Visual Mathematics,
Illustrated by the TI-92 and the TI-89

Springer-Verlag France S.A.R.L

George C. DORNER, Jean Michel FERRARD, Henri LEMBERG

Visual Mathematics,
Illustrated by the TI-92 and the TI-89

 Springer

Pr. George C. DORNER
William Rainey Harper College
Palatine, Illinois, USA

Dr. Jean Michel FERRARD
Professeur en classe PC
Lycée Jean Perrin
Lyon, France

Dr. Henri LEMBERG
Professeur en classe PC
Collège Stanislas
Paris, France

© Springer-Verlag France 2000

ISBN 978-2-287-59685-8   ISBN 978-2-8178-0201-5 (eBook)
DOI 10.1007/978-2-8178-0201-5

SPIN : 10749046

Preface

This book is unlike other mathematics books.

Some basic old topics and themes and some quite advanced, new concepts and themes of mathematics are covered. There is a constant concern for rigor and precision in the presentation of the material. A mathematics student will see here the integration of many themes which run throughout the history of mathematics and which are still the subject of intensive research. The emphasis is on concrete results which may see application by applied mathematicians, computer scientists, engineers, and other scientists.

The topics covered are in classical analysis (dynamic systems, Fourier series, differential equations, function interpolation, etc.) and in linear algebra (orthogonality, eigenvalues and eigenvectors). The breadth of topics is difficult to describe simply. Perhaps "introductory applied functional analysis" would come close.

An unstated subtheme is how to evaluate or look at a useful function. Classical topics of solution of equations, representation, approximation, and polynomial, rational function, and trigonometric interpolation are treated. These subjects are revisited in several settings. Bézier curves, splines, wavelets and other topics of interest in computer science applications are included in this comprehensive introduction.

This is classical, "post-calculus" mathematics which is unified and integrated with the tools and ideas of linear algebra. Because of the breadth of material covered and the diversity of the intended audience it may be incumbent on the reader to dig out an idea or a term which is not detailed here. For example, we did not cite some of the famous theorems of calculus, such as Rolle's Theorem or the Intermediate Value Theorem, even though they are frequently used. Similarly, a property of compact sets may be called upon or the idea of a basis may be used even though the definitions are not given in this book. On occasion there is a forward-reference in the book which may usually be skipped on first reading.

Certainly, the history of this mathematics is not "linear" and probably the same may be said about the learning. We attempted to present not a treatise but a useful introduction to many topics, illustrated by concrete examples.

So far, this preface reads like many others. The mathematics is classical, rigorous, and comprehensive. Then, why is it not like the others?

Most of the concepts which are developed are developed and illustrated by formal calculations using the TI-89 and TI-92 graphic and symbolic calculators from Texas Instruments. These calculators readily provide graphic displays and formal computations of a computer algebra system (CAS) which belie their portability and low cost. Little is sacrificed and much is gained by use of these machines.

Numerous programs, all very brief, are used to give examples, to illustrate significant points, and even to point toward extensions of the theory. This is what makes the breadth of material covered more accessible and what renders it more concrete. The reader may "see" and almost "touch" the mathematical themes studied which, without this clarification, may remain abstract theoretical constructions. While no list of problems or exercises accompany the text, the reader with calculator in hand may explore, experiment, and play with new ideas until they are comprehended.

We are in fact persuaded that from now on it is no longer possible to learn and to understand mathematics as one did not so long ago, before the daily and intensive use of computers. Today's formal calculators are true "pocket computers" dedicated to mathematics which permit both the professor and the student to renew their approach to understanding this science. For the former, it renders illustrations and demonstrations more accessible. For the latter, it provides immediate visual experiences and leads to better understanding. We want you to take part in our experiences. We are convinced of the benefits, and we definitely believe that using the calculator in this way yields a "less is more" result.

This work is thus really a book "unlike the others": it is a book of concrete mathematics.

Finally, the book differs not only in its conception and execution but also in its production. The book first appeared in France as:

Mathématiques concrète, illustrées par la TI-92 et la TI-89
J.M. Ferrard - H. Lemberg
Springer-Verlag France 1998 ISBN 2-287-59647-X

This English version appears as the result of a trans-oceanic collaboration via the internet between mathematics professors with similar interests and almost identical philosophies about the use of technology in teaching and learning of mathematics. The authors became collaborators and friends through this technology as a result of their common interests in mathematics, technology, and pedagogy.

We accept responsibility for any errors which may appear in this work, but, as another innovation, we attribute them to lost bits and bytes which linger somewhere over the Atlantic.

Contents

Discrete Dynamical Systems

In this chapter we will study sequences defined by recurrence relations of the form $x_{n+1} = f(x_n)$. This is a topic which has an interesting history and which has seen rapid development in recent years. Its study requires little in the way of mathematical preparation, and there are even interesting applications. An introduction to the popular topics of fractals and chaos properly belongs to this field. But our investigations may prove to be extremely delicate and challenging, even for a function as simple as $f : x \mapsto x^2 + c$, where c is a real or complex parameter.

A "discrete dynamical system" will be any pair (X, f), where X is a set and f is a function from X into itself. The study of such a system involves consideration of the behavior of the sequences defined by choosing some $x_0 \in X$ and computing $x_{n+1} = f(x_n)$: Does such a sequence converge? If not, in what manner does it diverge? The goal in this study is to determine the eventual behavior of the sequence.

1. Dynamical Systems in $\mathbb{R}$

We'll start off our study with one of the simpler cases: the study of dynamical systems of the real line. These are sequences defined by recurrence relations of the form $\begin{cases} x_0 \in \mathbb{R} \\ x_{n+1} = f(x_n) \end{cases}$, where $f : \mathbb{R} \to \mathbb{R}$.

Most often, f will be a regular or C^1 function, that is, it will be continuous and will have a continuous derivative. Even in this case the study of dynamical systems is fairly complex.

1. 1 The Logistic Model of Verhulst

We begin by describing a historic model which gave birth to the theory of discrete dynamical systems.

Let p_n be a measure of a population of individuals at some instant n. The rate of growth of the population is defined by the ratio $\dfrac{p_{n+1} - p_n}{p_n}$. If we suppose that the rate of growth r is constant, the law governing the growth of the population is particularly simple, since then $p_{n+1} = (1+r)p_n$. This represents a geometric sequence with ratio $1+r$. Thus $p_n = (1+r)^n p_0$.

However, this model is not very realistic. In a standard "predator-prey" model, the rate of growth depends on the ratio between the size of the population at a given instant and the maximum population possible. This is the hypothesis that the Belgian mathematician François Verhulst made around 1845. He thus proposed the following equation:

$$p_{n+1} = p_n + r p_n (1 - p_n)$$

Amazingly, this simple quadratic equation gives rise to a dynamical system (using $f(x) = x + rx(1 - x) = -rx^2 + (1 + r)x$) which has very unexpected behavior and is definitely not easy to handle!

First we explain some generalities for what follows. In order to study the sequence (x_n) defined by the recurrence relation $\begin{cases} x_0 \in I \\ x_{n+1} = f(x_n) \end{cases}$, one must determine an interval I which is stable or invariant under f. (This just means that $f(I) \subseteq I$). A brief investigation shows that the curve representing f is a parabola turning toward the negative y axis. The maximum of f is obtained at $x = \dfrac{r+1}{2r}$ and has the value $\dfrac{(r+1)^2}{4r}$. The function vanishes at 0 and at $\dfrac{1+r}{r}$. For an interval $[0, \alpha]$ to be stable under f requires that $\dfrac{(r+1)^2}{4r} \leq \dfrac{r+1}{r}$ and thus that $0 \leq r \leq 3$. Hence, if we choose $x_0 \in I = [0, 1 + \frac{1}{r}]$, we are assured that for all $n \geq 1, x_n \in I$. On the basis of this first experience, we will take different values $r = 1/2, 1, 5/2, 3$ and trace the first elements of the sequence starting from the initial value $x_0 = 0.3$.

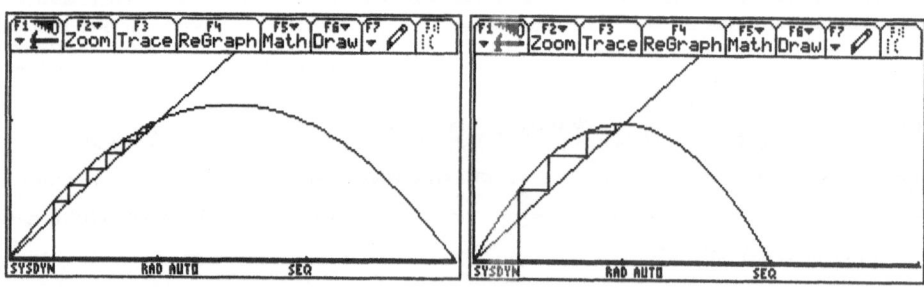

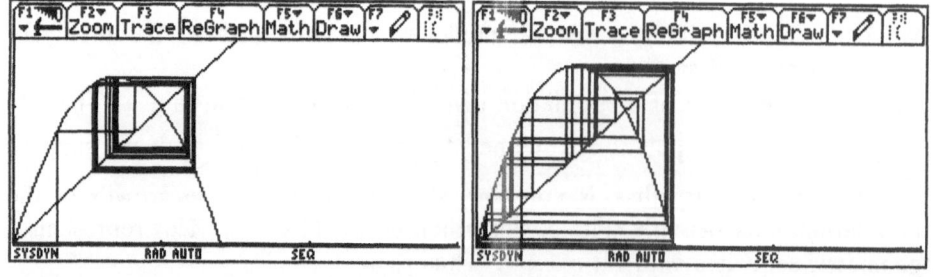

The study of various cases seems relatively complicated and strongly dependent on the values of r. Thus, we first pose this simple question: Does the sequence (x_n) converge? If yes, to what value?

Here is a general theorem which lets us show the existence of the limit of the sequence.

Theorem 1: (Fixed Point Theorem). *Let f be a real valued function defined on an interval $I \subseteq \mathbb{R}$ such that there exists a constant $0 \le M < 1$ such that for every $(x, y) \in I^2$:*

$$|f(x) - f(y)| \le M|x - y|$$

For an arbitrary $x_0 \in I$, the sequence (x_n) defined by $\begin{cases} x_0 \in I \\ x_{n+1} = f(x_n) \end{cases}$ converges to a unique point a such that $f(a) = a$. (We say that a is a fixed point of f).

Proof : we show that the sequence (x_n) is a Cauchy sequence in $\mathbb{R}$.

$$|x_{n+1} - x_n| = |f(x_{n+1}) - f(x_n)| \le M|x_n - x_{n-1}| \le \ldots \le M^n|x_1 - x_0| = \alpha M^n$$

Thus, if $m > n \ge N$:

$$|x_m - x_n| \le \sum_{k=n}^{m-1} |x_{k+1} - x_k| \le \alpha \sum_{k=n}^{m-1} M^k = \alpha \frac{M^n}{1 - M}$$

Then $0 \le M < 1$, for every $\varepsilon > 0$, there exists $N \in \mathbb{N}$ such that if $m > n \ge N$:

$$|x_m - x_n| \le \alpha \frac{M^n}{1 - M} < \varepsilon$$

Now $\mathbb{R}$ is complete, so the sequence (x_n) converges to $a \in \mathbb{R}$. This is a fixed point of f, $f(a) = a$, by the uniqueness of the limit and because f is continuous. Finally, the point a is unique. For suppose that there are two fixed points $a \ne b$ of f. Using the fact that $0 \le M < 1$, it follows that:

$$|a - b| = |f(a) - f(b)| \le M|a - b| < |a - b|$$

which can't be true.

This theorem is is a very important utility in analysis. One may see it used to prove the existence of solutions in problems as varied as the solution of numeric equations, the Cauchy-Lipschitz theorem for the solution of differential equations, or the local inverse theorem. But it also gives the key to the local study of sequences defined by recurrence relations.

Intuitively, an "attracting point" is any point a which "attracts" a sequence (x_n). This means "if x_n approaches a, x_{n+1} is closer to a". Adapting the preceding proof, we obtain:

Proposition 1: *Let f be a function of class C^1 on an interval I and a a fixed point of f such that $|f'(a)| < 1$. Then, there is a neighborhood $\mathcal{N}$ of a, such that for all $x_0 \in \mathcal{N}$, the sequence defined by $x_0 \in I$ and $x_{n+1} = f(x_n)$ converges to a.*

Proof : Since $|f'(a)| < 1$ and since f' is continuous, there exists $0 < \lambda < 1$ and a neighborhood $\mathcal{N}$ of a in I, such that for every $x \in \mathcal{N}, |f'(x)| \leq \lambda$.
Choosing $x \neq a$ and using the Mean Value Theorem, there exists $t \in [x, a]$ (or $[a, x]$) such that:

$$|f(x) - a| = |f(x) - f(a)| = |f'(t)||x - a| \leq \lambda|x - a| < |x - a|$$

(The point a is intuitively attracting.) By an immediate recurrence, it follows:

$$|x_n - a| = |f(x_{n-1}) - a| \leq \lambda|x_{n-1} - a| \leq \ldots \leq \lambda^n|x_0 - a|$$

The sequence (x_n) converges to a since $0 < \lambda < 1$.

We have thus shown that every fixed point a such that $|f'(a)| < 1$ is attracting.

The opposite of an attracting point will be called repelling. A repelling point is truly "repulsive"! In fact:

Proposition 2: *If we suppose that the sequence (x_n) defined by the recurrence relation $\begin{cases} x_0 \in I \\ x_{n+1} = f(x_n) \end{cases}$, with f of class C^1 on I, converges to a fixed point b of f such that $|f'(b)| > 1$, then there exists M such that for all $n \geq M, x_n = b$.*

Proof : We argue by contradiction and utilize again the Mean Value Theorem. Suppose that the (x_n) converges to b, each term being different from b. Then:

$$\forall \varepsilon > 0, \exists\, M \in \mathbb{N} \mid n \geq M \Rightarrow |x_n - b| < \varepsilon$$

We know that $|f'(b)| > 1$ and that f' is continuous. There is thus an open neighborhood $\mathcal{N}$ of b such that for every $x \in \mathcal{N}, |f'(x)| > 1$. Using the Mean Value Theorem, there exists N_0 such that, for every $n \geq N_0$:

$$\exists\, \theta_n\ /|x_{n+1} - b| = |(x_n - b)f'(\theta_n)| > |x_n - b|$$

Starting from N_0, the sequence $(|x_n - b|)$ is positive and strictly increasing. Thus, it can't tend to 0. We have thus shown that every fixed point b such that $|f'(b)| > 1$ is repelling.

In summary, for a function of class C^1, the behavior at a fixed point a depends on the value of $|f'(a)|$ relative to 1, but we still don't know what happens when $|f'(a)| = 1$.

The fixed points of $f : x \mapsto x + rx(1 - x)$ are 0 and 1. and $f'(0) = r + 1, f'(1) = 1 - r$. Since $r > 0$, 0 is a repelling point and 1 is attracting if and only if $r < 2$.

But even for $0 < r < 2$, the study of the sequence (x_n) doesn't reveal itself very simply as we show by studying the following graphs.

For $r = 0.75$, the sequence appears to increase monitonically to 1.

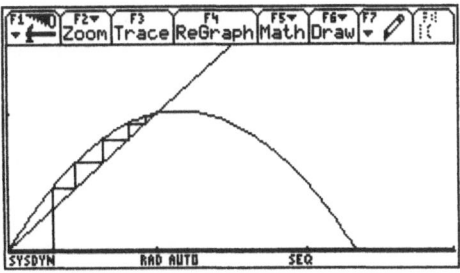

For $r = 1.75$, a first approximation gives an alternating convergence (like a "snail"), which is confirmed by zooming in on the point 1.

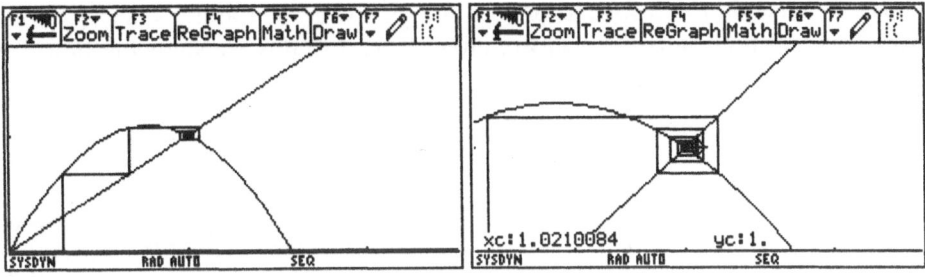

We observe the same phenomenon for $r = 2$:   apparent, but snail-slow convergence. This seems to be confirmed by a zoom into the neighborhood of 1.

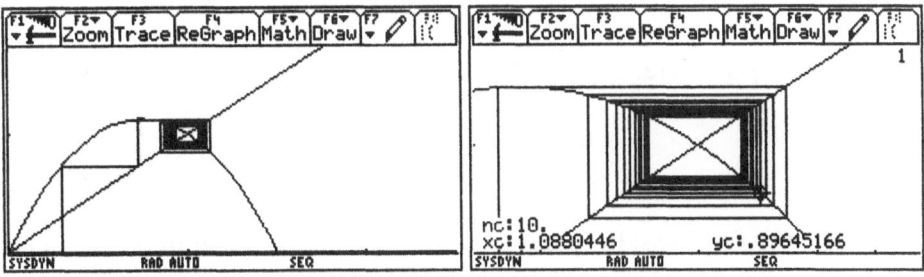

1. 2 The case of convergence

The sequence (x_n) behaves itself quite well in some cases.
In fact:

• if $0 < x_0 < 1$ and if $0 < r \leq 1$, the sequence (x_n) is increasing and converges to 1. We show by induction that $0 < x_n \leq 1$. This is true for x_0 and if it is satisfied by x_n, then $x_{n+1} = x_n + r x_n (1 - x_n) < 2 - x_n^2 \leq 1$. Moreover $x_{n+1} - x_n = r x_n (1 - x_n) > 0$ which shows that the sequence is increasing. The sequence (x_n) thus converges to the only fixed point possible which is $a = 1$.

• If $0 < x_0 < 1$ and if $1 < r \leq 2$, the sequence (x_n) still converges to 1, but it is no longer monotonic. We show this for the extreme case, where $r = 2$.
An elementary study of the function $f : x \mapsto 3x - 2x^2$ gives the following table of values:

x	0	1/2	3/4	1	9/8
$f(x)$	$\nearrow$	1	$\nearrow$ 9/8	$\searrow$ 0	17/32

Thus if $1/2 < x_0 < 1$, then $1 < x_1 < 9/8$ and $1/2 < x_2 < 1$, etc. One easily shows by recursion that for all for tout $n \in \mathbb{N}$, $0 < x_{2n} < 1$ and $1 < x_{2n+1} < 9/8$. We show that both subsequences are monotonic. For this, we study the function $h = f \circ f : x \mapsto -x(2x - 3)(4x^2 - 6x + 3)$. The graph of $k : x \mapsto \dfrac{h(x)}{x}$ is sketched below.

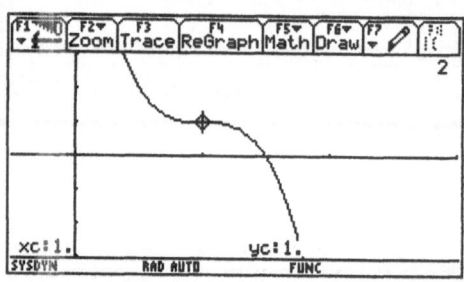

The graph of the function k for $x > 0$. The function is strictly decreasing and $k(1) = 1$. (This is logical since $f(1) = 1$.)

The function k is decreasing and $k(1) = 1$. Thus:

$$\begin{cases} 0 < x_{2n} < 1 \Rightarrow k(x_{2n}) > 1 \\ 1 < x_{2n+1} < 9/8 \Rightarrow k(x_{2n+1}) < 1 \end{cases}$$

which signifies that the subsequence (x_{2n}) is increasing and that the subsequence (x_{2n+1}) is decreasing. These two subsequences converge to a fixed point of $f \circ f$, and this function can't have two fixed points 0 and 1. The two subsequences thus converge to 1.

By contrast, when $2 < r \leq 3$, the study of our sequence is otherwise quite complicated.

1.3 Cycles

When $r > 2$, the point 1, which was attracting, has become repelling. First, recall what we obtain experimentally in the case where $r = 2.5$ and $r = 3$.

For $r = 2.5$, we have plotted the first 50 elements of the sequence. The image is not very clear. It is necessary to plot many more points.

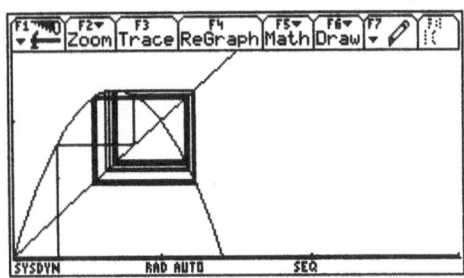

For $r = 3$, and after plotting the 100 first points similarly and zooming around 1, it still seems difficult to predict the behavior of the sequence.

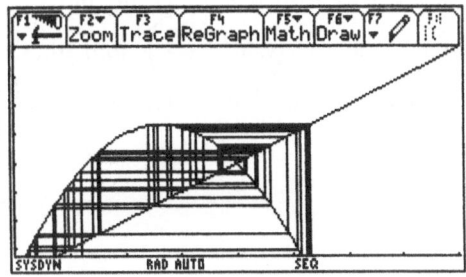

Obviously, our sequence doesn't seem to converge. But, while not regular, it may yet have an internal structure to be discovered. This is sometimes the case. We may find "cycles", that is, values for which the the sequence (x_n) oscillates periodically.

The demonstration is delicate and depends on the value of r. However, in the extreme case where $r = 3$, it is possible to make a precise analysis of the situation and to determine the explicit manner in which x_n is expressed as a function of x_0.

In fact, with $r = 3$, consider $f : x \mapsto 4x - 3x^2$. We know that $0 < x_0 < 1 + \frac{1}{r} = \frac{4}{3}$ and that for all $n \in \mathbb{N}, x_n$ satisfies the same inequality. Then we set:

$$x_0 = \frac{4}{3} \sin^2(\varphi), \ 0 \leq \varphi \leq \frac{\pi}{2}$$

or:

$$\varphi = \mathrm{Arcsin}\left(\sqrt{\frac{3x_0}{4}}\right)$$

In this case:

$$x_1 = 4x_0 - 3x_0^2 = \frac{16}{3} \sin^2(\varphi) \cos^2(\varphi) = \frac{4}{3} \sin^2(2\varphi)$$

and an obvious recursion shows that:

$$x_n = \frac{4}{3}\sin^2(2^n\varphi)$$

We say that the sequence (x_n) is *periodic* with period $p > 0$ if for all $n \in \mathbb{N}, x_{n+p} = x_n$.

It is easy to determine the values φ for which our sequence is periodic with period p:

$$x_{n+p} = x_n \iff \frac{4}{3}\sin^2(2^{n+p}\varphi) = \frac{4}{3}\sin^2(2^n\varphi)$$

$$\iff 2^{n+p}\varphi = 2^n\varphi + k\pi$$

$$\iff \varphi = \frac{k\pi}{2^n(2^p - 1)}$$

With $k = 2, n = 0, p = 3$, the initial value is $x_0 = \frac{4}{3}\sin^2\left(\frac{2\pi}{7}\right)$ and the sequence is of period 3. The graph is simple!

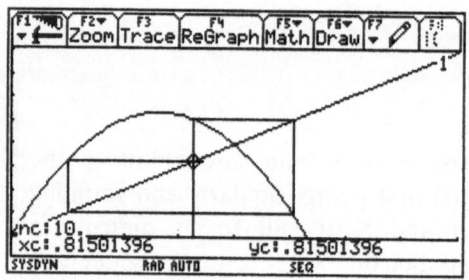

With $k = 4, n = 2, p = 3$, the initial value is $x_0 = \frac{4}{3}\sin^2\left(\frac{\pi}{7}\right)$ and the sequence is of period 3. The sketch is the same as above.

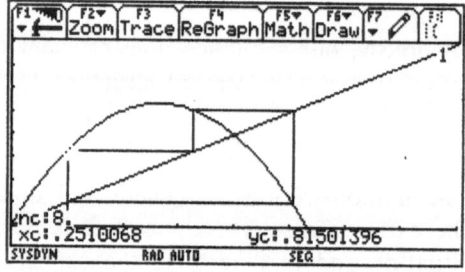

With $k = 21, n = 0, p = 7$, the value initial is $x_0 = \frac{4}{3}\sin^2\left(\frac{21\pi}{127}\right)$ and the sequence is of period 7.

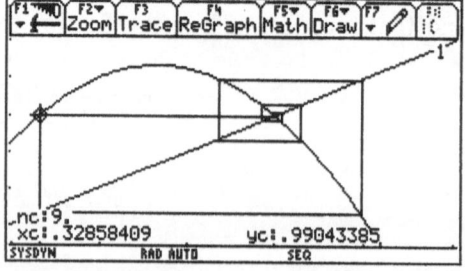

The precise theoretical determination of cycles in the case where $2 < r < 3$ is much more delicate. We need also a new notion, that of *limit cycle*.

Definition: Let (x_n) be a sequence defined by x_0 and $x_{n+1} = f(x_n), n \geq 1$. A limit cycle is any k-tuple of points $(a_1, a_2, \ldots, a_k)$ such that there are subsequences $(x_{n_1}, x_{n_2}, \ldots, x_{n_k})$ which tend respectively to $(a_1, a_2, \ldots, a_k)$. and for each $1 \leq i \leq k-1$, $f(a_i) = a_{i+1}$ and $f(a_k) = a_1$.

For example:

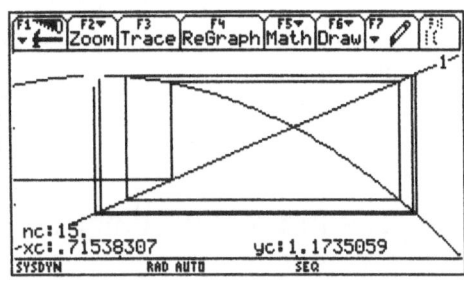

for $r = 2.25$, there may be no cycle, but the sequence seems to approach the points 0.715 and 1.173.

The following program attempts to determine such limit cycles. The idea is drawn from an article by Martin Gardner published in Scientific American. We calculate in parallel the elements x_n and x_{2n}, until we obtain the equality $x_n = x_{2n}$ which will be eventually be arrived at if our sequence is periodic. In this case, if p is the period, to determine the cycle is the same as to identify the list $\{x_n, x_{n+1}, \ldots, x_{n+p-1}\}$, thus comparing x_n to x_{n+j} $(j > 0)$ until equality is obtained.

```
:cycles()
:Func
:Local x,y,n,l
:rand()→x              a random value in [0,1] x₀
:x→y                   x₀ is preserved in y
:f(x)→x                we calculate following the term x₁
:f(f(y))→y             then the term x₂
:1→n
:While x*y and n≤500   a loop on n
:f(x)→x: f(f(y))→y     we calculate here xₙ and x₂ₙ
:n+1→n
:EndWhile              end the loop
:If x*y: Return        we haven't found a cycle
:x→l                   if we have found a cycle, we calculate
:f(y)→y                the first term
:While x*y             we loop until it recurs
:augment(l,{y})→l      we put the elements of the cycle in the list
:f(y)→y
:EndWhile
:l                     show the list l
:EndFunc
```

Calculation of limit cycles for different values of r. For $r = 2.75$, the function cycles() gives an empty list. Either there is no limit cycle, or the number of test terms is insufficient to determine one.

```
F1    F2    F3    F4    F5    F6
      Algebra Calc Other PrgmIO Clean Up
■ DelVar r                              Done
■ f(x)                          x - r·x·(x - 1)
■ 2.5 → r : cycles()
   {.7012378944  1.224996169  .535947556}
■ 2.75 → r : cycles()                      {}
■ 2.83 → r : cycles()
  {6829938681  1.295726349  .2113257523}
2.83→r:cycles()
SYSDYN         RAD AUTO        SEQ  5/30
```

1. 4 The bifurcation diagram

It seems that for certain values of r (for example $r = 2.57$), there is no limit cycle. All the same, to predict the length of an eventual limit cycle is a delicate question. On the other hand, the calculator is a marvellous tool for visualizing this phenomenon.

For that, we plot the pairs (r, x), where r varies from 0 to 3 and where x is one of the points of the cycle corresponding to r. (We already know that for $r < 2$, the graph will not be very interesting, since our sequence will then converge.) But, to keep the calculations from being too long, we will only plot, for each value of r, the terms x_{100} to x_{200} of the sequence.

Here is a little program which realizes this idea.

```
:bifurc()
:Prgm
:Local r,j,n,x
:ClrDraw:ClrGraph:FnOff          prepare the screen to plot
:100→n
:For r,xmin,xmax,(xmax-xmin)/238  on all the pixels on abcissa
:rand()→x                        store a random number a
:For j,1,n                       loop for n = 100 points
:f(x,r)→x                        calculate f(xₙ)
:End                             (r is defined as a parameter of f)
:For j,n,200                     plot the points x₁₀₀ to x₂₀₀
:f(x,r)→x
:PtOn r,x
:EndFor                          end of the inner loop
:EndFor                          end of the outer loop
:EndPrgm
```

and its result:

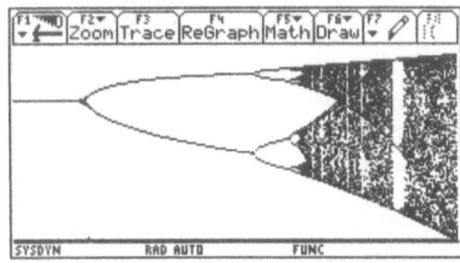

It's easy to see the points where the cycles divide in two. The dark part corresponds to chaos (numerous divisions or bifurcations), and the clear part corresponds to a lull.

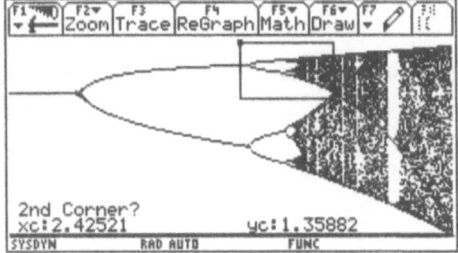

The Feigenbaum set has a "fractal" structure, which means that when we zoom in on part of the plot, we get the same structure as in the original figure. We pick the zoom rectangle at the second bifurcation.

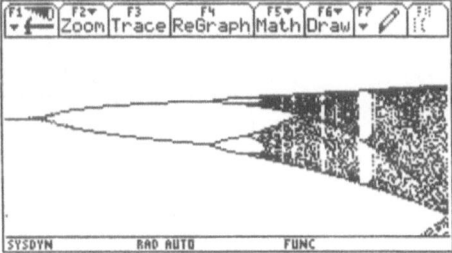

This second graph is similar at every point to the first. If we continue to zoom, we get the same structure. But note the length of calculation and the time to plot!

1. 5 The Feigenbaum constant

In the mid 70's, Mitchell Feigenbaum was the first to study the bifurcation phenomenon which we saw in the previous paragraph. He did this with only the help of a simple calculator! (It was *not* a TI, but perhaps you, too, may become famous if you master your calculator!)

Feigenbaum showed that if we denote by (b_k) the sequence of points where the bifurcations take place, then the sequence:

$$\delta_k = \frac{b_{k+1} - b_k}{b_{k+2} - b_{k+1}}$$

converges to a constant (called the Feigenbaum constant in his honor) $d \approx 4.66692016091029\ldots$

This constant is "universal" in the sense that it is independent of the recurrence treatment, when it is of the form $f : x \mapsto rx^2 g(x)$ when:

- f is of class C^1 on $[0,1]$.
- f has a maximum x_0 on $[0,1]$ such that $f''(x_0) \neq 0$.
- f is monotone on $[0, x_0[$ and on $]x_0, 1]$.
- f satisfies: $\dfrac{f^{(3)}(x)}{f'(x)} - \dfrac{3}{2}\left(\dfrac{f''(x)}{f'(x)}\right)^2 < 0, \forall\, x \in [0,1]$.

(As an example, we could use $f : x \mapsto rx^2 \sin(\pi x)$.)

We are going to redo Feigenbaum's experience, abandoning the Verhulst model and studying a very similar system given by the function $f_a : x \mapsto ax(1-x)$. These two models are in fact equivalent; we'll show this later.

We easily verify that f_a has a maximum on $[0,1]$ with the value $a/4$ at $x = 1/2$. For the usual reasons in the definition of the sequence, $(x_0 \in [0,1], x_{n+1} = f_a(x_n) \in [0,1])$, the study of the dynamics of the system is restricted to $0 < a \le 4$.

The fixed points of f_a are the solutions of the equation $f_a(x) = x$, which are 0 and $\dfrac{a-1}{a}$.

- at $x = 0$, we have $f'(0) = a$. Thus:

a	derivative	type of fixed point
$0 \le a < 1$	$|f'(0)| < 1$	attracting
$a = 1$	$|f'(0)| = 1$	?
$a > 1$	$|f'(0)| > 1$	repelling

- at $x = \dfrac{a-1}{a}$, we have $f'(x) = 2 - a$. and:

a	derivative	type of fixed point
$0 \le a < 1$	$1 < f'(x) \le 2$	repelling
$1 < a < 2$	$0 < f'(x) < 1$	attracting (monotone sequence)
$2 < a < 3$	$-1 < f'(x) < 0$	attracting (the snail sequence)
$3 < a < 4$	$-2 \le f'(x) < -1$	repelling

We will concentrate on the values of $a > 1$ and on the second fixed point. The value $a = 2$ is particularly interesting, since in this case (where the fixed point is $x = 1/2$) the variations of the sequence change: when a passes this value, the monotone convergent sequence is transformed into a sequence which is still convergent, but very slowly. Likewise, we see that $f_2'(1/2) = 0$. Such a fixed point will be called a *super attracting fixed point*, since the convergence of the sequence then seems extremely rapid). The point $x = 1/2$ (the only super attracting point of f_a) is very important for our study. We will come back to it later.

When a passes the value 3, the fixed point $\dfrac{a-1}{a}$ changes from attracting status to repelling status. We have seen experimentally in the Verhulst model that the sequence doesn't necessarily fall into "chaos", but that we could obtain a certain structure, namely cycles. Here it is easy to determine the cycles of order 2. In fact, these cycles are the fixed points of the map of $x \mapsto f_a \circ f_a$, which will be denoted f_a^2. Thus we have:

$$f_a(f_a(x)) = x \iff -a^3x^4 + 2a^3x^3 - (a^2 + a^3)x^2 + (a^2 - 1)x = 0$$

$$\iff x(x - \frac{a-1}{a})(-a^3x^2 + (a^2 + a^3)x - (a^2 - a)) = 0$$

whose solutions are the known fixed points 0 and $\dfrac{a-1}{a}$ and two others:

$$x_1 = \frac{a + 1 + \sqrt{a^2 - 2a - 3}}{2a}, \quad x_2 = \frac{a + 1 - \sqrt{a^2 - 2a - 3}}{2a}$$

Here is the preceding calculation automated.

Interesting solutions only exist for $a > 3$ and, for $a = 3$, we get the fixed point $\dfrac{a-1}{a}$.

In summary, for $a > 3$, the function $f_a^2 : x \mapsto f_a \circ f_a(x)$ has 4 global fixed points : the two fixed points of f_a and two new fixed points x_1, x_2 which may be determined as above; at $a = a_1 = 3$, the fixed point loses its stability and gives birth to a cycle of order 2.

What happens at our critical point $x = 1/2$? The solutions of $f_a^2(1/2) = 1/2$ are $2, 1 + \sqrt{5}, 1 - \sqrt{5}$. The important information here is that for $a = 1 + \sqrt{5}, y_2 = 1/2$. At y_2, a fixed point of f_a^2, we have $(f_a^2)'(y_2) = -a^2 + 2a + 4$ and the equation:

$$|(f_a^2)'(y_2)| = 1 \iff |-a^2 + 2a + 4| = 1$$

gives, for $a > 3$, only the solution $a = a_2 = 1 + \sqrt{6}$. For this value, the fixed point x_2 becomes unstable, and we could meet new bifurcations when a passes this value.

It now suffices to repeat this process: the attracting fixed points of f_a^4 give birth to a cycle of order 4, then those of f_a^8 has a cycle of order 8, etc. we thus determine two interesting sequences:

• a sequence (s_k) of super attracting points of $f_a, f_a^2, \ldots, f_a^{2^k}, \ldots$ for which the critical point $x = 1/2$ is a fixed point.
• a sequence (a_k) of values of a at which there is a bifurcation. (The fixed point passes from attracting to repelling there.)

We are now ready to study experimentally the Feigenbaum constant. The sequence (a_k) of bifurcation points is more and more difficult to determine. On the other hand the sequence (s_k) of super attracting points is, itself, distinctly more accessible. Thus, we will determine experimentally the eventual limit:

$$\lim_{k \to +\infty} \frac{s_{k-1} - s_{k-2}}{s_k - s_{k-1}}$$

where s_k is the value of the parameter a for which the critical point $1/2$ is fixed point of $f_a^{2^k}$. We know already that $s_1 = 2, s_2 = 1 + \sqrt{5}$. The calculator gives $s_3 \approx 3.498561699$. Evidently the complexity of the equation to be solved grows enormously with k.

In order to avoid the calculator work involved in solving all the equations which come up, we will use Newton's Method, whose convergence is particularly rapid but not always assured. (We'll come back to this last point later in the chapter, when this problem itself also gives rise to a dynamical system!)

Newton's Method is an algorithm which studies the sequence defined by:

$$\begin{cases} a_0 \in \mathbb{R} \\ a_{n+1} = a_n - \dfrac{g(a_n)}{g'(a_n)} \end{cases}$$

If this sequence converges, it converges to a solution of the equation $g(a) = 0$, thus to a fixed point of $f_a^{2^k}$. In fact, when the initial value (or "guess") a_0 is not very far from the solution to be found (this is almost a paradox...), one is assured of very rapid convergence of (a_n) to its limit.

Now, for our attack on Feigenbaum's constant, let $k \geq 1$ be fixed and $g : a \mapsto f_a^{2^k}(1/2) - 1/2$. We will define several sequences:

• The sequence (x_k) of iterates of the critical point is defined by:

$$\begin{cases} x_0 = 1/2 \\ x_{k+1} = ax_k(1 - x_k) \end{cases}$$

For $N > 1$, the element x_N corresponds to $f_a^N(1/2)$; it only remains to calculate $x_N - 1/2$ for an appropriate N to obtain $g(a)$.

- Each x_k is a function of a, so let $x_k = x_k(a)$. To calculate the denominator $g'(a)$, it suffices to calculate $\dfrac{dg}{da}(x_k)$.

We will then define the two sequences:

$$\begin{cases} x_0 = 1/2 \\ x_{k+1} = ax_k(1 - x_k) \end{cases}, \qquad \begin{cases} y_0 = 0 \\ y_{k+1} = x_k(1 - x_k) + ay_k(1 - 2x_k) \end{cases}$$

- To determine a good start for our algorithm which will assure convergence, we note that the Feigenbaum constant, d, is defined by the limit of the sequence d_k:

$$d_k = \frac{s_{k-1} - s_{k-2}}{s_k - s_{k-1}}$$

Thus $s_{k+1} = s_k + \dfrac{s_k - s_{k-1}}{d_k}$.

We will take as the initial value $s_{k+1.0} = s_k + d_k(s_k - s_{k-1})$ and we'll stop at the first j such that $\left| \dfrac{s_{k.j+1} - s_{k.j}}{s_{k.j}} \right| < \varepsilon$, where $\varepsilon > 0$ is a given precision. We put then $s_{k+1} = s_{k.j}$.

Here is the corresponding program, divided into two functions.

```
:iter(a,n)                       calculate f_a^{2^{n-1}}
:Func
:Local x,y,u,i
:0.5→x                           initial conditions for x
:0→y                             for y
:x→u
:For i,1,2^(n-1)
:a*x*(1-x)→x                     next x
:u*(1-u)+a*(1-2*u)*y→y           next y
:x→u
:EndFor
:x-0.5→x                         calculate g(x)
:Return {x,y }                   return the pair (x, y)
:EndFunc
```

```
:feigb()                                    set digits first
:Func
:Local 1,11,m,a,b,d,n                       results stored in l
:{2,approx(1+√(5))}→1                       initial values of the list
:dim(1)→m
:1[m]+(1[m]-1[m-1])/4→a
:For n,3,8                                   calculate 6 values. For each n
:iter(a,n)→11                                $s_{n.0}$
:a-11[1]/(11[2])→b                           $s_{n.1}$
:While abs((b-a)/a)>1E-10
:b→a                                         calculate the following values $s_{n.k}$
:iter(a,n)→11
:a-11[1]/(11[2])→b
:EndWhile
:augment(1,{b})→1                            add an element to the result
:dim(1)→m
:(1[m-1]-1[m-2])/(1[m]-1[m-1])→d             calculate $d_n$
:1[m]+(1[m]-1[m-1])/d→a                      initialize for the next time
:EndFor
:EndPrgm
```

An experimental study of the Feigenbaum constant. The preceding program lets us calculate effectively the elements which approach d_n which converges to d.

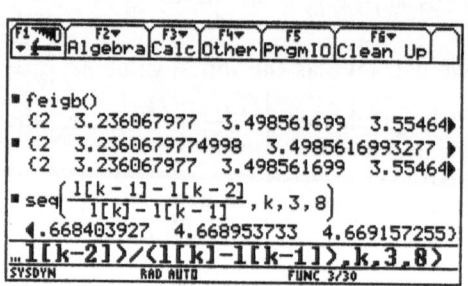

1. 6 Study of cycles

There is a simple graphic way to study the phenomena of cycles. In fact, when the dynamical system is defined by the function f, the fixed points correspond to solutions of the equation $f(x) = x$. The eventual cyclic points of order k correspond to solutions of the equation $f^k(x) = x$, (f^k designates the composition of f with itself k times), if there are no fixed points of order less than k.

Do there exist fixed points of every order? We will answer mathematically, but first here are two examples which show how the response to this question depends on the "context".

The fixed points of $f(x) = 3.6x(1-x)$ are the points of intersection of the graph of this curve and of the bisector of the first quadrant, the line $y = x$. We represent the sequence of iterates as f^k, for $k = 2, 3, \ldots, 6$ and $k = 10$. Note that there is no point of period 3. Why?

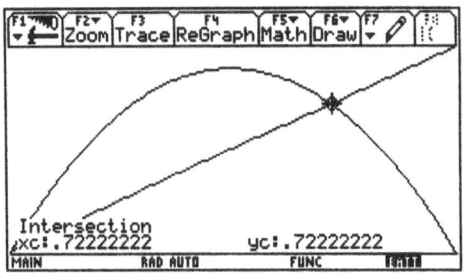

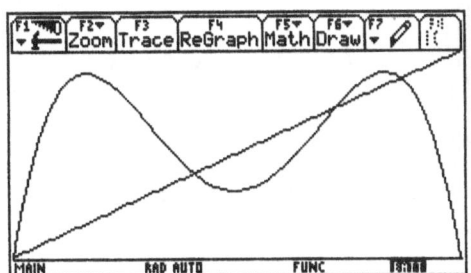

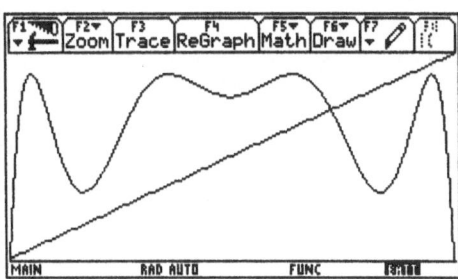

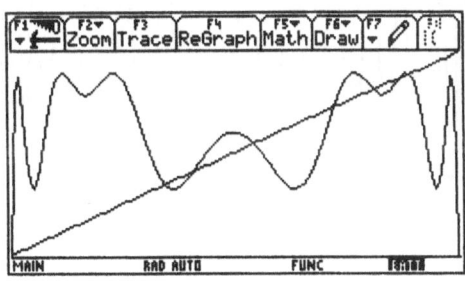

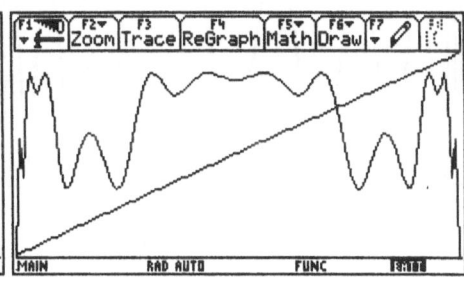

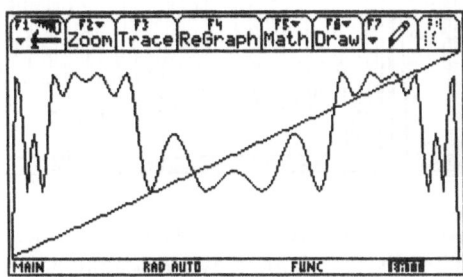

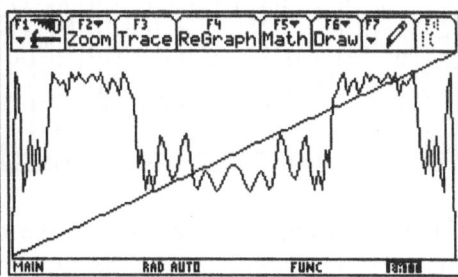

Starting the same study again with $f(x) = 3.99027x(1-x)$. the situation is completely different and is chaotic.

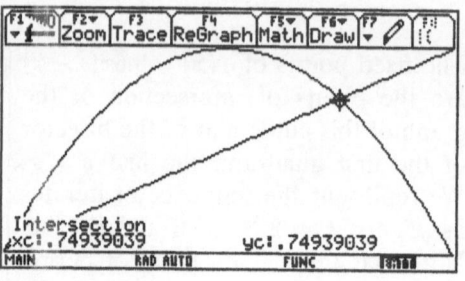

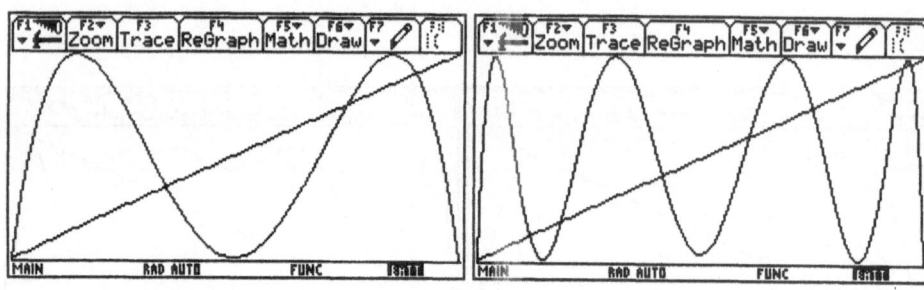

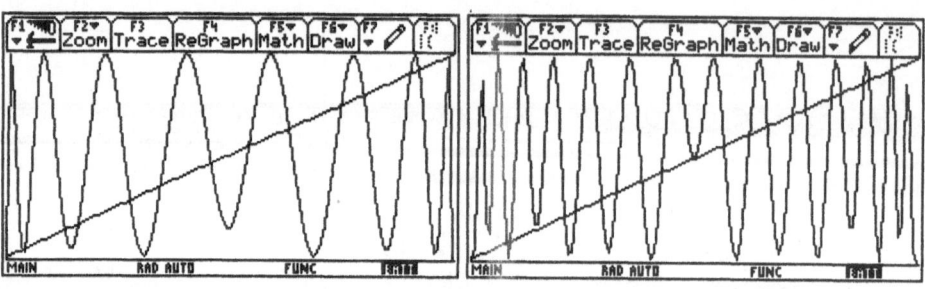

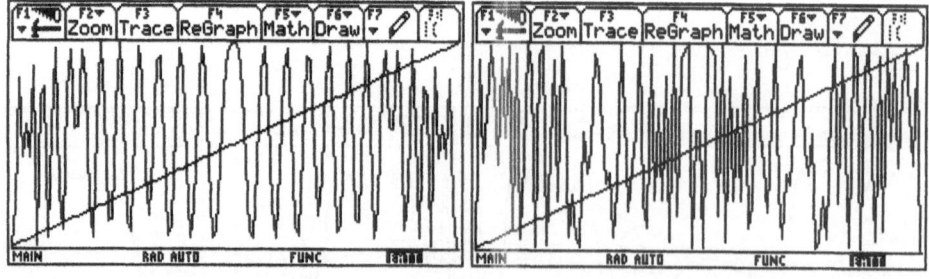

In these examples we could test the existence of cycles of certain dynamical systems. Is there a general result. which tells us whether every system has a

cycle, or, when a system has a cycle of period n, whether it has a cycle of any lesser period?
Of course, there are not general answers to these questions, but there are at least two remarkable results.

The first is the celebrated theorem of Sharkovskii from 1964.
First define a new order on the natural numbers $\mathbb{N}$, by:
- for the odd integers: $3 \rhd 5 \rhd 7 \rhd \ldots \rhd \ldots$
- add on the integers of the form 2^k times an odd integer:

$$3 \rhd 5 \rhd 7 \rhd \ldots \rhd 2 \cdot 3 \rhd 2 \cdot 5 \rhd \ldots \rhd 2^2 \cdot 3 \rhd 2^2 \cdot 5 \rhd \ldots$$
$$2^n \cdot 3 \rhd 2^n \cdot 5 \rhd \ldots \rhd 2^{n+1} \cdot 3 \rhd 2^{n+1} \cdot 5 \rhd \ldots$$

- finally we add on the powers of 2:

$$3 \rhd 5 \rhd 7 \rhd \ldots \rhd 2^n \cdot 3 \rhd 2^n \cdot 5 \rhd \ldots \rhd 2^{n+1} \rhd 2^n \rhd \ldots 2^2 \rhd 2 \rhd 1$$

Thus we have defined an order on $\mathbb{N}$ (the Sharkovskii order), and and we state the theorem:

Theorem 2:(Sharkovskii's Theorem) *Let $f : I \subseteq \mathbb{R} \to \mathbb{R}$ be a continuous function. If f has a point of period n, then f has a point of period k for all $n \rhd k$.*

We won't give a proof of the theorem here since that exceeds the scope of this book. (See the article by M. Misiurewicz cited in the Bibliography for references and an interesting discussion. The proof is also in the book on Chaotic Dynamic Systems by R. Devaney.) On the other hand, we propose to show you the result of Li and York on the existence of periodic points:

Theorem 3: *Let $f : \mathbb{R} \to \mathbb{R}$ be a continuous function. Suppose there exists a point a such that:*
- $f^3(a) \le a < f(a) < f^2(a)$, or
- $f^3(a) \ge a > f(a) > f^2(a)$.

then f has points of every period.

Proof : We may suppose that we have $f^3(a) \le a < f(a) < f^2(a)$. The second case may be deduced by symmetry with respect to the line $y = x$. We put $I_1 = [a, f(a)]$ and $I_2 = [f(a), f^2(a)]$.We then have

$$I_2 \subset f(I_1), \qquad I_1 \cup I_2 \subset f(I_2)$$

We break up the proof into 3 lemmas.
Lemma 1. Let I, J be two closed intervals such that $J \subset f(I)$. There exists an interval $K \subset I$ such that $f(K) = J$, $f(\text{int}(K)) = \text{int}(J)$ and $f(\partial K) = \partial J$, where $\text{int}(K)$ designates the interior of K and ∂K is the boundary of K.

Proof : Put $J = [b_1, b_2]$. There exist $a_1, a_2 \in I$ such that $f(a_1) = b_1$ and $f(a_2) = b_2$. We may suppose that $a_1 < a_2$. Let:

$$x_1 = \sup\{x \in [a_1, a_2] \ / \ f(x) = b_1\}$$

By continuity of f, $f(x_1) = b_1$. In the same manner, if:

$$x_2 = \inf\{x \in [x_1, a_2] \ / \ f(x) = b_2\}$$

then $f(x_2) = b_2$.
Thus we have $f(\{x_1, x_2\}) = \{b_1, b_2\}$ and $f(]x_1, x_2[\cap \partial J) = \emptyset$. Thus:

$$f(\mathrm{int}([x_1, x_2])) = \mathrm{int}(J) =]b_1, b_2[$$

We will say that an interval J is f-covered by an interval I if $J \subset f(I)$. We denote this $I \to J$.

Lemma 2. Let I be a closed interval f-covered by itself. Then f has a fixed point in I.

Proof : Let $I = [a, b]$. By the preceding lemma, there exists interval $K = [x_1, x_2]$ such that $I = f(K)$. Then we have, either $f(x_1) = a \leq x_1$ and $f(x_2) = b \geq x_2$, or $f(x_1) = b > x_1$ and $f(x_2) = a < x_2$. In the two cases, it only remains to apply the Intermediate Value Theorem.

Lemma 3. Let $J_0 \to J_1 \to \ldots \to J_n = J_0$, be a sequence of sets, a "chain", where, for $k \in \{0, 1, \ldots, n-1\}$, $J_{k+1} \subset f(J_k)$. Then, there exists a fixed point x_0 of f^n such that $f^k(x_0) \in J_k$ for all $k \in \{0, 1, \ldots, n\}$.

Proof : The proof is by induction. Put:

$$(\mathcal{H}_j) : \exists \ K_j \subset J_0 \ / \ \forall \ 1 \leq i \leq j, f^i(K_j) \subset J_i, f^i(\mathrm{int}(K_j)) \subset \mathrm{int}(J_i), f^j(K_j) = J_j$$

$(\mathcal{H}_1)$ is verified by Lemma 1.

Suppose that $(\mathcal{H}_{k-1})$ is true. The interval K_{k-1} exists. Then:

$$f^k(K_{k-1}) = f(f^{k-1}(K_{k-1})) = f(J_{k-1}) \supset J_k$$

By Lemma 1, there exists an interval $K_k \subset K_{k-1}$ such that $f^k(K_k) = J_k$, with $f^k(\mathrm{int}(K_k)) = \mathrm{int}(J_k)$. By recursion, the other properties of $(\mathcal{H}_k)$ are verified.

For $k = n$, we have $f^n(K_n) = J_0$. By Lemma 2, f^n has a fixed point $x_0 \in K_n \subset J_0$. Since $x_0 \in K_n, f^i(x_0) \in J_i$ for all $0 \leq i \leq n$.

Proof of the theorem.
Suppose that $f(a) = b > a, f^2(a) = f(b) = c > f(a) = b, f^3(a) = f(c) \leq a$. Let $I_1 = [a, b], I_2 = [b, c]$. then I_2 is f-covered by I_1, and I_1 and I_2 are f-covered by I_2.

We have $I_2 \subset f(I_2)$ which means that f has a fixed point (Lemma 2).

We show that f has a fixed point of period n for all $n \geq 2$. Let the loop of length n start with I_1 and repeat I_2 $(n-1)$ times. $(I_1 \to I_2 \to I_2 \ldots \to I_2 \to I_1)$. By Lemma 3, there exists $x_0 \in I_1$ such that $f^n(x_0) = x_0$ and for all $1 \leq j \leq n-1, f^j(x_0) \in I_2$.

Suppose there exists $k < n$ such that $f^k(x_0) = x_0$. In this case, $x_0 = f^k(x_0) \in I_2$. Thus $x_0 \in I_1 \cap I_2 = \{b\}$. We show that this is impossible.

- for $n = 2$, $f^2(b) = f^2(x_0) = x_0 = b$, in contradiction with $f^2(b) = f^3(a) \leq a$.
- for $n \geq 3$, we have $f^2(b) = f^2(x_0) \in I_2$, in contradiction with $f^2(b) = f^3(a) \leq a$.

Thus for all $1 \leq j \leq n-1$, $f^j(x_0) \neq x_0$ and x_0 is of period n.

2. Newton's Method in ℝ

Newton's Method which we used earlier is itself particularly fruitful for research into solutions of equations and of non-linear systems.

Let f be a continuous map from ℝ into ℝ. We want to find an approximate solution to the equation $f(x) = 0$. The intermediate value theorem assures us that a sufficient condition that f vanishes on an interval $[a, b]$ is $f(a)f(b) \leq 0$. We may then us a method of bisection to determine a zero of f in n steps with a precision of $\dfrac{b-a}{2^n}$.

If f is of class C^1, we prefer, in the general case, to use a sequence (x_n) defined by the recurrence relation:

$$\begin{cases} x_0 \in I \\ x_{n+1} = x_n - \dfrac{f(x_n)}{f'(x_n)} \end{cases}$$

Of course, this sequence is not always defined. But, if f' doesn't vanish (at least locally), it is obvious that determining a zero of f is equivalent to determining a fixed point of $g : x \mapsto x - \dfrac{f(x)}{f'(x)}$.

The sequence (x_n) has a geometric interpretation:

The line of slope $f'(x_n)$ passing through the point $(x_n, f(x_n))$ has the equation $y = \dfrac{x - x_n}{f'(x_n)} + f(x_n)$. It strikes the x-axis at $x = x_n - \dfrac{f(x_n)}{f'(x_n)} = x_{n+1}$.

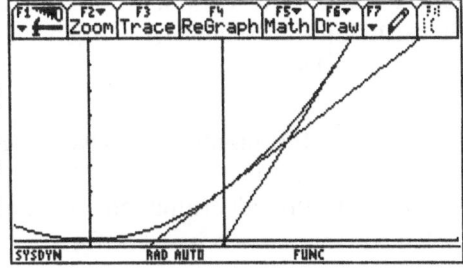

We plotted the tangent a point on this curve x_n, then its point of intersection with the x-axis (x_{n+1}), then the point of intersection of the line with equation $x = x_{n+1}$ with the curve, then the tangent and its point of intersection with the x-axis x_{n+1}...

Convergence of Newton's Method is only assured locally, if at all. In fact:

Theorem 1: Let $a < b$ and f be a function of class C^2 of $[a, b]$ with real values. Suppose there exists $x \in [a, b]$ such that $f(x) = 0$ and $f'(x) \neq 0$. Then there exists $\varepsilon > 0$ such that for all $x_0 \in [x - \varepsilon, x + \varepsilon]$, the Newton sequence defined by:

$$\begin{cases} x_0 \in [x - \varepsilon, x + \varepsilon] \\ x_{n+1} = x_n - \dfrac{f(x)}{f'(x_n)} \end{cases}$$

is defined and converges to x as n tends to infinity.

Proof : Since $f(x) = 0$, we may write:

$$x_{n+1} - x = x_n - \frac{f(x_n)}{f'(x_n)} - x + \frac{f(x)}{f'(x_n)}$$
$$= \frac{(x_n - x)f'(x_n) - f(x_n) + f(x)}{f'(x_n)}$$

Now:
- the function f' is continuous and $f'(x) \neq 0$. Then there exists $\alpha > 0$ and $K > 0$ such that, for all $x \in [x - \alpha, x + \alpha]$: $|f'(x)| > K$.
- the function f is of class C^2, so Taylor's formula with remainder in x gives:

$$f(x) = f(x_n) + (x - x_n)f'(x_n) + \int_{x_n}^{x} (x - t)f''(t)dt$$

and there exists $M > 0$ such that $\displaystyle\sup_{x \in [x - \alpha, x + \alpha]} |f''(x)| \leq M$. Thus:

$$|(x_n - x)f'(x_n) - f(x_n) + f(x)| \leq \left| \int_{x_n}^{x} (x - t)f''(t)dt \right| \leq \frac{M}{2}|x_n - x|^2$$

and

$$|x_{n+1} - x| \leq \frac{M}{2K}|x_n - x|^2$$

If we put $a_n = \dfrac{M}{2K}|x_n - x|$, then $a_{n+1} \leq a_n^2$, when $x_n \in [x - \alpha, x + \alpha]$.
Take $\varepsilon < \min(\alpha, (2K)/M)$. Then if $|x_n - x| \leq \varepsilon$, we have $|x_{n+1} - x| \leq \varepsilon$.
By recursion, the sequence (x_n) is well defined and the sequence (a_n) satisfies $a_n \leq a_0^{2^n}$, which tends to 0 as n tends to infinity, since $|x - x_0| \leq \varepsilon$ (thus $0 \leq a_0 < 1$).
We will note the rapidity of convergence of this method.

We will test this algorithm on the function $f : x \mapsto x^3 - x$, whose roots are $\{-1, 0, 1\}$.

The function g associated with the the Newton sequence is then:

$$g : x \mapsto \frac{2x^3}{3x^2 - 1}$$

- for all $x_0 > \dfrac{1}{\sqrt{3}}$, the sequence (x_n) converges to 1. In fact, studying the variations of g, we verify that $x_1 > 1$ and that for all $x > 1, g(x) < x$. The sequence $(x_n)_{n \geq 1}$ is then decreasing, bounded below by 1 and thus converges to the only limit which is possible which is 1.

- for reasons of symmetry, for all $x_0 < -\dfrac{1}{\sqrt{3}}$, the sequence (x_n) converges to -1.

- for x_0 in a neighborhood of 0, the theorem shows moreover that we are assured of the convergence of the sequence (x_n) to 0.

But, the picture is scrambled when x_0 is "far" from 0.

When $x_0 = 0.44720$, the sequence (x_n) tends to 0.

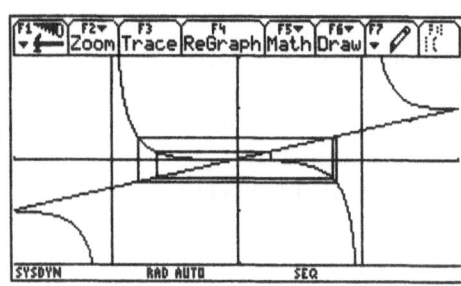

But, when $x_0 = 0.44725$, it tends to -1.

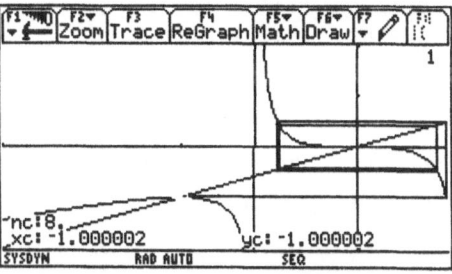

and when $x_0 = 0.44730$, it tends to 1.

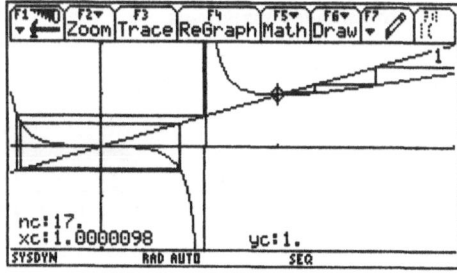

The next program, studies the convergence of the sequence (x_n), following the values of $x_0 \in [\text{xmin}, \text{xmax}]$.

```
:racines()
:Prgm
:Local x,i,j,l0,l1,l2,ll0,ll1,ll2
:{}→l0:{}→l1:{}→l2
:For i,xmin,xmax,(xmax-xmin)/238    for all the pixels on TI-92+
:i→x
:For j,1,10                         calculate from 10
:g(x)→x                            terms of the sequence
:EndFor
:If abs(x+1)<0.001 Then            if (xₙ) tends to −1
:augment(l2,{i})→l2                add i to l2
:ElseIf abs(x)<0.001 Then          if (xₙ) tends to 0
:augment(l0,{i})→l0                add i to l0
:Else                              otherwise
:augment(l1,{i})→l1                add i to l1
:EndIf
:EndFor
:FnOff :PlotsOff
:newList(dim(l0))→ll0              for the graph statistics
:Fill -0.5,ll0                     l0 is plotted at the bottom of the screen
:newList(dim(l1))→ll1
:Fill 0.5,ll1                      l1 is plotted at the top of the screen
:newList(dim(l2))→ll2              l2 is plotted at the middle
:NewPlot 1,1,l1,ll1,,,,2           plot !
:NewPlot 2,1,l0,ll0,,,,3
:NewPlot 3,1,l2,ll2,,,,1
:EndPrgm
```

And here is the resulting graph. The first graph corresponds to xmin=−0.6, xmax=−0.4. The second one is zoomed in on the first.

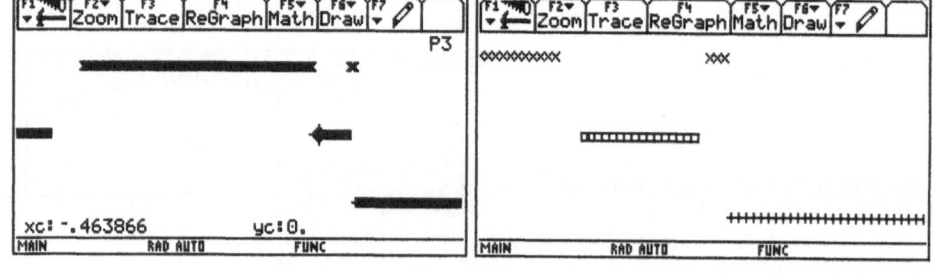

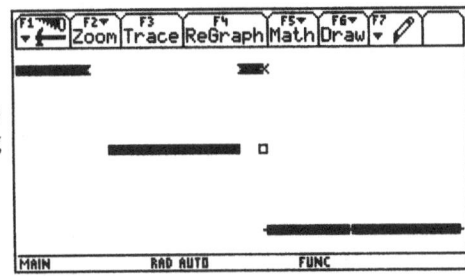

We then recalculated the values of x_0 in the window defined by the preceding zoom, $x \in [-0.475, -0.425]$.

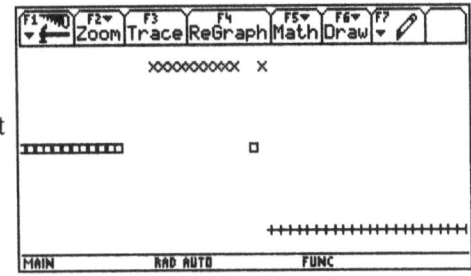

And a new zoom. Here we again get a fractal structure

The behavior of Newton's Method is thus different from that of the preceding dynamical systems. But the fractal structure discussed previously returns equally as much. And simple almost elementary questions (does the sequence converge or no?) give birth to complex problems with no answer known even today.

We will return to Newton's Method during our study of dynamical systems in the plane. Again, the algorithm will reveal numerous surprises.

3. Equivalence between different systems

The two dynamical systems which we have studied are, in fact, equivalent. This means that study of the logistic model of Verhulst is equivalent to study of the sequence defined by the recurrence $\begin{cases} x_0 \\ x_{n+1} = ax_n(1 - x_n) \end{cases}$

In fact, if we have:
$$\begin{cases} p_{n+1} = p_n + rp_n(1 - p_n) \\ x_{n+1} = ax_n(1 - x_n) \end{cases}$$

We may use the identification:

$$x_n = \frac{r}{1+r}p_n, \quad a = r + 1$$

since then:

$$x_{n+1} = \frac{r}{r+1}p_{n+1} = \frac{r}{r+1}(p_n + rp_n(1 - p_n))$$

$$= rp_n - \frac{r^2}{r+1}p_n^2$$

and

$$x_{n+1} = ax_n(1 - x_n) = (r+1)\frac{r}{r+1}p_n\left(1 - \frac{r}{r+1}p_n\right)$$

$$= rp_n - \frac{r^2}{r+1}p_n^2$$

We show likewise that these two systems are equivalent to a system defined by $u_{n+1} = 1 - au_n^2$. We will use this fact when we study the Hénon attractor.

We set:

$$u_n = \frac{r}{a}\left(x_n - \frac{1}{2}\right), \quad a = \frac{r(r-2)}{4}$$

in this case:

$$u_{n+1} = 1 - au_n^2 = 1 - a\frac{r^2}{a^2}\left(x_n - \frac{1}{2}\right)^2$$

$$= -\frac{r^2}{a}x_n^2 + \frac{r^2}{a}x_n + 1 - \frac{r^2}{4a}$$

and

$$u_{n+1} = \frac{r}{a}\left(x_{n+1} - \frac{1}{2}\right) = \frac{r}{a}\left(rx_n(1 - x_n) - \frac{1}{2}\right)$$

$$= -\frac{r^2}{a}x_n^2 + \frac{r^2}{a}x_n - \frac{r}{2a}$$

and $1 - \frac{r^2}{4a} = -\frac{r}{2a}$ for $4a = r(r-2)$.

4. Dynamical systems in the plane

We leave our study of dynamical systems in $\mathbb{R}$, that is, of the sequences defined by a relation of the form $x_{n+1} = f(x_n)$, with f a function from $\mathbb{R}$ into $\mathbb{R}$. We will now be interested in dynamical systems of the complex plane, that is, of sequences defined by the same type of recurrence, but this time with $f : \mathbb{C} \to \mathbb{C}$.

4. 1 Julia sets

To continue with the quadratic functions, we begin with $f : z \mapsto z^2 + c$, where $c \in \mathbb{C}$.

Take the (very) particular case where $c = 0$. Let (z_n) be the sequence defined by:

$$\begin{cases} z_0 \in \mathbb{C} \\ z_{n+1} = z_n^2 \end{cases}$$

It is clear that if:

- $|z_0| < 1$, the sequence (z_n) tends to 0.
- $|z_0| = 1$, the sequence (z_n) remains on the unit circle: for all $n, |z_n| = 1$.
- $|z_0| > 1$, the sequence (z_n) doesn't have a limit ($\lim_{n \to +\infty} |z_n| = \infty$).

We may consider that here there are two attracting points: 0 and infinity, and an entire set, the unit circle, stable under the map f, for which, we don't know if the sequence converges, diverges, has cycles, etc.

For $f_c : x \mapsto z^2 + c$, we will adopt a similar terminology in defining the set $E_c = \{z_0 \in \mathbb{C} \ / \ \lim_{n \to +\infty} |z_n| = \infty\}$ (the set of initial values of the sequence (z_n) for which the sequence "escapes" to infinity). We will call the boundary of E_c a "Julia set".

It is easy to propose an algorithm which allows us to visualize the Julia set corresponding to any complex value c. The idea is quite simple:

We choose z_0 and calculate the terms of the sequence (z_n). Consider that the sequence (z_n) has escaped if after N iterations, the point z_n is outside the disc centered at 0 with radius 2. In this case, we illuminate the pixel corresponding to z_0. Unfortunately, the graphic representation from this algorithm immediately inspired by the definition of E_c is much too long to plot. Also, it is preferable to use the following algorithm, less precise, but sufficient to visualize our set.

The Julia set of the system defined by $f_c : z \mapsto z^2 + c$ is the boundary of the points z_0 such that $\lim_{n \to +\infty} |z_n| = \infty$. We may determine it by calculating the preimages of z_0 instead of its images.

The preimage of $w \in \mathbb{C}$ by f_c is just a solution of the equation $z^2 + c = w$. This equation has two solutions, and we may choose one at random, which we denote z_{-1}. We repeat the process with z_{-1} and we thus define a sequence (z_{-n}), which approaches the Julia set.

We show that for the function $f_0 : z \mapsto z^2$: $|z_0| < 1$ or $|z_0| > 1$. One has $\lim_{n \to +\infty} |z_{-n}| = 1$ for $|z_{-n}| = |z_0|^{1/n}$. The general case for f_c, $c \neq 0$ is easily deduced.

The following program puts this algorithm to work.

```
:julia(c,z0)                              Julia for c ∈ ℂ, with initial term z₀
:Prgm
:Local w,n,l,a,b,ex,ey
:ClrDraw:ClrGraph:FnOff
:z0→w                                     initialization
:(xmax+xmin)/2→a                          abscissa of center
:(ymax+ymin)/2→b                          ordinate of center
:xmax-xmin→ex                             scale for the y-axis
:ymax-ymin→ey                             scale for the x-axis
:For n,1,10                               calculate 10 terms without plotting
:cZeros(z^2+c-w,z)→l                      find the square
:when(rand()<.5,l[1],l[2])→w              random choice
:EndFor
:For n,11,10000                           we plot 10 000 points
:cZeros(z^2+c-w,z)→l                      find the square root
:when(rand()<.5,l[1],l[2])→w              random choice
:PtOn (real(w)-a)/ex,(imag(w)-b)/ey       plot a point
:EndFor
:EndPrgm
```

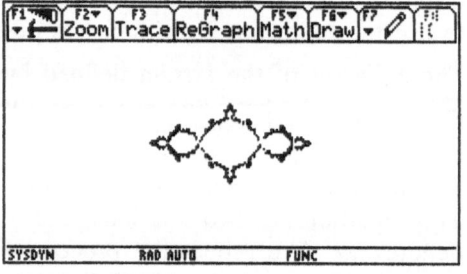

Here are various Julia sets. First, we take $c = -1$.

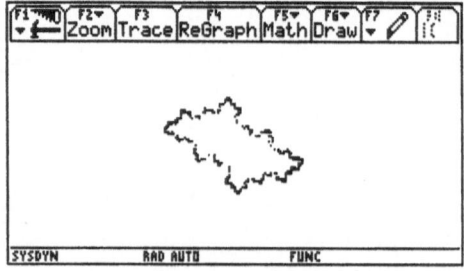

Then, we choose $c = -1/2 + i/2$.

Finally, we have "Douady's rabbit" for $c = 0.12 + 0.74i$. Adrien Douady is the principal mathematician to have studied Julia sets and their dictionary, the Mandelbrot Set. Sadly, the calculator plot is quite fuzzy and somewhat time-consuming.

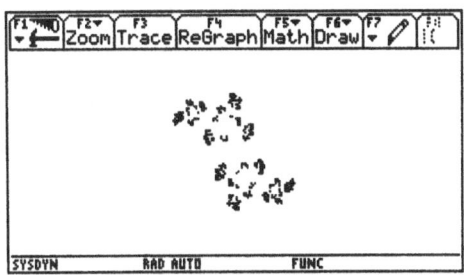

It is particularly interesting to study the fractal structure of Julia sets. For each value of c, the corresponding Julia set is characterized by invariance with respect to the function f_c and its inverse; that is how we plotted it. This invariance is based on a quadratic, and thus a non-linear, transformation. Each Julia set thus contains copies of itself, but deformed copies. Each of Douady's rabbits contains plenty of little rabbits!

A more precise study of Julia sets demands the use of more powerful mathematical utilities, such as functions of a complex variable. Thus, we will be content with a sketch.

4. 2 The Mandelbrot Set

We don't want to leave Julia sets without saying a word about the celebrated Mandelbrot Set, certainly the most popular and best known mathematical object among the public at large. Who hasn't seen the astonishing images, and the sometimes tormenting convolutions of sketches repeating to infinity. But the Mandelbrot Set is not just a simple decorative object. It certainly has many applications in mathematics, but also in practically all scientific domains: physics, biology, chemistry, even economics. Also, we can't mention the theory of complexity without citing this set. In spite of its importance we will stay on a modest level and be content to visualize it.

The Mandelbrot Set is a "map" (or a dictionary) of fractal curves. To plot the Mandelbrot Set bounded by Julia curves of the form $f_c : z \mapsto z^2 + c$, the algorithm involves considering, for each value of c corresponding to each pixel of the screen of the calculator, whether the sequence $(f^n(c))$ escapes or not. By "escape", we mean that it leaves the disk of radius 2.

The following program puts this idea to work.

```
:mandelb()
:Prgm
:Local a,b,c,i,j,k,x,y,z
:2.5/238→a                      for all TI-92+ pixels
:2./102→b
:For i,1,238                    also for the TI-92+
:For j,1,102
:a*i-2→x
:b*j→y
:x+i*y→c                        c is the pixel
:c→z
:0→k
:While abs(z)≤2 and k≤10        while z does not escape
:k+1→k
:z^2+c→z                        calculate the next element
:EndWhile
:If abs(z)≤2 Then               if z does not escape
:PtOn {x,x},{y,-y}              plot the pixel(+ symmetry)
:EndIf
:EndFor
:EndFor
:EndPrgm
```

Here is the Mandelbrot Set bound with the Julia curves. Then we show a zoom to a subset of this set. Note that the time to generate this screen is several hours!

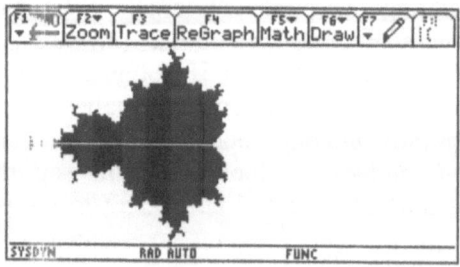

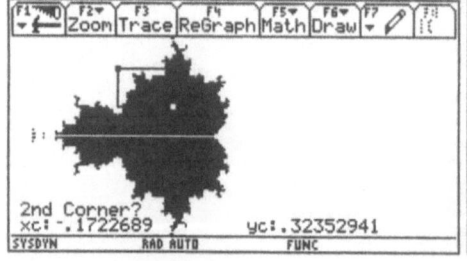

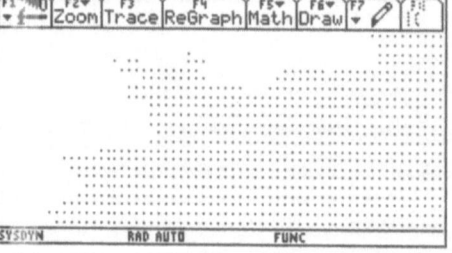

The properties of the Mandelbrot Set are numerous, intricate, and complex. They are still the object of much research. Among others we cite these results: the Mandelbrot Set bounded by Julia curves is connected (Douady-Hubbard), its boundary has 2 as its fractal dimension (Shishikura), which seems to explain the complexity we get after several zooms.

4. 3 Revisiting Newton's Method

In the preceding paragraph we studied Newton's method in $\mathbb{R}$ and we discussed its fractal structure. Newton's Method is adaptable to dimensions $n \geq 1$. It thus allows us to obtain a numerical solution of a system of non-linear equations.

Let $f : \mathbb{R}^n \to \mathbb{R}^n$, with the system $f(x) = 0$. We suppose that f is differentiable on a convex domain $C \subseteq \mathbb{R}^n$, and we denote by df_x the differential of f at x.

Proposition 1: If df_x exists for all $x \in C$, where C is convex in $\mathbb{R}^n$, and if there exists a $\gamma > 0$ such that for all $(x, y) \in C^2$:

$$||df_x - df_y|| \leq \gamma ||x - y||$$

then for all $(x, y) \in C^2$:

$$||f(x) - f(y) - df_y(x - y)|| \leq \frac{\gamma}{2}||x - y||^2$$

Proof : The function $\varphi : t \in [0,1] \mapsto f(y + t(x - y))$ is differentiable on $[0, 1]$ and, by the rule for composition of differentials, for all $t \in [0, 1]$:

$$\varphi'(t) = df_{y+t(x-y)}(x - y)$$

Thus, for all $t \in [0, 1]$:

$$
\begin{aligned}
||\varphi'(t) - \varphi'(0)|| &= ||df_{y+t(x-y)}(x - y) - df_y(x - y)|| \\
&\leq ||df_{y+t(x-y)} - df_y||\, ||x - y|| \\
&\leq \gamma t ||x - y||^2
\end{aligned}
$$

Moreover:

$$
\begin{aligned}
||f(x) - f(y) - df_y(x - y)|| &= ||\varphi(1) - \varphi(0) - \varphi'(0)|| = ||\int_0^1 (\varphi'(t) - \varphi'(0))dt|| \\
&\leq \int_0^1 ||(\varphi'(t) - \varphi'(0))||dt \leq \gamma ||x - y||^2 \int_0^1 t\,dt \\
&= \frac{\gamma}{2}||x - y||^2
\end{aligned}
$$

We have then the:

Theorem 1: Let C be an open convex subset of $\mathbb{R}^n$ and let $f : C \to \mathbb{R}^n$ be differentiable on C. We suppose that:

a) there exists $\gamma > 0$ such that for all $(x, y) \in C^2$, $\|df_x - df_y\| \leq \gamma \|x - y\|$

b) there exists $\beta > 0$ such that for all $x \in C, df_x^{-1}$ exists and $\|df_x^{-1}\| \leq \beta$.

Then there exists $\alpha > 0$ such that for all x_0 such that $\|df_{x_0}^{-1}(f(x_0))\| \leq \alpha$, the sequence defined by x_0 and the recurrence relation:

$$x_{n+1} = x_n - df_{x_n}^{-1}(f(x_n))$$

is well defined and converges to a point ζ satisfying $f(\zeta) = 0$.

Proof : By hypothesis b), df_x^{-1} exists for all $x \in C$. Thus, the sequence (x_n) is well defined.

Moreover, $\|x_1 - x_0\| \leq \|df_{x_0}^{-1}(f(x_0))\| \leq \alpha$.

Let $n \geq 1$. By the definition of the elements x_n, we have:

$$\|x_{n+1} - x_n\| = \|df_{x_n}^{-1}(f(x_n))\| \leq \beta \|f(x_n)\|$$
$$= \beta \|f(x_n) - f(x_{n-1}) - df_{x_{n-1}}(x_n - x_{n-1})\|$$

Now, by the preceding proposition:

$$\|x_{n+1} - x_n\| \leq \frac{\beta\gamma}{2} \|x_n - x_{n-1}\|^2 = C \|x_n - x_{n-1}\|^2$$

A recursion immediately shows that for all $n \geq 1$:

$$\|x_n - x_{n-1}\| \leq C^{2^n} \|x_1 - x_0\|^{2^n - 1} \leq C^{2^n} \alpha^{2^n - 1}$$

We choose then α in a manner that $C\alpha < 1$. In this case, the sequence (x_n) is a Cauchy sequence since:

$$\|x_{m+1} - x_n\| \leq \sum_{k=n}^{m} \|x_{k+1} - x_k\| \leq \sum_{k=n}^{m} C^{2^k} \alpha^{2^k - 1}$$

This last quantity tends to 0 as the remainder of the series converges. Moreover:

$$\|x_{n+1} - x_0\| \leq \sum_{k=0}^{n} \|x_{k+1} - x_k\| \leq \sum_{k=0}^{n} C^{2^k} \alpha^{2^k - 1} \leq \frac{1}{\alpha} \frac{1}{1 - C\alpha} = r$$

The space $\mathbb{R}^n$ is complete, so the sequence (x_n) converges to a point ζ. We will show that $f(\zeta) = 0$.

>From a), we have:

$$\|df_{x_n} - df_{x_0}\| \leq \gamma \|x_n - x_0\| \leq \gamma r$$

thus:

$$||df_{x_n}|| \leq ||df_{x_0})|| + \gamma r = K$$

Since $f(x_n) = -df_{x_n}(x_{n+1} - x_n)$, we get that $||f(x_n)|| \leq K||x_{n+1} - x_n||$. This implies that $\lim_{n \to +\infty} f(x_n) = 0$, and since f is continuous: $f(\zeta) = 0$.

It is possible to visualize the convergence of a sequence toward a root of the equation $f(x) = 0$, when $f : \mathbb{C} \mapsto \mathbb{C}$ by Newton's Method. For example, for $f : z \mapsto z^3 - 1$, we obtain $g(z) = z - \dfrac{f(z)}{f'(z)} = \dfrac{2z^3 + 1}{3z^2}$. We define the sequence (z_n) by its real and imaginary parts as in the following screen:

The real part of $g(z)$ and and its imaginary part are stored in the variables u1 and u2.

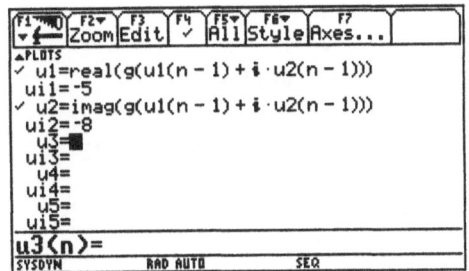

This is an example using a sequence which converges to 1.

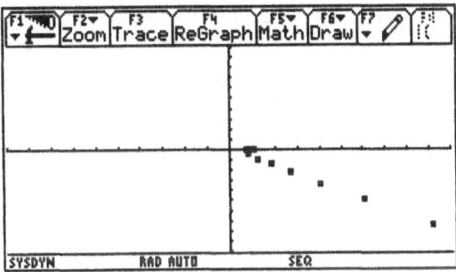

And here is a zoom in the neighborhood of 1.

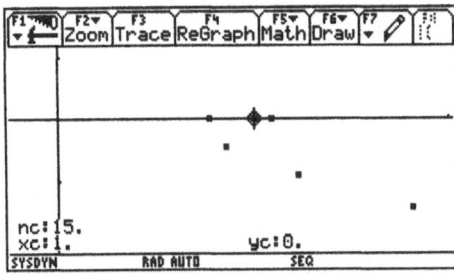

The equation $f(z) = 0$ has three roots $1, j, j^2$, the so-called "cube roots of unity."(We may take $j = -\frac{1}{2} + \frac{i\sqrt{3}}{2}$.) The convergence of Newton's Method is only assured locally. We could thus pose the same question as in the real case: for a function of the initial value z_0, to which root of $f(z) = 0$, will the sequence (z_n) converge? Will it here also have a fractal structure? The following program, adapted from that of the real case, will attempt to determine

34   Discrete Dynamical Systems

a graphic response. We say "attempt", since the program takes a very (very) long time to run. Its graphic results are also somewhat deceiving. This is due to the resolution on the screen of the calculator, which is relatively gross with respect to that of a computer.

```
:racinec()
:Prgm
:Local x,y,i,j,k,r
:{}→l0:{}→l1:{}→l2               a complex list
:For i,xmin,xmax,Δx              for each pixel
:For j,ymin,ymax,Δy
:approx(i+i*j)→x                 $x_0$
:g(x)→y
:While abs((y-x)/x)>0.001
:y→x                             calculate $x_n$
:g(x)→y                          until it approaches
:EndWhile                        a root
:If abs(x-1)<0.001 Then          determination of the root
:augment(l2,i+i*j)→l2            1
:ElseIf abs(x+1/2+i*√(3)/2)<0.001 Then
:augment(l0,i+i*j)→l0            $j$
:ElseIf abs(x+1/2-i*√(3)/2)<0.001 Then
:augment(l1,i+i*j)→l1            $j^2$
:EndIf
:EndFor
:EndFor
:FnOff :PlotsOff
:real(l1)→s1                     for the plot statistics
:imag(l1)→ss1                    real and imaginary part
:real(l0)→s0
:imag(l0)→ss0
:real(l2)→s2
:imag(l2)→ss2
:NewPlot 1,1,s1,ss1,,,,2         plot 3 lists
:NewPlot 2,1,s0,ss0,,,,3         use ×, +, box
:NewPlot 3,1,s2,ss2,,,,1         for $1, j, j^2$
:EndPrgm
```

We launched the program on $[-1.2, 1.2]$ $[-1.2, 1.2]$. Indeed, the screen was found to be saturated with illuminated pixels. We show a zoom onto $[-1.2, -1.01] \times [-1.2, 0.1]$.

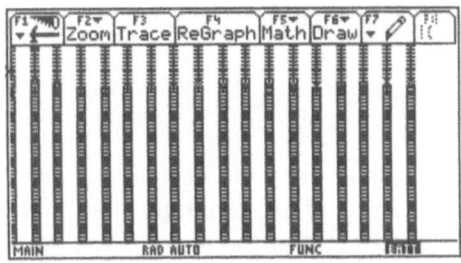

We decided to zoom some more, as indicated on the first screen. On the second, the sequences which converge to each of the roots of the equation $z^3 = 1$ are all jumbled. We obtain a fractal structure, though it is difficult to distinguish here.

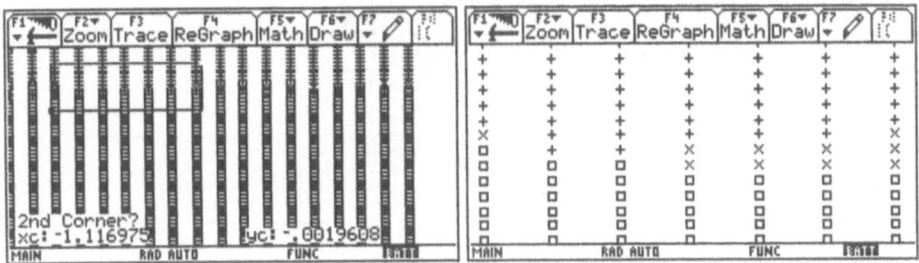

4. 4 The Hénon attractor

The Hénon attractor is a dynamical system defined in the plane $\mathbb{R}^2$ by the map:

$$f_{a,b} : (x, y) \mapsto (y + 1 - ax^2, bx)$$

where a and b are two real parameters. Our problem consists of studying the sequence of pairs (x_n, y_n) defined by:

$$\begin{cases} (x_0, y_0) \in \mathbb{R}^2 \\ x_{n+1} = y_{n+1} + 1 - ax_n^2 \\ y_{n+1} = bx_n \end{cases}$$

We note that it is possible to decompose f geometrically into three maps:
- a "folding", $f_1 : (x, y) \mapsto (x, y + 1 - ax^2)$. For example the x-axis is found to transform into a parabola turned toward the negative y-axis.
- a contraction, $f_2 : (x, y) \mapsto (bx, y)$ (when $|b| < 1$).
- a symmetry, $f_3 : (x, y) \mapsto (y, x)$ with respect to the line $y = x$.

As in all systems studied up to now, the dynamics depend on the values of two parameters a, b and and on the initial value of the sequence. We give a prime example, the historic example treated by Michel Hénon in 1976.

The Hénon attractor, for $b = 0.3$ and $a = 1.4$. The definition the two sequences and of the plot window are on the following screens.

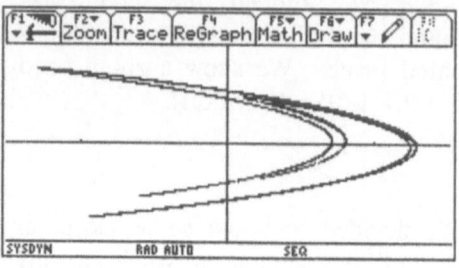

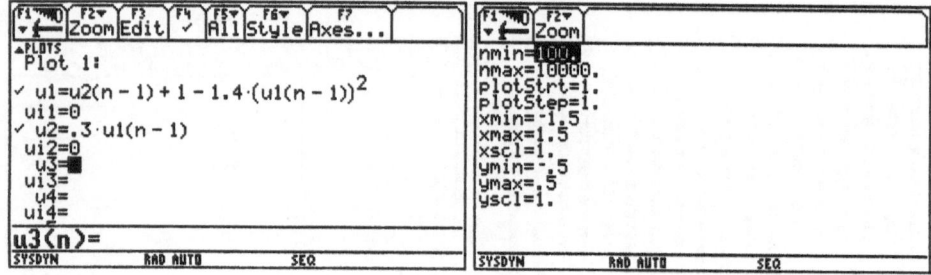

This first experience shows that the point $(0, 0)$ is attracted to the set laid out above, that is, that the sequence (x_n, y_n) "converges", not to a point, but to a set of points. This is nothing to be surprised at: we are familiar with the same phenomenon through our study of Julia sets. Not all points are attracted; for example, try the plot of the trajectory of point $(0, 1)$. It will leave the screen very swiftly, and a look at the table of values will show you that the series diverges.

For the initial value of $(0, 1)$, the table of values on the calculator shows that the sequence (x_n, y_n) seems to escape to infinity.

n	u1	u2			
3.	-2.15	.45			
4.	-5.022	-.645			
5.	-34.95	-1.506			
6.	-1710.	-10.48			
7.	-4.1e6	-513.1			
8.	-2.e13	-1.2e6			
9.	-8.e26	-7.e12			
10.	-8.e53	-2.e26			

n=3.

In spite of this there exists a region of the plane, determined experimentally by Hénon, for which all the points are attracting. This region is the quadrilateral Q defined by the points $P_1 = (-1.33, 0.42)$, $P_2 = (1.32, 0.133)$, $P_3 = (1.245, -0.14)$, $P_4 = (-1.06, -0.5)$. This region Q is therefore stable under $f_{1.4.0.3}$.

In fact this region gradually shrinks in correlation with application of f. In fact, while f is not a linear transformation of the plane (it is quadratic), we may locally replace it by its differential $\begin{pmatrix} -2ax & 1 \\ b & 0 \end{pmatrix}$, giving the determinant with value $-b$. We know that for a linear map, the modulus of the determinant, here $|b|$, represents a magnification factor for the map. Here $b = 0.3$. At the n-th step, the area of a quadrilateral will be multiplied by $(0.3)^n$. The Hénon attractor thus has "zero area".

A second property is that this attractor is dependent on initial conditions. In our first study, when the sequence (x_n) was defined by its first term x_0 and the converging recurrence relation $x_{n+1} = x_n + r x_n(1 - x_n)$, a small variation of x_0 had no influence on the convergence of the sequence. It is totally different here: a seemingly insignificant difference between two initial values will involve important differences after only a few terms.

The two sequences $(u1, u2)$ and $(u3, u4)$ are defined by the same recurrence relation, but the initial condition of the first is $(0,0)$ and that of the second is $(10^{-5}, 0)$. We see a noticeable difference appear after thirty terms.

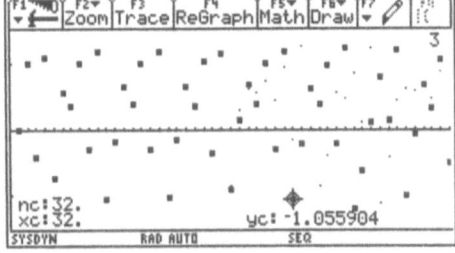

On this graph we plotted two sequences as a function of time n. The Trace option of the calculator lets us follow the differences..

A third property of our attractor resides in its fractal structure: a zoom onto a subset lets us retrieve another curve of the same type. The Hénon attractor consists of an infinite number of parabolic curves, each one bending among the others.

We remark that this attractor is also a generalization in dimension dimension 2 of the quadratic dynamical system studied previously in this chapter, since for $b = 0$, we obtain $x_{n+1} = 1 - ax_n^2$. When $b \neq 0$, do we get the Feigenbaum phenomenon? The answer is yes!

The fixed points of the Hénon transformation are given by the solutions of the system:

$$\begin{cases} x &= 1 + y - ax^2 \\ y &= bx \end{cases}$$

whose solutions are (x_1, y_1) and (x_2, y_2) defined by:

$$\begin{cases} x = \dfrac{b - 1 \pm \sqrt{(b - 1)^2 + 4a}}{2a} \\ y = bx \end{cases}$$

There are the calculations done by the machine.

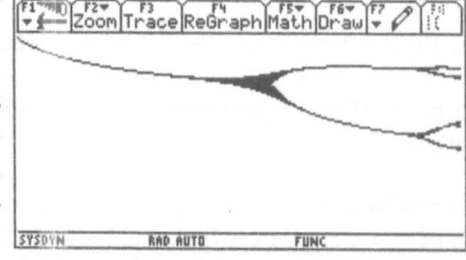

The discriminant $(b - 1)^2 + 4a > 0$ (so there is only one fixed point) if and only if $a < a_0 = -\dfrac{(b - 1)^2}{4}$. For $b = 0.3$ this gives $a_0 = -0.1225$. There is no fixed point for $a < a_0$ and when a exceeds this value, there are two fixed points (x_1, y_1) and (x_2, y_2), the first, with $\left(x_1 = \dfrac{b - 1 + \sqrt{(b - 1)^2 + 4a}}{2a} \right)$, being attracting.

We know that the differential of $f_{a,b}$ is the matrix $\begin{pmatrix} -2ax & 1 \\ b & 0 \end{pmatrix}$ whose eigenvalues are:

$$\lambda_{1,2} = -ax \pm \sqrt{a^2 x^2 + b}$$

For the fixed point x_1, we may find that the second eigenvalue is -1 when $a = a_1 = \dfrac{3}{4}(b - 1)^2$. This signifies the passage from a stable (attracting) fixed point to an unstable (repelling) fixed point. Thus, substituting $b = 0.3$, we get $a_1 = 0.3675$. At this point, there is a doubling of the period.

The Feigenbaum diagram adapted for the Hénon attractor. We get the structure of the doubling of unstable points. However not everything occurs just as in dimension 1. For $a \approx 1.08$, there are in fact two attractors, but the calculator can't show us this phenomenon!

We have only dented the surface of the subject of this chapter and of Hénon attractors in particular. But we have many other topics to explore with our calculator...

2

Differential Equations

A number of concrete problems of physics, biology, chemistry, ecology, economics... may be reduced to the study of functions whose derivatives satisfy certain relations: the movement of a mass particle, the process of evolution of a population, laws of market finance... We assume that the reader has had a first exposure to both applications and the solution of such problems. Here we give a detailed review of the subject of differential equations from a mathematician's point of view, using the graphic calculator as both a pedagogical aid and as a "solver."

An equation connecting the derivatives of an unknown function is called a differential equation. Attempting to solve such an equation by finding the unknown function leads the mathematician to give a formal framework of the subject and to construct a theory adapted to its solution.

1. Definition of the problem

Let $n \in \mathbb{N}$, be an open set of $\mathbb{R}^{n+2}$, and let F be a map from Ω into $\mathbb{R}$. A differential equation of order n is any equation of the form:

$$(E) \qquad F(x, y, y', \ldots, y^{(n)}) = 0$$

where the unknown is $y : I \to \mathbb{R}$.

If I is an interval, an I-solution of (E), is any function f of class $C^n(I, \mathbb{R})$, satisfying for all $x \in I$:

$$F(x, f(x), f'(x), \ldots, f^{(n)}(x)) = 0$$

The integer n is called the order of the differential equation (E) and the curves representing the solutions of (E) are called integral curves of (E).

Remarks.
1. Solving a differential equation thus consists of determining the set of solution pairs (I, f) formally when it is possible, otherwise graphically or with numerical tables.
2. We could generalize the definition above in the case of functions $F : \Omega \to \mathbb{C}$, with Ω an open set of $\mathbb{R} \times \mathbb{C}^{n+1}$. The solutions are then the pairs (I, f), with $f : I \to \mathbb{C}$.

For example, $2xy' + 5y + \cos(x) = 0$ is a differential equation of order 1, and $3y^2y'' - y^{(3)}y'^2(1+x^2) = 0$ is a differential equation of order 3.

Solving any differential equation is an arduous task and may even be impossible. In general, we don't know how to express the solution of most differential equations formally, even when such a solution exists. We may barely and only in certain cases guarantee the existence of a solution and more often only that of a local solution, a solution on an interval containing a point of interest. That is why the scientist must often simplify the hypotheses of a very complex problem in order to fit it into a mathematical framework.
We begin our theory with the most simple equations, those for which an answer to all the preceding questions is known: the linear differential equations of order 1.

2. Linear equations of first order

The equation:
$$(E): \qquad a(x)y' + b(x)y = c(x)$$
where a, b, c are continuous on an interval $J \subseteq \mathbb{R}$, with real values, is called a linear differential equation of first order.
The equation $(H):$ $a(x)y' + b(x)y = 0$ is called the associated homogeneous equation.

We say that the equation (E) is normalized if $a(x) = 1$, for all $x \in J$. We will **always** seek to normalize an equation, dividing by the function a, and restricting ourselves to the interval of continuity of the functions a, b, c.
For example, consider the equation $(E): xy' + (\cos x)y = e^x$. The functions a, b, c are continuous on $\mathbb{R}$. Normalizing this equation results in the equation $y' + \dfrac{\cos(x)}{x}y = \dfrac{e^x}{x}$. The new functions $a = 1, b = \dfrac{\cos(x)}{x}$ and $c = \dfrac{e^x}{x}$ are continuous on $\mathbb{R}^{>0}$ and $\mathbb{R}^{<0}$. Thus, we seek the I-solutions for $I \subseteq \mathbb{R}^{>0}$ and for $I \subseteq \mathbb{R}^{<0}$.

The equation (E) is not equivalent to the two previous equations. The solutions of (E) must thus be deduced from the solutions of these two equations.

Theorem 1: Denote by (E) the equation: $y' + b(x)y = c(x)$ and by (H) the equation: $y' + b(x)y = 0$, where b and c are continuous on an interval $I \subseteq \mathbb{R}$.

a) The set of solutions of (H) is an $\mathbb{R}$-vector space S_H which is of dimension 1.

b) The set of solutions S_E of (E) is an affine space with direction S_H.

Proof: a) It is easy to verify that S_H has the structure of a vector space, since S_H is non-empty (the function which is identically zero is there) and for $(f, g) \in S_H^2$, $\lambda \in \mathbb{R}$, then:

$$(\lambda f + g)'(x) + b(x)(\lambda f + g)(x) = \lambda(f'(x) + b(x)f(x)) + (g'(x) + b(x)g(x)) = 0$$

Since b is continuous on I, we may denote, for $x \in I$:

$$B : x \mapsto \int_{x_0}^{x} b(t)dt$$

where $x_0 \in I$.   For every differentiable function f on I, we denote $g(x) = e^{B(x)} f(x)$; then $f(x) = e^{-B(x)} g(x)$. The function g is differentiable and:

$$f'(x) = e^{-B(x)} g'(x) - b(x) e^{-B(x)} g(x) = e^{-B(x)} g'(x) - b(x) f(x)$$

Now we have:

$$f \in \mathcal{S}_H \Longleftrightarrow \forall\, x \in I, f'(x) + b(x) f(x) = 0$$
$$\Longleftrightarrow \forall\, x \in I, e^{-B(x)} g'(x) = 0$$
$$\Longleftrightarrow \forall\, x \in I, g'(x) = 0 \Longleftrightarrow \exists\, C \in \mathbb{R}, \forall\, x \in I, g(x) = C$$

Thus:

$$\mathcal{S}_H = \{ f : x \mapsto C e^{B(x)}, \ C \in \mathbb{R} \}$$

b) To say that $\mathcal{S}_E$ is an affine space with direction $\mathcal{S}_H$ means that:

$$\mathcal{S}_E = f_0 + \mathcal{S}_H$$

where f_0 is a particular solution of (E).
Equivalently, this means: every solution of (E) is the sum of a particular solution of (E) and of the general solution of (H).
In fact, if f_1 is a solution of (E), the linearity of differentiation implies that $(f_1 - f_0) \in \mathcal{S}_H$:

$$(f_1 - f_0)'(x) + b(x)(f_1 - f_0)(x) = f_1'(x) + b(x) f_1(x) - f_0'(x) - b(x) f_0(x) = 0$$

thus $\mathcal{S}_E \subseteq \mathcal{S}_H + f_0$.
Conversely, if $g \in \mathcal{S}_H$ and f_0 is a particular solution of (E), for the same reason of linearity $g + f_0$ is a solution of (E), so $\mathcal{S}_H + f_0 \subseteq \mathcal{S}_E$.

Remark: We note that, when b and c are continuous on I, the solutions of (E), as well as those of (H), are defined on the entire interval I.

Theorem 2: Let b and c be two continuous functions with real values on an interval $I \subseteq \mathbb{R}$. For every $(x_0, y_0) \in I \times \mathbb{R}$, there is a unique solution f of $(E) : y' + b(x)y = c(x)$ satisfying the initial condition $f(x_0) = y_0$.

Proof: We know that the general solution f of (E) may be written:

$$f : x \mapsto C e^{B(x)} + f_0(x)$$

where B is an anti-derivative of b on I, f_0 is a particular solution of (E) and C is a real constant. Then:

$$y_0 = Ce^{B(x_0)} + f_0(x_0) \Leftrightarrow C = (y_0 - f(x_0))e^{-B(x_0)}$$

which verifies our theorem.

Let's now give some examples of the use of the calculator to solve first order differential equations.

Here are some examples of first order differential equations. The equations are normalized and all of the functions involved are continuous on $\mathbb{R}$. The solutions are thus defined on all of $\mathbb{R}$.

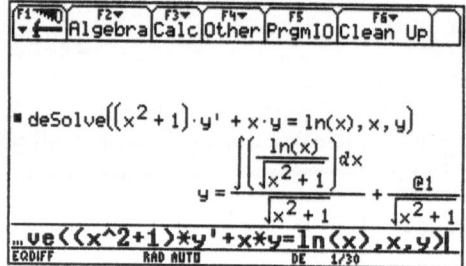

This is an example of a non-normalized first order differential equation. But the function $x \mapsto x^2 + 1$ doesn't vanish on $\mathbb{R}$, so the normalized equation is equivalent. The solution found is thus defined on $\mathbb{R}^{>0}$ as the natural logarithm function.

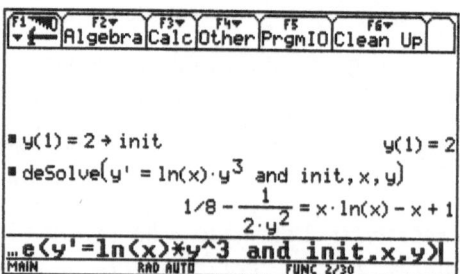

Now we have an example of a first order differential equation with an initial condition. It remains to express y as a function of x with the aid of the command solve. This equation is of order 1, but it is not linear.

2. 1 Non-normalized equation

We study the case of a non-normalized equation.

First example:
$$(E): \quad 2x(1+x)y' + (1+x)y = 1$$

The normalized equation is:

$$(E'): \quad y' + \frac{1}{2x}y = \frac{1}{2x(1+x)}$$

By the preceding theorem, we are assured of the existence of I-solutions for $I =]-\infty, -1[$, $I =]-1, 0[$, and $I =]0, +\infty[$ for (E) and (E'). In order that the calculator can determine the solutions of (E'), the mode must be placed in Complex Format RECTANGULAR .

We store the normalized differential equation in eqd and we require the solution by $x > 0$.

Now consider $x < -1$. The calculator was working in $\mathbb{C}$. That is why i appears in the result obtained.

And finally we look at $x \in]-1, 0[$. We remark that this solution is identical to the preceding one.

Thus, (E') has solutions on:

- $I_0 =\;] -\infty, -1[,\quad f_0 : x \mapsto \dfrac{C_0}{\sqrt{-x}} - \dfrac{\ln\left(\frac{\sqrt{-x}-1}{\sqrt{-x}+1}\right)}{2\sqrt{-x}} = \dfrac{C_0}{\sqrt{-x}} - \dfrac{\text{Argth}\sqrt{-x}}{\sqrt{-x}}$

- $I_1 =\;] -1, 0[,\quad f_1 : x \mapsto \dfrac{C_1}{\sqrt{-x}} - \dfrac{\ln\left(\frac{\sqrt{-x}-1}{\sqrt{-x}+1}\right)}{2\sqrt{-x}} = \dfrac{C_1}{\sqrt{-x}} - \dfrac{\text{Argth}\sqrt{-x}}{\sqrt{-x}}$

- $I_2 =\;] 0, +\infty[,\quad f_2 : x \mapsto \dfrac{C_2}{\sqrt{x}} + \dfrac{\text{Arctan}\sqrt{x}}{\sqrt{x}}$

Now we pass to the equation (E).

Let (I, f) be a solution of (E). The restriction of this solution to I_1 is a solution of (E'). It is thus f_1. Likewise, the solution on I_2 is a solution of (E'). This is thus f_2. Hence, if there is an I-solution f of (E), with $I_1 \subset I$, I contains a neighborhood of 0. Now f must be continuous and differentiable at $x = 0$. We must therefore study an extension of f_1 and f_2 to 0 by continuity, and we must assure that the function obtained is differentiable.

Since:

- f_1 has a limit at 0^+ if and only if $C_1 = 0$. In that case, $f_1(0) = 1$.
- f_2 has a limit at 0^- if and only if $C_2 = 0$. In that case, $f_2(0) = 1$.

Thus

$$f : x \mapsto \begin{cases} \dfrac{\text{Arctan}\sqrt{x}}{\sqrt{x}} & x > 0 \\ 1 & x = 0 \\ -\dfrac{\text{Argth}\sqrt{-x}}{\sqrt{-x}} & -1 < x < 0 \end{cases}$$

This function f is differentiable at $x = 0$, since a Taylor expansion at 0 gives:

$$\frac{\text{Arctan}\sqrt{x}}{\sqrt{x}} = 1 - \frac{1}{3}x + o(x)$$

$$-\frac{\text{Argth}\sqrt{-x}}{\sqrt{-x}} = 1 - \frac{1}{3}x + o(x)$$

One must replace x by $-x$ to obtain the result of the second Taylor expansion.

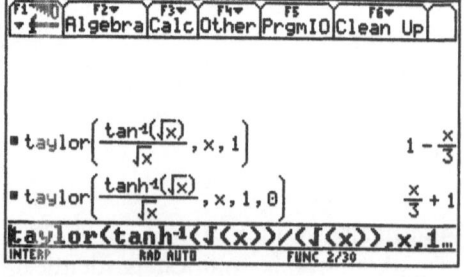

We remark that for $x = 0$, the equation (E) is satisfied since $f(0) = 1$.

Finally, there exists an I-solution for every interval I containing $]-1, +\infty[$, which is:

$$f : x \mapsto \begin{cases} \dfrac{\text{Arctan } \sqrt{x}}{\sqrt{x}} & x > 0 \\ 1 & x = 0 \\ -\dfrac{\text{Argth } \sqrt{-x}}{\sqrt{-x}} & -1 < x < 0 \end{cases}$$

We now consider the same issue at $x = -1$. But this is much simpler, since if $x = -1$ in (E), we obtain the contradiction $0 = 1$! Thus there is no solution of (E) on any interval containing -1.

Second example:

$$(e^x - 1)y' + (e^x + 1)y = 3 + 2e^x$$

We make the calculator consider separately the cases $x > 0$ and $x < 0$ ($x = 0$ is the only root of the equation $x \mapsto e^x - 1 = 0$). We remark that the two solutions are of the same form:

$$x \mapsto \frac{2e^{2x} + xe^x + 3}{(e^x - 1)^2} + \frac{C_i e^x}{(e^x - 1)^2}, \quad (i = 1, 2)$$

We have thus determined the I-solutions on $I_1 = \mathbb{R}^{>0}$ and $I_2 = \mathbb{R}^{<0}$. We determine the I-solutions for an interval I containing $x = 0$. Now:

We obtain the Taylor expansion of the numerator, then of the denominator of the solution about 0 and of order 3.

A Taylor expansion of order 3 about 0 of the solution (stored in f) gives:
• for the numerator:

$$(C + 5) + (C + 5)x + \frac{10 + C}{2}x^2 + \frac{C + 19}{6}x^3 + o(x^3)$$

- for the denominator:

$$x^2 + x^3 + o(x^3)$$

Since the denominator has a term of degree 2, the quotient

$$\frac{(C+5) + (C+5)x + \frac{10+C}{2}x^2 + \frac{C+19}{6}x^3}{x^2 + x^3}$$

has a limit at 0 if and only if $C = -5$. We then obtain:

$$\frac{\frac{5}{2}x^2 + \frac{7}{3}x^3}{x^2 + x^3}$$

and

For $C = -5$, the function obtained has a continuous and differentiable extension at 0.

Finally, replacing x by 0 in (E) gives $y(0) = 5/2$, which we have obtained above. In conclusion, there is a unique solution of (E) on every interval I containing 0 (and thus on all $\mathbb{R}$) which is $x \mapsto \dfrac{\frac{5}{2}x^2 + \frac{7}{3}x^3}{x^2 + x^3}$.

3. Non-linear first order equations

A first order differential equation, of the form $g(x, y, y') = 0$, must first be normalized, that is, taken to the form $y' = \varphi(x, y)$. In fact, there are no general theorems assuring the existence of a solution of equations of the form $g(x, y, y') = 0$. And, even in its normal form, the existence of one or several solutions of the normalized equation is not assured.

Let U be an open set of $\mathbb{R} \times \mathbb{R}$, $\varphi : U \to \mathbb{R}$ a continuous map, and $(x_0, y_0) \in U$. A solution of the Initial Value Problem (IVP):

$$(E) \quad \begin{cases} y' = \varphi(x, y) \\ y(x_0) = y_0 \end{cases}$$

is any pair (I, f) where I is an interval containing x_0 and $f : I \to \mathbb{R}$ is a solution of (E) such that $f(x_0) = y_0$.

Let U be an open set of $\mathbb{R} \times \mathbb{R}$, $\varphi : U \to \mathbb{R}$ a continuous map, and $(x_0, y_0) \in U$. A maximal solution of the Initial Value Problem:

$$(E) \quad \begin{cases} y' = \varphi(x, y) \\ y(x_0) = y_0 \end{cases}$$

is any solution (I, f) such that for every solution (I_1, f_1) of (E) which satisfies:

$$\begin{cases} I \subset I_1 \\ \forall\ x \in I, f_1(x) = f(x) \end{cases}$$

we have $I_1 = I$.

To say that (I, f) is a maximal solution of the IVP thus means that there does not exist a strict extension of the solution f of the maximal IVP to an interval I_1 containing the interval I.

Let U an open set of $\mathbb{R} \times \mathbb{R}$ and $\varphi : U \to \mathbb{R}$ a continuous map. We say that φ is locally Lipschitz with respect to the second variable if for every $(x_1, y_1) \in U$, there is a neighborhood V_1 of x_1, a neighborhood W_1 of y_1, and a real number $C > 0$ such that, for every $(x, y_2, y_3) \in V_1 \times W_1^2$:

$$|f(x, y_2) - f(x, y_3)| \leq C|y_2 - y_3|$$

We have then the following fundamental theorem:

Theorem 1:(Cauchy-Lipschitz). Let U be an open set of $\mathbb{R} \times \mathbb{R}$, $\varphi : U \to \mathbb{R}$ a continuous map, locally Lipschitz with respect to the second variable, and $(x_0, y_0) \in U$. Then, there is a unique maximal solution (I, f) of the IVP

$$(E) \quad \begin{cases} y' = \varphi(x, y) \\ y(x_0) = y_0 \end{cases}$$

Moreover, I is an open interval and every solution of (E) is the restriction of this maximal solution.

Remark: The condition required of φ is very general. In practice, the conditions satisfied by φ are more restrictive. The proof of the Cauchy-Lipschitz theorem is quite technical, so we only show a weaker version which is sufficient in most cases.

Theorem 2:(Cauchy-Lipschitz, weak version). Let $U = [a, b] \times \mathbb{R}$, $\varphi : U \to \mathbb{R}$ be a continuous map which satisfies:

There exists $C > 0$ such that, for every $x \in [a, b]$, for every $(y_1, y_2) \in \mathbb{R}^2$, $|f(x, y_1) - f(x, y_2)| \leq C|y_1 - y_2|$.

Then, for some $(x_0, y_0) \in [a, b] \times \mathbb{R}$, there is a unique solution $([a, b], f)$ of the IVP:

$$(E) \quad \begin{cases} y' = \varphi(x, y) \\ y(x_0) = y_0 \end{cases}$$

Remark: In its weakened version the Cauchy-Lipschitz theorem assures a maximal solution on all $[a, b]$ (those for which $(x_0, y_0) \in [a, b] \times \mathbb{R}$), and no more local solutions. The condition required of φ in this theorem is referred to as a "Lipschitz condition".

The proof of this second form of the Cauchy-Lipschitz theorem depends on the construction of a sequence of functions converging to the desired solution. This sequence is defined through use of the following:

Proposition 1:(I, f) is a solution of the IVP (E) if and only if we have:

$$\begin{cases} \forall\, x \in I, (x, f(x)) \in U \\ \forall\, x \in I, f(x) = y_0 + \displaystyle\int_{x_0}^{x} \varphi(t, f(t))dt \end{cases}$$

Proof: In fact, if (I, f) is a solution of the IVP, then f' is continuous, since $f'(x) = \varphi(x, f(x))$ and for every $x \in I$:

$$f(x) = f(x_0) + \int_{x_0}^{x} f'(t)dt = y_0 + \int_{x_0}^{x} \varphi(t, f(t))dt$$

Conversely, if:

$$\begin{cases} \forall\, x \in I, (x, f(x)) \in U \\ \forall\, x \in I, f(x) = y_0 + \displaystyle\int_{x_0}^{x} \varphi(t, f(t))dt \end{cases}$$

then, $f(x_0) = y_0$, f has a continuous derivative on I and by differentiation $f'(x) = \varphi(x, f(x))$.

We move now to the proof of the Cauchy-Lipschitz theorem, weak version. Consider the space $E = C^0([a, b], \mathbb{R})$ of continuous functions on $[a, b]$ equipped with the uniform norm $\|f\| = \sup_{x \in [a,b]} |f(x)|$, and the operator T on E defined for each $f \in E$, by:

$$\forall\, x \in [a, b] \quad T(f)(x) = y_0 + \int_{x_0}^{x} \varphi(t, f(t))dt$$

We show that T is a strictly contracting map.

$$\forall\, x \in [a, b] \quad T(f)(x) - T(g)(x) = \int_{x_0}^{x} (\varphi(t, f(t)) - \varphi(t, g(t)))\, dt$$

and, by the hypotheses of the theorem:

$$(*) \qquad \forall \, x \in [a, b] \quad |T(f)(x) - T(g)(x)| \leq \left| \int_{x_0}^{x} C \|f - g\| dt \right|$$

thus:

$$\|T(f) - T(g)\| \leq C \|f - g\| \max(|b - x_0|, |x_0 - a|)$$

If we have $C \|f - g\| \max(|b - x_0|, |x_0 - a|) < 1$, we could use the fixed point theorem. However, we don't have that knowledge. Thus, we use an iteration method, an idea of the mathematician Picard. Considering the inequality $(*)$, we know that:

$$\forall \, x \in [a, b] \quad |T(f)(x) - T(g)(x)| \leq C \|f - g\| |x - x_0|$$

We show by recursion that for every $n \geq 1$:

$$\forall \, x \in [a, b] \quad |T^n(f)(x) - T^n(g)(x)| \leq C^n \|f - g\| \frac{|x - x_0|^n}{n!}$$

This is true for $n = 1$, and if we suppose that it is true for $n - 1$, then:

$$|T^n(f)(x) - T^n(g)(x)| \leq \int_{x_0}^{x} C |T^{n-1}(f)(t) - T^{n-1}(g)(t)| dt \leq C^n \|f - g\| \frac{|x - x_0|^n}{n!}$$

This last quantity tends to 0 (it is the general term of a convergent series) and is thus strictly less than 1 after some value of n. Thus, there exists $n \in \mathbb{N}$ such that:

$$\frac{C^n}{n!} \max(|b - x_0|^n, |x_0 - a|^n) < 1$$

which implies that T^n is a contraction map and therefore has a unique fixed point f. This fixed point is likewise a fixed point of T, since $T^n(f) = f$ implies that $T^{n+1}(f) = T^n(T(f)) = T(f)$. Thus $T(f)$ is a fixed point of T^n. By uniqueness, $T(f) = f$.

We may easily show that for every $f \in E$, $T(f)$ is a function of class C^1, since f and φ are continuous. Thus, differentiating, for every $x \in [a, b]$:

$$f(x) = T(f)(x) = y_0 + \int_{x_0}^{x} \varphi(t, f(t)) dt \Rightarrow f'(x) = \varphi(x, f(x))$$

the initial condition $y_0 = f(x_0)$ is satisfied.

The converse has been demonstrated in the preceding proposition.

If we analyze the proof, we actually construct by recursion a sequence (f_n) of functions which converge to the solution sought. In fact, we have put:

$$\begin{cases} f_0(x) = y_0 \\ f_{n+1}(x) = y_0 + \int_{x_0}^{x} \varphi(t, f_n(t)) dt \end{cases}$$

The calculator lets us look at this sequence, which is called Picard's method of successive approximations:

We define $f_0(t) = 0$ and for every natural number n

$f_{n+1}(t) = \int_0^x (1 + f^2(t))dt$.

This corresponds to the differential equation $y' = 1 + y^2$.

We have calculated the first four iterations. The first three terms of the fourth result remind us of the Taylor expansion of the tangent function.

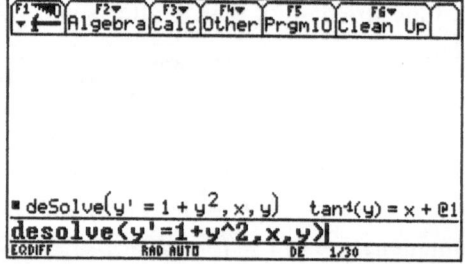

Verification : the general solution of the equation $y' = 1 + y^2$ is $y = \tan(x+C)$, which satisfies $f(0) = 0$ is actually $\tan(x)$.

We have graphed the tangent function and the preceding approximations, polynomials of degree 15 and 31 on $[-\pi, \pi]$. It appears that the approximation is correct in a neighborhood of 0. Even so, this method only gives one branch of the tangent function by reason of its discontinuity at the points $(2k + 1)\pi/2$.

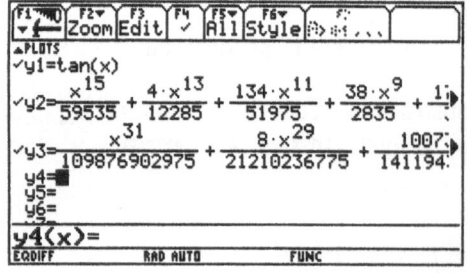

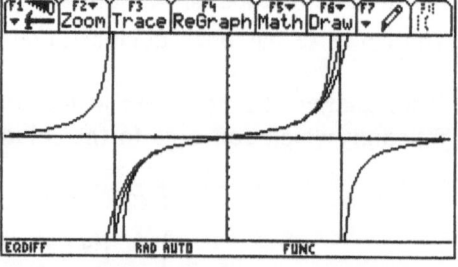

Three solutions of the same equation $y' = 1 + y^2$ with three different initial conditions at $x = 0$ (indicated by the little circles). Here we obtain three distinct branches of the tangent function.

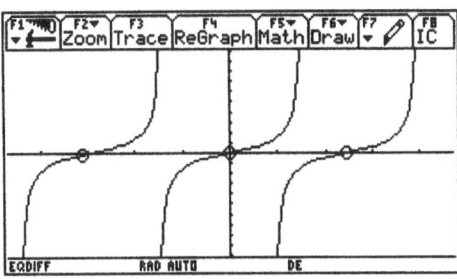

3. 1 Examples of the Cauchy-Lipschitz theorem

First example: Solve $y' - \dfrac{y}{x} + y^2 + \dfrac{1}{x^2} = 0$, for $x > 0$.

Letting the calculator work is enough to guide our demonstration.

The calculator shows that $x = 0$ poses a problem. Nonetheless it determines a solution in the implicit form $f(x, y) = 0$, and then we determine y as a function of x

It only remains to formalize precisely the work done in the preceding screen.

Let $(x_0, y_0) \in \mathbb{R}^{+*} \times \mathbb{R}$. The map $x \mapsto \dfrac{y}{x} - y^2 - \dfrac{1}{x^2}$ is of class C^1 on $\mathbb{R}^{>0} \times \mathbb{R}$. We may apply the Cauchy-Lipschitz theorem, either in its general form or in its weak form on any compact set of $\mathbb{R}^{>0} \times \mathbb{R}$ containing (x_0, y_0). We are thus assured of the existence and of the uniqueness of a maximal solution of the IVP at (x_0, y_0). We know then that the solution is of the form $f_a : x \mapsto \dfrac{\ln x + a + 1}{x(\ln x + a)}$.

The equation $f_a(x_0) = y_0$ has as a solution:

$$a_0 = \frac{1}{x_0 y_0 - 1} - \ln x_0$$

- if $x_0 y_0 = 1$, a_0 is not defined and the solution of the differential equation is $x \mapsto \dfrac{1}{x}$ which is defined on $\mathbb{R}^{>0}$

- if $x_0 y_0 \neq 1$, the solution is not continuous at $x = e^{-a_0}$. The restriction of this solution to one of the intervals, $]0, e^{-a_0}[$ or $]e^{-a_0}, +\infty[$ which contains x_0 is the

maximal solution we seek. We thus obtain two types of solutions of the form
$$\frac{1}{x} + \frac{1}{x \ln x + a_0 x}.$$

Here are the graphs of several solutions of the differential equation using different initial conditions. The graphs form a partition of the half plane $x > 0$. Note that no two such graphs ever meet: otherwise we would have a contradiction of the Cauchy-Lipschitz theorem at the point of intersection.

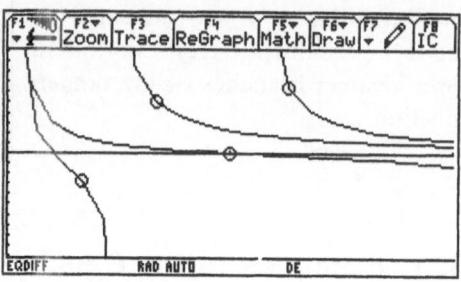

The slope field lets us visualize the graphs of many solutions as a function of the initial condition.

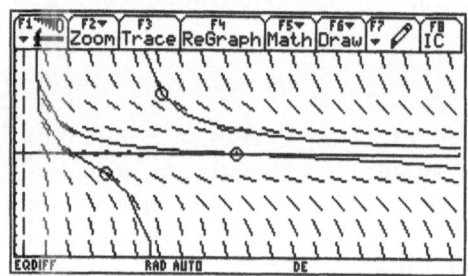

Second example: Solve $y' = \text{ch}(x + y)$.

Let the calculator guide us (if that is possible...)

The machine can't solve every problem without help. We propose a linear change of variable $z = x + y$. It then knows how to solve the equivalent differential equation.

In fact, if we put $z(x) = x + y(x)$, then $z'(x) = 1 + y'(x)$ and the equation $y' = \text{ch}(x + y)$ is equivalent to $z' = \text{ch}(z) + 1$.

Solving for z as a function of x doesn't pose a problem. We must always pay attention to the sign in the argument of the logarithm. That is what determines the interval for the definition of the maximal solution.

The general solution determined by the calculator is thus:

$$y_C : x \mapsto \ln\left(-\frac{x + 2C + 2}{x + 2C}\right) - x$$

The function $(x, y) \mapsto \mathrm{ch}(x + y)$ is of class C^1 on $\mathbb{R}^2$. The Cauchy-Lipschitz theorem assures us of the existence and of the uniqueness of the maximal solution of the IVP, for all $(x_0, y_0) \in \mathbb{R}^2$. We may now determine the maximal interval of definition of the solution.

The constant C is determined by $y_0 = \ln\left(-\dfrac{x_0 + 2C + 2}{x_0 + 2C}\right) - x_0$; thus, let:

$$C_0 = -\frac{x_0 e^{x_0 + y_0} + x_0 + 2}{2(e^{x_0 + y_0} + 1)}$$

The condition imposed so that we may express z, then y, as a function of x is $-\dfrac{2}{x + 2C} - 1 > 0$. A quick study of the function $g : x \mapsto -\dfrac{2}{x + 2C} - 1$ shows that g only remains positive on the interval $]-2 - 2C, -2C[$. This is the domain of the definition of the maximal solution.

In summary, for each $(x_0, y_0) \in \mathbb{R}^2$, there is a unique maximal solution of the IVP defined on $]-2 - 2C_0, -2C_0[$, with $C_0 = -\dfrac{x_0 e^{x_0 + y_0} + x_0 + 2}{2(e^{x_0 + y_0} + 1)}$. This solution is:

$$x \mapsto \ln\left(-\frac{x + 2C_0 + 2}{x + 2C_0}\right) - x$$

Here are the graphs of several solutions for different initial conditions. We distinguish, using the slope field, the "lines" which each graph may not cross, and the domain of definition of each solution.

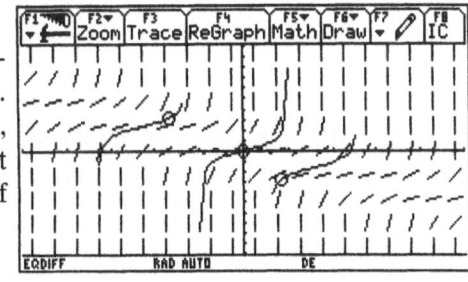

Third example: Solve $y' = y^2 - x$.

It is not fruitful here to use the function Desolve for the formal solution of the differential equation. This equation is well known as a Ricatti equation. Its solution may not be expressed in terms of the usual functions. Only a theoretical study, based on the Cauchy-Lipschitz theorem, will permit us to study and to graph the maximal solutions.

In fact, the map $(x, y) \mapsto y^2 - x$ is of class C^1 on $\mathbb{R}^2$. The Cauchy-Lipschitz theorem assures us, for all $(x_0, y_0) \in \mathbb{R}^2$, the existence and uniqueness of a maximal solution of the IVP, defined on an open interval.

We use our calculator to graph some solutions.

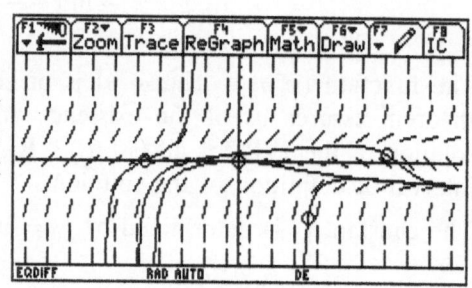

There are the graphs of several solutions. Some solutions are monotone; some seem to be defined on a strict subinterval of $\mathbb{R}$.

The curve representing the set of solutions of the equation $y'(x) = 0$, (the parabola with equation $x = y^2$), is called the zero isocline. It is a particularly important curve. Every solution which crosses this parabola at x_1 has a horizontal tangent at this point. Moreover, this curve partitions the plane into two regions: at every point of the interior of the first, each solution curve will be decreasing (it has a negative derivative); At every point of its exterior every solution curve will be increasing (it has a positive derivative).

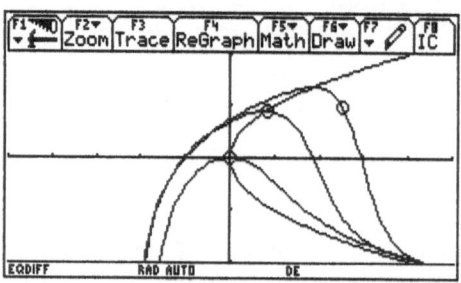

We have graphed the equation of $x = y^2$ and 3 solutions. Each solution in the exterior of the parabolic region is increasing and otherwise is decreasing.

This curve differentiates the solutions of our differential equation: those which don't intersect the parabola and those which do touch it.

Let $I =]a, b[$ be the interval of definition of the maximal solution f such that $f(x_0) = y_0$.

1. First we study the zeros of the derivative f', and we show that they are isolated. Let α be a zero of f'.
Since $f'(x) = f^2(x) - x$, the function f is of class C^2 and $f''(x) = 2f(x)f'(x) - 1$. Thus, $f''(\alpha) = -1$. Thus, there exists a neighborhood centered on α, $]\alpha - \delta, \alpha + \delta[$, contained in I, where f' is decreasing. Since $f'(\alpha) = 0$, $f'(x) > 0$ for $x \in]\alpha - \delta, \alpha[$ and $f'(x) < 0$ for $x \in]\alpha, \alpha + \delta[$. This proves that the zeros of f' are isolated points.

2. We show now that f' does not vanish more than once.
Suppose that f' has at least a second zero $\beta > \alpha$. Then f' is positive to the right of α and negative to the left of β. By continuity, the intermediate value assures us of the existence of another zero $\alpha < \gamma < \beta$. By recursion, we thus construct a sequence (x_n) of zeros of f' contained in $[\alpha, \beta]$. By a compactness argument, this sequence has a subsequence (x_{n_k}) which converges to some $x_0 \in [\alpha, \beta]$. By the continuity of f', we have $f'(x_0) = 0$, contradicting the fact that the zeros of f' are isolated, since x_0 is not isolated from the subsequence (x_{n_k}).
The demonstration for $\beta < \alpha$ is identical.

3. We now move on to the study of the variations among solutions of the same differential equation. We will distinguish several cases as a function of the initial condition.

- $y_0^2 - x_0 < 0$.

a) Suppose that there is an $x_1 > x_0$ such that $f'(x_1) = 0$. Then, the preceding work shows that $f'(x) > 0$ for $x < x_1$. But since $f'(x_0) = y_0^2 - x_0 < 0$, this implies that f' vanishes between x_0 and x_1. This is a contradiction. Thus, for every $x > x_0$, $f'(x) < 0$ and the function f is decreasing on $[x_0, b[$.
We show that $b = +\infty$. Suppose the contrary, that b is finite. The function f does not have a limit at b, since if it had one, say, ℓ, we could similarly define the solution f' at b by $f'(b) = \ell^2 - b$. This contradicts the maximality of the interval of definition of f. (Recall that this is an open interval.) Thus $\lim_{x \to b^-} f(x) = -\infty$, which implies that: $\lim_{x \to b^-} f'(x) = +\infty$, contradicting its sign.
Thus, $b = +\infty$.
Hence, f is decreasing on $[x_0, +\infty[$. Suppose that $\lim_{x \to +\infty} f(x) = \ell$. In this case,
$\lim_{x \to +\infty} f'(x) = -\infty$, and:

$$\lim_{x \to +\infty} f(x) = \lim_{x \to +\infty} \left(y_0 + \int_{x_0}^{x} f'(t)dt \right) = -\infty$$

Again we have a contradiction. By definition: $\lim_{x \to +\infty} f(x) = -\infty$.

b) We show that f' vanishes at a point of $]a, x_0[$. Otherwise, we know that f' remains negative on $]a, x_0[$. Hence, for all $x \in]a, x_0]$, $x = f^2(x) - f'(x) \geq 0$. Thus $a \geq 0$.
The function f is decreasing on $]a, x_0[$. If it has a limit at a, we could extend it to a, f, and then f', contradicting the maximality of the interval of definition. Thus $\lim_{x \to a^+} f(x) = -\infty$ and $\lim_{x \to a^+} f'(x) = +\infty$ in contradiction with its sign. Thus,

f' vanishes for at least one point at x_1 of the interval $]a, x_0[$, and only one, by the study of its zeros. We likewise know that for all $x \in]a, x_1[, f'(x) > 0$. We show that a is finite. If we suppose otherwise, (that is $a = -\infty$), then $\lim\limits_{x \to -\infty} f'(x) = +\infty$ and:

$$\lim_{x \to +\infty} f(x) = \lim_{x \to +\infty} \left(y_0 + \int_{x_0}^{x} f'(t)dt \right) = -\infty$$

Thus, there is a point $x_2 \in]-\infty, x_1[$ such that for $x < x_2, f(x) < 0$. Thus, for $x < x_2$:

$$\int_{x}^{x_2} \frac{f'(t)}{f^2(t)} dt = \frac{1}{f(x)} - \frac{1}{f(x_2)}$$

tends to $-\dfrac{1}{f(x_2)}$ when x tends to $-\infty$, and:

$$\int_{x}^{x_2} \frac{f'(t)}{f^2(t)} dt = \int_{x}^{x_2} \left(1 - \frac{t}{f^2(t)} dt \right) \geq \int_{x}^{x_2} dt = x_2 - x$$

which, itself, tends to $+\infty$. This is a contradiction and so a is finite.

If f has a limit at a^+, we could extend f and f', contradicting the maximality of the interval of definition of our solution. Thus, $\lim\limits_{x \to a^+} f(x) = -\infty$.

In summary, the interval of definition of the maximal solution is $]a, +\infty[$.

The case $y_0^2 - x_0 = 0$ may be treated in completely similar manner.

• $y_0^2 - x_0 > 0$.

The preceding reasoning at a, shows that a is finite and that $\lim\limits_{x \to a^+} f(x) = -\infty$. It suffices to repeat it.

If f' vanishes at a point of $]a, b[$, we fall back to the preceding arguments. We may thus suppose that f' does not vanish on $]a, b[$, that stays positive and thus that f is increasing on this interval. We show that b is finite. Suppose the contrary, $b = +\infty$.

Since $f^2(x) = x + f'(x) > x$, the function f tends to $+\infty$. Thus, there exists a point $\gamma > x_0$ such that for $x > \gamma, f(x) > 0$. Then we have, for $x > \gamma$:

$$\int_{\gamma}^{x} \frac{f'(t)}{f^2(t)} dt = \frac{1}{f(\gamma)} - \frac{1}{f(x)}$$

which tends to $\dfrac{1}{f(\gamma)}$ when x tends to $+\infty$. But, for $x > \gamma$:

$$f(x) = f(\gamma) + \int_{\gamma}^{x} f'(t)dt \geq f(\gamma) + (x - \gamma)f'(x_0)$$

with x_0 between x and γ, which shows that:

$$\lim_{t \to +\infty} \frac{t}{f^2(t)} = 0$$

Hence,:

$$\int_{\gamma}^{x} \frac{f'(t)}{f^2(t)} dt = \int_{\gamma}^{x} \left(1 - \frac{t}{f^2(t)}\right) dt$$

which tends to $+\infty$ when x tends to $+\infty$. This is a contradiction.

This is a family of solutions with the zero isocline, the curve on which the derivative of a solution vanishes. We distinguish various solutions which cross this parabola ($y'(x) = 0$) from other solutions. Once a solution gets inside the parabola, it stays inside forever. Graphically we get the results of our study.

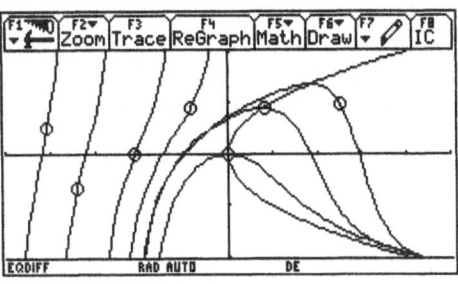

One of the lessons to be drawn from this example is that what is "obvious" graphically may not be so easy to justify mathematically.

4. Systems of differential equations of first order

A system of differential equations of first order is a system of the form:

$$\begin{cases} y'_1 &= \varphi_1(y_1, y_2, \ldots, y_n) \\ y'_2 &= \varphi_2(y_1, y_2, \ldots, y_n) \\ \vdots &= \vdots \\ y'_m &= \varphi_m(y_1, y_2, \ldots, y_n) \end{cases}$$

These systems are particularly important since the scalar differential equation order n is equivalent to such a system.
In fact, if $y^{(n)} = \psi(x, y, y', \ldots, y^{(n-1)})$ is such an equation, it is enough to write:

$$\begin{cases} y' &= y_1 \\ y'_1 &= y_2 \\ \vdots & \vdots \quad \vdots \\ y'_{n-2} &= y_{n-1} \\ y'_{n-1} &= \psi(x, y, y_1, y_2, \ldots, y_{n-1}) \end{cases}$$

For example, the second order equation $y'' = \psi(x, y, y')$ is equivalent to the system:

$$\begin{cases} y' = z \\ z' = \psi(x, y, z) \end{cases}$$

This is what one must do to graph the solutions of a system of equations on the calculator:

To graph the solution of the second order differential equation $y'' = -y' - y/2 + x/2$ with $y(0) = -3$, $y'(0) = 0$, we proceed as above. We only activate the graph of the first equation.

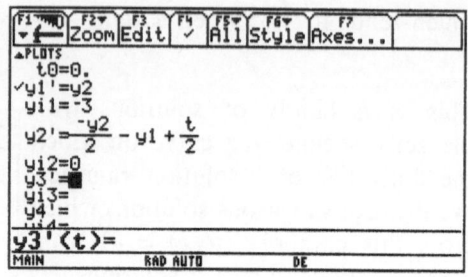

This is the graphical result for the above differential equation. The tic-mark is 1 on each axis.

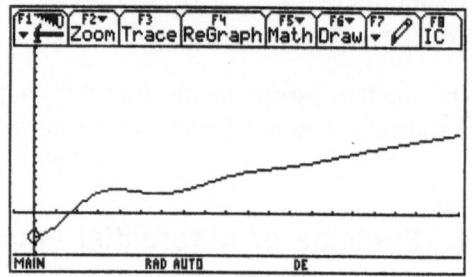

4. 1 Linear differential equations of first order

Let n be a positive integer. Any differential system of the form $X'(t) = A(t)X(t) + B(t)$, where B is a continuous function on $I \subseteq \mathbb{R}$ in $\mathbb{R}^n$, and A is continuous map of I into $\mathcal{M}_n(\mathbb{R})$ is called a first order linear differential system. Note that we have a matrix whose entries are continuous functions of a real variable. For example:

$$\begin{cases} x_1'(t) & = & \cos(t)x_1(t) + e^t x_2(t) + t^2 \\ x_2'(t) & = & 3x_1(t) + \ln(t)x_2(t) + 1 \end{cases}$$

is such a system. We could also write this in the matrix form: $X'(t) = A(t)X(t) + B(t)$, with:

$$X(t) = \begin{pmatrix} x_1 \\ x_2 \end{pmatrix}, \quad A(t) = \begin{pmatrix} \cos(t) & e^t \\ 3 & \ln(t) \end{pmatrix}, \quad B(t) = \begin{pmatrix} t^2 \\ 1 \end{pmatrix}$$

Remark: Here we treat only "square" linear systems: the matrix A belongs to $\mathcal{M}_n(\mathbb{R})$. The general case is not much more complicated.

Let A and B be two continuous maps defined on an interval I of $\mathbb{R}$, with values respectively in $\mathcal{M}_n(\mathbb{R})$ and $\mathbb{R}^n$. Let $t_0 \in I, Y_0 = \begin{pmatrix} y_1 \\ y_2 \\ \vdots \\ y_n \end{pmatrix} \in \mathbb{R}^n$. A solution of the

IVP at (t_0, Y_0) is any pair (J, F), where J is an interval of I and F is a function from J into $\mathbb{R}^n$ of class C^1 such that for all $t \in J, F'(t) = A(t)F(t) + B(t)$ and $F(t_0) = Y_0$.

The following theorem guarantees the existence and uniqueness of the solution of the IVP.

Theorem 1:(Cauchy-Lipschitz, linear case) Let I be an interval of $\mathbb{R}$, and let $A : I \to \mathcal{L}(\mathbb{R}^n)$ and $B : I \to \mathbb{R}^n$ be two continuous maps. The IVP problem, sometimes called the Cauchy problem in this book:

$$\begin{cases} X'(t) = A(t)X(t) + B(t) \\ X(t_0) = Y_0 \end{cases}$$

has a unique solution F defined on all of I.

Proof: The proof of this theorem has already been established in the case where I is compact. In fact, in this case, A is continuous on I, y is bounded and then it remains to apply the weak version of the Cauchy-Lipschitz theorem.

When I is not compact, let (K_n) be an increasing sequence of compact sets such that $\bigcup_n K_n = I$. (For example, if $I =]a, b]$, we could take $K_n = [a - 1/n, b]$). For each n, there exists a unique solution of the IVP F_n defined on K_n. Since $K_n \subset K_{n+1}$, F_{n+1} is an extension of F_n. Then we put, for every $t \in I$:

$$F(t) = F_n(t), \text{ if } t \in K_n$$

This definition is correct since if $t \in K_n$, then $t \in K_{n+j}$ for all $j \geq 1$ and F_{n+j} extends F_n. Finally, F is a solution of the IVP on (t_0, Y_0) (because the F_n are). This solution is unique since if there is another solution G, then for all n, the restriction of G to K_n will be F_n (by the uniqueness for the IVP on K_n), thus, for each $t \in I, F(t) = G(t)$.

We show that the structure of the set of solutions of a differential equation (E) $X'(t) = A(t)X(t) + B(t)$ is similar to that in the case of scalar equations.

Denote by $\mathcal{S}_E(I)$ the set of solutions of the equation (E) and by $\mathcal{S}_H(I)$ the set of solutions of the associated homogeneous equation (E) which is $(H) : X'(t) = A(t)X(t)$.

Proposition 1:

a) $\mathcal{S}_H(I)$ is an n dimensional vector space.

b) $\mathcal{S}_E(I)$ is an affine space with direction $\mathcal{S}_H(I)$.

(This is a geometric statement of the fact that if f_0 is a particular solution of (E), then any solution of (E) may be obtained as the sum of f_0 and some solution of (H): $\mathcal{S}_E = f_0 + \mathcal{S}_H$.)

Proof: The vector space structure of $\mathcal{S}_H(I)$ and the affine structure of $\mathcal{S}_E(I)$ may be shown as in the scalar case. The only difficulty here consists of determining the dimension of $\mathcal{S}_H(I)$.

For that, we use the, Cauchy-Lipschitz theorem. Let $t_0 \in I$ be fixed. The map:

$$\phi_{t_0} : \begin{array}{ccc} \mathcal{S}_H(I) & \to & \mathbb{R}^n \\ F & \to & F(t_0) \end{array}$$

which is clearly linear, is an isomorphism.

In fact, the Cauchy-Lipschitz theorem assures us of the bijectivity of ϕ_{t_0}, since for every $Y_0 \in \mathbb{R}^n$, there exists a unique $F \in \mathcal{S}_H(I)$ such that $F(t_0) = Y_0$. Thus, $\mathcal{S}_H(I)$ is of dimension n.

We will say that n functions $(F_1, F_2, \ldots, F_n)$ from $\mathcal{S}_H(I)$ form a basis of $\mathcal{S}_H(I)$ if the family $(F_1, F_2, \ldots, F_n)$ is a linearly independent set or if:

$$\forall\, t \in I, \sum_{k=1}^{n} \lambda_k F_k(t) = 0 \Rightarrow \lambda_1 = \lambda_2 = \ldots = \lambda_n = 0, \quad (\lambda_i \in \mathbb{R})$$

Now we have the following fundamental result:

Proposition 2: Let $(F_1, F_2, \ldots F_p)$ be elements of $\mathcal{S}_H(I)$. If there exists $t_0 \in I$ and p scalars $(\lambda_1, \lambda_2, \ldots, \lambda_p)$ such that $\sum_{k=1}^{p} \lambda_k F_k(t_0) = 0$, then:

$$\forall\, t \in I, \sum_{k=1}^{p} \lambda_k F_k(t) = 0$$

Proof: The function $\sum_{k=1}^{p} \lambda_k F_k(t)$ satisfies the initial condition $\sum_{k=1}^{p} \lambda_k F_k(t_0) = 0$, just like the function which is identically zero and which also appears in $\mathcal{S}_H(I)$. The Cauchy-Lipschitz theorem allows us to conclude the proof.

Recall that for each $1 \le i \le n$, F_i is is a vector of n scalar functions $F_i = \begin{pmatrix} F_{1,i} \\ F_{2,i} \\ \vdots \\ F_{n,i} \end{pmatrix}$.

In a similar manner, we have:

Proposition 3: The n functions $(F_1, F_2, \ldots, F_n)$ form a basis of $S_H(I)$ if and only if there exists $t_0 \in I$ such that the determinant:

$$\begin{vmatrix} F_{1.1}(t_0) & F_{2.1}(t_0) & \cdots & F_{n.1}(t_0) \\ F_{1.2}(t_0) & F_{2.2}(t_0) & \cdots & F_{n.2}(t_0) \\ \vdots & \vdots & \ddots & \vdots \\ F_{1.n}(t_0) & F_{2.n}(t_0) & \cdots & F_{n.n}(t_0) \end{vmatrix} \neq 0$$

This determinant is called the Wronskian of $(F_1, F_2, \ldots, F_n)$ and is denoted $W(F_1, F_2, \ldots, F_n)$.

This last proposition may also be demonstrated as follows:
$W(F_1, F_2, \ldots, F_n)$ is a differentiable function and by the rules for differentiation of determinants:

$$W'(F_1, F_2, \ldots, F_n) = \sum_{k=1}^{n} \begin{vmatrix} F_{1.1} & F_{2.1} & \cdots & F'_{k.1} & \cdots & F_{n.1} \\ F_{1.2} & F_{2.2} & \cdots & F'_{k.2} & \cdots & F_{n.2} \\ \vdots & & \ddots & & & \vdots \\ \vdots & & & \ddots & & \vdots \\ \vdots & & & & \ddots & \vdots \\ F_{1.n} & F_{2.n} & \cdots & F'_{k.n} & \cdots & F_{n.n} \end{vmatrix}$$

Using the fact that each function F_k is an element of $S_H(I)$ $(F'_k = AF_k)$ and subtracting for each determinant a linear combination of its other rows, it follows that:

$$W'(F_1(t), F_2(t), \ldots, F_n(t)) = \text{tr}(A(t))W(F_1(t), F_2(t), \ldots, F_n(t))$$

where $\text{tr}(A(t))$ designates the trace of the matrix $A(t)$. Thus, the Wronskian itself is the solution of a homogeneous linear system which satisfies, for all $t \in I$:

$$W(F_1(t), F_2(t), \ldots, F_n(t)) = W(F_1(t_0), F_2(t_0), \ldots, F_n(t_0))e^{\int_{t_0}^{t} \text{tr}(A(u))du} \neq 0$$

We may now determine the solutions of $S_E(I)$.

Theorem 2: Let $(F_1, F_2, \ldots, F_n)$ be a basis of $S_H(I)$. Denote by $\mathcal{W}$ the matrix:

$$\mathcal{W} = \begin{pmatrix} F_{1.1} & F_{2.1} & \cdots & F_{n.1} \\ F_{1.2} & F_{2.2} & \cdots & F_{n.2} \\ \vdots & \vdots & \ddots & \vdots \\ F_{1.n} & F_{2.n} & \cdots & F_{n.n} \end{pmatrix}$$

The general solution of the equation $(E):$ $X'(t) = A(t)X(t) + B(t)$ is of the form:

$$Y(t) = W(t) \int_{t_0}^{t} W^{-1}(u)B(u)du$$

Proof: We put $Y(t) = W(t)C(t)$ where C is a vector of functions to be determined. The function Y is differentiable and:

$$Y'(t) = W'(t)C(t) + W(t)C'(t)$$

From above, we know that $W'(t) = A(t)W(t)$. Thus:

$$Y'(t) = A(t)W(t)C(t) + W(t)C'(t)$$

The matrix W is invertible (its determinant is the Wronskian). We choose C such that:

$$C'(t) = W^{-1}(t)B(t)$$

or

$$C(t) = \int_{t_0}^{t} W^{-1}(u)B(u)du$$

Then we have:

$$Y'(t) = A(t)Y(t) + B(t)$$

which verifies that the formula of the theorem indeed gives a solution to the original non-homogeneous differential system.

The difficulty in the solution of the linear differential system (E) thus consists of determining the set $(F_1, F_2, \ldots, F_n)$, the basis of the space of solutions S_H of (H). There is no general method for this task. On the other hand, once a basis is found, it is easy to determine the solutions of (E) by virtue of the preceding proposition.

Example : Solve the system:

$$\begin{cases} x'(t) = \dfrac{(1+t^4)x(t) - 2t^2y(t)}{t(t^4 - 1)} + t \\ y'(t) = \dfrac{-2t^2x(t) + (1+t^4)y(t)}{t(t^4 - 1)} + t^2 \end{cases}$$

In terms of matrices this system may be written:

$$X'(t) = \begin{pmatrix} x'(t) \\ y'(t) \end{pmatrix} = \frac{1}{t(t^4 - 1)} \begin{pmatrix} 1+t^4 & -2t^2 \\ -2t^2 & 1+t^4 \end{pmatrix} X(t) + \begin{pmatrix} t \\ t^2 \end{pmatrix}$$

We may verify that the vectors: $\begin{pmatrix} 1/t \\ t \end{pmatrix}, \begin{pmatrix} t \\ 1/t \end{pmatrix}$ are two solutions of (H).

Moreover, their Wronskian has the value $\dfrac{1 - t^4}{t^2}$.   These solutions are independent on $]-\infty, -1[$, on $]-1, 0[$, on $]0, 1[$, and on $]1, +\infty[$. The solutions of (E) are then calculated directly by the machine using the formula of the theorem.

Define the matrix of solutions and calculate the Wronskian.

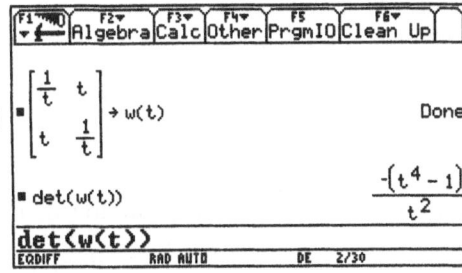

The solution of the general equation is calculated automatically using the formula shown in the last theorem. Here $t_0 = 0$.

4. 2 Differential systems with constant coefficients

When the matrix A is constant (that is, when its coefficients are real numbers independent of t), it is easy to determine a basis of $\mathcal{S}_H(I)$. The method of subsequently finding a general solution of the equation (E) has been explained above and is no different here.

Recall the notation:
(E) is the system $X'(t) = AX(t) + B(t)$, (H) is the associated homogeneous system $X'(t) = AX(t)$. With constant coefficients, $A = (a_{i,j}) \in \mathcal{M}_n(\mathbb{R})$, the set of real, square matrices. Hence:

$$\begin{cases} x_1'(t) &= a_{1,1}x_1(t) + a_{1,2}x_2(t) + \ldots + a_{1,n}x_n(t) + b_1(t) \\ x_2'(t) &= a_{2,1}x_1(t) + a_{2,2}x_2(t) + \ldots + a_{2,n}x_n(t) + b_2(t) \\ \quad \vdots & \quad \vdots \qquad\qquad\qquad\qquad \vdots \\ x_n'(t) &= a_{n,1}x_1(t) + a_{n,2}x_2(t) + \ldots + a_{n,n}x_n(t) + b_n(t) \end{cases}$$

For now, we occupy ourselves only with the homogeneous system $(H) : X'(t) = AX(t)$, where A is a square constant matrix.

Proposition 4: *Let A be a constant square matrix and let $(H) : X'(t) = AX(t)$. The function $e^{\lambda t}Y$ is a solution of (H) if and only if λ is an eigenvalue of A and Y is an associated eigenvector of λ.*

Proof: a) if λ is an eigenvalue of A and Y is an associated eigenvector, we have $AY = \lambda Y, Y \neq 0$. Then:

$$\begin{cases} d/dt(e^{\lambda t}Y) &= \lambda e^{\lambda t}Y \\ A(e^{\lambda t}Y) &= \lambda e^{\lambda t}Y \end{cases}$$

Thus, $e^{\lambda t}Y$ is a solution of (H).

b) Conversely, if $e^{\lambda t}Y$ is a solution of (H), with $Y \neq 0$:

$$\frac{d}{dt}e^{\lambda t}Y = Ae^{\lambda t}Y$$

let:

$$\lambda Y e^{\lambda t} = e^{\lambda t}AY$$

which implies that $A(Y) = \lambda Y$.

For each eigenvalue λ of a matrix, there are one or more linearly independent eigenvectors. We designate the set of all linear combinations of these eigenvectors as E_λ and call this set the eigenspace corresponding to λ. It is easy to see that E_λ is closed under addition and scalar multiplication and is thus a vector space.

Proposition 5: *Let λ be an eigenvalue of A and let $(Y_1, Y_2, \ldots, Y_k)$ be a basis of the eigenspace E_λ. Then the family of functions $(e^{\lambda t}Y_1, e^{\lambda t}Y_2, \ldots, e^{\lambda t}Y_k)$ is a linearly independent set of solutions of (H).*

Proof: We may see that each of these functions is a solution of (H), and since for all $1 \leq i \leq k$, at $t = 0$, $e^{\lambda 0}Y_i = Y_i$, the family is linearly independent.

We will now consider two special cases of interest which are readily solved.

Case a) The constant matrix A is a diagonal matrix or is at least diagonalizable.

Case b) The constant matrix A is a triangular matrix or is at least triangularizable. As will be seen, these two cases cover linear systems with constant coefficients completely.

4. 3 Case a) Diagonalizable differential systems

Definition: We say that $A \in \mathcal{M}_n(\mathbb{K})$ is diagonalizable in $\mathbb{K}$ if it is similar to a diagonal matrix D of $\mathcal{M}_n(\mathbb{K})$, that is, if there exists an invertible matrix P of $\mathcal{M}_n(\mathbb{K})$ such that $D = P^{-1}AP$.

Remark: Theorems from linear algebra tell us that (real) symmetric or (complex) Hermitian matrices are diagonalizable. For readers unfamiliar with these ideas it will be sufficient to consider diagonal matrices in a first reading of this material. These concepts are treated in later chapters, as well.

Theorem 3: Let $A \in M_n(\mathbb{R})$ be a *diagonalizable matrix*. Let $(\lambda_1, \lambda_2, \ldots, \lambda_p)$ be its distinct eigenvalues and for each $1 \leq j \leq p, let(Y_{j.1}, Y_{j.2}, \ldots Y_{j.k_j})$ be a basis of the eigenspace associated with λ_j.
The differential equation $(H) : X'(t) = AX(t)$ has as a basis of solutions the the functions:

$$e^{\lambda_1 t}(Y_{1.1}, Y_{1.2}, \ldots Y_{1.k_1}), e^{\lambda_2 t}(Y_{2.1}, Y_{2.2}, \ldots Y_{2.k_2}), \ldots, e^{\lambda_p t}(Y_{p.1}, Y_{p.2}, \ldots Y_{p.k_p})$$

Proof: We know already that $\mathcal{S}_H(\mathbb{R})$ is a vector space of dimension n. Moreover, for each $1 \leq j \leq p$, the family $e^{\lambda_j t}(Y_{j.1}, Y_{j.2}, \ldots Y_{j.k_j})$ is a linearly independent set generating a subspace V_j of $\mathcal{S}_H(\mathbb{R})$ with dimension k_j, and $\sum_{j=1}^{p} k_j = n$. Finally, the sum of the subspaces is a direct sum, since if $(X_1, X_2, \ldots, X_p)$ is family of p vectors, with $X_i \in V_i$, then the equation:

$$\sum_{j=1}^{p} \alpha_j X_j(t) = 0 \Rightarrow \sum_{j=1}^{p} \alpha_j X_j(0) = 0$$

which gives a linear combination of the eigenvectors of A.

In summary, there is a relatively simple method to determine the solutions of (H) (and then those of (E)) when the matrix A is diagonalizable:
• Determine the eigenvalues of A.
• For each eigenvalue, determine the associated eigenvectors.
• By the preceding theorem, the set all the eigenvectors for all the eigenvalues forms a basis of solutions of (H).
• Calculate the Wronskian $\mathcal{W}$ of the basis determined, then apply the formula:

$$Y(t) = \mathcal{W}(t) \int_{t_0}^{t} \mathcal{W}^{-1}(u)B(u)du$$

to determine the general solution of (E).

Example: Solve the system:

$$\begin{cases} x'(t) &=& 2x(t) - y(t) + z(t) + t \\ y'(t) &=& -x(t) + 2y(t) - z(t) + t^2 \\ z'(t) &=& x(t) - y(t) + z(t) + t^3 \end{cases}$$

The following screen shots show the steps outlined above:

Define the matrix A and calculate its eigenvalues. The matrix is real and symmetric, so it has three distinct eigenvalues and is diagonalizable in $\mathbb{R}$.

Then we determine the eigenvalues. It is better not to use the built-in functions
EigVl and EigVc which work in the real approximate mode and not in the
exact mode.

Calculate the basis of eigenvectors of
the three distinct eigenvalues determined above, using the definitions of
eigenvalue and eigenvector.

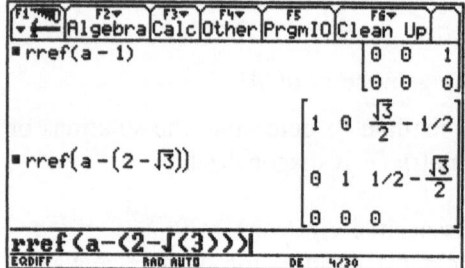

A basis of eigenvectors of A is thus:

$$e_1 = \begin{pmatrix} 1 \\ 1 \\ 0 \end{pmatrix}, \quad e_2 = \begin{pmatrix} \frac{1-\sqrt{3}}{2} \\ \frac{\sqrt{3}-1}{2} \\ 1 \end{pmatrix}, \quad e_3 = \begin{pmatrix} \frac{1+\sqrt{3}}{2} \\ -\frac{\sqrt{3}+1}{2} \\ 1 \end{pmatrix}$$

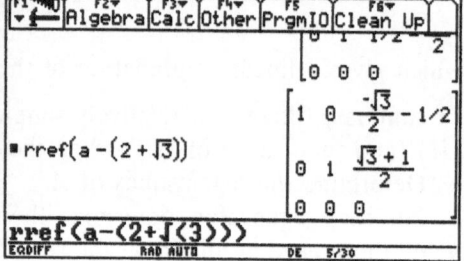

A basis for $\mathcal{S}_H(\mathbb{R})$ is thus:

$$X_1 = e^t \begin{pmatrix} 1 \\ 1 \\ 0 \end{pmatrix}, \quad X_2 = e^{(2-\sqrt{3})t} \begin{pmatrix} \frac{1-\sqrt{3}}{2} \\ \frac{\sqrt{3}-1}{2} \\ 1 \end{pmatrix}, \quad X_3 = e^{(2+\sqrt{3})t} \begin{pmatrix} \frac{1+\sqrt{3}}{2} \\ -\frac{\sqrt{3}+1}{2} \\ 1 \end{pmatrix}$$

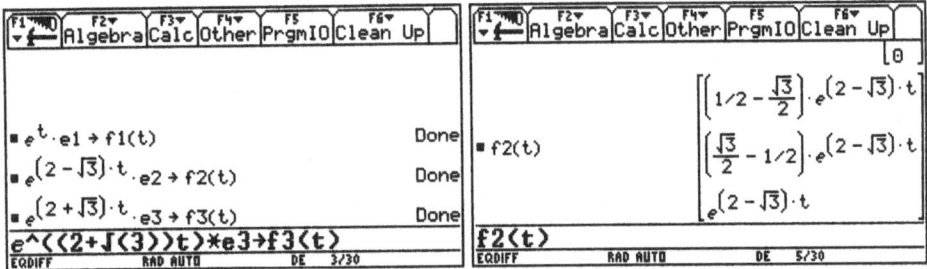

We let the calculator do the work of calculating the general solution of the equation (E) .

Define the Wronskian $\mathcal{W}(t)$, then calculate the general solution using the formula shown earlier.

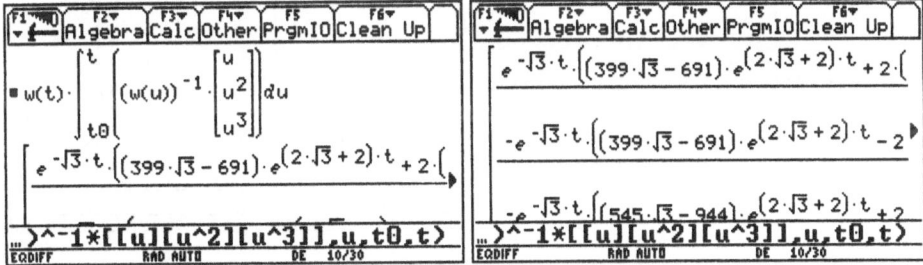

The calculator is thus shown to be a powerful aid in solving a differential system with constant coefficients which has distinct real eigenvalues.

Now suppose that the real matrix A has distinct complex eigenvalues. Everything works as before. In fact, if λ is a complex eigenvalue of A and x is an associated eigenvector in $\mathbb{C}^n$, the characteristic polynomial is real, since we know that $\overline{\lambda}$ is likewise an eigenvalue of A with $\overline{x}$ as its associated eigenvector:

$$A\overline{x} = \overline{Ax} = \overline{\lambda x} = \overline{\lambda}\,\overline{x}$$

Let $\lambda \in \mathbb{C}$ be a complex eigenvalue of A and let $(x_1, x_2, \ldots, x_k)$ be a basis of the associated subspace. The family $(x_1, x_2, \ldots, x_k)$ is independent over $\mathbb{R}$, and so is the family $(\overline{x_1}, \overline{x_2}, \ldots, \overline{x_k})$. The latter is a basis of the eigenspace associated with the eigenvalue $\overline{\lambda}$. Thus, the eigenvalues λ and $\overline{\lambda}$ give $2k$ complex valued functions independent of $S_H(\mathbb{R})$ which are:

$$e^{\lambda t}x_1, e^{\lambda t}x_2, \ldots, e^{\lambda t}x_k \text{ and } e^{\overline{\lambda}t}\overline{x_1}, e^{\overline{\lambda}t}\overline{x_2}, \ldots, e^{\overline{\lambda}t}\overline{x_k}$$

Their real parts and their imaginary parts are similarly solutions of (H) since the matrix A is real. We thus obtain as a solution basis:

$$\Re e(e^{\lambda t}x_1), \Re e(e^{\lambda t}x_2), \ldots, \Re e(e^{\lambda t}x_k) \text{ and } \Im m(e^{\lambda t}x_1), \Im m(e^{\lambda t}x_2), \ldots, \Im m(e^{\lambda t}x_k)$$

Example: Solve the system:

$$\begin{cases} x'(t) = -y(t) \\ y'(t) = x(t) \\ z'(t) = 2z(t) \end{cases}$$

Define the matrix A and calculate its eigenvalues. They are $2, i, -i$.

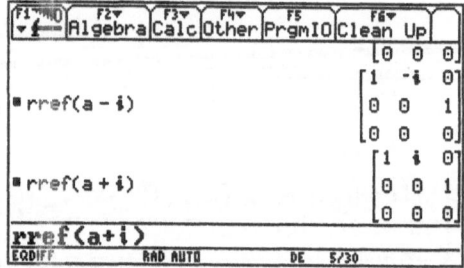

Calculate the complex eigenvectors of A.

A solution for the real eigenvalue $\lambda = 2$ is shown. Then we determine a complex valued solution of the system using the eigenvalue $\lambda = i$, then we find its real and imaginary parts.

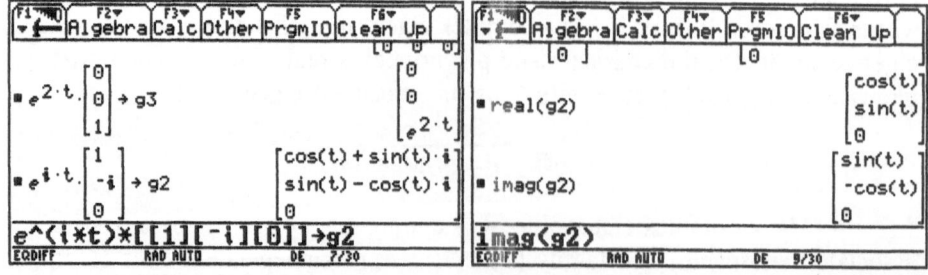

And we verify that both parts are indeed solutions.

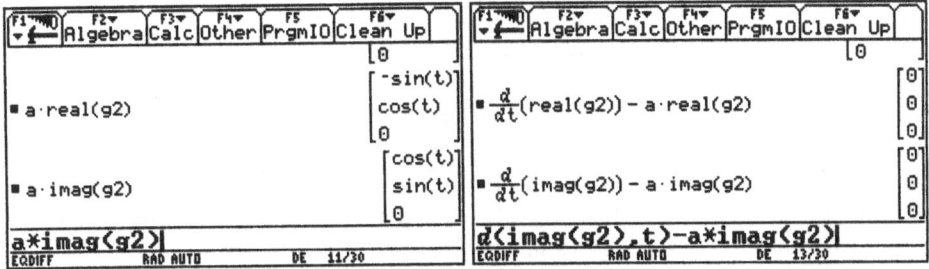

Thus, we have a basis, (g_1, g_2, g_3), for the solution space of the differential system.

4. 4 Case b) Triangularizable differential systems

Definition: We say that $A \in \mathcal{M}_n(\mathbb{K})$ is triangularizable in $\mathbb{K}$ if it is similar to a triagonal matrix T of $\mathcal{M}_n(\mathbb{K})$, that is, if there exists an invertible matrix P of $\mathcal{M}_n(\mathbb{K})$ such that $T = P^{-1}AP$.

Remark: This concept is also treated in chapter on eigenvalues.

We know from experience with Gauss reduction of linear systems that, working in the field $\mathbb{C}$, we could always remain in this case. Thus, we suppose that there is a triangular matrix T similar to A. In that case:

$$X'(t) = AX(t) + B(t) \Leftrightarrow X'(t) = PTP^{-1}X(t) + B(t)$$
$$\Leftrightarrow P^{-1}X'(t) = T(P^{-1}X(t)) + P^{-1}B(t)$$

Putting $Y(t) = P^{-1}X(t)$ and $C(t) = P^{-1}B(t)$, we obtain the following equivalent system:

$$Y'(t) = TY(t) + C(t)$$

which may be written:

$$\begin{cases} y_1'(t) &= \lambda_1 y_1(t) + c_1(t) \\ y_2'(t) &= t_{21}y_1(t) + \lambda_2 y_2(t) + c_2(t) \\ \vdots & \qquad \vdots \\ y_n'(t) &= t_{n1}y_1(t) + t_{n2}y_2(t) + \ldots + \lambda_n y_n(t) + c_n(t) \end{cases}$$

Theorem 4: Let $(H) : Y'(t) = TY(t)$ where $T \in \mathcal{M}_n(\mathbb{R})$ is a triangular matrix. Let $(\lambda_1, \lambda_2, \ldots, \lambda_p)$ be the distinct eigenvalues of T and $(r_1, r_2, \ldots, r_p)$ their multiplicities. A basis of the space $\mathcal{S}_H(\mathbb{R})$ is formed by the families:

$$(e^{\lambda_i t}P_{i,0}(t), e^{\lambda_i t}P_{i,1}(t), \ldots, e^{\lambda_i t}P_{i,r_i-1}(t)), \quad 1 \le i \le p$$

where $P_{i,k}(t)$ is a vector of $(\mathbb{R}_k[X])^n$. (Its components are polynomials of degree less than or equal to k).

Proof: The proof is by induction on n.
We know that $y_1'(t) = \lambda_1 y_1(t) \Rightarrow y_1(t) = \alpha_1 e^{\lambda_1 t}$, and α_1 is a constant, a polynomial of degree 0.
Consider a homogeneous system of order n.   We are interested in the first $(n-1)$ equations.   By the induction hypothesis, there is a basis of the solution space of this sub-system formed by the elements of the form $e^{\lambda_i t} P_{i,k}^*(t)$, $1 \le i \le n-1, 0 \le k \le r_i^* - 1$, where $P_{k,i}^*$ is a polynomial vector of dimension $(n-1)$ and r_i^* is the multiplicity of λ_i in this sub-system.

$$e^{\lambda_i t} P_{i,k}^*(t) = e^{\lambda_i t} \begin{pmatrix} p_1(t) \\ p_2(t) \\ \vdots \\ p_{n-1}(t) \end{pmatrix}$$

To obtain a solution of the system of order n, we must determine $y_n(t)$ which is defined by:

$$\frac{dy_n(t)}{dt} - \lambda_n y_n(t) = t_{n1} y_1(t) + \ldots + t_{n.n-1} y_{n-1}(t)$$

If we replace each $y_i(t), 1 \le i \le n-1$ by the right part of the last equation by $e^{\lambda_i t} P_{i,k}^*(t)$, we obtain a sum of elements of the form $a_j e^{\lambda_i t} p_j(t)$ where each p_j is a polynomial of degree less than or equal to k. This right-hand part of the equation may thus be written in the form $e^{\lambda_i t} g(t)$, with g a polynomial of degree less than or equal to k. Then the equation may be written:

$$\frac{dy_n(t)}{dt} - \lambda_n y_n(t) = e^{\lambda_i t} g(t)$$

• $\lambda_n \ne \lambda_i$.   In this case, there is a polynomial p of degree k such that $y_n(t) = e^{\lambda_i t} p(t)$. Thus, we obtain as a solution of the system the vector:

$$e^{\lambda_i t} \begin{pmatrix} p_1(t) \\ \vdots \\ p_{n-1}(t) \\ p(t) \end{pmatrix}$$

which is a polynomial vector of degree less than or equal to k. We thus obtain, starting with solutions of the sub-system of order $n-1$, a solution of the initial system.

• $\lambda_n = \lambda_i$.   In this case, there is a solution which is a polynomial of degree equal to $\deg(g) + 1$, and which is a solution of the equation:

$$\frac{dy_n(t)}{dt} - \lambda_i y_n(t) = e^{\lambda_i t} g(t)$$

We thus obtain a polynomial vector of degree $k+1$, denoted $Q_{k+1,i}$ starting from the polynomial vector $P_{k,i}^*$.

Applying the same procedure to each element of the basis $e^{\lambda_i t} P_{k,i}^*(t)$ of solutions of the sub-system, we obtain $n-1$ solutions of the system. A supplementary solution is none other than:

$$ e^{\lambda_n t} \begin{pmatrix} 0 \\ 0 \\ \vdots \\ 0 \\ 1 \end{pmatrix} $$

It remains to show the independence of the solutions which we have found.

The $(n-1)$ solutions obtained from the sub-system of order $(n-1)$ form an independent set, by the induction hypothesis and by the fact that adding a coordinate doesn't modify their independence. It remains to consider this last solution which we have found. To verify the independence of the functions, it suffices to verify their independence at one point, for example at $t=0$, thus with:

$$ \begin{pmatrix} 0 \\ \vdots \\ 0 \\ 1 \end{pmatrix}, \begin{pmatrix} V_1 \\ w_1 \end{pmatrix}, \begin{pmatrix} V_2 \\ w_1 \end{pmatrix}, \dots, \begin{pmatrix} V_{n-1} \\ w_{n-1} \end{pmatrix} $$

In this notation the V_i are $(n-1)$-component column vectors from the sub-system. Or:

$$ \beta_1 \begin{pmatrix} 0 \\ \vdots \\ 0 \\ 1 \end{pmatrix} + \sum_{i=2}^{n-1} \beta_i \begin{pmatrix} V_i \\ w_i \end{pmatrix} = 0 \Rightarrow \sum_{i=1}^{n-1} \beta_i V_i = 0 \text{ and } \beta_1 \begin{pmatrix} 0 \\ \vdots \\ 0 \\ 1 \end{pmatrix} = 0 $$

which entails that $\beta_1 = \beta_2 = \dots = \beta_{n-1} = 0$. It remains to verify that the solutions are indeed of the form described in the theorem proposed. This is implicit by the construction used.

In fact, the solution technique is much more simple with the aid of the calculator. It suffices to "cascade" the solutions of a triangular system. As an example, let's solve the system:

$$ \begin{cases} x' = 2x \\ y' = x + y \\ z' = x + y + z \end{cases} $$

Define the matrix A, thus the vector AX.

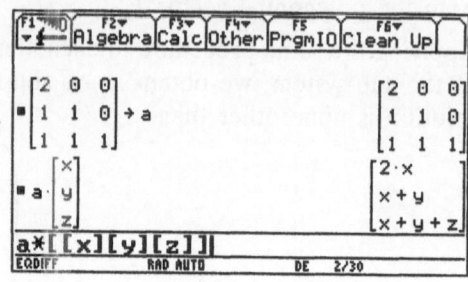

Call the function which solves formally the equation by integrating at each step the preceding solution.

The general solution is of the form:

$$\begin{cases} x(t) = c_1 e^{2t} \\ y(t) = c_1 e^{2t} + c_2 e^t \\ z(t) = 2c_1 e^{2t} + c_2 t e^t + c_3 e^t \end{cases}$$

Here is a complete example. Solve the system:

$$\begin{cases} x' = -2x - y + 2z + e^t \\ y' = -15x - 6y + 11z + e^{-t} \\ z' = -14x - 6y + 11z \end{cases}$$

Consider this matrix as a linear operator φ of $\mathbb{R}^3$ with respect to its canonical basis (e_1, e_2, e_3).

Enter the matrix A; calculate its characteristic polynomial. There is only one eigenvalue, 1. The matrix A is not diagonalizable. Thus, we treat it as triangularizable and determine the eigenvectors.

The dimension of the eigenspace E_1 is 1. It is generated by $e_1' = \begin{pmatrix} 1 \\ 1 \\ 2 \end{pmatrix}$.

We make a change of basis (e_1', e_2, e_3), then extract the 2×2 submatrix which corresponds to the projection onto the plane generated by (e_2, e_3). We then seek an eigenvector in this plane.

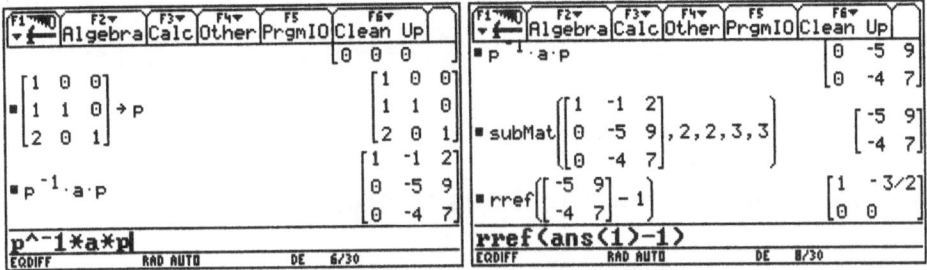

Such a vector is $e_2' = 3e_2 + 2e_3$, so in the basis we started with:

$$ e_2' = \begin{pmatrix} 0 \\ 3 \\ 2 \end{pmatrix} . $$

We pass to the basis (e_1', e_2', e_3). The matrix will be triangular, and and we could solve the differential system in cascade.

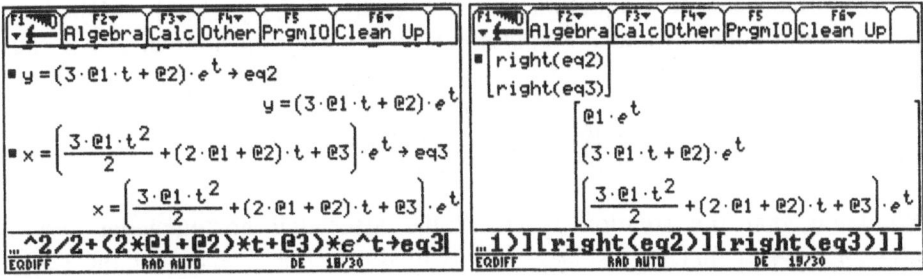

Here are some technical manipulations:

Finally:

To determine the final result, we multiply the result of the triangular system by P^{-1}.

$$\begin{bmatrix} \dfrac{3 \cdot @1 \cdot t^2}{2} + (2 \cdot @1 + @2) \cdot t + @3 \end{bmatrix} \cdot e^t$$

$$\begin{bmatrix} @1 \cdot e^t \\[6pt] \left(@1 \cdot t - \dfrac{@1}{3} + \dfrac{@2}{3} \right) \cdot e^t \\[10pt] \left(\dfrac{3 \cdot @1 \cdot t^2}{2} + @2 \cdot t - \dfrac{4 \cdot @1}{3} - \dfrac{2 \cdot @2}{3} + @3 \right) \cdot e^t \end{bmatrix}$$

`p^-1*[[@1*e^t][(3*@1*t+@2)*e^...`
EQDIFF   RAD AUTO   DE   20/30

5. Linear differential equations of order n

A linear differential equation of order n is one of the form:

$$a_n(t)y^{(n)}(t) + a_{n-1}y^{(n-1)}(t) + \ldots + a_1 y'(t) + a_0 y(t) = b(t)$$

where the $a_i, 1 \leq i \leq n$ and b are continuous functions on an interval I of $\mathbb{R}$. As agreed, we will only work with normalized equations of the form:

$$y^{(n)}(t) + a_{n-1}(t)y^{(n-1)}(t) + \ldots + a_1(t)y'(t) + a_0(t)y(t) = b(t)$$

The theory of differential systems which we are developing applies perfectly here since we know how to express an equation of order n with a system of equations of order 1. Here, it suffices to put $X'(t) = A(t)X(t) + B(t)$ with:

$$\begin{pmatrix} y' \\ y'' \\ \vdots \\ y^{(n)} \end{pmatrix} = \begin{pmatrix} 0 & 1 & 0 & \cdots & 0 \\ 0 & 0 & 1 & \cdots & 0 \\ \vdots & & \ddots & \ddots & \vdots \\ 0 & \cdots & \cdots & \cdots & 1 \\ -a_0 & -a_1 & \cdots & -a_{n-1} & -a_{n-1} \end{pmatrix} \begin{pmatrix} y \\ y' \\ \vdots \\ y^{(n-1)} \end{pmatrix} + \begin{pmatrix} 0 \\ \vdots \\ 0 \\ b \end{pmatrix}$$

5. 1 Linear equations of order 2 with constant coefficients

Every normalized linear equation of order 2, $y''(t) + ay'(t) + by(t) = c(t)$ where a, b are both constant and c a continuous function on an interval $I \subseteq \mathbb{R}$ thus becomes a 2×2 system:

$$(E): \begin{pmatrix} y' \\ y'' \end{pmatrix} = \begin{pmatrix} 0 & 1 \\ -b & -a \end{pmatrix} \begin{pmatrix} y \\ y' \end{pmatrix} + \begin{pmatrix} 0 \\ c \end{pmatrix}$$

We know that the solution space of the associated homogeneous system $(H): y''(x) + a(x)y'(x) + b(x)y(x) = 0$ is of dimension 2, and that the set of solutions of (E) is an affine space with the direction $\mathcal{S}_H(I)$. By the Cauchy-Lipschitz theorem, for each $t_0 \in I$, $(y_0, y_0') \in \mathbb{R}^2$, there is a unique solution $F = \begin{pmatrix} f \\ f' \end{pmatrix}$ such that $F(t_0) = (y_0, y_0')$; or again, there exists a unique real-valued function f such that $f(t_0) = y_0$ and $f'(t_0) = y_0'$.

To find a basis of $\mathcal{S}_H(I)$, we must determine the eigenvalues, and then the eigenvectors, of the matrix $A = \begin{pmatrix} 0 & 1 \\ -b & -a \end{pmatrix}$.

The characteristic polynomial of the matrix M and its zeros which depend on a and b

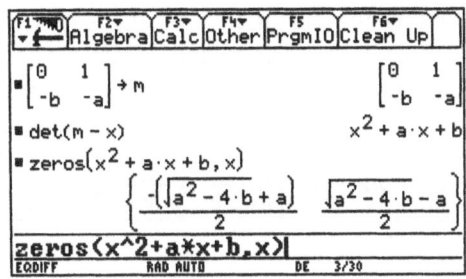

- If $a^2 - 4b \neq 0$, the matrix M has two distinct eigenvalues (real or complex). We then know how to determine a basis for $\mathcal{S}_H(I)$, starting with the eigenvalues and eigenvectors.
- If $a^2 = 4b$, the matrix M has one eigenvalue and is never diagonalizable. We know how to determine a basis from a triangular system $\mathcal{S}_H(I)$.

Example: Solve the differential system $y''(t) + 5y'(t) - 6y() = \cos(t)$.

The definition and the diagonalization of the matrix M

The eigenvectors are $\begin{pmatrix} 1 \\ 1 \end{pmatrix}$, associated with $\lambda = 1$, and $\begin{pmatrix} 1 \\ -6 \end{pmatrix}$, associated with $\lambda = -6$. We know that $F_1 = \begin{pmatrix} e^x \\ e^x \end{pmatrix}$ and $F_2 = \begin{pmatrix} e^{-6x} \\ -6e^{-6x} \end{pmatrix}$ form a basis of $\mathcal{S}_H$.

The definition of the basis of solutions and of the Wronskian.

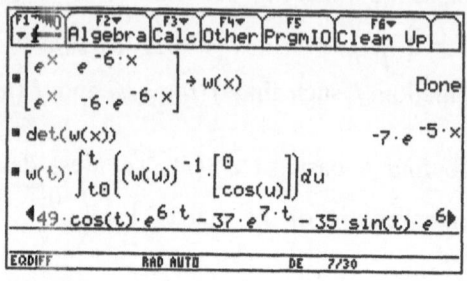

Here is the calculation of the solutions which is somewhat difficult to read. Next to it is a simplified version of the result. Only the first line is to be considered.

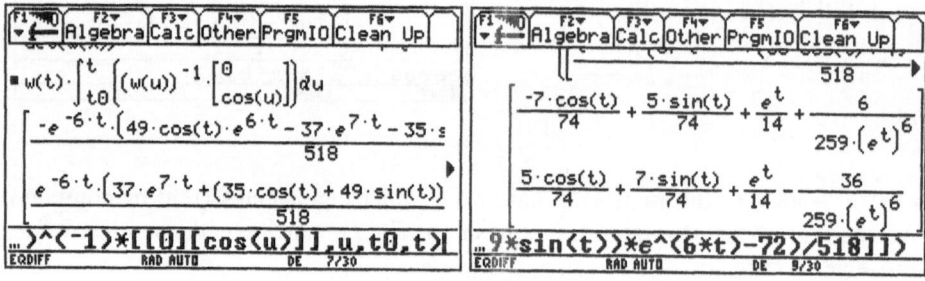

In fact, the calculator knows how to solve this form of differential equation.

Solving a linear differential equation of second order. The solution is given in its general form.

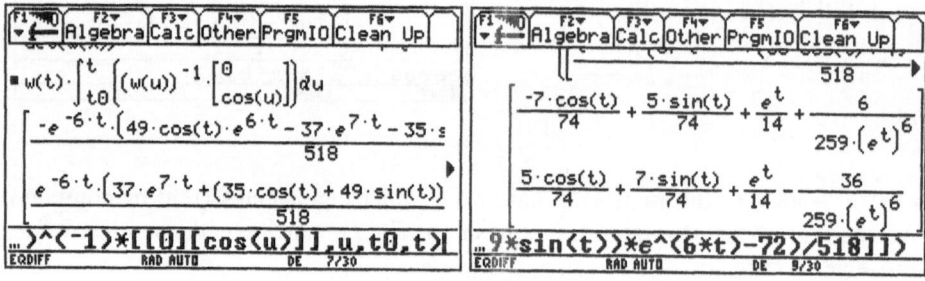

By contrast, the method developed above remains necessary in the case of differential equations with non-constant coefficients. For example, consider how to solve the equation $t^2 y''(t) + 6ty'(t) + 6y(t) = e^t$. This equation is not normalized. But it is a so-called Euler equation, and we look for solutions of the homogeneous equation of the form $t \mapsto t^\alpha$. In this case, we find t^{-2} and t^{-3} defined on $\mathbb{R}^{>0}$ and $\mathbb{R}^{<0}$.

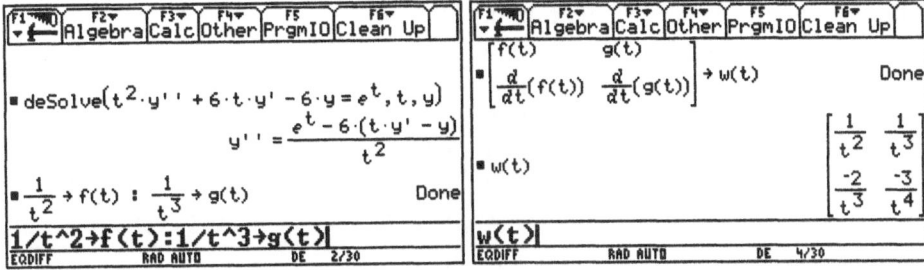

We start by determining both solutions of the homogeneous equation, then calculate the Wronskian. The method of calculation of solutions of the general system is "standard". Only the first line of this result is to be considered here.

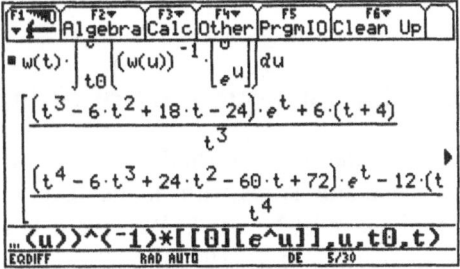

When we know one solution of a linear second order differential equation, there is a way to determine a second, linearly independent, solution which will lead to the Wronskian and to the complete solution of the equation.

Let $(H) : y''(x) + a(x)y'(x) + b(x)y(x) = 0$ and let f be a solution. We change the variable function by setting $y(x) = z(x)f(x)$. We then have: $y'(x) = z(x)f'(x) + z'(x)f(x)$ and $y''(x) = z(x)f''(x) + 2z'(x)f'(x) + z''(x)f(x)$. Substituting these expressions for y, y', y'' in (H) and taking into account the fact that f satisfies (H), it follows that:

$$f(x)z''(x) + 2a(x)f'(x)z'(x) = 0$$

which is a first order differential equation in z', and which we know how to solve. This is where we obtain another solution of H.

As an example, let's solve the equation $xy''(x) - xy'(x) - y(x) = 0$, knowing that $x \mapsto xe^x$ is a solution.

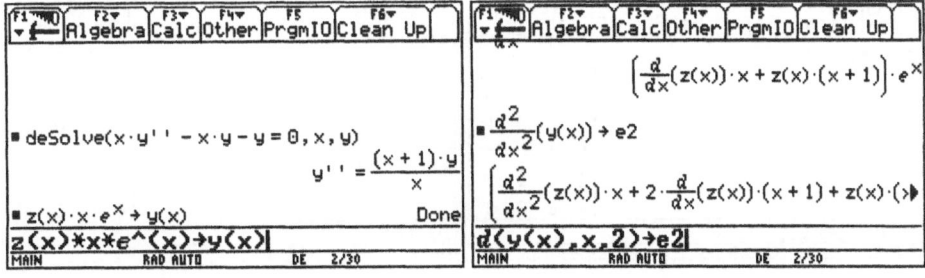

After defining $z(x)$, we calculate both derivatives, then substitute the expressions obtained in the differential equation. We integrate to obtain a solution z. But this function does not have an anti-derivative which is expressible in terms of familiar functions.

Finally, the general solution of the equation is, on an interval not containing 0:

$$y(x) = Axe^x + B \int_a^x \frac{e^{-t}}{t^2} dt$$

5. 2 Linear equations of order n with constant coefficients

The calculation of formal solutions of linear differential equations built into the calculator stops with 2. For every equation of higher order, the previous technique must be used. We give a last example.

Solve the equation $y^{(4)} - y = \cos(t)$.

First we consider the homogeneous equation $y^{(4)} - y = 0$. In terms of matrices we must solve:

$$\begin{pmatrix} y' \\ y'' \\ y^{(3)} \\ y^{(4)} \end{pmatrix} = \begin{pmatrix} 0 & 1 & 0 & 0 \\ 0 & 0 & 1 & 0 \\ 0 & 0 & 0 & 1 \\ 1 & 0 & 0 & 0 \end{pmatrix} \begin{pmatrix} y \\ y' \\ y'' \\ y^{(3)} \end{pmatrix}$$

Definition of the matrix of the system, calculation of the eigenvalues, and then the eigenvectors.

The matrix M is diagonalizable in $\mathbb{C}$ and a basis of eigenvectors respectively associated with $1, -1, i, -i$ is:

$$\begin{pmatrix} 1 \\ 1 \\ 1 \\ 1 \end{pmatrix}, \quad \begin{pmatrix} 1 \\ -1 \\ -1 \\ -1 \end{pmatrix}, \quad \begin{pmatrix} -1 \\ -i \\ 1 \\ i \end{pmatrix}, \quad \begin{pmatrix} 1 \\ -i \\ -1 \\ i \end{pmatrix}$$

The Wronskian matrix lets us determine a solution of the non-homogeneous equation.

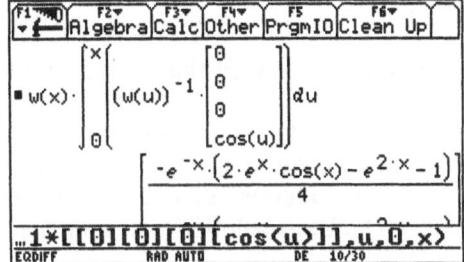

The Wronskian method to determine a particular solution.

Finally, the set of solutions of $y^{(4)} - y = \cos(t)$ is:

$$c_1 e^x + c_2 e^{-x} + c_3 \cos(x) + c_4 \sin(x) + \frac{\mathrm{ch}(x) - \cos(x)}{2}$$

6. Autonomous systems in the plane

Let U be an open set in the plane $\mathbb{R}^2$ and let $F : U \to \mathbb{R}^2$ be a map of class C^1 which associates $F(x,y) = (f(x,y), g(x,y))$ to $X = (x,y)$. An autonomous systemCs is a differential system of this type :

$$(E): \ X' = F(x,y) \Leftrightarrow \begin{cases} x' = f(x,y) \\ y' = g(x,y) \end{cases}$$

Thus, X is a vector function of the independent variable t, and $X' = \dfrac{dX(t)}{dt}$. The word "autonomous" means that the defining equations of the system are independent of the variable t. Of course, in most applications we think of the independent variable as time.

Now, if $X = \phi(t)$ is a (vector) solution of (E) on U, then for all $t_0, \psi(t) = \phi(t - t_0)$ is likewise a solution. In fact, the transformation $t \to t - t_0$ doesn't change the equation (E) since t does not appear. We say that the differential system is "translation invariant". Also, if ϕ is a solution of the IVP for (E) such that $\phi(0) = X_0$, then $\phi(t - t_0)$ is a solution of the IVP for (E) with $\phi(t_0) = X_0$. This second solution may be obtained by translating the first on the (t) axis.

A critical point (or, in mechanics problems, an equilibrium point) of a system is any point (x_0, y_0) such that $F(x_0, y_0) = 0$, or, in other words, where $\dfrac{dX(0)}{dt} = 0$.

We call the plane with coordinates (x,y) the "phase plane". Note again that x and y are functions of t. Graphing in the phase plane involves, not graphing x and y as functions of t, but graphing y against x — as the parametric curve $(x(t), y(t))$. Each solution curve in the phase plane is called a "trajectory" or sometimes an "orbit".

6. 1 Linear systems

Consider an autonomous system of the form:

$$\begin{cases} x' = ax + by + e \\ y' = cx + dy + f \end{cases} \Leftrightarrow \begin{pmatrix} x' \\ y' \end{pmatrix} = \begin{pmatrix} a & b \\ c & d \end{pmatrix} \begin{pmatrix} x \\ y \end{pmatrix}$$

where a, b, c, d, e, f are real constants. For obvious reasons, this is called a first order linear autonomous system with constant coefficients. In vector form, we may write it as $X'(t) = AX + B$, where X has components $x(t)$ and $y(t)$, B has components e and f, and A is the matrix of the defining equations.

It is easy to see finding critical point(s) of such a system involves solution of a system of linear algebraic equations. When there are non-zero solutions, the critical point is not at the origin. However, by making a translation of the dependent variables by e and f, we may always consider the "homogeneous" linear differential system X'=AX which has a critical point only at the origin. In what immediately follows we study the linear homogeneous autonomous systems in detail. They may be easily and completely analyzed, and understanding them leads to understanding of more general differential systems.

In a homogeneous linear autonomous system, then, the only critical point is 0. To study the trajectories of such a system, everything depends on the matrix $A = \begin{pmatrix} a & b \\ c & d \end{pmatrix}$, which we assume to be invertible. We use techniques of linear algebra to "reduce" the matrix to simpler form.

For a given square matrix, A, the "eigenvalues" are numbers for which we may find vectors X satisfying the equation $AX = \lambda X$, or $AX - \lambda X = 0$. This is a homogeneous algebraic linear system of a very familiar type. Such a system always has the vector 0 as a solution, and it is "well-known" that such a system only has additional, more interesting, solutions in case the matrix $A - \lambda I$ is non-singular or invertible. This happens when its determinant vanishes: $\det(A - \lambda I)=0$. In general, we will have two, possibly repeated and possibly complex eigenvalues in the simple systems we are investigating here.

Once we have found an eigenvalue, corresponding eigenvectors which satisfy the equation $AX = \lambda X$ may be found. This is easily done on a modern scientific calculator. The built-in functions yield only numeric answers, and it is better to use the definitions and other features of the calculator to find eigenvalues and eigenvectors in the simple examples we are considering.

An entire chapter is devoted to eigenvalues and eigenvectors later in the book. In what follows in this chapter, we assume many results which are basic to the linear algebra topic. The reader who has not studied eigenvalues of a matrix may wish to look ahead in this book or to other references. As will be seen, many applications and other topics in mathematics will be seen to depend on these characteristics of a matrix.

We denote by λ_1, λ_2 the eigenvalues of A. We assume that none of them are zero. This will be the case if the matrix A is invertible.

Now, consider these cases:
- λ_1, λ_2 are distinct real numbers.

If the matrix A has distinct eigenvalues it is diagonalizable and has two eigenvectors which are not proportional: they are linearly independent. It follows and is easily checked that then $X_1(t) = e^{\lambda_1 t} v_1$ and $X_2(t) = e^{\lambda_2 t} v_1$ are each solutions of (E), and that these two solutions are linearly independent as (function) vectors.

Differential equations theory now tells us that every solution of (E) may be written in the form $e^{\lambda_1 t} v_1 + e^{\lambda_2 t} v_2$, that is $x(t) = \alpha e^{\lambda_1 t}$ and $y(t) = \beta e^{\lambda_2 t}$. We could thus write:

$$y = \gamma x^{\lambda_2/\lambda_1}$$

In fact, the behavior of curves in the phase plane depends on the respective signs of λ_1 and λ_2.

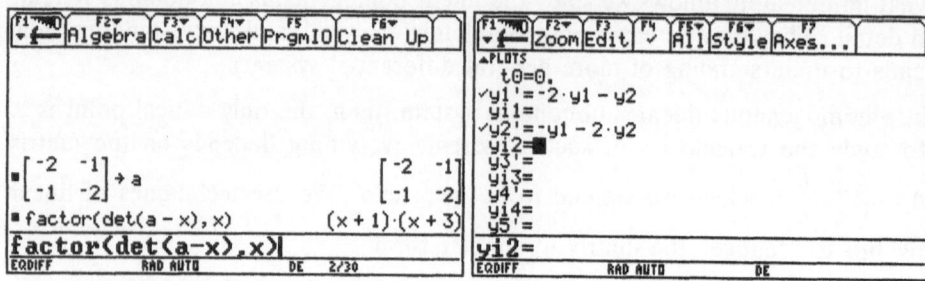

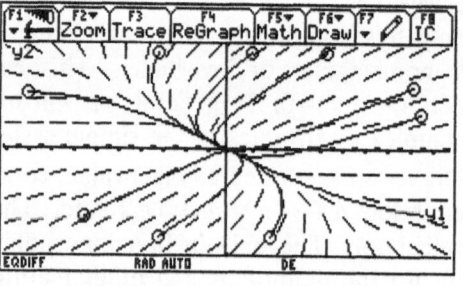

When the two eigenvalues are negative, the critical point is attracting or attractive. We say that it is stable (or is a "stable node"). In the phase plane, the curves are "attracted" to the origin. These are parabolas.

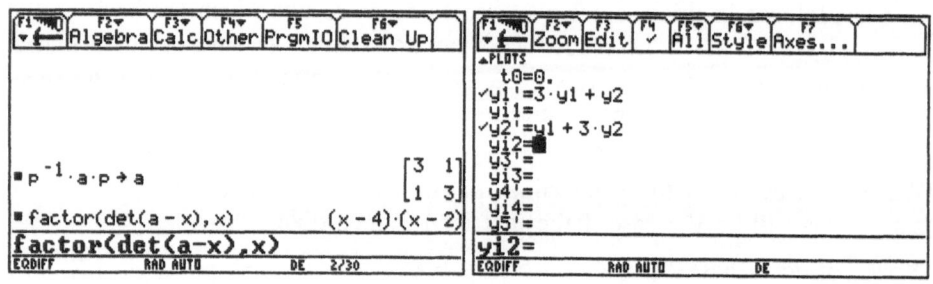

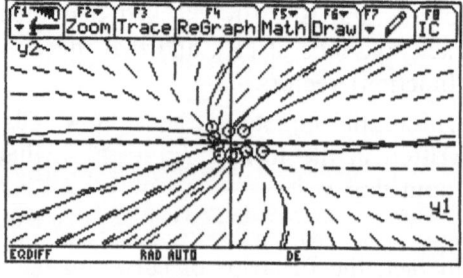

When both eigenvalues are positive, the critical point is repelling or repulsive (or is an "unstable node". In the phase plane, the curves are repelled by the origin. They are again parabolas.

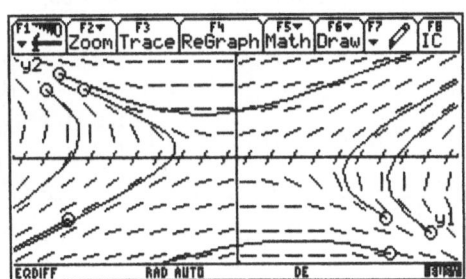

When the two eigenvalues are of opposite sign, the critical point is a saddle point (or unstable saddle). These are hyperbolas.

- $\lambda_1 = \lambda_2 = \lambda$.

If the matrix A is diagonalizable, it is then equal to λI. It is easy to verify that the curves in the phase plane are lines passing through the origin, which is an attracting point when $\lambda < 0$ and repulsive when $\lambda > 0$.

In the general case, the matrix A is not diagonalizable, but is triangularizable into the form $\begin{pmatrix} \lambda & 1 \\ 0 & \lambda \end{pmatrix}$. We know that in the basis (e_1, e_2) of triangularization, the solutions are of the form:

$$\begin{cases} x(t) = (c_0 + c_1 t)e^{\lambda t} \\ y(t) = c_1 e^{\lambda t} \end{cases}$$

Again, two cases are possible depending on the sign of λ.

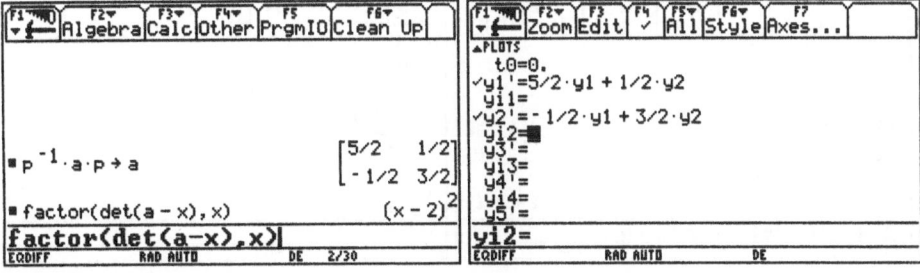

For $\lambda > 0$, the critical point is repulsive (unstable node).

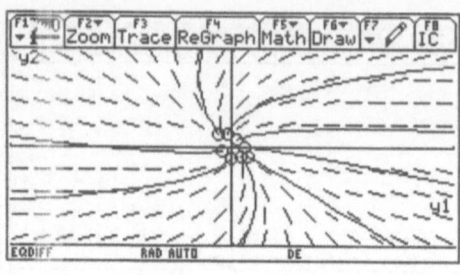

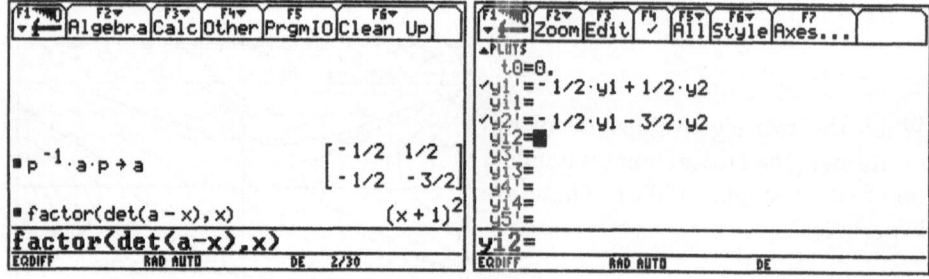

For $\lambda < 0$, the critical point is now attracting (a stable node).

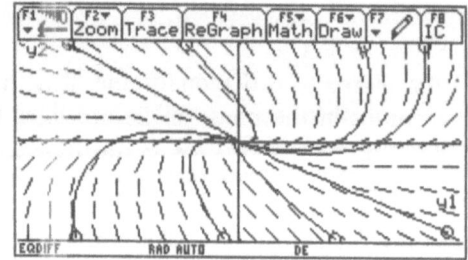

- λ_1 and λ_2 are complex conjugates $a \pm ib$.

We know then that the complex solutions of the differential system are of the form $e^{at \pm ibt}$, and that their combination gives the real solutions $e^{at}\cos(bt)$ and $e^{at}\sin(bt)$. The polar equation of the solution curves is of the type $\rho = Ce^{\frac{a}{b}\theta}$. Again, behavior of the solution depends on a and of its sign.

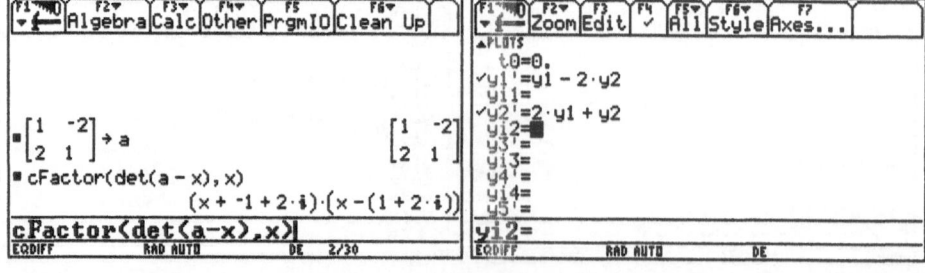

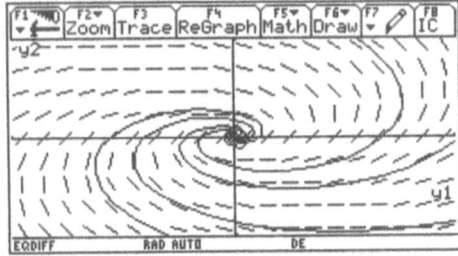

For $a > 0$, the curves are spirals. The critical point is called a focus and is repulsive.

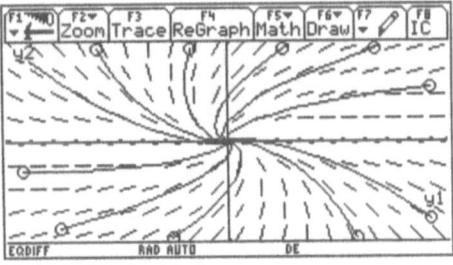

For $a < 0$, the curves are also spirals, but the critical point is attracting. It is again a focus.

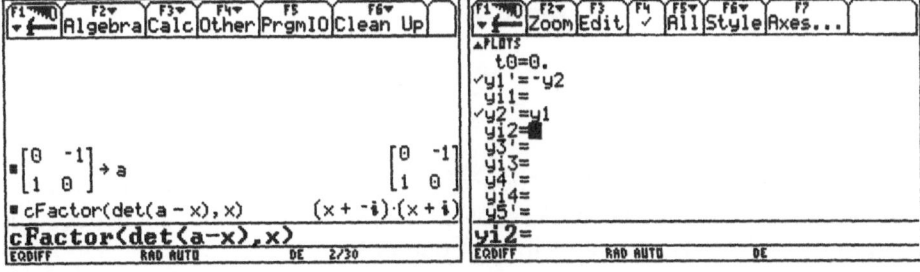

For $a = 0$, the curves are circles centered on the critical point (a center). This is only true with an orthonormalized basis. Otherwise, we obtain ellipses.

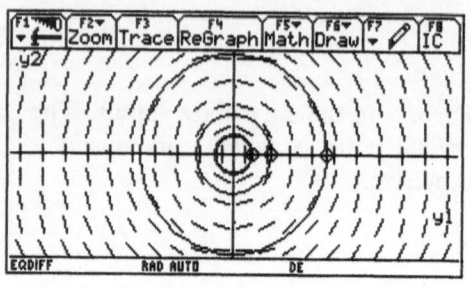

We remark that if $A = \begin{pmatrix} a & b \\ c & d \end{pmatrix}$, its characteristic polynomial $P_A(x) = \det(A - xI)$ is equal to $x^2 - \operatorname{tr}(A)x + \det(A)$. We thus have $\lambda_1 + \lambda_2 = \operatorname{tr}(A)$ and $\lambda_1 \lambda_2 = \det(A)$. The real or non-real nature and the sign of the eigenvalues of A thus depend on the trace and on the determinant of A.

We may then summarize the preceding results in the following diagram:

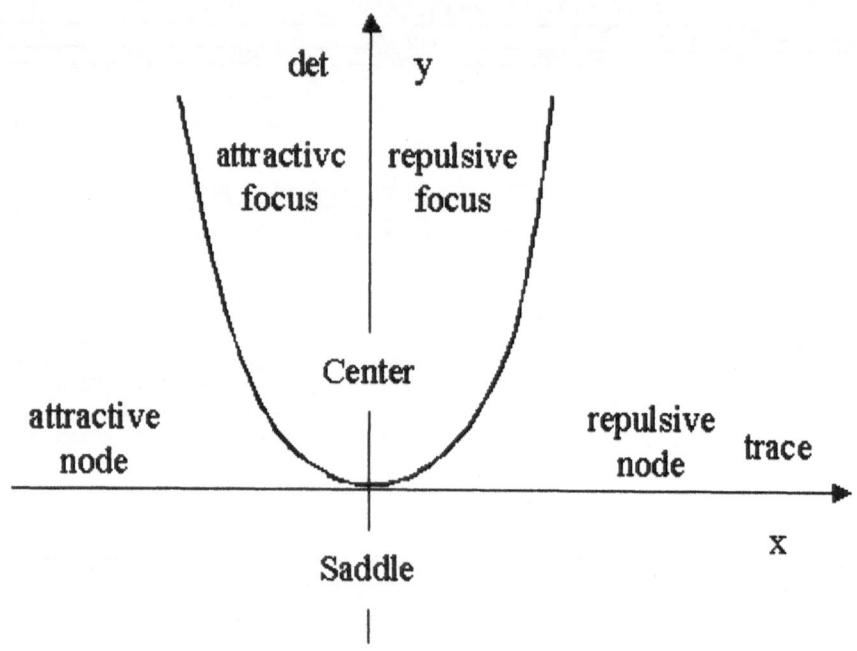

Is it possible to study the attraction of trajectories of an autonomous linear differential system qualitatively? For example, is it possible to deduce the behavior of all orbits, starting with just one of them ?

Before responding, let's be more precise:

Let $M = (x, y)$ be a solution of an autonomous linear differential system defined on $[0, +\infty[$. We say that this solution is *stable* if for all $\varepsilon > 0$, there exists $\delta > 0$ such that if $N = (x_1, y_1)$ is another solution of the system, then for all $t \geq 0$:

$$||M(0) - N(0)|| < \delta \Rightarrow ||M(t) - N(t)|| < \varepsilon$$

(This means that if the trajectories are close to each other at the initial point, they remain so on all $\mathbb{R}^{>0}$.)

The solution N is *asymptotically stable*, if in addition to the preceding condition we also have $\lim\limits_{t \to +\infty} (M(t) - N(t)) = 0$.

There is at least one theorem which assures of the stability of the solutions of a linear autonomous system.

Theorem 1: Let A be the matrix of an autonomous linear system.

a) we have stability of the system if and only if the eigenvalues of A all have a negative or zero real part.

b) we have asymptotic stability of a system if and only if the eigenvalues of A all have a strictly negative real part.

Proof: Note that it suffices to study the different cases which we have examined since in every case the solutions, whether real or complex, may be expressed in exponential form.

We consider the behavior of $t \mapsto e^{(a+ib)t}$ in a neighborhood of $+\infty$.

If $a < 0$, $\lim\limits_{t \to +\infty} e^{(a+ib)t} = 0$.

If $a = 0$, $e^{(a+ib)t}$ remains bounded.

If $a > 0$, $e^{(a+ib)t}$ diverges in a neighborhood of $+\infty$.

By linearity and because the null solution is stable, we have stability of all the solutions for which the real part of the eigenvalues of A is negative or zero. When the eigenvalues are pure imaginary (real part equal to zero), the trajectories of the solution obtained are ellipses, which signifies that the distance from the origin is conserved. The null solution is thus stable.

If the eigenvalues have a strictly negative real part, we have asymptotic stability ; it suffices to examine the four examples corresponding to this case.

6. 2 Non-linear systems

The above classification of autonomous linear systems is relatively simple and well known. What about non-linear autonomous systems? Of course, this question is much more difficult. The general case is impossible to treat, so we suppose that the vector function $F = (f, g)$ is in class C^1. We may then approximate F by its differential DF. Specifically, if we consider that the origin

$(0,0)$ is a critical point (we may always arrange this by a change of variable), this means that we may write:

$$F(x, y) = F(0,0) + DF_{(0,0)}(x, y) + h(x, y)$$

with:

$$\lim_{(x,y)\to(0,0)} \frac{\|h(x, y)\|}{\|(x, y)\|} = 0$$

or

$$\begin{pmatrix} f(x, y) \\ g(x, y) \end{pmatrix} = \begin{pmatrix} f(0,0) \\ g(0,0) \end{pmatrix} + \begin{pmatrix} \frac{\partial f}{\partial x}(0,0) & \frac{\partial g}{\partial x}(0,0) \\ \frac{\partial f}{\partial y}(0,0) & \frac{\partial g}{\partial y}(0,0) \end{pmatrix} \begin{pmatrix} x \\ y \end{pmatrix} + h(x, y)$$

We quite naturally wish to approximate the non-linear autonomous system, by its "linearization" $X' = AX$, with:

$$A = \begin{pmatrix} \frac{\partial f}{\partial x}(0,0) & \frac{\partial g}{\partial x}(0,0) \\ \frac{\partial f}{\partial y}(0,0) & \frac{\partial g}{\partial y}(0,0) \end{pmatrix}$$

Note the analogy of this "local linear approximation" with the differential of the beginning calculus course.

Now several questions may be raised: are the solutions of the original non-linear system "approximated" by the solutions of the linearized system? In what sense is the "approximation" made? Are the forms of the graphs conserved? If the critical point is attracting (or repelling), does it remain so in the linearized case?

We analyze some examples to establish a good foundation for understanding these questions.

First example

$$\begin{cases} x' = y \\ y' = -x - y^3 \end{cases}$$

The graph of solutions of the system in the phase plane appear to be asymptotic curves. In fact, the curves move toward and turn about the origin, the critical point, in a spiral. Proof of this result rests on a qualitative analysis of the system.

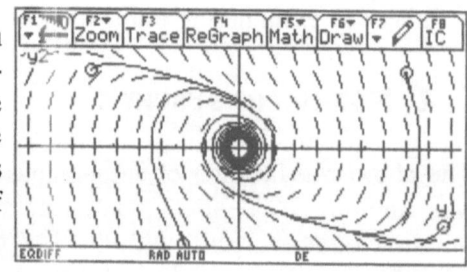

The linearized system is $(x' = y, y' = -x)$. The eigenvalues are the complex numbers $\pm i$. The trajectories are circles centered at the origin. The behavior of solutions is not preserved in the linearization.

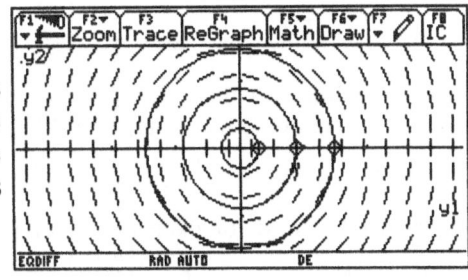

Second example:

$$\begin{cases} x' = y \\ y' = -x - 4y - x^2 \end{cases}$$

The linearized system is $(x' = y, y' = -x - 4y)$. The matrix of the system has two real negative eigenvalues $-2 \pm \sqrt{3}$. The critical point at the origin is a stable node.

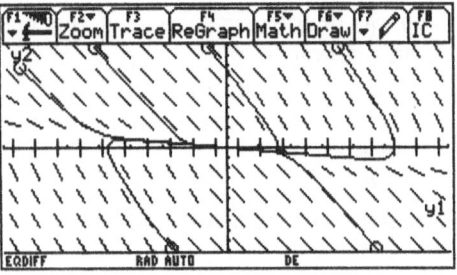

The graph of solutions in the phase plane appears to have the critical point $(0,0)$ as an attracting point. The stability as well as the form of the solutions seem to be conserved.

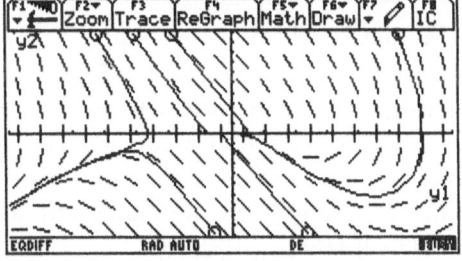

We have the following theorem about autonomous non-linear systems:

Theorem 2: Let (E) be an autonomous differential system of class C^1 and let (L) be the linearized autonomous system with the matrix A. We suppose that the origin $(0,0)$ is a critical point.

If the eigenvalues of A have strictly negative real parts, both systems behave similarly in a neighborhood of the origin.

Proof of this theorem appeals to the notion of the exponential of a matrix, a general notion which we have not developed, but which we now introduce in the particular case of 2×2 matrices. We refer to the chapter dedicated to the reduction of operators for more complete coverage.

Let $A = \begin{pmatrix} a & b \\ c & d \end{pmatrix}$. Two cases may be considered.

Case 1: A has two distinct eigenvalues $\lambda_1 \neq \lambda_2$, either both real or complex conjugates. Then, there is an invertible matrix P (real or complex) such that $A = PDP^{-1}$, with $D = \begin{pmatrix} \lambda_1 & 0 \\ 0 & \lambda_2 \end{pmatrix}$.

For each $t \in \mathbb{R}$ and for each $k \in \mathbb{N}$, $(tA)^k = P(tD)^k P^{-1}$.

It follows that, $(tD)^k = \begin{pmatrix} t^k \lambda_1^k & 0 \\ 0 & t^k \lambda_2^k \end{pmatrix}$, and the following series of matrices is convergent:

$$\sum_{k=0}^{+\infty} \frac{t^k D^k}{k!} = \begin{pmatrix} \sum_{k \geq 0} \frac{t^k \lambda_1^k}{k!} & 0 \\ 0 & \sum_{k \geq 0} \frac{t^k \lambda_2^k}{k!} \end{pmatrix} = \begin{pmatrix} e^{t\lambda_1} & 0 \\ 0 & e^{t\lambda_2} \end{pmatrix}$$

which we denote e^{tD}. We call the matrix $e^{tA} = Pe^{tD}P^{-1}$ the "exponential of A", and denote it e^{tA}.

Case 2: A has only one eigenvalue λ, which must be real. Then it may be verified that there exists an invertible matrix P such that $A = PTP^{-1}$, with $T = \begin{pmatrix} \lambda & 1 \\ 0 & \lambda \end{pmatrix}$. Thus $T = \lambda I + N$ with $N = \begin{pmatrix} 0 & 1 \\ 0 & 0 \end{pmatrix}$ and $N^2 = 0$. For all real t, and for every natural number k, and using the binomial theorem: $(tT)^k = \lambda^k I + k\lambda^{k-1} N$. and

$$\sum_{k=0}^{+\infty} \frac{t^k T^k}{k!} = \begin{pmatrix} \sum_{k \geq 0} \frac{t^k \lambda^k}{k!} & \sum_{k \geq 1} \frac{\lambda^{k-1} t^k}{(k-1)!} \\ 0 & \sum_{k \geq 0} \frac{t^k \lambda^k}{k!} \end{pmatrix} = \begin{pmatrix} e^{t\lambda} & t e^{t\lambda} \\ 0 & e^{t\lambda_2} \end{pmatrix}$$

is a matrix which we denote e^{tT} for this case. We then call the matrix $e^{tA} = Pe^{tT}P^{-1}$ the "exponential of A", and we denote it e^{tA} in either case.

Here are two remarks which we will use in our proof.

Remark 1. Let a be the larger of the real parts of the eigenvalues of A. For every $a' > a$, there is a constant $K > 0$ such that for all $t > 0$, $\|e^{tA}\| \leq Ke^{ta'}$. Since all norms on $\mathbb{R}^2$ are equivalent, we choose an algebraic norm (which thus satisfies $\|AB\| \leq \|A\| \|B\|$).

Remark 2. The solution of the homogeneous autonomous system $X' = AX$ may be written in the form $F(t) = e^{tA}C$, where C is a constant vector in $\mathbb{R}^2$. This vector is determined by the initial condition $F(t_0) = X_0$, In this case, the solution is:

$$F(t) = e^{(t-t_0)A} X_0$$

In the same manner, every solution of a non-homogeneous system $X' = AX + B$, with $B = B(t, X)$ may be written:

$$F(t) = e^{tA}C + \int_{t_0}^{t} e^{(t-u)A} B(u, F(u)) du$$

It suffices to differentiate this function to verify this assertion.

Proposition 1:(*Gronwall's Lemma*) *Let f and u be two real-valued functions which are defined and continuous on an interval I. Let $C > 0$ be a real positive number. We suppose that $u > 0$ and that for all $t_0, t \in I, t > t_0$:*

$$f(t) \leq C + \int_{t_0}^{t} f(s)u(s)ds$$

Then, for every $t_0, t \in I, t > t_0$:

$$f(t) \leq C \exp\left(\int_{t_0}^{t} u(s)ds\right)$$

This lemma is important since it is often the basis of the qualititative study of differential equations.

Proof: We denote $F(t) = \displaystyle\int_{t_0}^{t} f(s)u(s)ds$. Our hypotheses may be written $f(t) \leq C + F(t)$. The function F is differentiable and $F'(t) = u(t)f(t)$. Because $u(t) \geq 0$, it follows that:
$F'(t) = u(t)f(t) \leq Cu(t) + u(t)F(t)$ or $F'(t) - u(t)F(t) \leq Cu(t)$. We have then:

$$F'(t) - u(t)F(t) \leq Cu(t)$$

$$\Leftrightarrow (F'(t) - u(t)F(t)) \exp(-\int_{t_0}^{t} u(s)ds) \leq Cu(t)\exp(-\int_{t_0}^{t} u(s)ds)$$

$$\Leftrightarrow \left(F(t)\exp(-\int_{t_0}^{t} u(s)ds)\right)' \leq -C\left(\exp(-\int_{t_0}^{t} u(s)ds)\right)'$$

Integrating over the interval $[t_0, t]$, it follows, since $F(t_0) = 0$:

$$F(t)\exp(-\int_{t_0}^{t} u(s)ds) \leq C\left(1 - \exp\left(\int_{t_0}^{t} u(s)ds\right)\right)$$

or

$$F(t) \leq C\exp\left(\int_{t_0}^{t} u(s)ds\right) - C$$

We finish the proof with the relation $f(t) \leq C + F(t)$.

We pass now to the proof of Theorem 2.

An autonomous system may be written locally in the neighborhood of an isolated critical point (the origin) in the form $X'(t) = AX(t) + B(t, X(t))$, with $B = o(X)$ in a neighborhood of 0.

Since the eigenvalues of A are strictly negative, we choose a real negative number $-\rho$ which majorizes them. By the preceding Remark 1, there is a constant $K > 0$ such that for all $t \geq 0$:

$$\|e^{tA}\| \leq Ke^{-\rho t}$$

By Remark 2, every solution of a non-linear system may be written in the form:

$$F(t) = e^{tA}F(0) + \int_0^t e^{(t-u)A}B(u, F(u))du$$

Using these two facts, it follows that:

$$||F(t)|| \le Ke^{-\rho t}||F(0)|| + K\int_0^t e^{-\rho(t-u)}||B(u, F(u))||du$$

Let $\varepsilon > 0$. We know that $B = o(F)$ in a neighborhood of 0. Thus, there is a δ such that $||F|| < \delta$ implies that $||B(u, F(u))|| < \varepsilon||F(u)||/K$. Since $F(0) = 0$, there is a neighborhood of 0 on which $||F(u)|| < \delta$. Then we obtain:

$$e^{\rho t}||F(t)|| \le K||F(0)|| + \varepsilon\int_0^t e^{\rho u}||F(u)||du$$

By Gronwall's Lemma we deduce that:

$$||F(t)|| \le K||F(0)||e^{(\varepsilon-\rho)t}$$

which ends our proof.

Third example: the Lotka-Volterra equation

The Lotka-Volterra equation is the historic model of the study of predator-prey relationships in populations.

Let $x(t)$ be a population of sheep and let $y(t)$ be a population of wolves at the instant t. We suppose that the populations and hence the functions x and y satisfy the equations:

$$\begin{cases} x'(t) = (a - by(t))x(t) \\ y'(t) = (cx(t) - d)y(t) \end{cases}$$

where a, b, c, d are positive constants. The critical points (or equilibrium populations) are determined by:

$$\begin{cases} a - by = 0 \\ cx - d = 0 \end{cases} \Rightarrow x = \frac{d}{c}, \quad y = \frac{a}{b}$$

We make a change of origin to place the critical point at $(0, 0)$ by putting $y_1 = x - \dfrac{d}{c}, y_2 = y - \dfrac{a}{b}$. We obtain:

$$\begin{cases} y_1'(t) = -by_2(t)(y_1(t) + d/c) \\ y_2'(t) = cy_1(t)(y_2(t) + a/b) \end{cases}$$

Here is a graph of solutions for the particular values $a = 1, b = 2, c = 3, d = 1$.

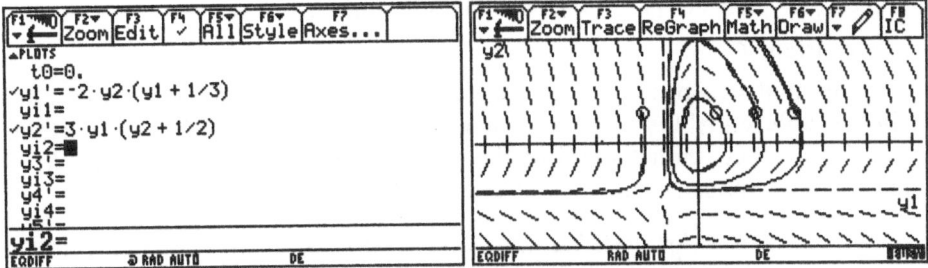

The linearized system then, using variables convenient for the calculator, is:

$$\begin{cases} y_1'(t) = -\frac{bd}{c} y_2(t) \\ y_2'(t) = \frac{ca}{b} y_1(t) \end{cases} \quad \text{or} \quad \begin{pmatrix} y_1' \\ y_2' \end{pmatrix} = \begin{pmatrix} 0 & -\frac{bd}{c} \\ \frac{ca}{b} & 0 \end{pmatrix} \begin{pmatrix} y_1 \\ y_2 \end{pmatrix}$$

Its eigenvalues, $\pm i \sqrt{ad}$, are pure imaginary. The trajectories in the phase plane are ellipses centered on the critical point.

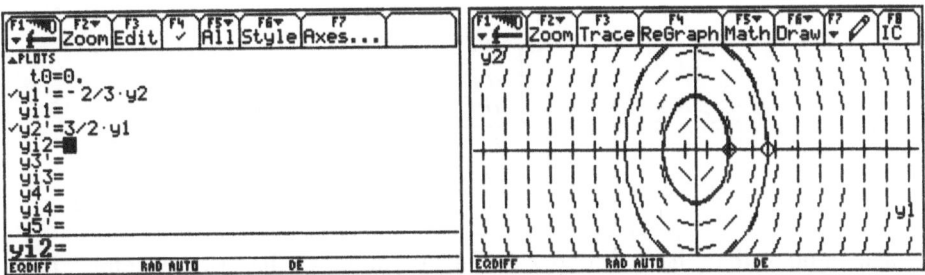

The two graphs don't seem much different in a neighborhood of the critical point. We will try to demonstrate that this is the case.

It is possible to give explicit formal solutions of the Lotka-Volterra equation. In fact, by writing $x'(t) = \dfrac{dx}{dt}, y'(t) = \dfrac{dy}{dt}$ and eliminating dt from the two equations, it follows that:

$$(cx - d)y\,dx + (by - a)x\,dy = 0$$

Since x and y are strictly positive, dividing by xy, this equation is equivalent to:

$$\left(c - \frac{d}{x}\right) dx + \left(b - \frac{a}{y}\right) dy = 0$$

an exact equation which may be integrated to:

$$\ell(x, y) = cx - d \ln x + by - a \ln y = C$$

The solution curves are thus given implicitly by the equation $\ell(x, y) = C$.

We remark that: $\lim\limits_{x\to+\infty}\ell(x,y)=\infty,\ \lim\limits_{y\to+\infty}\ell(x,y)=\infty$ and that $\lim\limits_{x\to0}\ell(x,y)=+\infty,\ \lim\limits_{y\to0}\ell(x,y)=+\infty$. The solution curves are thus bounded for every constant C.

We study their curvature. The Lotka-Volterra equations show that x and y are infinitely differentiable. We could then calculate $x''(t)$ and $y''(t)$ as functions of x,x',y,y', then, with the aid of these equations, as functions of x,y. After some rather tedious calculations, it follows that:

$$x'y'' - x''y' = \left(a(cx-d)^2 + d(a-by)^2\right)xy$$

which is always positive because a and d are.

Thus, since the curvature is positive, we obtain convex closed curves in the plane.

Moreover:

$$\begin{cases} \dfrac{\partial\ell}{\partial x} = c - \dfrac{d}{x} \\[2mm] \dfrac{\partial\ell}{\partial y} = b - \dfrac{a}{y} \end{cases}$$

At the critical point, the function ℓ is stationary and:

$$\begin{cases} \dfrac{\partial^2\ell}{\partial x^2} = \dfrac{d}{x^2} \\[2mm] \dfrac{\partial^2\ell}{\partial y^2} = \dfrac{a}{y^2} \\[2mm] \dfrac{\partial^2\ell}{\partial x\partial y} = 0 \end{cases}$$

which implies that the function ℓ has a minimum at the critical point. Thus, the latter is interior to a domain bounded by each solution curve.

When the time t increases, it appears that neither x', nor y' vanish. At the initial point $(x(0),y(0))$, $x'(0)<0$ and $y'(0)>0$. The curve is then described in the positive direction. By continuity, every such curve is described in the positive direction. Since the curve is closed, the fluctuation of populations is periodic: for some numbers of sheep or wolves, neither of the two populations can either decrease nor increase indefinitely. The population is a periodic function of time. Here is a model which the ecologists will surely ignore!

On this screen, we have graphed the respective evolutions of these populations of sheep (the light curve) and wolves (the bold curve) as functions of time. We see clearly the growth and decrease of each population with respect to the other.

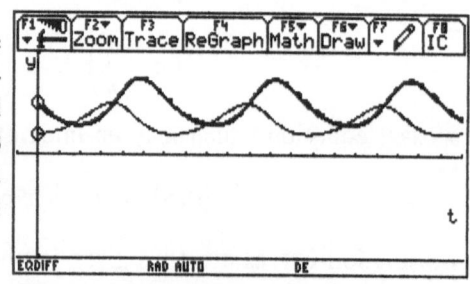

Fourth example: the harmonic oscillator

A point mass m situated on the real line is attracted to the origin O by a force proportional to its distance from it. We denote by $x(t)$ the distance from the origin at t. The fundamental equation of the dynamics of the particle may be written: $mx''(t) = -kx(t)$, where k is a positive constant. We will simplify - or "normalize" - this equation by setting $m = 1$, and we may suppose that $k = 1$. We transform the second order equation to the autonomous system:

$$\begin{cases} x' = y \\ y' = -x \end{cases}$$

We have already studied this second order linear equation, and we know that its solutions are of the form:

$$\begin{cases} x(t) = \alpha \cos(t - t_0) \\ y(t) = -\alpha \sin(t - t_0) \end{cases}$$

The solution curves are circles centered on the origin and with radius α. These circles are described in a clockwise or negative direction.

The solution circles of the harmonic oscillator system.

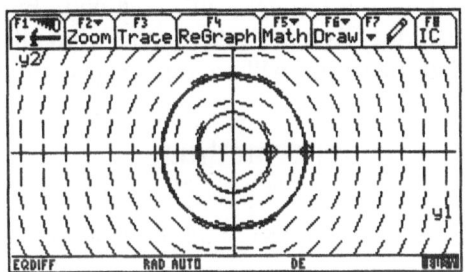

The origin is a stable critical point: as one may trace very easily, the curve remains in a neighborhood of this point.

We suppose now that there is friction, proportional to the speed and opposing the motion. The equation becomes $x''(t) = -x(t) - Cx'(t)$, with $C > 0$. This transforms into an autonomous equation:

$$\begin{cases} x' = y \\ y' = -x - Cy \end{cases}$$

This linear autonomous system has for its matrix $A = \begin{pmatrix} 0 & 1 \\ -1 & -C \end{pmatrix}$. The eigenvalues depend on the sign of $C^2 - 4$. If we suppose that C is less than 2, A has two complex eigenvalues $-\dfrac{C}{2} \pm Bi$, where B is a positive real number. The solution curves are spirals converging to the origin.

The graph of a solution curve with $C = 0.2$ is shown here.

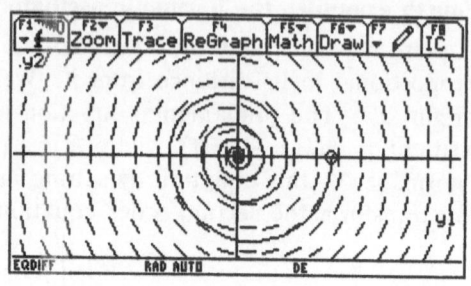

Thus, a small amount of friction completely perturbs the behavior of the solution curves. We call this an unstable oscillator.

7. Numeric solutions

The two most common numerical algorithms for solving differential equations are supplied on the TI graphic calculators to graph solutions: Euler's method and the RK (Runge-Kutta) method. Euler's method clearly graphs more rapidly than the Runge-Kutta method but also less precisely. We will study both algorithms and will analyze their respective performances.

7. 1 Euler's method

This is the simplest of all the numeric methods for solving differential equations. If the initial value problem differential equation (IVP) is $x' = \varphi(t, x(t))$, $x(a) = x_0$ on an interval $[a, b]$, the algorithm calculates and graphs a sequence (x_i) of points defined by:

$$x_{i+1} = x_i + h\varphi(t_i, x_i)$$

with: $h = \dfrac{b - a}{N}, t_i = a + ih, (0 \leq i \leq N)$. h is called the step size. N is a positive integer specifying the number of steps. Here we have chosen a constant step size for reasons of simplicity, but it may also vary as a function of i $(0 \leq i \leq N))$.

In fact, because $\varphi(t, x(t)) = x'(t)$, we have $x_{i+1} = x_i + hx'(t_i)$ which is an approximation of the formula of finite growth $x(t_{i+1}) = x(t_i) + hx'(t^*_i)$. Here t^*_i is a value in the appropriate subinterval, a so-called "mean value."

The following simple program puts this idea to work. It is not optimal. We could just as well plot starting from the lower bound of the interval $[a, b]$, but that algorithm comes with the calculator.

```
:euler(a,b,n,in)              [a,b], in : initial value
:Prgm
:Local h,i,x,y
:(b-a)/n→h                    h is the step
:in→x                         x takes the initial value
:ClrGraph
:For i,0,n                    we loop on [a,b]
:x+h*f(a+i*h,x)→y             the next term
:Line a+i*h,x,a+(i+1)*h,y     graph of the segment (tₙ, xₙ), (tₙ₊₁, xₙ₊₁)
:y→x                          for the next term
:EndFor
:EndPrgm
```

We define the function $f(x,y)$ on the command screen before launching the program. In the following example, we have chosen $f(x,y) = 1 + y/x$.

Here is an example using the preceding program for the equation $x'(t) = 1 + x(t)/t$ on $[1,3]$, $x(1) = 1$. What is the solution of this differential equation?

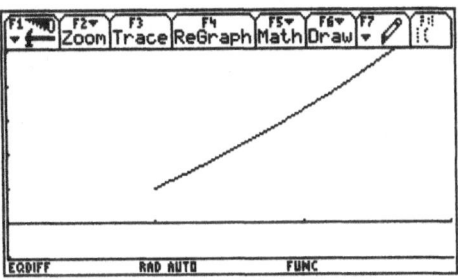

Observe the graph for $n = 10$ on $[1,3]$ superimposed on the graph of the solution $y = x(1+\ln(x))$ on the interval $[0,3]$. The graphs start to diverge toward the middle of the interval. The step $h = 0.2$ is pretty large. A smaller step slows the graphing but improves the precision.

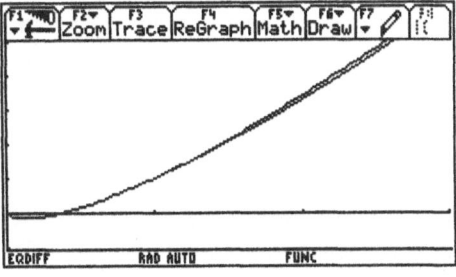

Here is the same type of program, but with a numeric emphasis. We calculate the values of the sequence (x_i) defined by the Euler algorithm and we save it in a list l. We then calculate the values of the exact solution, (saved in y1) at the same points, and we calculate the maximum of the difference. This lets us evaluate numerically the error committed by the Euler algorithm.

```
:eultest(a,b,n,in)          [a, b], in : initial value
:Func
:Local h,i,x,y,l,ll,r
:(b-a)/n→h                  h is the step
:in→x                       x takes the initial value
:{x}→l                      tore in the list l
:For i,0,n-1                a loop on [a, b]
:x+h*f(a+i*h,x)→y           next term
:augment(1,{y})→1           next element in the list l
:y→x
:EndFor
:seq(y1(a+i*h),i,0,n)→ ll   same thing for the function solution
:max(abs(1-11))→ r          r = the maximum of the difference the results
:EndFunc
```

Here is a test for several values of n(10, 50, 100, 200, 1000), on the interval $[0.2, 1]$. It seems that the error is of the order of h.

```
F1▼  F2▼  F3▼  F4▼   F5    F6▼
  Algebra Calc Other PrgmIO Clean Up

■ eultest(.2, 1, 10, y1(.2))          .147399
■ eultest(.2, 1, 50, y1(.2))          .031488
■ eultest(.2, 1, 100, y1(.2))         .015872
■ eultest(.2, 1, 200, y1(.2))         .007968

EQDIFF          RAD AUTO        FUNC 4/30
```

If we repeat on a larger interval $[0.2, 3]$, the rounding errors due to the calculations of the sequence (x_i) add up. The error is approximately 10 times greater.

```
F1▼  F2▼  F3▼  F4▼   F5    F6▼
  Algebra Calc Other PrgmIO Clean Up

■ eultest(.2, 3, 50, y1(.2))          .372635
■ eultest(.2, 3, 100, y1(.2))         .191131
■ eultest(.2, 3, 200, y1(.2))         .096781
eultest(.2,3,200,y1(.2))
EQDIFF          RAD AUTO        FUNC 3/30
```

It is easy to evaluate a bound for the error committed by Euler's method in the case where the function f satisfies the condition of the Cauchy-Lipschitz theorem "weak version" ($|f(t, x) - f(t, y)| \leq K|x - y|$, the linear case for example), even when we don't know the exact solution of the differential equation $x'(t) = f(t, x(t)), x(t_0) = x_0$. In fact, we know that the solution $x(t)$ satisfies the integral equation:

$$x(t) = x_0 + \int_{t_0}^{t} f(u, x(u))du$$

Thus, if we write $t_k = a + kh, 0 \leq k \leq n$, we have:

$$x(t_{k+1}) = x(t_k) + \int_{t_k}^{t_{k+1}} f(u, x(u)) du$$

The evaluation error is then:

$$E = \max_{0 \leq k \leq n} |e_k|, \text{ with } e_k = x(t_k) - x_k$$

Let's define:

$$\varepsilon_k = x(t_{k+1}) - x(t_k) - h f(t_k, x(t_k))$$

If we suppose that the function x is of class C^2 on the interval $[t_0, t_0 + T]$ (thus majorized by the constant C), we could write, using Taylor's formula:

$$\varepsilon_k = \int_{t_k}^{t_{k+1}} (x'(u) - x'(t_k)) \, du$$

$$= \int_{t_k}^{t_{k+1}} (t_{k+1} - u) \, x''(u) du$$

where:

$$|\varepsilon_k| \leq h \int_{t_k}^{t_{k+1}} |x''(u)| du \leq Ch^2$$

Hence:

$$e_{k+1} = e_k + h \left(f(t_k, x(t_k)) - f(t_k, x_k) \right) + \varepsilon_k$$

thus:

$$|e_{k+1}| \leq (1 + hK)|e_k| + |\varepsilon_k|$$

We show by recursion that:

$$e_k \leq (1 + hK)^k e_0 + Ch^2 \frac{(1 + hk)^k - 1}{hK} = Ch^2 \frac{(1 + hk)^k - 1}{hK}$$

when $e_0 = 0$. When k is small, we use the approximation $(1 + hK)^k \approx 1 + khK$ to obtain $e_k \sim Ckh^2 \leq Cnh^2$. Hence, $h = \dfrac{b - a}{n}$ gives:

$$\max_{0 \leq k \leq n} e_k = O\left(\frac{1}{n}\right)$$

Thus, we see that $\lim_{h \to 0} \max(e_k) = 0$ and this assures us of the convergence of Euler's method. But note the following fact: the hypotheses we placed on f are only practical in the case of linear differential equations. They are generally unrealistic otherwise.

7. 2 The Runge-Kutta method

Euler's method is not sufficiently accurate to successfully solve many differential equations. To do so we use a more precise method, which however requires many more calculations. We now state definitions for this problem.

Let $\begin{cases} x'(t) = f(t, x(t)) \\ x(t_0) = x_0 \end{cases}$ be the IVP, where f is a continuous function of $[t_0, t_0 + T] \times \mathbb{R}$ satisfying a Cauchy condition.

Let N be a natural number, let $\sigma = (t_0 < t_1 < \ldots < t_N = t_0 + T)$ be a subdivision (or partition) of the interval $[t_0, t_0 + T]$ with a variable step, and let:

$$h_n = t_{n+1} - t_n, \quad h = \max_{0 \le n < N} h_n$$

The method of approximate calculation which we are going to describe is of the form:

$$(*) \quad \begin{cases} x_{n+1} = x_n + h_n \Phi(t_n, x_n, h_n), n \ge 0 \\ x_0 \in \mathbb{R} \end{cases}$$

where we will suppose that Φ is a continuous function on $[t_0, t_0 + T] \times \mathbb{R} \times [0, a]$ with real values which only depend on f.

We will thus generalize the work of the previous paragraph, when Euler's method corresponds to $\Phi(t, x, h) = f(t, x)$ and where Φ is independent of h.

With Euler's method, we saw that the error e_n was proportional to h. The method proposed in this paragraph gives an error proportional to $h^p, p > 1$, under the condition that the solution of the differential equation is regular or sufficiently smooth.

As with Euler's method, we direct our attention to a study of the error and not just to the problem of convergence.

The method $(*)$ is said to be of order $p > 0$ if there is a constant $K > 0$ only depending on x and on Φ such that:

$$\sum_{n=0}^{N-1} |x(t_{n+1}) - x(t_n) - h_n \Phi(t_n, x(t_n), h_n)| \le K h^p$$

for every solution x of the IVP of class C^{p+1} on $[t_0, t_0 + T]$.

Thus, if the method is of order p, the error E may be shown to be majorized by $K h^p$.

Before announcing a theorem giving a necessary and sufficient condition that a method $(*)$ be of order p, we put:

$$f_0(t, x) = f(t, x)$$

$$f_1(t, x) = \frac{\partial f}{\partial t}(t, x) + \frac{\partial f}{\partial x}(t, x) f(t, x)$$

$$\vdots = \vdots$$

$$f_k(t, x) = \frac{\partial f_{k-1}}{\partial t}(t, x) + \frac{\partial f_{k-1}}{\partial x}(t, x) f(t, x)$$

We could easily show by recurrence that if x is a solution of the differential equation $x'(t) = f(t, x(t))$ and if f is sufficiently differentiable, then $x^{(k+1)}(t) = f_k(t, x(t)) = \dfrac{d^k}{dt^k} f(t, x(t))$.

Theorem 1: Let f be of class C^p on $[t_0, t_0 + T] \times \mathbb{R}$. We suppose that the functions $\Phi, \dfrac{\partial \Phi}{\partial h}, \ldots, \dfrac{\partial^p \Phi}{\partial h^p}$ exist and are continuous on $[t_0, t_0 + T] \times [0, a]$. Then the method $(*)$ to be of order p if and only if:

$$\Phi(t, x, 0) = f(t, x)$$

$$\frac{\partial \Phi}{\partial h}(t, x, 0) = \frac{1}{2} f_1(t, x)$$

$$\vdots \qquad\qquad \vdots$$

$$\frac{\partial^{p-1} \Phi}{\partial h^{p-1}}(t, x, 0) = \frac{1}{p} f_{p-1}(t, x)$$

Proof: Let $\varepsilon_n = x(t_{n+1}) - x(t_n) - h_n \Phi(t_n, x(t_n), h_n)$, and:

$$\Psi_k(t, x) = \frac{1}{(k+1)!} f_k(t, x) - \frac{1}{k!} \frac{\partial^k \Phi}{\partial h^k}(t, x, 0)$$

(We remark that the conditions of the theorem are equivalent to $\Psi_k(t, x) = 0, (0 \le k \le p - 1)$).
By Taylor's formula, there is a $c_n \in]t_n, t_{n+1}[and a \lambda_n \in]0, h_n[$ such that:

$$\varepsilon_n = \sum_{k=0}^{p-1} h_n^{k+1} \Psi_k(t_n, x(t_n)) + \frac{h_n^{p+1}}{(p+1)!} x^{(p+1)}(c_n) - \frac{h_n^{p+1}}{(p)!} \frac{\partial^p \Phi}{\partial h^p}(t_n, x(t_n), \lambda_n)$$

If the conditions of the theorem are satisfied, then by the preceding remark:

$$|\varepsilon_n| \le C h_n^{P+1} \le C h^p h_n$$

and:

$$\sum_{n=0}^{N} |\varepsilon_n| \le C T h^p$$

Conversely, suppose that the conditions of the theorem are not satisfied. Then there exists a smallest $k < p$ such that $\Psi_k(t, x) \ne 0$. In the case where for all $n, h_n = h$, we have:

$$\varepsilon_n = h^{k+1} \Psi_k(t_n, x(t_n)) + O(h^{k+2})$$

and:

$$\sum_{n=0}^{N} |\varepsilon_n| = h^k \sum_{n=0}^{N} h |\Psi_k(t_n, x(t_n))| + O(h^{k+1})$$

But if the method is of order p:

$$0 = \lim_{h \to 0} \frac{1}{h^k} \sum_{n=0}^{N} |\varepsilon_n| = \int_{t_0}^{t_0+T} |\Psi_k(u, x(u))| du$$

Thus for all $t \in [t_0, t_0 + T]$ and for every solution x of the IVP, $\Psi_k(t, x(t)) = 0$. Hence by the Cauchy-Lipschitz theorem, for each (t, y) there is a solution x of for all $t \in [t_0, t_0 + T]$, for all y, $\Psi_k(t, y) = 0$. This is a contradiction, so the theorem is proved.

The Runge-Kutta methods of the form $(*)$ are given by this general definition: Let q be a natural number, let $(a_{i,j})_{1 \le i,j \le q}$ be q^2 positive real numbers, and let $(c_1, c_2, \ldots, c_q)$ be real numbers.
For every $0 \le n \le N$, and for every $1 \le i \le q$, we put $t_{n,i} = t_n + c_i h_n$. Then:

$$x_{n,i} = x_n + h_n \sum_{j=1}^{q} a_{i,j} f(t_{n,j}, x_{n,j})$$

and:

$$x_{n+1} = x_n + h_n \sum_{j=1}^{q} b_j f(t_{n,j}, x_{n,j})$$

or, in a more general form:

$$\begin{cases} x_i = x + h \sum_{j=1}^{q} a_{i,j} f(t + c_j h, x_j), 1 \le i \le q \\ \Phi(x, t, h) = \sum_{j=1}^{q} b_j f(t + c_j h, x_j) \end{cases}$$

The most common Runge-Kutta method is given by the formula:

$$\begin{cases} x_{n,1} = x_n \\ x_{n,2} = x_n + \frac{h_n}{2} f(t_n, x_{n,1}) \\ x_{n,3} = x_n + \frac{h_n}{2} f(t_n + \frac{h_n}{2}, x_{n,2}) \\ x_{n,4} = x_n + \frac{h_n}{2} f(t_n + \frac{h_n}{2}, x_{n,3}) \end{cases}$$

and:

$$x_{n+1} = x_n + h_n \left(\frac{1}{6} f(t_n, x_n) + \frac{1}{3} f\left(t_n + \frac{h_n}{2}, x_{n,2}\right) \right.$$
$$\left. + \frac{1}{3} f\left(t_n + \frac{h_n}{2}, x_{n,3}\right) + \frac{1}{6} f(t_{n+1}, x_{n,4}) \right)$$

This is (almost) the method used by the calculator when we ask it to graph the solutions of differential equations with the RK option. The method implemented there is in fact an acceleration of the method shown here which reduces the number of calculations. Nevertheless, we now better understand why the graph with the option EULER is more rapid but less precise than with the option RK.

Here is a second example showing this contrast. The differential equation to be solved is:

$$y' = \frac{1}{1000}y(100 - y), \quad y(0) = 10$$

whose solution is:

$$y = \frac{100e^{x/10}}{9 + e^{x/10}}$$

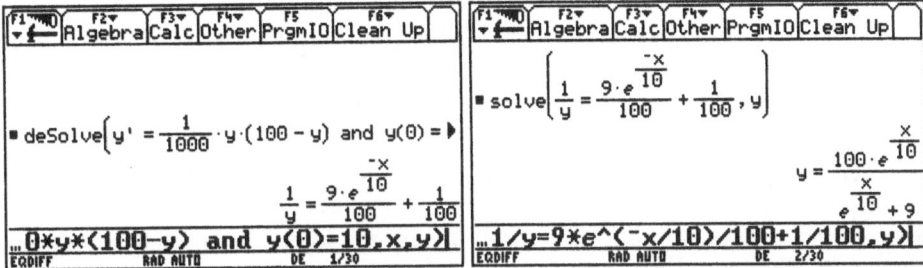

We construct two tables, allowing us to store the values calculated by each of the methods, Runge Kutta and Euler, for comparison. The fourth column holds the absolute value of the difference of the two values.

t	rk	euler	diff
	c1	c2	c3	c4
17	16.	35.4831	34.4536	1.02954
18	17.	37.8048	36.7119	1.09284
19	18.	40.1834	39.0353	1.14803
20	19.	42.6083	41.4151	1.1932
21	20.	45.0689	43.8414	1.22753
22	21.	47.5545	46.3035	1.25099
23		50.0524	48.7898	1.26256

We then defined two statistical graphs:
- We first graphed the pairs $(t, RK(t))$ with plus signs,
- Second, we graphed the pairs $(t, Euler(t))$ with dots.

We have added the graph of the exact solution calculated above.

This shows the graph of the exact solution and of the two approximate solutions. In this window, [-10,100] by [-5,120], we don't see any difference. The step size was 1.

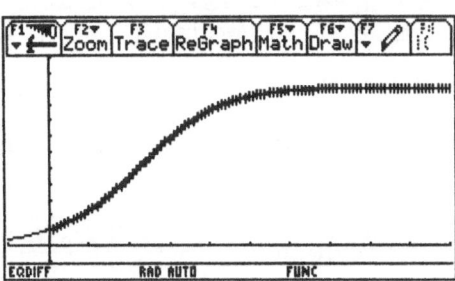

Zooming in, we see clearly the difference. From the top of the screen, the curves graphed are Runge Kutta, the exact solution and Euler.

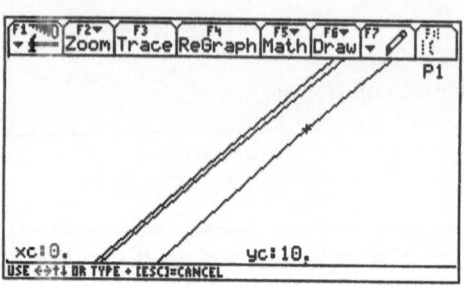

8. The Laplace transformation

The Laplace transformation will allow us to solve differential equations "operationally" using only algebraic techniques.

Let f be a function defined on $\mathbb{R}$, which is zero on $\mathbb{R}^{<0}$. The Laplace transform of f is the function defined by:

$$F(s) = \int_0^{+\infty} f(t)e^{-st}dt$$

for the values of s for which this integral converges. We may say that the function f(t) has been transformed from the t-domain to the s-domain.

The mapping or assignment $f \mapsto F$ may also be written as $\mathcal{L}(f)(s) = F(s)$ or just $\mathcal{L}(f) = F$.

Some functions don't have Laplace transforms. For example, the Laplace transformation of $t \mapsto e^{t^2}$, is not defined since the defining integral does not converge for any value of s. For another example, the Laplace transform of 1 only exists for $s > 0$.

In order to focus on basic information about Laplace transforms, we suppose that our work here is defined for a class of functions for which $\mathcal{L}(f)$ exists on $\mathbb{R}^{\geq 0}$.

Here are some computed examples of Laplace transforms:

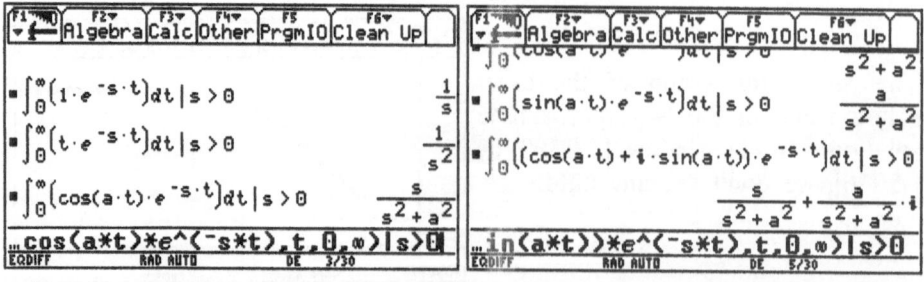

Now we show some other properties of this transformation.

Proposition 1: 1. The transformation $\mathcal{L}$ is linear, that is, for any (f, g) and any real λ:

$$\mathcal{L}(f + \lambda g) = \mathcal{L}(f) + \lambda \mathcal{L}(g)$$

2. For any $n \in \mathbb{N}$, $\mathcal{L}(t^n) = \dfrac{n!}{s^{n+1}}$

3. $\mathcal{L}(e^{-at}) = \dfrac{1}{s+a}$ $(s + a > 0)$

4. $\mathcal{L}(f(at)) = \dfrac{1}{a}\mathcal{L}(f)(\dfrac{s}{a}), a > 0$

5. $\mathcal{L}(f(t-a)) = e^{-as}\mathcal{L}(f)(s), a > 0$

6. $\mathcal{L}(e^{-at}f(t)) = \mathcal{L}(f)(s+a)$

Proof: We prove only a few of these properties. The first follows directly from the linearity of the integral, the second by induction, the third by the change of variable $u = a + s$.

4. $\mathcal{L}(f(at)) = \displaystyle\int_0^{+\infty} f(at)e^{-st}dt = \dfrac{1}{a}\int_0^{+\infty} f(u)e^{-\frac{s}{a}u}du$, with $u = at$.

5. $\mathcal{L}(f(t-a)) = \displaystyle\int_0^{+\infty} f(t-a)e^{-st}dt = \int_{-a}^{+\infty} f(u)e^{-s(a+u)}du$ with $t - a = u$. We obtain the desired result since f vanishes on $\mathbb{R}^{<0}$.
The last property is obvious.

Proposition 2: Let f be a function of class C^1 on $\mathbb{R}^{/ge0}+$. We suppose that, for all $s > 0$, $\lim\limits_{t \to +\infty} f(t)e^{-st} = 0$. Then:

$$\mathcal{L}(f')(s) = s\mathcal{L}(f)(s) - f(0)$$

Proof: It is sufficient to integrate by parts:

$$\int_0^{+\infty} f'(t)e^{-st}dt = \left[f(t)e^{-st}\right]_0^{+\infty} + s\int_0^{+\infty} f(t)e^{-st}dt$$

If f is of class C^k and if all the derivatives of f satisfy the hypotheses of the proposition, it follows that:

$$\mathcal{L}(f^{(k)})(s) = s^k\mathcal{L}(f)(s) - s^{k-1}f(0) - s^{k-2}f'(0) - \ldots - f^{(k-1)}(0)$$

Proposition 3: Let $F(t) = \displaystyle\int_0^t f(u)du$. Then:

$$\mathcal{L}(F)(s) = \dfrac{1}{s}\mathcal{L}(f)(s)$$

Proof: This follows from the preceding proposition and the fact that $F'(t) = f(t)$ with $F(0) = 0$.

There are many more useful properties of this transform. In particular, the Laplace transformation takes a particular combination called the convolution of two functions into the product of the transforms. However, for this chapter on differential equations, we won't need that application.

Now we give an example of how the Laplace transformation may be used to solve certain differential equations. The idea is a simple use of the preceding propositions.

For example, let's solve the following differential system on $\mathbb{R}^{\geq 0}$:

$$\begin{cases} x'(t) + y'(t) = x(t) - y(t) + 3e^{2t} \\ x''(t) + y'(t) = 2e^{2t} \end{cases}$$

with the initial conditions $x(0) = 0, x'(0) = 1, y(0) = -1$.

We apply the Laplace transformation to each of the two equations. Using the properties shown for the transform of a derivative $(\mathcal{L}(f')(s)) = s\mathcal{L}(f)(s) - f(0)$ and $\mathcal{L}(f'')(s) = s^2\mathcal{L}(f)(s) - sf(0) - f'(0))$ and with the transform of an exponential function, it follows that:

$$\begin{cases} sX + sY = X - Y + \dfrac{3}{s-2} - 1 \\ s^2 X + sY = \dfrac{2}{s-2} \end{cases}$$

Now use the calculator to do the necessary algebra:

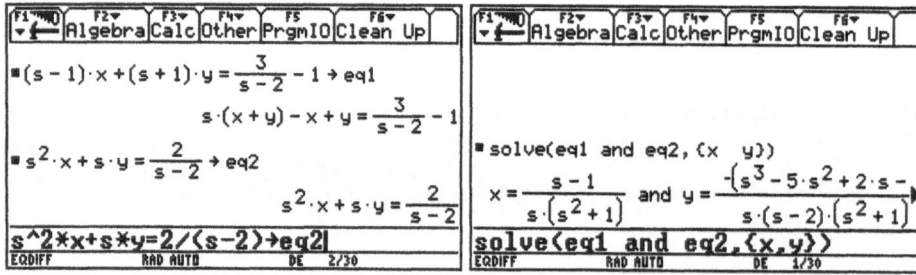

Here are the definitions of the two equations and their solution. We decompose each result into simple components. It only remains to look up the result in a table of transforms.

Referring to the results on the first two screens, we observe that:

$$\mathcal{L}(\cos t) = \frac{s}{s^2+1}, \qquad \mathcal{L}(\sin t) = \frac{1}{s^2+1}.$$

$$\mathcal{L}(e^{2t}) = \frac{1}{s-2}, \qquad \mathcal{L}(1) = \frac{1}{s}.$$

By linearity of the Laplace transformation, we obtain:

$$\begin{cases} \mathcal{L}(\sin t + \cos t - 1) = X = \mathcal{L}(x(t)) \\ \mathcal{L}(\sin t - \cos t + e^{2t} - 1) = Y = \mathcal{L}(y(t)) \end{cases}$$

If the Laplace transform is a one to one transformation, we may identify the arguments in each of the above equations. This is the case as we now show: Let E be a subspace of the vector space of functions generated by $P(t) \cos t$ and $Q(t) \sin t$, where P and Q are any polynomials. Each element of E may be written in the form $P(t) \cos t + Q(t) \sin t$. Since, for every natural number n:

$$I_n = \int_0^{+\infty} t^n e^{it} e^{-st}\, dt = \int_0^{+\infty} t^n e^{(i-s)t}\, dt$$

$$= \left[\frac{t^n}{i-s} e^{(i-s)t} \right]_0^{+\infty} + \frac{n}{s-i} \int_0^{+\infty} t^{n-1} e^{(i-s)t}\, dt$$

so, for $s > i$ we may write:

$$I_n = \frac{n}{s-i} I_{n-1}$$

We may compute that $I_0 = \dfrac{1}{s-i}$, and we also obtain:

$$I_n = \frac{n!}{(s-i)^n} = \frac{n!(s+i)^{n+1}}{(s^2+1)^n}$$

Taking the real part and the imaginary part, we see that the transform of every element of E may be written as a rational fraction $R(s)$. If $R(s) = 0$, for all $s > 0$ it follows that R is identically zero, by uniqueness of the decomposition into simple components using partial fraction decomposition. This proves that $\mathcal{L}$ is one to one (or injective) on E and this allows us conclude that the solution of the differential system is:

$$\begin{cases} x(t) = \cos t + \sin t - 1 \\ y(t) = -\cos t + \sin t + e^{2t} - 1 \end{cases}$$

With the foregoing brief overview of the Laplace transform, we close this chapter on differential equations. Many more applications to this subject may be made with the modern graphic calculator, both for solution and illustration of the mathematical theory.

3

Fourier analysis

The theory conceived and developed by Joseph Fourier in the early nineteenth century has since had many applications in areas as various as those of the telephone, radio, television, communications, and virtually every field in which vibrations or oscillations occur. This theory is still alive and is even being reborn today under the name of "wavelet theory". Few technical areas are unaffected by applications of Fourier theory. The starting point for Fourier theory is one which is found constantly in the history of mathematics: to attempt to approximate functions by simpler ones. Here we try to represent "arbitrary" functions in terms of trigonometric polynomials that are as easy to manipulate as the usual polynomials of school mathematics. In some ways, the Fourier expansion of a function resembles the representations proposed in the next chapter about interpolation of the values of functions where we use approximation by the method of least squares. It is also the case that many developments of mathematical analysis have historical beginnings in this field. This chapter will be both an introduction the the mathematical theory and to some of the computational nuances of Fourier analysis.

1. Fourier series

In this chapter f designates a 2π periodic function ($\forall x \in \mathbb{R}, f(x+2\pi) = f(x)$), defined on $\mathbb{R}$, with complex or real values. We are going to address the following question:
Is there a trigonometric series, that is a series of the form

$$\sum_{n \geq 0}(a_n \cos(nx) + b_n \sin(nx))$$

such that for all real x:

$$f(x) = \sum_{n \geq 0}(a_n \cos(nx) + b_n \sin(nx)) \ ?$$

Definition: Let f be a 2π periodic function, integrable on all compact subsets of $\mathbb{R}$. The complex numbers $(c_n(f))_{n \in \mathbb{Z}}$ defined by:

$$c_n(f) = \frac{1}{2\pi}\int_0^{2\pi} f(t)e^{-int}dt$$

are called the complex Fourier coefficients of f. The series:

$$SF(f)(x) = \sum_{n=-\infty}^{+\infty} c_n(f)e^{inx}$$

is called the Fourier series of f.

Remark: This series may not converge for some values of x, or may not even converge for any real x. in addition, $SF(f)(x)$ may possibly converge to a value different from $f(x)$. We will see later what conditions we will need to impose on f to obtain some kind of reasonable and useful convergence.

Here f is the 2π periodic function such that $f(x) = x$ on $[0, 2\pi]$. For the calculator, n does not represent an integer (in the first result shown). We therefore calculate the Fourier coefficients of f with the help of the integer variable @n1. It remains to determine $c_0(f)$ individually.

For a function which is defined piecewise, we must help the calculator calculate the Fourier coefficients. Here we do so for a "folded parabola".

Remarks: We may show easily that:

1. Since the functions $x \mapsto f(x)$ and $x \mapsto e^{inx}$ are 2π periodic, for every $a \in \mathbb{R}$ we have:

$$c_n(f) = \frac{1}{2\pi} \int_a^{2\pi+a} f(t)e^{-int}\,dt$$

2. When f is a real valued function, we may write $e^{int} = \cos nt + i\sin nt$, and define:

$$a_n(f) = \frac{1}{\pi} \int_0^{2\pi} f(t)\cos(nt)\,dt,\ (n \geq 0) \quad b_n(f) = \frac{1}{\pi} \int_0^{2\pi} f(t)\sin(nt)\,dt\ (n \geq 1)$$

We then have the following relationships: $a_0(f) = 2c_0(f)$ and for all $n \geq 1$:

$$\begin{cases} c_n(f) = \dfrac{1}{2}(a_n(f) - ib_n(f)) \\ c_{-n}(f) = \overline{c_n(f)} \end{cases}$$

or:

$$\begin{cases} a_n(f) = c_n(f) + c_{-n}(f) \\ b_n(f) = i(c_n(f) + c_{-n}(f)) \end{cases}$$

Moreover, when f is even, we obtain, for all $n \geq 0$:

$$a_n(f) = \frac{2}{\pi} \int_0^\pi f(t) \cos(nt) dt, \qquad b_n(f) = 0$$

and when f is odd:

$$b_n(f) = \frac{2}{\pi} \int_0^\pi f(t) \sin(nt) dt, \qquad a_n(f) = 0$$

3. When f is T periodic, its Fourier coefficients $(c_n(f))$ are defined by the formula:

$$c_n(f) = \frac{1}{T} \int_0^T f(t) \exp\left(\frac{-2i\pi t}{T}\right) dt$$

The following program allows for the definition of f of period T and calculation of its nth Fourier coefficient. The coefficient of order 0 is calculated by a simple integral. The result is stored in a variable 1.

```
:fourierf()
:Prgm
:Local f,a,c,p,k
:Dialog
:Text "Definition of f"
:Request "f",f                                              f
:Request "Period",t                                         t is the period
:DropDown "parity :",{"even","odd","none"},c                even, odd, no parity
:EndDlog
:expr(f)→f:expr(t)→t                                        transformation of f and t
:If c=1 Then                                                if f is even
:4/t*∫(f*cos(2*π*@n1/t*x),x,0,t/2)→l                        one calculates $a_n$
:Elseif c=2 Then                                            if f is odd
:4/t*∫(f*sin(2*π*@n1/t*x),x,0,t/2)→l                        one calculates $b_n$
:Else                                                       else
:1/t*∫(f*e^(i*2*π*@n1/t*x),x,0,t)→l
:EndIf
:EndPrgm
```

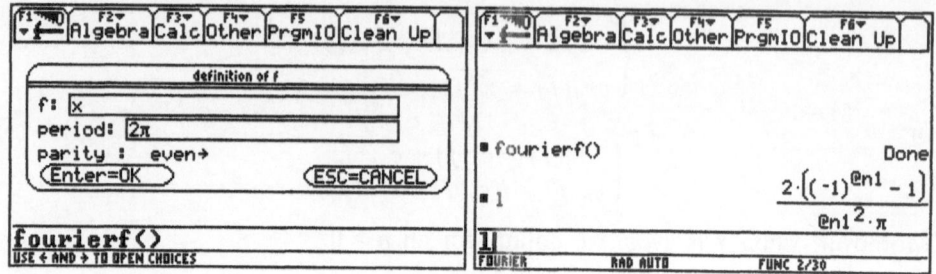

1.1 Convergence of Fourier series

We now consider the problem of convergence of the Fourier series $SF(f)$ of f.

Proposition 1:(Lebesgue's Lemma). Let f be defined and integrable on a compact interval $[a, b]$. Then:

$$\lim_{x \to \pm\infty} \int_a^b f(t)e^{itx} dt = 0$$

Proof: a) We first show this result for a step function φ. In this case, there is a partition $\sigma = (a = t_0 < t_1 < \ldots < t_n = b)$ of $[a, b]$ such that for all $1 \le k \le n$, the restriction of φ to the interval $]t_{k-1}, t_k[$ is a constant c_k. We then have:

$$\int_a^b \varphi(t)e^{itx} dt = \sum_{k=1}^n \int_{t_{k-1}}^{t_k} \varphi(t)e^{itx} dt$$

$$= \sum_{k=1}^n \int_{t_{k-1}}^{t_k} c_k e^{itx} dt = \frac{1}{ix} \sum_{k=1}^n \left(e^{ixt_{k-1}} - e^{ixt_k}\right) c_k$$

Therefore:

$$\left| \int_a^b \varphi(t)e^{itx} dt \right| \le \frac{2}{|x|} \sum_{k=1}^n |c_k| = \frac{A}{|x|}$$

so the result follows for step functions.

b) In the general case, with f integrable on $[a, b]$, and for any $\varepsilon > 0$, there exists a step function φ such that, for all $x \in [a, b]$,

$$\int_a^b |f(t) - \varphi(t)| dt < \varepsilon$$

Therefore, for all real x:

$$\left| \int_a^b f(t)e^{itx} dt - \int_a^b \varphi(t)e^{itx} dt \right| = \left| \int_a^b (f(t) - \varphi(t))e^{itx} dt \right| < \varepsilon$$

For this step function φ, also by part a), there exists X such that for all $|x| > X$, we have:

$$\left| \int_a^b \varphi(t)e^{itx} dt \right| < \varepsilon$$

Thus, since we can write:

$$\left| \int_a^b f(t)e^{itx} dt \right| = \left| \int_a^b (f(t) - \varphi(t) + \varphi(t))e^{itx} dt \right| \leq \left| \int_a^b (f(t) - \varphi(t))e^{itx} dt + \int_a^b \varphi(t)e^{itx} d \right|$$

the triangle inequality gives, for $|x| > X$:

$$\left| \int_a^b f(t)e^{itx} dt \right| < 2\varepsilon$$

We may conclude that the sequence of Fourier coefficients of a periodic integrable function tends to 0, when $n \to \pm\infty$.

Theorem 1:(Dirichlet's Theorem). Let f be a real or complex valued, 2π periodic function, integrable on any compact subset of $\mathbb{R}$. Let x be a point such that the limits $f(x+0)$ and $f(x-0)$ exist. If at this point f has both a right and a left derivative, then the Fourier series $SF(f)(x)$ converges to $\frac{1}{2}(f(x+0) + f(x-0))$.

In particular, $SF(f)(x)$ converges to $f(x)$ at any point x where f is continuous and differentiable. (So we say that regular functions have convergent Fourier series).

Proof: Note that

$$SF_n(f)(x) = \sum_{k=-n}^{n} c_k(f)e^{ikx} = \frac{a_0}{2} + \sum_{k=1}^{n} (a_k(f)\cos(kx) + b_k(f)\sin(kx))$$

By definition of the Fourier coefficients, we may write:

$$SF_n(f)(x) = \frac{1}{\pi} \int_{-\pi}^{\pi} \left(\frac{1}{2} + \sum_{k=1}^{n} (\cos(kx)\cos(kt) + \sin(kx)\sin(kt)) \right) f(t) dt$$

$$= \frac{1}{\pi} \int_{-\pi}^{\pi} \left(\frac{1}{2} + \sum_{k=1}^{n} \cos k(x - t) \right) f(t) dt$$

$$= \frac{1}{\pi} \int_{-\pi}^{\pi} \frac{\sin(n + 1/2)(x - t)}{2\sin\frac{x-t}{2}} f(t) dt$$

by an immediate calculation of the second sum.

Indeed, using the sum for a geometric sequence:

$$\frac{1}{2} + \sum_{k=1}^{n} \cos(ku) = \frac{1}{2} + \Re \sum_{k=1}^{n} e^{iku}$$

$$= \frac{1}{2} + \Re \frac{e^{iu} - e^{i(n+1)u}}{1 - e^{iu}} \quad \text{(sum of geometric sequence)}$$

$$= \frac{1}{2} + \cos\left(\frac{n+1}{2}\right) \frac{\sin(nu/2)}{\sin(u/2)} = \frac{\sin(n+1/2)/2}{\sin(u/2)}$$

(Check this using the calculator if it is not clear!)
The change of variable $u = t - x$ gives:

$$SF_n(f)(x) = \frac{1}{2\pi} \int_{-\pi-x}^{\pi-x} f(x+u) \frac{\sin(n+1/2)u}{\sin(u/2)} du$$

and, since the function being integrated is 2π periodic, by partitioning the interval $[-\pi, \pi]$ into $[-\pi, 0]$ and $[0, \pi]$, and by making the change of variable $v = -u$ in the first integral obtained, we obtain:

$$SF_n(f)(x) = \frac{1}{2\pi} \int_{0}^{\pi} (f(x+u) + f(x-u)) \frac{\sin(n+1/2)u}{\sin(u/2)} du$$

We notice that for $f = 1$, we have $a_0 = 2, a_n = b_n = 0$, for all $n \geq 1$, so that the last equality translates into:

$$\int_{0}^{\pi} \frac{\sin(n+1/2)u}{\sin(u/2)} du = \pi$$

Thus:

$$SF_n(f)(x) - \frac{1}{2}(f(x+0) + f(x-0)) =$$

$$\frac{1}{2\pi} \int_{0}^{\pi} (f(x+u) + f(x-u) - f(x+0) - f(x-0)) \frac{\sin(n+1/2)u}{\sin(u/2)} du$$

This last expression may be written as the sum of the two integrals:

$$\frac{1}{2\pi} \int_{0}^{\pi} (f(x+u) - f(x+0)) \frac{\sin(n+1/2)u}{\sin(u/2)} du + \frac{1}{2\pi} \int_{0}^{\pi} (f(x-u) - f(x-0)) \frac{\sin(n+1/2)u}{\sin(u/2)} du$$

We know that f has a righthand derivative at x, which means that the limit:

$$\lim_{u \to 0^+} \frac{f(x+u) - f(x+0)}{u} \quad \text{exists}$$

In a neighborhood of $0, \sin(u/2) \sim u/2$, and we have:

$$\lim_{u \to 0^+} \frac{f(x+u) - f(x+0)}{\sin(u/2)} \text{ exists}$$

and the function $\dfrac{f(x+u) - f(x+0)}{\sin(u/2)}$ is integrable on the interval $[0, \pi]$. Lebesgue's Lemma assures us that the limit of the first integral is 0.

By the same reasoning, the function $\dfrac{f(x-u) - f(x-0)}{\sin(u/2)}$ is integrable on the interval $[0, \pi]$ and Lebesgue's Lemma ensures us that the second integral converges to 0.

The following program, which is just an adaptation of the preceding program, allows us to calculate the real partial sums (in cosine and sines) of order n of the Fourier series of f.

```
:fouriern()
:Prgm :Local f,a,c,p,k
:delar l,ll
:Dialog
:Text "Definition of f"                          function f
:Request "f",f
:Request "Period",t
:DropDown "parity :",{"even","odd","none"},c
:Request "Number of coeffs.",n                    number of coefficients
:EndDlog
:expr(f)→f:expr(t)→t
:expr(n)→n
:If c=1 Then                                      if f is even, calculation of aₙ
:seq(4/t*∫(f*cos(2*π*k/t*x),x,0,t/2),k,0,n)→l
:l[1]/2+Σ(l[k+1]*cos(2*π*k*x/t),k,1,n)→sf(x)
:ElseIf c=2 Then                                  if f is odd, calculation of bₙ
:seq(4/t*∫(f*sin(2*π*k/t*x),x,0,t/2),k,0,n)→l
:Σ(l[k]*sin(2*π*k*x/t),k,1,n)→sf(x)
:Else                                             else, the general case
:seq(2/t*∫(f*cos(2*π*k/t*x),x,0,t),k,0,n)→l
:seq(2/t*∫(f*sin(2*π*k/t*x),x,0,t),k,0,n)→ll
:l[1]/2+Σ(l[k+1]*cos(2*π*k*x/t)+
  ll[k]*sin(2*π*k*x/t),k,1,n)→sf(x)
:EndIf
:EndPrgm                                           result in sf(x)
```

We launch the preceding program for the π periodic function $x \mapsto |\sin x|$. Then we ask for the partial sums of order 3 and (not shown) 5.

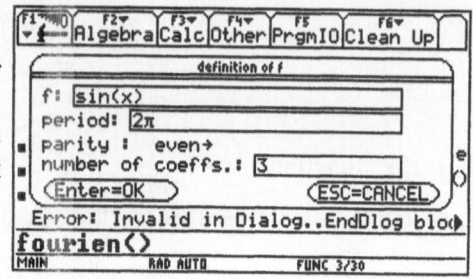

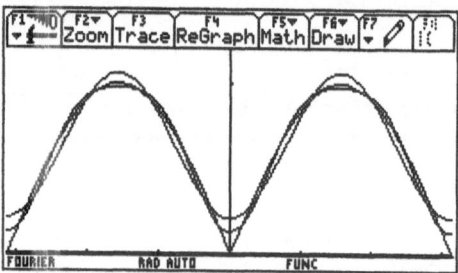

After entering the results in the function editor, we graphed the function f and its first two Fourier approximations. We notice that the convergence doesn't seem to be optimal at the point $x = 0$. At this point f is continuous, but not differentiable.

Now consider the same type of layout for the "signum" function on the interval $[-\pi, \pi]$

$$f : x \mapsto \begin{cases} -1 & \text{if } x \in [-\pi, 0] \\ 1 & \text{if } x \in [0, \pi] \end{cases}$$

We launch the preceding program for this function. We ask for the partial sums of orders 3 to 9. We have graphed the approximations 3 and 5, then 7 and 9 in the next two screenshots.

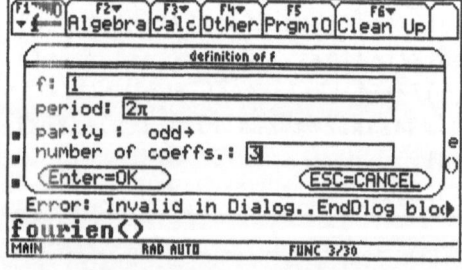

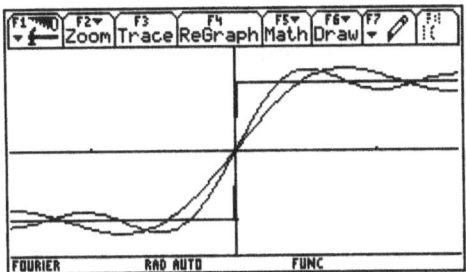 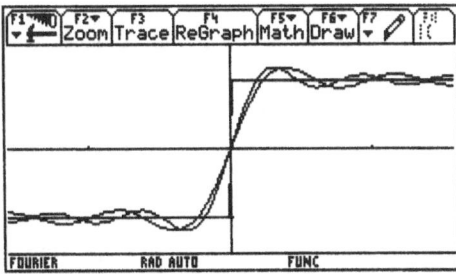

It is interesting to zoom on a neighborhood of 0, a point of discontinuity of f. Although approximations are all equal to 0 at $x = 0$, they have to "climb" rapidly to 1. In fact, they "overshoot" 1, making a bump on the graph that never disappears for a partial sum of any order. This is known as Gibbs phenomenon.

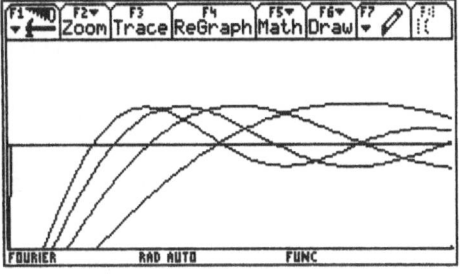

1. 2 The Gibbs phenomenon

Let f be the step function:

$$f : x \mapsto \begin{cases} -\frac{\pi}{4} & \text{if } x \in [-\pi, 0] \\ \frac{\pi}{4} & \text{if } x \in [0, \pi[\end{cases}$$

extended as a 2π periodic function on all $\mathbb{R}$.

Let's determine the partial sum of order $2n - 1$, $SF_{2n-1}(f)(x)$ of the Fourier series of f. Since f is odd, all Fourier coefficients $a_n(f) = 0$. Moreover, for $n \geq 1$:

$$b_n(f) = \frac{2}{\pi} \int_0^\pi \frac{\pi}{4} \sin(nt) dt = \frac{1}{2} \left(\frac{1 - (-1)^n}{n} \right)$$

$$SF_{2n-1}(f)(x) = \sum_{k=1}^{2n-1} b_k(f) \sin kx$$

$$= \sum_{k=1}^{n} \frac{\sin(2k-1)x}{2k-1} = \sum_{k=1}^{n} \int_0^x \cos(2k-1)t\,dt$$

$$= \int_0^x \left(\sum_{k=1}^{n} \cos(2k-1)t \right) dt$$

$$= \frac{1}{2} \int_0^x \frac{\sin 2nt}{\sin t}\,dt$$

Dirichlet's Theorem insures us that $SF_{2n-1}(f)(x)$ converges to $\dfrac{\pi}{4}$ for $0 < x < \pi$ and to 0 for $x = 0$.

Let's study the variation of this function. Differentiation yields:

$$SF'_{2n-1}(f)(x) = \frac{\sin 2nx}{2 \sin x}$$

so, using elementary calculus facts, $SF_{2n-1}(f)$ has a maximum at points with x-values:

$$x_k = \frac{(2k+1)\pi}{2n}$$

and a minimum at points with x-values:

$$x_k = \frac{k\pi}{n}$$

Thus, the y coordinate of the first maximum is equal to:

$$y_n = \int_0^{\pi/2n} \frac{\sin 2nt}{2 \sin t}\,dt$$

Let's determine its limit as n goes to infinity. We may write:

$$2y_n = \int_0^{\pi/2n} \frac{\sin 2nt}{t}\,dt + \int_0^{\pi/2n} \sin 2nt \left(\frac{1}{\sin t} - \frac{1}{t} \right) dt$$

$$= \int_0^\pi \frac{\sin s}{s}\,ds + \int_0^{\pi/2n} \sin 2nt \left(\frac{1}{\sin t} - \frac{1}{t} \right) dt$$

But :

$$\left| \int_0^{\pi/2n} \sin 2nt \left(\frac{1}{\sin t} - \frac{1}{t} \right) dt \right| \le \int_0^{\pi/2n} \left| \frac{1}{\sin t} - \frac{1}{t} \right| dt \le \frac{C}{2n}$$

by continuity of the function $t \mapsto \dfrac{1}{\sin t} - \dfrac{1}{t}$ in a neighborhood of 0. Thus:

$$\lim_{n\to+\infty} y_n = \frac{1}{2} \int_0^\pi \frac{\sin s}{s}\,ds \approx 0.9259 > \frac{\pi}{4}$$

The first maximum is that of $SF_{2n-1}(f)$. We will not show this here. On the other hand, we have proved that the maximum of the function SF_{2n-1} tends to a limit strictly greater than $\frac{\pi}{4}$.

Here is an approximate value of the limit of the sequence (y_n). The first maximum is that of $SF_{2n-1}(f)$ on $[0,\pi]$. We have proven that the bump observed graphically will always persist.

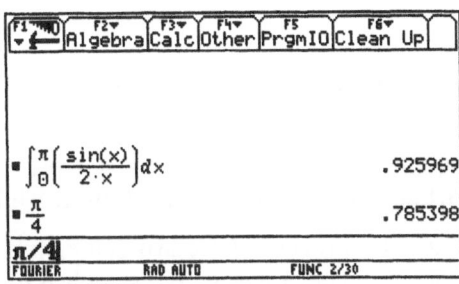

1. 3 Cesáro summability

Dirichlet's Theorem guarantees convergence of the Fourier series of f to $f(x)$ at any point of continuity of f, provided that we have at least a left and a right derivative at this point.

We now introduce Cesaro or (C,1), summability which is a "generalization" of the idea of convergence of a sequence or a series. We will apply it to Fourier series, and, among other things, this will allow us to "erase" the Gibbs phenomenon.

Definition: Let (u_n) be a sequence of real or complex numbers. We put:

$$S_n = \sum_{k=0}^{n} u_k \ (n \geq 0), \qquad \sigma_n = \frac{1}{n} \sum_{k=0}^{n-1} S_k \ (n \geq 1)$$

If the sequence (σ_n) tends to a limit σ, we say that the series $\sum u_k$ is Cesaro summable or (C,1) summable to σ.

We may easily show that if the series $\sum u_k$ converges to U, its Cesaro sum is equal to U.

Indeed, if $\lim_{n\to+\infty} \sum_{k=0}^{n} u_k = U$, we may write:

$$\forall\, \varepsilon > 0 \ \exists\, N, \quad \text{such that} \quad n \geq N \Rightarrow |S_n - U| < \varepsilon$$

We may then likewise write:

$$|\sigma_n - U| = \frac{1}{n} \left| \sum_{k=0}^{n-1} (S_k - U) \right|$$

$$\leq \frac{1}{n} \left| \sum_{k=0}^{N-1} (S_k - U) \right| + \frac{1}{n} \sum_{k=N}^{n-1} |(S_k - U)|$$

$$\leq \frac{C_N}{n} + \frac{n-N}{n}\varepsilon \leq \frac{C_N}{n} + \varepsilon$$

We obtain the announced result when n goes to infinity because $\lim\limits_{n \to +\infty} \dfrac{C_N}{n} = 0$.
The converse of this property is false. To show this, it suffices to take the sequence (u_n) defined by $u_n = (-1)^n$. The series $\sum u_n$ has no limit, since partial sums all equal either 0 or 1, while the sequence (σ_n) tends to 0.

Thus, the process of Cesaro summability is a true generalization of the usual concept of convergence. Let's apply it to the Fourier series. If $SF_n(f)$ represents the nth partial sum of the Fourier of f, put:

$$\sigma_n(f) = \frac{1}{n} \sum_{k=0}^{n-1} SF_k(f)$$

Theorem 2:(Cesaro). *Let f be a real or complex valued, 2π periodic function, integrable on any compact subset of $\mathbb{R}$. Let x be a point such that the limits $f(x+0)$ and $f(x-0)$ exist. Then $\sigma_n(f)(x)$ converges to $\frac{1}{2}(f(x+0) + f(x-0))$. In particular, $\sigma_n(f)(x)$ converges to $f(x)$ at each point of continuity of f.*

Proof: Recall that for all $n \geq 0$:

$$SF_n(f)(x) = \frac{a_0(f)}{2} + \sum_{k=1}^{n} (a_k(f)\cos(kx) + b_k(f)\sin(kx))$$

and that:

$$\sigma_n(f)(x) = \frac{1}{n} \sum_{k=0}^{n-1} SF_k(f)(x)$$

By applying the first calculations made during the proof of the theorem of Dirichlet, it follows that:

$$\sigma_n(f)(x) = \frac{1}{2n\pi} \int_0^{\pi} (f(x+u) + f(x-u)) \sum_{k=0}^{n-1} \frac{\sin(k+1/2)u}{\sin u/2} du$$

$$= \frac{1}{2n\pi} \int_0^{\pi} (f(x+u) + f(x-u)) \frac{\sin^2 nu/2}{\sin^2 u/2} du$$

(the second line comes from the calculation of the preceding sum).

When f is the constant function equal to 1, we get:

$$\int_0^\pi \frac{\sin^2 nu/2}{\sin^2 u/2} du = n\pi$$

Thus, if one puts $y = \frac{1}{2}(f(x+0) + f(x-0))$:

$$\sigma_n(f)(x) - y = \frac{1}{2n\pi} \int_0^\pi (f(x+u) + f(x-u) - 2y) \frac{\sin^2 nu/2}{\sin^2 u/2} du$$

Let $\varepsilon > 0$ be given. There exists $\delta > 0$ (depending on ε and x) such that if $0 < u < \delta$, then $|f(x+u) + f(x-u) - f(x+0) - f(x-0)| < \varepsilon$. In this case:

$$\frac{1}{2n\pi} \left| \int_0^\delta (f(x+u) + f(x-u) - 2y) \frac{\sin^2 nu/2}{\sin^2 u/2} du \right| \le \frac{\varepsilon}{2n\pi} \int_0^\delta \frac{\sin^2 nu/2}{\sin^2 u/2} du < \frac{\varepsilon}{2}$$

and:

$$\frac{1}{2n\pi} \left| \int_\delta^\pi (f(x+u) + f(x-u) - 2y) \frac{\sin^2 nu/2}{\sin^2 u/2} du \right| \le$$

$$\frac{1}{2n\pi \sin^2(\delta/2)} \int_0^\delta |f(x+u) + f(x-u) - 2y| \, du = \frac{C_\delta}{2n\pi}$$

The proof is completed by allowing n go to infinity in this last inequality.

Remark: If f is continuous on the closed interval $[a, b]$, it is uniformly continuous. In this case, with $y = f(x)$, the number δ above is independent of x and depends only ε. The constant C_δ (that depends on x) is bounded by:

$$\frac{1}{2n\pi \sin^2(\delta/2)} \int_{-\pi}^\pi |f(u)| du + 2M\pi$$

which no longer depends on x. Thus, the convergence of $\sigma_n(f)$ to f is uniform on any compact subset by the continuity of f.

So the Cesaro sum of f converges to f when the function is a continuous function. But, we must keep in mind that this convergence does not deal with the Fourier series of f itself, rather with an "average" of the Fourier series.

This result should be considered in parallel with the Weierstrass Approximation Theorem which deals similarly with approximation by algebraic polynomials (see the next chapter on interpolation). Indeed, we have the following result:

Theorem 3: Let f be a 2π periodic, real valued continuous function. For any $\varepsilon > 0$, there exists a trigonometric polynomial P, that is to say a function of the form $\sum_{k=0}^n a_k \cos kx + b_k \sin kx$, such that: $\sup_{x \in [-\pi, \pi]} |f(x) - P(x)| < \varepsilon$

This may also be expressed by saying that the trigonometric polynomials are dense in the set of continuous functions on an interval.

The next program allows us to calculate the nth Cesaro partial sum. It is similar to the preceding program that determines partial sums of Fourier $SF_N(f)(x)$. Only the calculation of $\sigma_n(f)(x)$ is new; it is based on the following observations:

$$\sigma_n(f)(x) = \frac{1}{n} \sum_{k=0}^{n-1} SF_k(f)(x)$$

$$= \frac{1}{n} \sum_{k=0}^{n-1} \left(\frac{a_0(f)}{2} + \sum_{j=1}^{k} a_j(f) \cos jx + b_j(f) \sin jx \right)$$

$$= \frac{a_0(f)}{2} + \sum_{j=0}^{n-1} \frac{n-j}{n} \left(a_j(f) \cos jx + b_j(f) \sin jx \right)$$

```
:cesaro()
:Prgm
:Local f,a,c,p,k
:DelVar l,ll
:Dialog
:Text "Definition of f"
:Request "f",f
:Request "Period",t
:DropDown "parity :",{"even","odd","none"},c
:Request "no. (C,1) partial sums",n
:EndDlog
:expr(f)→f:expr(t)→t
:expr(n)→n
:If c=1 Then
:seq(4/t*∫(f*cos(2*π*k/t*x),x,0,t/2),k,0,n)→l
:l[1]/2+Σ((n-k+1)/(n+1)*l[k+1]*cos(2*π*k*x/t),k,1,n)→sigf(x)
:ElseIf c=2 Then
:seq(4/t*∫(f*sin(2*π*k/t*x),x,0,t/2),k,1,n)→l
:Σ((n-k+1)/(n+1)*l[k]*sin(2*π*k*x/t),k,1,n)→sigf(x)
:Else
:seq(2/t*∫(f*cos(2*π*k/t*x),x,0,t),k,0,n)→l
:seq(2/t*∫(f*sin(2*π*k/t*x),x,0,t),k,0,n)→ll
:l[1]/2+Σ((n-k+1)/(n+1)*(l[k+1]*cos(2*π*k*x/t)+
  ll[k]*sin(2*π*k*x/t)),k,1,n)→sigf(x)
:EndIf
:EndPrgm
```

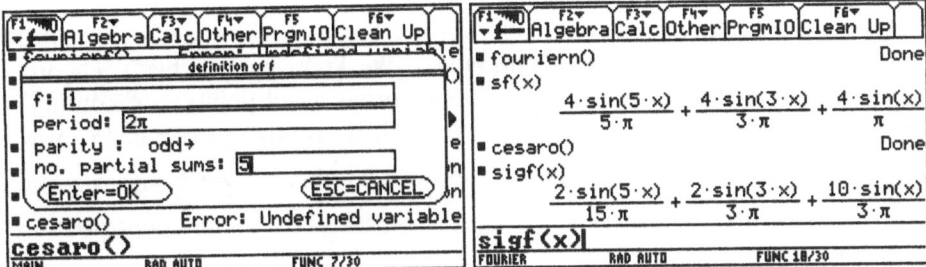

Here is the graph of the periodic step function f and its 5th Fourier and Cesaro sums. The partial sum of the Fourier series better approaches f, but the convergence of the Cesaro sum is more regular but not faster or better!

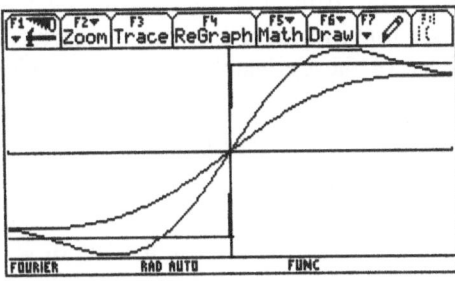

The graph of the odd periodic function equal to x on $[-1, 1]$ and its 5th Fourier and Cesaro sums. The convergence is excellent on half of the interval. The Cesaro sums do not seem to be much improvement when compared to Fourier sums.

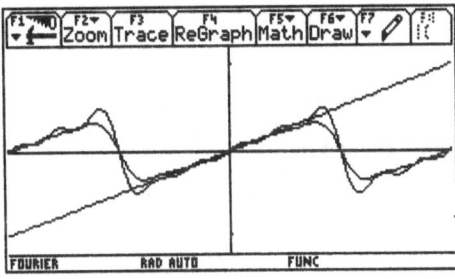

Look at the graph of the periodic odd function equal to 1 on $[0, 1]$ and the 7th Fourier and Cesaro sums. We see what the averaging of Cesaro summability means. Oscillations of the Fourier sums have disappeared, even if the Cesaro convergence is not very good (because of discontinuities).

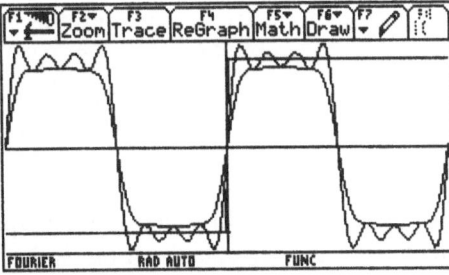

2. Acceleration of convergence of Fourier series

Let f be a piecewise continuous, 2π-periodic function whose derivatives up to order n are likewise piecewise continuous. We know by Dirichlet's Theorem that $f(x)$ is equal to the sum of its Fourier series at all points x of continuity of f. When the Fourier coefficients approach 0 "slowly" (for example, when they are of order $\frac{1}{n}$), many terms are required to obtain a good approximation of f by its Fourier series. This introduces the risk of errors and poor precision in calculations.

There is an acceleration method for the convergence of the Fourier series similar to what exists for slowly converging numerical series. It consists of extracting from f a function g which is a piecewise polynomial with the same discontinuities as f and whose derivatives possess the same discontinuities as those of the derivatives of f. Moreover we require that for all $1 \leq i \leq n$:

$$f^{(i)}(-\pi+0) - g^{(i)}(-\pi+0) = f^{(i)}(-\pi-0) - g^{(i)}(-\pi-0)$$

This assures us that $\varphi(x) = f(x) - g(x)$ will be of class C^n. Then, if let $(c_n(\varphi))$ designate the Fourier coefficients of φ, we have:

$$f(x) = g(x) + \sum_{n \in \mathbb{Z}} c_n(\varphi)e^{inx}$$

Since the function φ is of class C^n, we know that coefficients $c_k(\varphi)$ are "infinitely small of order $\frac{1}{k}$".

Indeed, integrating by parts, and for φ of class C^1 on $[0, 2\pi]$ we have:

$$c_n(\varphi') = \frac{1}{2\pi} \int_0^{2\pi} \varphi'(x)e^{-inx}dx = in \cdot \frac{1}{2\pi} \int_0^{2\pi} \varphi(x)e^{-inx}dx = inc_n(\varphi)$$

So, if φ is of class C^k, repeating this argument gives that for all $n \in \mathbb{Z}$:

$$|c_n(\varphi)| = \left| \frac{1}{(in)^k} c_n(\varphi^{(k)}) \right| \leq \frac{M_k}{n^k}$$

Here is how to build g and φ for a specific example for a general function f. Define the 2π periodic function f_0, by:

$$f_0(x) = \frac{\pi - x}{2}, \quad x \in]0, 2\pi[, \qquad f_0(0) = f_0(2\pi) = 0$$

This is known as a sawtooth wave.

These shots show the graph of the sawtooth wave function f_0 defined above, the definition of f and, below, the calculation of Fourier coefficients which are stored in the list l:

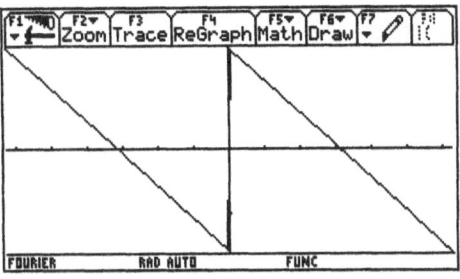

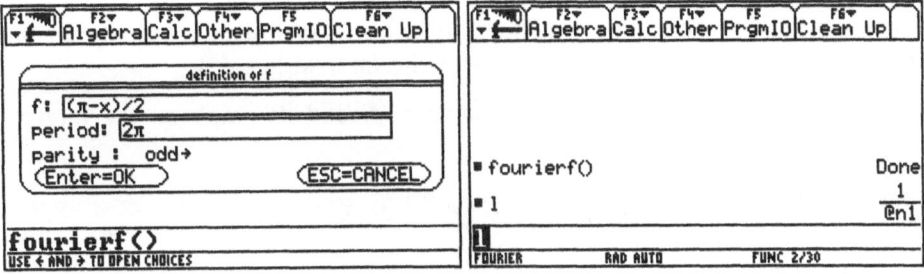

The Fourier series of f is equal to $\displaystyle\sum_{n=1}^{+\infty} \frac{\sin nx}{n}$. Dirichlet's Theorem assures us that:

$$f_0(x) = \sum_{n=1}^{+\infty} \frac{\sin nx}{n}, \quad 0 < x < 2\pi$$

On the interval $[-\pi, \pi]$, the function f_0 has a unique discontinuity at $x = 0$, a jump discontinuity of height π. Thus, for all points x_0, the translated $\tilde{f}_0(x) = f_0(x - x_0)$ will possess the same jump of π at x_0.

Now put:

$$f_1(x) = a_1 + \int_0^x f_0(t)\,dt$$

Since the convergence of the series defining f_0 is uniform on all compact subsets of $\mathbb{R}$, we may write:

$$f_1(x) = a_1 + \int_0^x \left(\sum_{n=1}^{+\infty} \frac{\sin nt}{n} \right) dt$$

$$= a_1 + \sum_{n=1}^{+\infty} \int_0^x \frac{\sin nt}{n}\,dt$$

$$= a_1 + \sum_{n=1}^{+\infty} \frac{1}{n^2} - \sum_{n=1}^{+\infty} \frac{\cos nx}{n^2}$$

The first series in this sum has a known sum of $\frac{\pi^2}{6}$, so we choose the constant $a_1 = -\frac{\pi^2}{6}$. Then:

$$f_1(x) = \sum_{n=1}^{+\infty} \frac{\cos nx}{n^2}$$

Moreover, f_1 is 2π periodic and, for all $x \in [0, 2\pi]$:

$$f_1(x) = \frac{\pi^2}{12} - \frac{(\pi - x)^2}{4}$$

This polynomial function is continuous on $\mathbb{R}$ and its derivative f_0 has a discontinuity at 0 with jump π (There are other points of discontinuity on the real line, but we are only concerned with the interval $[-\pi, \pi]$).

Below we see the definition of the function f_1 and its graph on the interval $[-\pi, \pi]$.

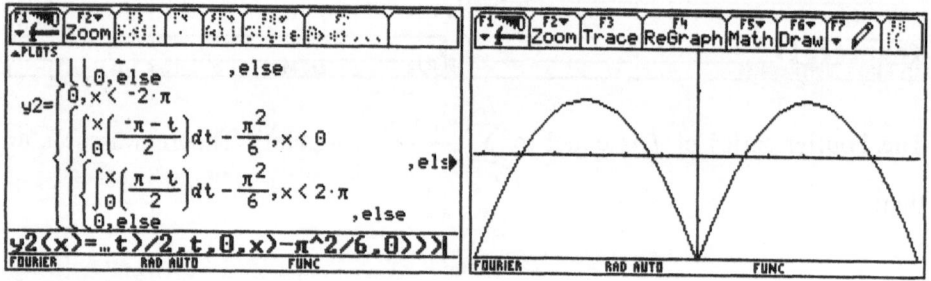

It remains only to resume "eliminating discontinuities" by recursion. When f_k has already been defined, we define:

$$f_{k+1}(x) = a_{k+1} + \int_0^x f_k(t)\,dt$$

The constant a_{k+1} is chosen to cancel the constant term in the previous integral, or:

$$\pi a_{k+1} + \int_0^\pi \left(\int_0^x f_k(t)\,dt \right) dx = 0$$

The function f_{k+1} is 2π periodic and continuous on $\mathbb{R}$, as are its derivatives, except for $f_{k+1}^{(k+1)} = f_0$ which has a jump discontinuity of π at $x = 0$. Thus, the function $\widetilde{f_{k+1}}(x) = f_{k+1}(x - x_0)$ and its derivatives, are continuous for all $x \in [-\pi, \pi]$, except that $\widetilde{f_{k+1}}^{(k+1)}$ has one discontinuity of height π at x_0.

We denote:
$(x_{0.1}, x_{0.2}, \ldots, x_{0.k_0})$ as the points of discontinuity of f,
$(x_{1.1}, x_{1.2}, \ldots, x_{1.k_1})$ as the points of discontinuity of f', $\ldots$

$(x_{n.1}, x_{n.2}, \ldots, x_{n.k_n})$ as the points of discontinuity of $f^{(n)}$ on the interval $[-\pi, \pi]$. Likewise, denote, for all $0 \leq l \leq n$ and $1 \leq j \leq k_l$:

$$f^{(l)}(x_{l.j} + 0) - f^{(l)}(x_{l.j} - 0) = s_{l.j}$$

as the discontinuity jump of $f^{(l)}$ at the point $x_{l.j}$. Finally, we arrive at a way to define the desired function g:

$$g(x) = \sum_{i=1}^{k_0} \frac{s_{0.i}}{\pi} f_0(x - x_{0.i}) + \sum_{i=1}^{k_1} \frac{s_{1.i}}{\pi} f_1(x - x_{1.i}) + \ldots + \sum_{i=1}^{k_n} \frac{s_{n.i}}{\pi} f_n(x - x_{n.i})$$

The function g has the properties that we required above. Indeed:
• g is discontinuous at the points $(x_{0.1}, x_{0.2}, \ldots, x_{0.k_0})$, with the same jumps as those of f. In fact, for all $1 \leq j \leq k_1$:

$$g(x_{0.j} + 0) - g(x_{0.j} - 0) = \frac{s_{0.j}}{\pi}(f_0(x - x_{0.j} + 0) - f_0(x - x_{0.j} - 0)) = s_{0.j}$$

• for all $1 \leq l \leq n$, for all $1 \leq j \leq k_l$, the derivative $g^{(l)}$ is discontinuous at the points $(x_{l.1}, x_{l.2}, \ldots, x_{l.k_l})$ with the same jumps as those of $f^{(l)}$. The proof is the same.
• at all other points, g and its derivatives are continuous, because f and its derivatives are.
Thus, the function $\varphi = f - g$ is of class C^n on $[-\pi, \pi]$.

We note that it is easy to calculate the Fourier series of g since:

$$f_0(x - x_{0.i}) = \sum_{n=1}^{+\infty} \frac{\sin(n(x - x_{0.i}))}{n}, f_1(x - x_{1.i}) = \sum_{n=1}^{+\infty} \frac{\cos(n(x - x_{1.i}))}{n^2}, \ldots$$

As a concrete example, let f be the odd, 2π-periodic function which is equal to $x \mapsto \cos x$ on $]0, \pi]$ and is zero at $x = 0$.
Below is the definition of the function f and the calculation of its Fourier coefficients. It remains to simplify the expression obtained.

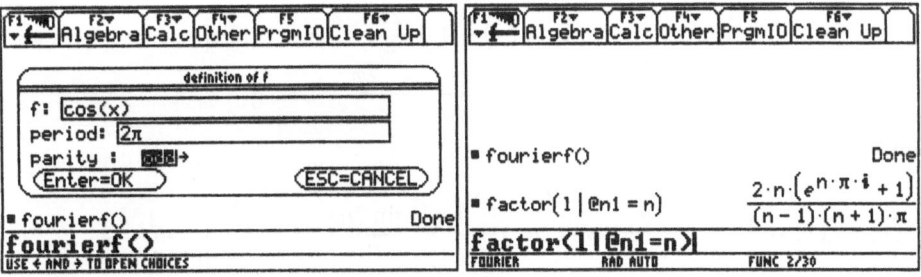

Thus, by simplifying the Fourier expansion, we may write for all $x \in]0, \pi[$:

$$\cos x = \frac{8}{\pi} \sum_{n=1}^{+\infty} \frac{n}{4n^2 - 1} \sin(2nx)$$

This last sum equals 0 at $x = 0$. And, for $x \in]-\pi, 0[$:

$$\cos(x + \pi) = -\cos x = \frac{8}{\pi} \sum_{n=1}^{+\infty} \frac{n}{4n^2 - 1} \sin(2nx)$$

The function f has three points of discontinuity on $[-\pi, \pi]$: $(-\pi, 0, \pi)$. Since our function is 2π periodic, the jumps at $-\pi$ and at π are the same and are equal to 2. Because of the periodicity of f_0, we have $f_0(x + \pi) = f_0(x - \pi)$. In the formula for g, it is sufficient to use only one of the two points $-\pi, \pi$. Indeed, the jump of g at $-\pi$, for example, is equal to 2, the same as at π. When we calculate $\varphi = f - g$, by suppressing the discontinuity at $-\pi$, we also suppress the discontinuity at π!
We therefore put :

$$g(x) = \frac{2}{\pi} f_0(x + \pi) + \frac{2}{\pi} f_0(x)$$

so that:

$$g(x) = \begin{cases} \dfrac{2}{\pi} \left(\dfrac{\pi}{2} - x \right) & \text{if } x \in]0, \pi[\\ \dfrac{2}{\pi} \left(-\dfrac{\pi}{2} + x \right) & \text{if } x \in]-\pi, 0[\end{cases}$$

But $f_0(x) = \displaystyle\sum_{n=1}^{+\infty} \frac{\sin nx}{n}$. Therefore:

$$f_0(x + \pi) = \sum_{n=1}^{+\infty} \frac{\sin n(x + \pi)}{n} = \sum_{n=1}^{+\infty} (-1)^n \frac{\sin nx}{n}$$

and:

$$g(x) = \frac{2}{\pi} \sum_{n=1}^{+\infty} (-1)^n \frac{\sin nx}{n} + \frac{2}{\pi} \sum_{n=1}^{+\infty} \frac{\sin nx}{n} = \frac{4}{\pi} \sum_{n=1}^{+\infty} \frac{\sin 2nx}{2n}$$

Finally:

$$f(x) - g(x) = \begin{cases} \cos(x) - 1 + \dfrac{2x}{\pi} & \text{if } x \in]0, \pi[\\ -\cos(x) + 1 - \dfrac{2x}{\pi} & \text{if } x \in]-\pi, 0[\end{cases}$$

and:

$$f(x) - g(x) = \frac{8}{\pi} \sum_{n=1}^{+\infty} \frac{n}{4n^2 - 1} \sin(2nx) - \frac{4}{\pi} \sum_{n=1}^{+\infty} \frac{\sin 2nx}{2n} = \frac{8}{\pi} \sum_{n=1}^{+\infty} \frac{1}{4n(4n^2 - 1)} \sin 2nx$$

So we started with convergence of order $\dfrac{1}{n}$ and finish with convergence of order $\dfrac{1}{n^3}$.

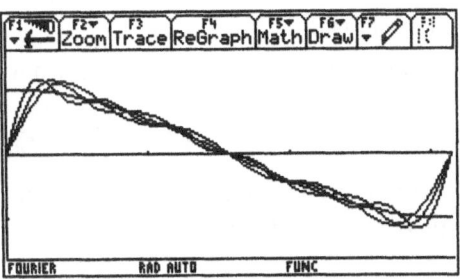

The graph of $\cos x$ on $[0, \pi]$, as well as partial sums of order $6, 8$ and 12 of the Fourier series of f. We again see the Gibbs phenomenon studied earlier.

The Gibbs phenomenon has disappeared (since the function $f : x \mapsto \cos x - 1 + 2x/\pi$ is continuous on $[0, \pi]$) and the convergence of the partial sums of the Fourier series is rapid: we can't see the difference between the graph of f and the "accelerated" partial sums of order 6 and 8 that are graphed.

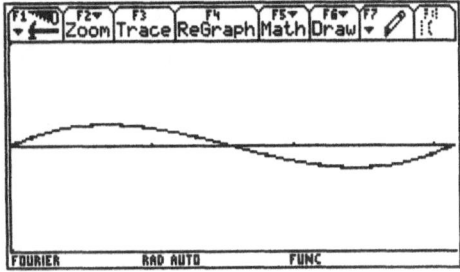

3. Hilbert analysis

Let E be the vector space of complex valued functions defined on $]-\pi, \pi[$ which are integrable on this interval. We extend each element of E to all of $\mathbb{R}$ by 2π periodicity and at $\pm\pi$ so that $f(\pi) = f(-\pi)$. We may thus define the Fourier series of any element of E.

The map:

$$(f, g) \to\; <f, g> = \frac{1}{2\pi} \int_{-\pi}^{\pi} f(t)\overline{g(t)}dt$$

defines a semi-norm:

$$\|f\|_2 =\; <f, f>^{1/2} = \left(\frac{1}{2\pi} \int_{-\pi}^{\pi} f(t)\overline{g(t)}dt \right)^{1/2}$$

Indeed, $\|f\|_2$ is "not quite" a norm since $\|f\|_2 = 0$ does not imply that $f = 0$. Let E_0 be the set of all functions in E such that $\|f\|_2 = 0$. (For example, E_0

contains all functions which differ from the zero function at a finite number of points.) To obtain a norm, we pass to the quotient vector space E/E_0, whose elements are the classes $[f + E_0], f \in E$. We will work with these equivalence classes, thereby "factoring out" the elements of E_0. In this manner, $(E, \| \ \|_2)$ is a preHilbert or inner product space, and we call $\| \ \|_2$ as defined above the Hilbert norm.

The family $(e^{inx})_{n \in \mathbb{Z}}$ is an orthonormal family of E, since for all $n \in \mathbb{Z}$:

$$< e^{inx}, e^{imx} > = \delta_{n,m}, \text{and } \|e^{inx}\|_2 = 1$$

The Fourier coefficients $(c_n(f))_{n \in \mathbb{Z}}$ of a function f of this family are also defined with respect to this family as:

$$c_n(f) = \frac{1}{2\pi} \int_{-\pi}^{\pi} f(t)e^{-int} dt = < f, e^{inx} >$$

Finally, recall that:

$$SF_n(f)(x) = \sum_{k=-n}^{n} c_k(f)e^{ikx} \in \mathcal{P}_n$$

where $\mathcal{P}_n$ denotes the vector space of trigonometric polynomials:

$$\mathcal{P}_n = \left\{ \sum_{k=-n}^{n} a_k e^{ikx} \| \ (a_k)_{|k| \leq n} \in \mathbb{C}^{2n+1} \right\}$$

The next theorem is quite similar to the ones covered in the next chapter on orthogonality. It should remind you of the Pythagorean theorem.

Theorem 1: Let $f \in E$. If we define the distance from f to $\mathcal{P}_n$ by:

$$d(f, \mathcal{P}_n) = \inf\{ \ \|P - f\|_2 \| \ P \in \mathcal{P}_n\}$$

then this distance is actually obtained for a unique polynomial of $\mathcal{P}_n$, which is $SF_n(f)$.

Proof: If $P \in \mathcal{P}_n$, then:

$$\|P - f\|_2^2 = \|P - SF_n(f) + SF_n(f) - f\|_2^2$$
$$= \|P - SF_n(f)\|_2^2 + \|SF_n(f) - f\|_2^2 + 2\Re < P - SF_n(f), SF_n(f) - f >$$

But, for all $-n \leq k \leq n$:

$$< SF_n(f) - f, e^{ikx} > = < SF_n(f), e^{ikx} > - < f, e^{ikx} > = 0$$

Therefore, $< P - SF_n(f), SF_n(f) - f > = 0$ and:

$$\|P - f\|_2^2 = \|P - SF_n(f)\|_2^2 + \|SF_n(f) - f\|_2^2$$

(which is just the Pythagorean relation for the triangle $(f, P, SF_N(f))$).
The minimal value of $\|P - f\|_2$ is obtained when $\|P - SF_n(f)\|_2 = 0$.

Proposition 1:(Bessel inequality). Let $f \in E$, and let $(c_n(f))$ be the sequence of Fourier coefficients of f. For all $n \geq 0$:

$$\sum_{k=-n}^{n} |c_n(f)|^2 \leq \|f\|_2^2 = \frac{1}{2\pi} \int_{-\pi}^{\pi} |f(t)|^2 dt$$

The series $\sum |c_n(f)|^2$ is convergent and:

$$\sum_{k=-\infty}^{+\infty} |c_n(f)|^2 \leq \|f\|_2^2 = \frac{1}{2\pi} \int_{-\pi}^{\pi} |f(t)|^2 dt$$

Proof: Since $< SF_n(f) - f, SF_n(f) >= 0$, it follows that:

$$\|SF_n(f) - f\|_2^2 = \|SF_n(f)\|_2^2 + \|f\|_2^2 - 2\Re < SF_n(f), f >= \|f\|_2^2 - \|SF_n(f)\|_2^2$$

Thus, for all $n \geq 0, \|SF_n(f)\|_2^2 \leq \|f\|_2^2$, which is the theorem.

Remark: If we wish to use the "real" form $(a_n(f))$ and $(b_n(f))$ of Fourier coefficients, the Bessel inequality becomes: for all $n \geq 0$,

$$\frac{|a_0(f)|^2}{2} + \sum_{k=1}^{n}(|a_k(f)|^2 + |b_k(f)|^2) \leq \frac{1}{\pi} \int_{-\pi}^{\pi} |f(t)|^2 dt$$

In fact, when n tends to infinity this inequality becomes an equality. Indeed:

Proposition 2:(Parseval's Theorem) Let $f \in E$ and let $(c_n(f)_n)$ be the sequence of Fourier coefficients of f. Then:

$$\sum_{k=-\infty}^{+\infty} |c_n(f)|^2 = \|f\|_2^2 = \frac{1}{2\pi} \int_{-\pi}^{\pi} |f(t)|^2 dt$$

Proof: The proof is divided in two parts.
• Assume that f is continuous on $[-\pi, \pi]$ and satisfies $f(\pi) = f(-\pi)$ (which ensures the continuity of f on $\mathbb{R}$ by a 2π periodic extension).
We know then that the sequence $(\sigma_n(f))$ is uniformly convergent to f on $[-\pi, \pi]$.
But, for all $n \in \mathbb{N}, \sigma_n(f) \in \mathcal{P}_n$. We therefore have:

$$\|SF_n(f) - f\|_2^2 \leq \|\sigma_n(f) - f\|_2^2 = \frac{1}{2\pi} \int_{-\pi}^{\pi} |\sigma_n(f)(t) - f(t)|^2 dt$$

$$\leq \sup_{x \in [-\pi, \pi]} |\sigma_n(f)(t) - f(t)|^2 \to 0$$

But $||SF_n(f) - f||_2^2 = ||f||_2^2 - ||SF_n(f)||_2^2$, which, when n tends to the infinity, gives:

$$\sum_{k=-\infty}^{+\infty} |c_n(f)|^2 = ||f||_2^2$$

• In the general case, we use the density of the continuous functions on $[-\pi, \pi]$ in the space of functions whose square is integrable on this interval, using the Hilbert norm. That is, if $f \in E$ and $\varepsilon > 0$, there exists $\varphi \in \mathbb{C}^0$, with $\varphi(-\pi) = \varphi(\pi)$ such that $||f - \varphi||_2 < \varepsilon$.

But $||SF_n(f) - SF_n(\varphi)||_2 \le ||f - \varphi||_2 < \varepsilon$, and:

$$||f - SF_n(f)||_2 \le ||f - SF_n(\varphi)||_2 + ||SF_n(\varphi) - SF_n(f)||_2 + ||SF_n(f) - f||_2 < 3\varepsilon$$

Examples.

1. Let f be the 2π-periodic function equal to $x \to x$ on $]-\pi, \pi[$. Let the calculator determine its Fourier coefficients, then apply the Parseval formula shown above.

Here are calculations of the nth Fourier coefficient and the norm $||f||_2^2$ of f:

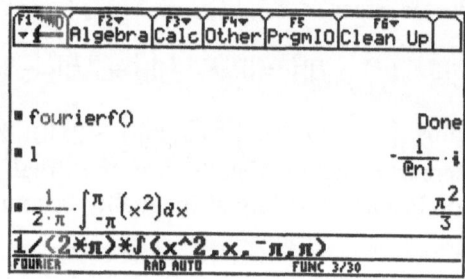

The Parseval formula gives:

$$||f||_2^2 = \frac{\pi^2}{3} = \sum_{n \in \mathbb{Z}^*} \frac{1}{n^2} = 2\sum_{n=1}^{+\infty} \frac{1}{n^2}$$

Therefore:

$$\sum_{n=1}^{+\infty} \frac{1}{n^2} = \frac{\pi^2}{6}$$

2. Let f be the 2π-periodic odd function equal to 1 on $]0, \pi]$.

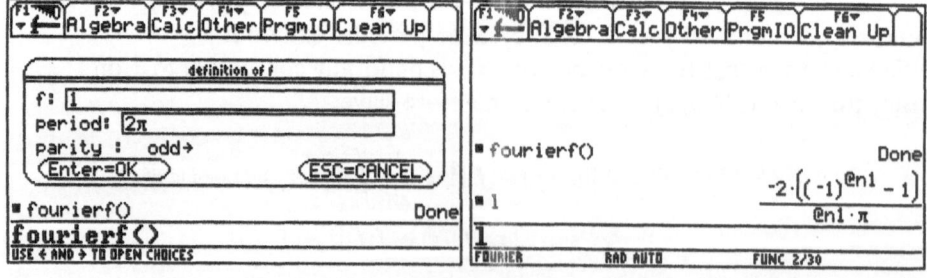

Therefore, for all $n \geq 0, a_n(f) = b_{2n}(f) = 0$ and $b_{2n+1}(f) = \dfrac{4}{\pi(2n+1)}$. Thus:

$$\frac{1}{2\pi} \int_{-\pi}^{\pi} |f(t)|^2 dt = 1$$

Therefore:

$$1 = \frac{1}{2} \sum_{n=1}^{+\infty} |b_n(f)|^2 = \frac{8}{\pi^2} \sum_{n=0}^{+\infty} \frac{1}{(2n+1)^2}$$

Therefore :

$$\sum_{n=0}^{+\infty} \frac{1}{(2n+1)^2} = \frac{\pi^2}{8}$$

These last two results are confirmed by the calculator. The other formulas are built in to the machine.

4. Discrete Fourier Transform

Suppose that f is a 2π-periodic function which is regular enough for us to know that its Fourier series converges simply to f at all point of continuity. Suppose also that we only know some equally distributed values $f\left(\dfrac{2k\pi}{N}\right)_{0 \leq k \leq N-1}$.

We nevertheless want to determine the Fourier coefficients $c_n(f)$ of f. Since we know only N values of f, we will try to calculate the N coefficients c_n, $-N/2 \leq n \leq N/2 - 1$.

We propose two methods for the approximate calculation of $c_n = \dfrac{1}{2\pi} \int_0^{2\pi} f(t)e^{-int}dt$.

The trapezoidal method
When we seek to calculate an approximate value of the previous integral by the trapezoidal rule, we calculate:

$$c_n' = \frac{1}{N} \sum_{k=0}^{N-1} y_k \exp\left(\frac{-2ikn\pi}{N}\right) = \sum_{k=0}^{N-1} y_k \omega_N^{-nk}$$

with $y_k = f\left(\dfrac{2k\pi}{N}\right)$ and $\omega_N = \exp(\dfrac{2i\pi}{N})$.

By polynomial interpolation

We may also determine the trigonometric interpolation polynomial of f at the points $\left(\dfrac{2k\pi}{N}\right)$, in the form:

$$p(t) = \sum_{n=-N/2}^{N/2-1} c_{n.N} e^{int}$$

We then have to solve the $N \times N$ system with unknowns $c_{-N/2.N}, \ldots, c_{N/2-1.N}$:

$$\sum_{n=-N/2}^{N/2-1} c_{n.N} \exp\left(\dfrac{2ikn\pi}{N}\right) = y_k, \qquad 0 \le k \le N-1$$

or:

$$\sum_{n=-N/2}^{N/2-1} c_{n.N} \omega_N^{nk} = y_k, \qquad 0 \le k \le N-1$$

But, it is possible to solve this system formally because its matrix is a Vandermonde matrix. Indeed:

$$\sum_{n=-N/2}^{N/2-1} c_{n.N} \omega_N^{nk} = \sum_{j=N/2}^{N-1} c_{j-N.N} \omega_N^{k(j-N)} = \sum_{j=N/2}^{N-1} c_{j-N.N} \omega_N^{jk}$$

By setting:

$$Y_n = \begin{cases} c_{n,N} & 0 \le n \le N/2 - 1 \\ c_{n-N.N} & N/2 \le n \le N-1 \end{cases}$$

it follows that:

$$\sum_{n=0}^{N-1} Y_n \omega_N^{nk} = y_k$$

By using the fact that ω_N is an N-th root of 1, it follows, for all p such that $0 \le p \le N-1$ that:

$$\sum_{k=0}^{N-1} y_k \omega_N^{-kp} = \sum_{k=0}^{N-1} \sum_{n=0}^{N-1} Y_n \omega_N^{k(n-p)}$$

$$= \sum_{n=0}^{N-1} Y_n \sum_{k=0}^{N-1} \omega_N^{k(n-p)}$$

But:

$$\sum_{k=0}^{N-1} \omega_N^{k(n-p)} = \begin{cases} N & \text{if } p = n \\ 0 & \text{else} \end{cases}$$

Therefore:

$$\sum_{k=0}^{N-1} y_k \omega_N^{-kp} = NY_p \Rightarrow Y_p = \frac{1}{N}\sum_{k=0}^{N-1} y_k \omega_N^{-kp}$$

Notice that we land on the same coefficients as with the trapezoidal method. For either method, the approximated values of the Fourier coefficients $c_n(f)$ for $-N/2 \leq n < N/2$ are:

$$c_{n,N} = \begin{cases} Y_n & \text{if } 0 \leq n < N/2 \\ Y_{n+N} & \text{if } -N/2 \leq n < 0 \end{cases}$$

with $Y_n = \dfrac{1}{N}\displaystyle\sum_{n=0}^{N-1} y_k \omega_N^{-nk}$.

The map that, associates the points $(Y_n)_{0 \leq n \leq N-1}$ to points $(y_k)_{0 \leq k \leq N-1}$ is called the Discrete Fourier Transform or DFT. It is a one-to-one linear map, since its matrix is the invertible Vandermonde matrix $(\omega_N^{ij})_{0 \leq i,j \leq N-1}$.

We close this section with some remarks.
1. There is an even more surprising relationship between the approximate and the exact Fourier coefficients.
If the function f is of class C^1 piecewise, we know that the series of its Fourier coefficients is absolutely convergent ($\sum |c_n|$ converges). We have then, for all $t \in [0, 2\pi]$:

$$f(t) = \sum_{n=-\infty}^{+\infty} c_n e^{int}$$

By grouping:

$$f\left(\frac{2k\pi}{N}\right) = \sum_{n=-\infty}^{+\infty} c_n \omega_N^{nk} = \sum_{n=0}^{N-1}\left(\sum_{p=-\infty}^{+\infty} c_{n+pN}\right)\omega_N^{nk}$$

and, since the Discrete Fourier Transform is one-to-one:

$$c_{n,N} = \sum_{p=-\infty}^{+\infty} c_{n+pN}$$

This relationship allows us to estimate the error made by replacing c_n by $c_{n,N}$ since:

$$c_{n,N} - c_n = \sum_{p \neq 0} c_{n+pN}, (\text{with} - N/2 \leq n < N/2)$$

2. If f is a trigonometric polynomial of degree P, meaning that $f(t) = \sum_{n=-P}^{P} c_n e^{int}$, and if $N \geq 2P + 1$, the approximate coefficients are in fact exact.   Indeed, in the difference $(c_{n.N} - c_n)$ above, only the coefficients $c_{n+N}, c_{n+2N}, \dots, c_{n-N}, c_{n-2N}, \dots$ intervene, and these are null as soon as $N \geq 2P + 1$, since we have $-N/2 \leq n < N/2$.

5. Fast Fourier Transform

The Discrete Fourier Transform is an interesting method for determining the Fourier coefficients of a function. Unfortunately its calculation time will often be prohibitive, especially because practical applications may require many hundreds of data points. The calculation of each element $Y_n = \dfrac{1}{N} \sum_{k=0}^{N-1} y_k \omega_N^{-nk}$ requires $2N$ multiplications for $y_k \omega_N^{-nk}$, and repeating this N times gives a calculation of order $O(N^2)$.

In the 1960's, two American mathematicians, J.W. Cooley and J.W. Tukey, found an algorithm which is much more rapid and which since has become famous due to its numerous applications. It is known as the "Fast Fourier Transform"(FFT). This algorithm is based on the method of "divide and conquer": to treat a problem, divide it into two problems of the same type and process each of its sub-problems..., but before that, again divide each sub-problem. The gain is important since at each step the data to be processed will be divided by 2.

5. 1 Principle of the FFT

Suppose that $N = 2\ell$. In the calculation of $Y_k = \dfrac{1}{N} \sum_{j=0}^{N-1} y_j \omega_N^{-nj}$, regroup the even terms and the odd terms:

$$\begin{cases} P_k = \dfrac{1}{\ell}(y_0 + y_2 \omega_N^{-2k} + \dots + y_{N-2} \omega_N^{-(N-2)k}) \\ I_k = \dfrac{1}{\ell}(y_1 + y_3 \omega_N^{-2k} + \dots + y_{N-1} \omega_N^{-(N-2)k}) \end{cases}$$

We then have:

$$Y_k = \frac{1}{2}(P_k + \omega_N^{-k} I_k)$$

Likewise, we have:

$$P_{k+\ell} = P_k, \quad I_{k+\ell} = I_k, \quad \omega_N^{-(k+\ell)} = -\omega_N^{-k}$$

Thus, to calculate Y_k, it suffices, for each $k \in \{0, 1, \dots, \ell - 1\}$, to:

- calculate P_k and $\omega_N^{-k} I_k$
- add these two results to obtain $Y_k = \dfrac{1}{2}(P_k + \omega_N^{-k} I_k)$.
- deduce that $Y_{k+\ell} = \dfrac{1}{2}(P_k - \omega_N^{-k} I_k)$.

The computation cost is on the order of $2\ell^2$ multiplications, or about $\dfrac{N^2}{2}$.

So we divide the data, as well as the cost of processing, by two. But why stop after such good progress? By supposing that ℓ is even, we may use again the same algorithm on each sequence (P_k) and (I_k). In the case where $N = 2^p$, we thus continue down to the Discrete Fourier Transform of order 2:

$$
\begin{cases}
Y_0 = \dfrac{1}{2}(y_0 + y_1) \\
Y_1 = \dfrac{1}{2}(y_0 - y_1)
\end{cases}
$$

Proposition 1: Let $N = 2^p$. The number of multiplications required by the FFT is of order $N \ln_2 N$.

Proof: Let α_p be the number of complex multiplications necessary in the FFT algorithm when $N = 2^p$ data are given. The calculation of P_k and I_k each costs α_{p-1} multiplications. We have then $2^{p-1} - 1$ multiplications for ω_N^{-k}. Thus:

$$
\begin{cases}
\alpha_1 = 0 \\
\alpha_p = 2\alpha_{p-1} + 2^{p-1} - 1
\end{cases}
$$

We have $\dfrac{\alpha_p}{2^p} = \dfrac{\alpha_{p-1}}{2^{p-1}} + \dfrac{1}{2} - \dfrac{1}{2^p}$, which by telescoping gives:

$$
\alpha_p = (p - 2)2^{p-1} + 1
$$

This is a cost of order $N \ln_2 N$, clearly more advantageous than that of N^2 in the earlier, naive method.

Comparative cost table of the two algorithms, DFT and FFT. We note the huge gain for large values of n.

DATA	n	n^2	nln(n)
	c1	c2	c3
1	8	64	1.66355E1
2	32	1024	1.10904E2
3	1024	1048576	7.09783E3
4	32768	1073741824	3.40696E5
5	1048576	109951162...	1.45363E7
6	1.07374E9	1.15292E18	2.23278E10
7			

r6c1=1073741824.

MAIN   RAD AUTO   DE

5. 2 Programming the FFT

Algorithms of the "divide and conquer" class lend themselves well to recursive programming. The FFT is no exception.

To write a recursive function is straightforward; to solve the problem, we treat two cases:
- the final case (to give the result)
- if we are not in the final case, we call the procedure.

The simplest example is that the calculation of $n!$ that can be written recursively as:

```
fac(n)
if n=0 the result is 1
otherwise the result is n * fac(n - 1)
```

Let's calculate for instance 3!. We call fac(3). The result of this call is 3*fac(2). The call of fac(2) gives 2*fac(1). The call of fac(1) is 1*fac(0), which finally gives the result 1. Going back, we get fac(3)=1*2*3.

In similar manner, here is pseudo-code for a recursive version of the Cooley-Tukey algorithm.

To obtain the FFT of a vector y of $N = 2^p$ points,
put $\omega = e^{2i \pi / N}$
If $n = 1$ return y else
   1. let $m = N/2$.
   2. call the FFT with $m = N/2$ points of y with even coordinates
   and ω^2 (store the result in X)
   3. call the FFT with the $m = N/2$ points of
   y with odd coordinates and ω^2 (store the result in Y)
EndIf
Let $u = 1$
For k from 1 to $m - 1$
   1. store $(X(k) + u * Y(k))/2$ in $Z(k)$
   2. store $(X(k) - u * Y(k))/2$ in $Z(k + m)$
   3. let $u = u * \omega$
EndFor
Return the vector Z
End of procedure

But recursion demands a lot of memory space, for each call requires storage of all the variables of the preceding calls. And despite its 640 Kb, the calculator will strain to execute a recursive procedure for a large number of points. Thus, we have opted for programming an iterative procedure.

```
:fft(l)                          list l of data
:Func
:Local n,m,z,z0,i,j,k,h,tmp
:dim(l)→n                        n is the length of l
:n/2→m                           m = n/2
:While m≥1                       while m ≥ 1
:1→z                             from 1
:e^(-iπ/m)→z0                    ω
:For k,1,m
:For h,1,n/(2*m)
:k+2*(h-1)*m→i                   define vectors Pₖ, Iₖ
:i+m→j
:(l[i]-l[j])*z→tmp
:l[i]+l[j]→l[i]                  put them in l
:tmp→l[j]
:EndFor
:z*z0→z
:EndFor
:m/2→m                           again with m/2
:EndWhile                        end of while
```

```
:1→j
:For i,1,n                       for i from 1 to n
:If j>i Then                     if j > 1
:l[j]→tmp                        exchange l(i) and l(j)
:l[i]→l[j]
:tmp→l[i]
:EndIf
:n/2→m                           m takes the value n/2
:While m≥2 and j>m               determination
:j-m→j                           of the right place for
:m/2→m                           l(j)
:EndWhile
:j+m→j
:EndFor
:l                               return l
:EndFunc
```

5. 3 Applications of the FFT

The FFT may be used to calculate an interpolation polynomial. We will take the example of Chebishev polynomials. The reader may refer to the chapter on orthogonality for further details and properties.

The Chebishev polynomials $(T_n)_{n \geq 0}$ are defined for all $n \geq 0$ by by $T_n(\cos \theta) = \cos(n\theta)$. Thus, $T_0(x) = 1, T_1(x) = x, T_2(x) = 2x^2 - 1, \ldots$ These functions form a basis of $\mathbb{R}[X]$, the vector space of real polynomials. Therefore, if $P \in \mathbb{R}[X]$ has a degree less or equal to n, there are unique real numbers $(a_0, a_1, \ldots, a_n)$ such that:

$$P(x) = \sum_{j=0}^{n} a_j T_j(x)$$

For $0 \leq k \leq n$, let $x_k = \cos\left(k\dfrac{\pi}{n}\right)$, and suppose that we are given $n + 1$ points $(y_0, y_1, \ldots, y_n)$. The Fast Fourier Transform will help us to determine the interpolation polynomial P which interpolates or "fits", the (y_k) at the given points (x_k), using the Chebishev polynomials as a basis. To do this, we must solve the system:

$$y_k = \sum_{j=0}^{n} a_j T_j(\cos(k\pi/n)) = \sum_{j=0}^{n} a_j \cos(jk\pi/n), \quad 0 \leq k \leq n$$

By writing $\cos x$ in terms of e^{ix} and e^{-ix}, these equations may be written equivalently as:

$$(*) \qquad y_k = \frac{1}{2}\sum_{j=0}^{n} a_j \omega_{2n}^{jk} + \frac{1}{2}\sum_{j=1}^{n} a_{-j} \omega_{2n}^{-jk} = \sum_{j=-n}^{n} c_j \omega_{2n}^{jk}$$

with $\omega_n = e^{2i\pi/n}$ and:

$$c_j = \begin{cases} \dfrac{a_j}{2} & \text{if } 0 < j \leq n \\ a_0 & \text{if } j = 0 \\ \dfrac{a_{-j}}{2} & \text{if } -n \leq j \leq -1 \end{cases}$$

Extend the sequence $(y_0, y_1, \ldots, y_n)$ to a sequence $(y_0, y_1, \ldots, y_{2n-1})$ by setting:

$$y_{2n-k} = y_k, \quad 1 \leq k \leq n - 1$$

The system $(*)$ then becomes:

$$y_k = \sum_{j=-n}^{n} c_j \omega_{2n}^{jk}, \quad 0 \leq k \leq 2n - 1$$

Inspired by the DFT, we seek to invert this system by putting:

$$u_p = \sum_{j=0}^{2n-1} y_j \omega_{2n}^{-jp}, \quad 0 \leq p \leq n$$

or:

$$u_p = \sum_{j=0}^{2n-1} \sum_{l=-n}^{n} c_l \omega_{2n}^{(l-p)j} = \sum_{l=-n}^{n} c_l \sum_{j=0}^{2n-1} \omega_{2n}^{(l-p)j}$$

The last sum is equal to $2n$ if $l = p[2n]$ and 0 otherwise, which gives:

$$u_p = 2nc_p, \quad (0 \leq p \leq n-1), \quad u_n = 4nc_n$$

>From the relationships between (c_n) and (a_n), it follows that:

$$a_k = \frac{1}{n\alpha_k} \sum_{j=0}^{2n-1} y_j \omega_{2n}^{-jk}, 0 \leq k \leq n$$

with $\alpha_j = 1$, for $1 \leq j \leq n-1$ and $\alpha_0 = \alpha_n = 2$.

In practice, we will use the following algorithm:
- Calculate $y_{2n-k} = y_k, \quad 1 \leq k \leq n-1$
- Use the FFT algorithm to pass from $(y_0, y_1, \ldots, y_{2n-1})$ to $(Y_0, Y_1, \ldots, Y_{2n-1})$
- Obtain the result: $a_k = Y_k, (1 \leq k \leq n-1), a_0 = \dfrac{Y_0}{2}, a_n = \dfrac{Y_n}{2}$.

:tcheb(1)	*l* is the list of given points
:Func	
:Local 11,11,k,p,n	
:dim(1)→p	
:seq(1[k],k,p-1,2,-1)→11	add new points
:augment(1,11)→1	new list
:fft(1)→11	FFT
:left(11,p)→11	take only $(n+1)$ points
:11[1]/2→11[1]	modify the first one
:11[p]/2→11[p]	and the last one
:11	
:EndFunc	

Here are two examples of the use of the preceding program, calculating the coefficients of the interpolation polynomial on the points $\cos(k\pi/n)$ in terms of the basis of Chebishev polynomials for 5 and 9 points.

```
F1    F2▼    F3▼   F4▼   F5     F6▼
     Algebra Calc Other PrgmIO Clean Up


■ tcheb({1  2  3  4  5})
            {12  -2·√2-4  0  2·√2-4  0}
■ tcheb({-1  0  -2  0  -3  0  3  0  4})
     {-1/2  -5·√2-5  9  5·√2-5  -5  5·√2
tcheb({-1,0,-2,0,-3,0,3,0,4})
MAIN         RAD AUTO        FUNC 2/30
```

Proposition 2: Let $(y_0, y_1, \ldots, y_n)$ be $(n+1)$ real numbers. There is a unique trigonometric (interpolation) polynomial $Q(t) = \sum_{k=0}^{n} a_k \cos kt$ such that, for all $0 \leq j \leq n$:

$$y_j = Q(\frac{j\pi}{n})$$

Proof: Indeed, since $P(x) = \sum_{j=0}^{n} a_j T_j(x)$, for $x = \cos t$, it follows that:

$$Q(t) = P(\cos t) = \sum_{j=0}^{n} a_j \cos(jt)$$

We have $y_j = Q(\frac{j\pi}{n})$. The preceding formulas allow us to finish the proof.

6. An introduction to wavelets

To develop a 2π-periodic function as a Fourier series is the same as decomposing it in terms of the Hilbert basis $(e^{inx})_{n \in \mathbb{Z}}$. But what is a more precise geometric meaning of the nth Fourier coefficient $c_n(f)$? To answer this question, it is preferable to work with the real form of the Fourier series, that is to say with the form $\frac{a_0}{2} + \sum_{n \geq 1}(a_n(f) \cos(nx) + b_n(f) \sin(nx))$. We recall that:

$$a_n(f) = \frac{1}{\pi} \int_{-\pi}^{\pi} f(t) \cos(nt)dt, \quad b_n(f) = \frac{1}{\pi} \int_{-\pi}^{\pi} f(t) \sin(nt)dt$$

Each of these coefficients contributes to the series development, and the contribution is the value of the integral of the product of f with the sine and cosine functions, functions that oscillate more and more rapidly with n.

Computing the Fourier coefficients of $f(x) = |\sin(x)|$, involves the shaded areas with, for example, the functions $\cos(x)$, $\cos(2x)$, $\cos(3x)$, $\cos(4x)$ and $\cos(10x)$. The latter are the dotted graphs. The bold graph is the product of $f(x)$ by each of these functions. The shaded area is the coefficient.

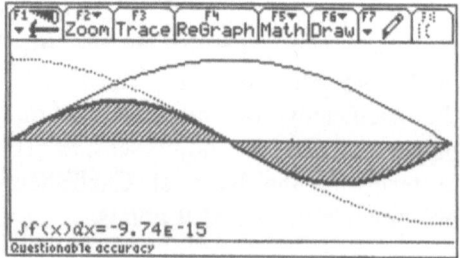

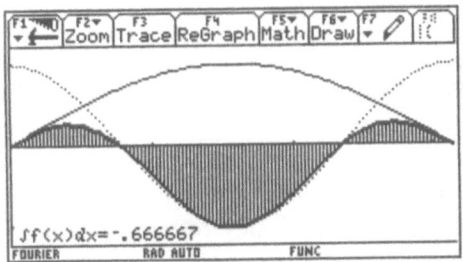 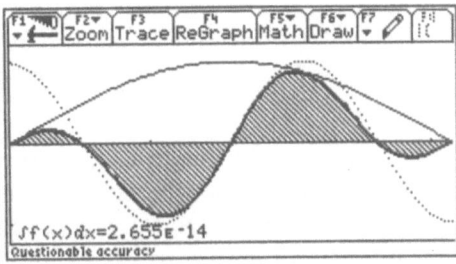

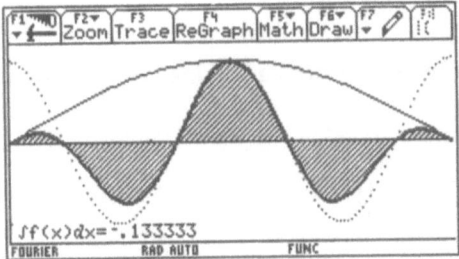 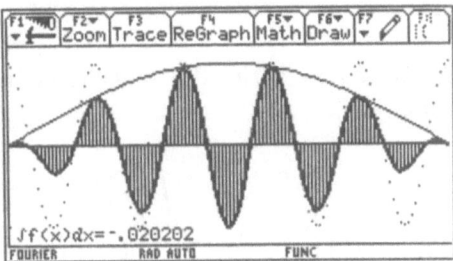

Here is the decomposition of $f(x) = |\sin(x)|$ with the function $\sin(5x)$, used in finding $b_n(f)$. The net contribution is zero. We see how the increasingly rapid oscillation of each $\cos(nx)$ and $\sin(nx)$ function makes its contribution to the development of the Fourier series.

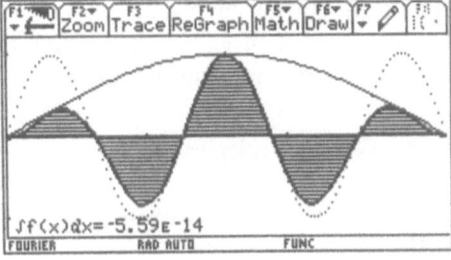

When the function f is periodic and regular (class C^1), the Fourier series expansion is an excellent tool for analysis of the function f. This is only the case because its Fourier coefficients $(c_n(f))$ tend rapidly to 0. Thus, a limited number of calculations and the utilization of the FFT will be sufficient to rebuild the function f. On the other hand, when f is more irregular, the necessary calculation times lengthen enormously.

What about a function defined on $\mathbb{R}$ which is not periodic? It is obviously impossible to decompose it into a Fourier series, since such an expansion gives a periodic function, which will not be able to represent our original non-periodic function. If one wishes to represent an arbitrary function with the help of the exponential, it will be necessary to use functions $x \mapsto e^{i\lambda x}$ for every $\lambda \in \mathbb{R}$. This family is not countable, so one we will not be able to make a series expansion. Therefore, it will be necessary to use a continuous analog, an "integral representation" for f. This is the role of the Fourier transform, a map that assigns to any function f which is integrable on $\mathbb{R}$, the function defined for $\zeta \in \mathbb{R}$ by:

$$\hat{f}(\zeta) = \int_{-\infty}^{+\infty} f(x)e^{-2i\pi\zeta x}dx$$

In practice, the Fourier transformation is as important as the Fourier series development of periodic functions. Indeed, we may show that, if f is continuous and if the functions f and $\hat{f}$ are integrable on $\mathbb{R}$, one may retrieve f from $\hat{f}$ with the help of an inverse transformation:

$$f(x) = \int_{-\infty}^{+\infty} \hat{f}(\zeta)e^{2i\pi x\zeta}d\zeta$$

We also may verify the Plancherel formula, a formula analogous to that of Parseval:

$$\int_{-\infty}^{+\infty} |f(x)|^2 dx = \int_{-\infty}^{+\infty} |\hat{f}(\zeta)|^2 d\zeta$$

But we will not study the Fourier transform in detail here. We will only observe that, as in the periodic case, the Fourier transform does is not such a good a tool for the study of f when the former possesses irregularities (points of discontinuity or nondifferentiability), because these are usually "smoothed out" by the integration involved. Moreover, to determine $\hat{f}$, it is necessary to know f at each point of $\mathbb{R}$, which is never the case in practice, since data points for f are only acquired one by one.

6. 1 Gabor windows

To try to reply to questions raised by Fourier transforms, Dennis Gabor proposed in the 1940's "to open windows", that is to say, to truncate the function f to an interval $[-a, a]$. The simplest manner is to study the function $f_a = f \cdot \chi_{[-a,a]}$, where $\chi_{[-a,a]}$ represents the characteristic function of the interval. (Recall that $\chi_{[-a,a]}(x)$ has the value 1 on the interval and the value 0 outside.)
Other non-rectangular windows are used similarly, such as the triangular window or the Hanning window.

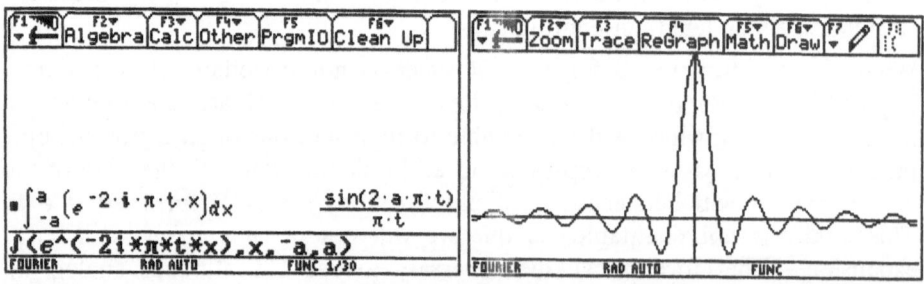

Above, we have calculation of the Fourier transform of $\chi_{[-a,a]}$ and its graph, the first example of a window. The transform tends to 0 slowly and has a large amplitude in a neighborhood of 0. This is due to the discontinuities of χ. This is why one uses other, more regular, window functions as follow.

Here is the graph of the triangle function $(1 - |x|/a)$ on the interval $[-a, a]$. Its Fourier transform is relatively simple to calculate. This transform is positive and tends to 0 more rapidly than the previous one, but retains a large amplitude near 0.

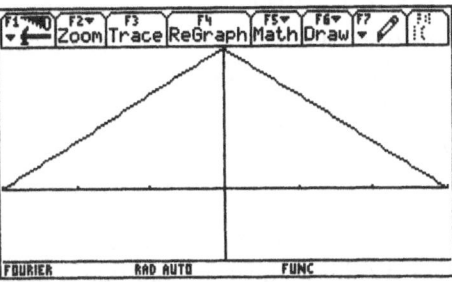

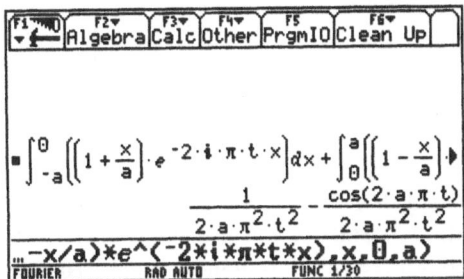

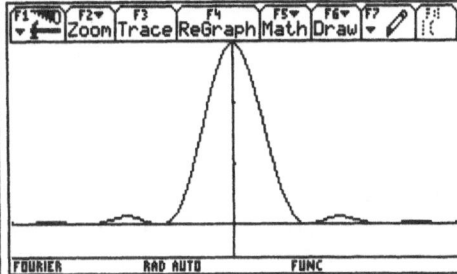

Observe the graph of the Hanning function $1/2 + 1/2\cos(2\pi x/a)$ on the interval $[-a, a]$. Its Fourier transform is also simple to calculate.

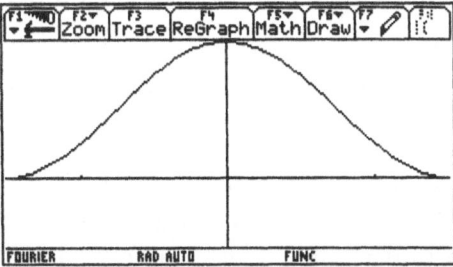

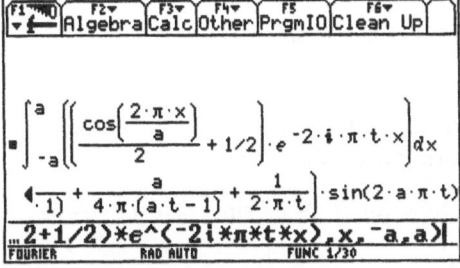

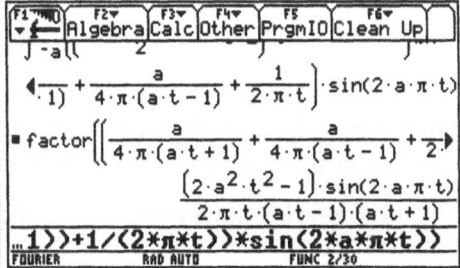

This shows the graph of the Fourier transform of the Hanning function shown in a neighborhood of 0. This transform tends to 0 more rapidly than the two previous ones, and its amplitude at 0 is somewhat reduced.

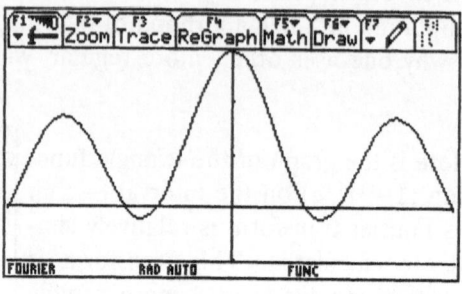

In general, we will denote a window function by w. Gabor's idea was to "slide" these windows ahead on the graph of the function f. In order to do that, it is sufficient to use the translation of w defined by $w_a : x \mapsto w(x - a)$.

We then replace the Fourier transform $\hat{f}$ of f by:

$$c_{\zeta,a}(f) = \int_{-\infty}^{+\infty} f(t)\overline{w}(x - a)e^{-2i\pi\zeta x}dx$$

We thus obtain a set of coefficients $(c_{\zeta,a}(f))_{\zeta,a}$ associated with the function f. Each scalar $c_{\zeta,a}(f)$ provides information based on the behavior of f on a neighborhood of a for the frequency ζ. Gabor was able to demonstrate formulas similar to those for Fourier series and transforms, including the possibility of retrieving the function f from its coefficients $(c_{\zeta,a}(f))_{\zeta,a}$, as well as the analog in dimension 2 of the Plancherel formula.

What is the difference between the Fourier and the Gabor analysis? If f is a function with compact support (in other words, f is zero outside a bounded closed interval), its reconstruction from its Fourier transform $\hat{f}$ will require enormously expensive calculations. For how can we reconstitute the null part of f from $\hat{f}(\zeta)$? It will require very many calculations which must also telescope so as to obtain a value close to 0.

With the Gabor analysis, if f is null on $[x_0 - a, x_0 + a]$ and if the support of w is in $[-1, 1]$, then, for x on a neighborhood of x_0:

$$c_{\zeta,a}(f) \approx \int_{x-1}^{x+1} f(t)w(t - a)e^{-2i\pi\zeta t}dt = 0$$

On the other hand, if f oscillates greatly on a neighborhood of x_0, the coefficients $c_{\zeta,a}(f)$ will be important on this neighborhood of x_0.

We illustrate this phenomenon with the function $f : x \mapsto \sin(2x) + 0.9\cos(5x) + 0.2\cos(42x)$.

This is the graph of the function f on $[-2\pi, 2\pi]$.

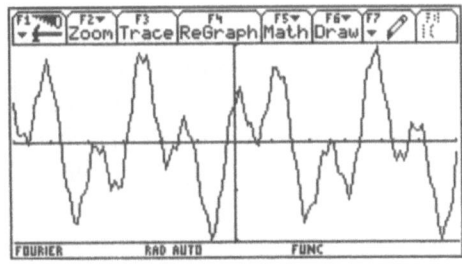

We see a zoom on the graph of f (the normal graph), that of $w(t - 5)\cos(2\sqrt{2}\pi x)$, where w is the Hanning function (dotted), and their product (bold), as well as the calculation of a Gabor coefficient around a strong oscillation of f (shaded).

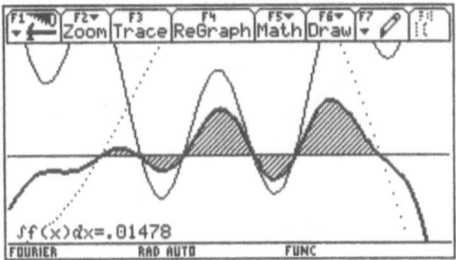

The disadvantage of the Gabor analysis is that it works only on fixed length windows. How do we process, then, functions whose variations are abrupt? (For example, a signal representing a burst of sound!). An answer is given by Morlet windows, and this really begins the theory of wavelets.

6. 2 The Morlet wavelets

Jean Morlet proposed to improve Gabor's idea by constructing windows "in accordion style". These windows resemble small waves that pass by and whose amplitude damps off, thus their name of wavelets.

Let's start from a so-called mother function $\Psi(x) = e^{-x^2/2}\cos(5x)$ and construct a family of functions $\psi_{a,b}(x) = \frac{1}{\sqrt{a}}\Psi\left(\frac{x-b}{a}\right)$, for $b \in \mathbb{R}, a > 0$ (called the "daughter wavelets").

The Morlet coefficients $c_{a,b}(f)$ are then defined by:

$$c_{a,b}(f) = \int_{-\infty}^{+\infty} f(t)\psi_{a,b}(t)dt$$

This is the Morlet wavelet on $[-3, 3]$. We give here a first translation-dilation: the graph of $\psi((x+3)/2)$ on the interval $[-9, 4]$. Then we show the graph of $\psi(4x - 9)$ on $[0, 4]$. The y-axis shown is $[-2, 2]$.

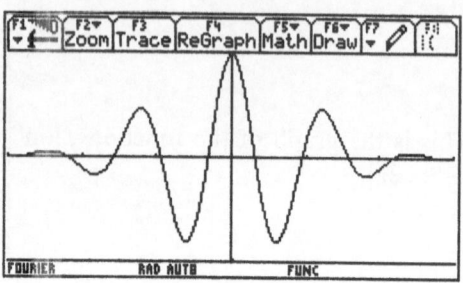

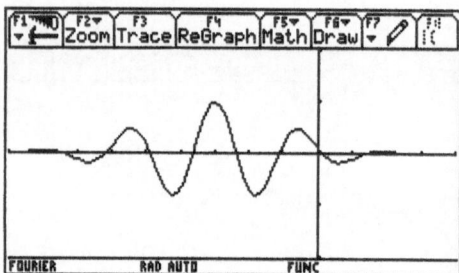

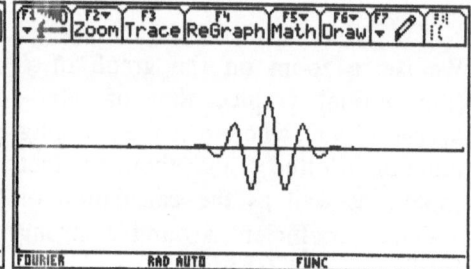

Compare the graph of the function f, the Morlet wavelet (dotted), and the product of the two functions (bold). The value of the wavelet coefficient is about 1.128. The contribution of $\psi_{1.0}$ will be relatively important.

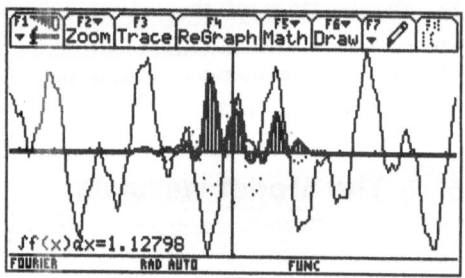

Here is the graph of the function f, as well as the dilated Morlet wavelet ($\psi(4x)$) (dotted) and the product of the two functions (bold). The value of the wavelet coefficient is clearly insignificant, roughly $2 \cdot 10^{-4}$. The contribution of $\psi_{1.0}$ will be negligible.

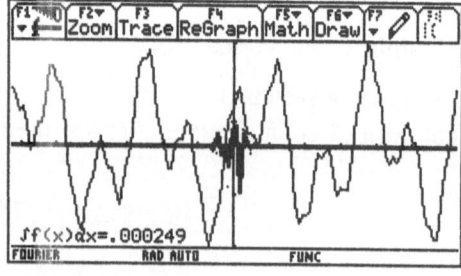

Here is a zoom to a neighborhood of 0 of the preceding graph, on the part where $f(x) = \psi(4x)$ is more important.

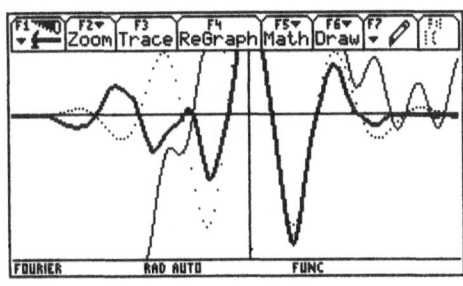

Of course, as in the theory developed by Fourier and then by Gabor, the main two theorems (reconstitution of f from coefficients $c_{a,b}(f)$ and Plancherel's theorem) exist in the wavelet framework.

If you try the examples described above, you will notice that, even numerically, the times of calculations for wavelet coefficients are extremely long. It is necessary therefore to find a more rapid means of calculating them. Moreover, in practice, we only rarely are given a function which is given by an equation. More often we will have a discrete data sample, the function f being only an interpolation of these data. This is the subject of the following paragraph.

6. 3 The multi-resolution analysis

Let V_0 be the set of constant functions with support in the interval $[0, 1[$ (that is, they vanish outside of this interval). For all $p \in \mathbb{N}$, let $\sigma_p = \left(0, \dfrac{1}{2^p}, \dots, \dfrac{k}{2^p}, \dots, \dfrac{2^p - 1}{2^p}\right)$ be a (uniform) partition of $[0, 1[$. We denote by V_p the set of step functions on σ_p, null outside of the interval $[0, 1[$ (therefore, the points of σ_p are possible discontinuity points.)

An element of V_2, a step function on $\sigma_2 = \left(0, \dfrac{1}{4}, \dfrac{1}{2}, \dfrac{3}{4}, 1\right)$.

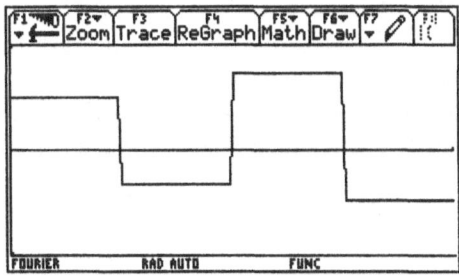

Thus, a sample of 2^p values $(x_0, x_1, x_2, \dots, x_{2^p-1})$ will be identified with the function of V_p defined by:

$$f(x) = \sum_{k=0}^{2^p-1} x_k \chi_{\left[\frac{k}{2^p}, \frac{k+1}{2^p}\right[}$$

Proposition 1: The sets (V_p) are real vector spaces satisfying the following properties:

a) $V_0 \subset V_1 \subset \dots V_p \subset V_{p+1} \subset \dots$

b) For all $p \in \mathbb{N}$, for all $i \in \{0, \dots, 2^p - 1\}$ $(f \in V_p) \Rightarrow (t \mapsto f(2t - i) \in V_{p+1})$

c) Let ϕ be the function defined by:

$$\phi(x) = \begin{cases} 1 & \text{if } x \in [0,1[\\ 0 & \text{otherwise} \end{cases}$$

If $p \in \mathbb{N}$ and $q \in \{0, \dots, 2^p - 1\}$, let $\phi_{p.q}$ be the function defined by $\phi_{p.q}(x) = \phi(2^p x - q)$. Then, for each $p \in \mathbb{N}$, the family $(\phi_{p.q})_{0 \le q \le 2^p - 1}$ is a basis of V_p (which thus is of dimension 2^p).

d) $V = \bigcup_{p \in \mathbb{N}} V_p$ with the scalar product

$$< f, g >= \int_0^1 f(t)g(t)t$$

is a real pre-Hilbert space, and for all $p \in \mathbb{N}, V_p$ is a Euclidean space.

Proof: a) This property is evident, since the subpartition σ_{p+1} is finer than σ_p. (It contains, among others, all points of σ_p.)

b) This property follows from the following result.

c) Let $p \in \mathbb{N}$ and $q \in \{0, \dots, 2^p - 1\}$; we note that $\phi_{p.q} = \phi(2^p x - q)$ is not zero if and only if $0 \le 2^p x - q < 1$, if and only if $\dfrac{q}{2^p} \le x < \dfrac{q+1}{2^p}$. Moreover, in this case it is constant and equal to 1. The function $\phi_{p.q}$ is therefore the indicator function (or characteristic function) of the interval $[\dfrac{q}{2^p}, \dfrac{q+1}{2^p}[$. The family of the $(\phi_{p.q})_{0 \le q \le 2^p - 1}$ is thus linearly independent and clearly spans V_p.

d) As an increasing union of vector spaces, V is a vector space. It is easy to verify that $< f, g >= \int_0^1 f(t)g(t)dt$ is a scalar product. If $< f, f >= 0$ then f is identically zero except on the possible points of discontinuity. Of course, if we use equivalence classes of functions by identifying such functions which are "almost null" we obtain a pre-Hilbert space. Similarly, V_p is finite dimensional and is a Euclidean space with the the indicated scalar product.

For each natural number p, define W_p as the orthogonal complement of V_p in V_{p+1}, or:

$$V_{p+1} = V_p \oplus^{\perp} W_p$$

Thus, by recursion, for all $n \in \mathbb{N}^*$,:

$$V_n = V_0 \oplus^{\perp} W_0 \oplus^{\perp} \dots \oplus^{\perp} W_{n-1}$$

This formula will allow us to decompose the data sample representing our function into a "coarse" part (represented by V_0), and into "fine" parts (represented by $W_0, W_1, \ldots, W_{n-1}$) which have increasingly "sharp" detail. The sequence V_n constructed above using the characteristic function of the unit interval as a "scaling function" is an example of what is called a multiresolution analysis.

The explanation that we just gave is general. To illustrate it, and to calculate wavelet coefficients, we will use this particular case: the Haar wavelets. They have been known since the beginning of the century, but their properties do not allow us to use them much in practice. On the other hand, they are simple enough to to be easily manipulated and understood. In other applications, one may wish to begin with "smoother" father or mother functions, but this may be at the cost of more computations and less clarity.

We use as a so-called father wavelet, or scaling function, the characteristic function of the closed unit interval:

$$\phi(x) = \begin{cases} 1 & \text{if } x \in [0, 1] \\ 0 & \text{otherwise} \end{cases}$$

The mother wavelet of Haar is defined by:

$$\psi(x) = \begin{cases} 1 & \text{if } x \in [0, 1/2[\\ -1 & \text{if } x \in [1/2, 1[\\ 0 & \text{otherwise} \end{cases}$$

You may verify that $\psi(x) = \phi(2x) - \phi(2x - 1)$.
If $p \in \mathbb{N}$ and $q \in \{0, \ldots, 2^p - 1\}$, let $\psi_{p.q}$ be the function defined by $\psi_{p.q}(x) = \psi(2^p x - q)$.

Proposition 2: For all $p \in \mathbb{N}$, the family $(\psi_{p.q})_{0 \leq q \leq 2^p - 1}$ is an orthogonal basis of W_p.

Proof:
- dim $W_p =$ dim $V_{p+1} -$ dim $V_p = 2^{p+1} - 2^p = 2^p$
- the family $(\psi_{p.q})_{0 \leq q \leq 2^p - 1}$ has cardinality 2^p and consists of functions which are not identically zero, so it forms a basis of W_p if and only if these functions are independent. In order to show that they are, it is sufficient to show that its elements are pairwise orthogonal.

The support of $\psi_{p.q}$ is included in $[\frac{q}{2^p}, \frac{q+1}{2^p}[$.
We obtain immediately $< \psi_{p.q}, \psi_{p.q'} >= 0$ if $q \neq q'$.
Without loss of generality, we may suppose $p > p'$. The function $\psi_{p'.q'}$ is constant on all intervals of the type $[\frac{j}{2^{p'+1}}, \frac{j+1}{2^{p'+1}}[$, and the integral of $\psi_{p.q}$ is null when its support is included in such an interval. From this, it follows that $< \psi_{p.q}, \psi_{p'.q'} >= 0$, which ends our demonstration.

Note finally that $< \psi_{p.q}, \psi_{p.q} >= \dfrac{1}{2^p}$.

We may now calculate the wavelet coefficients associated with a function f_p defined on the interval $[0, 1[$ by a sampling on $\left(0, \dfrac{1}{2^p}, \ldots, \dfrac{2^p-1}{2^p}\right)$. In order to do that, set:

$$f_p = \sum_{q=0}^{2^p-1} f_{p,q+1} \phi_{p,q}$$

Since $V_p = V_{p-1} \oplus W_{p-1}$, it follows that $f_p = f_{p-1} + d_{p-1}$ with:

$$\begin{cases} f_{p-1} = \displaystyle\sum_{q=0}^{2^{p-1}-1} f_{p-1,q+1} \phi_{p-1,q} \\ d_{p-1} = \displaystyle\sum_{k=0}^{2^{p-1}-1} d_{p-1,k} \psi_{p-1,k} \end{cases}$$

For every $j \in \{0, \ldots, 2^{p-1} - 1\}$:

$$< f_p, \psi_{p-1,j} > = \frac{d_{p-1,j}}{2^{p-1}} = \sum_{q=0}^{2^p-1} s_{p,q+1} < \phi_{p,q} \psi_{p-1,j} >$$

By considering the support of $\phi_{p,q}$ and $\psi_{p-1,j}$, we get: $< \phi_{p,q} \psi_{p-1,j} > = 0$ if $q \notin \{2j, 2j+1\}$, then:

$$< \phi_{p,2j}, \psi_{p-1,j} > = \int_0^1 \phi(2^p t - 2j) \psi(2^{p-1} t - j) dt = \frac{1}{2^p}$$

and that:

$$< \phi_{p,2j+1} \psi_{p-1,j} > = \int_0^1 \phi(2^p t - 2j - 1) \psi(2^{p-1} t - j) dt = -\frac{1}{2^p}$$

Finally:

$$d_{p-1,j} = \frac{f_{p,2j} - f_{p,2j+1}}{2}$$

On the other hand, for each $j \in \{0, \ldots, 2^{p-1} - 1\}$:

$$< f_p, \phi_{p-1,j} > = \frac{f_{p-1,j+1}}{2^{p-1}} = \sum_{q=0}^{2^p-1} f_{p,q+1} < \phi_{p,q} \phi_{p-1,j} >$$

and a calculation identical to the preceding one gives:

$$f_{p-1,j} = \frac{f_{p,2j} + f_{p,2j+1}}{2}$$

This allows us to pass from a sample of size 2^p to a sample of size 2^{p-1} using the details above. This may be done in a very simple and extremely rapid way since we use only sums and differences. (The program may be easier to understand than the previous explanation!)

`:ond(1)`	*l is list of data*
`:Func`	
`:Local ls,ld,k,j,r,p`	
`:dim(1)→p`	*p is the dimension of the list l*
`:ln(p)/(ln(2))→p`	*then the power of 2*
`:{}→r`	*result list*
`:For j,p-1,0,-1`	
`:seq((1[2*k+1]+1[2*k+2])/2,k,0,2^j-1)→ls`	*calculation of sums*
`:seq((1[2*k+1]-1[2*k+2])/2,k,0,2^j-1)→ld`	*calculations of differences*
`:ls→l`	*sums are saved in l*
`:augment(ld,r)→r`	*differences in the result*
`:EndFor`	
`:augment(1,r)→r`	*we add the last element*
`:EndFunc`	

Two examples of rapid calculation of the wavelet coefficients on some given data.

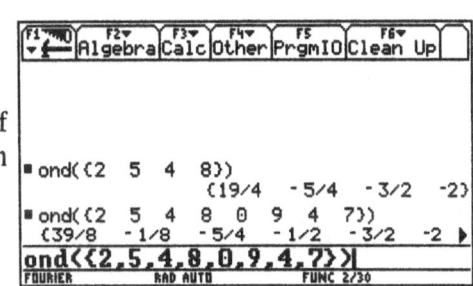

Let's quickly explain how the preceding program works with the first example, $\{2, 5, 4, 8\}$.

• during the first pass `ls` contains the averages of $(2, 5)$ and $(4, 8)$, $((7/2, 6))$, while `ld` contains the averages of these pairs, that is, $(-3/2, -2)$. Thus, r contains $(-3/2, -2)$ and l is $(7/2, 6)$.

• during the second pass `ls` contains the average of $(7/2, 6)$, that is, $(19/4)$ and `ld` contains the average of this pair, that is, $(-5/4)$. Thus, r contains $(-5/4, -3/2, -2)$ and l is $(19/4)$.

Wavelets may be used, among other things, to compress data. Indeed, if our sampled function is relatively smooth, we would presume that consecutive sample values will be close. The contribution to the wavelet coefficients stemming from two consecutive difference values of the sample will therefore be small. Specifying a precision ε and keeping only the wavelet coefficients greater than ε in absolute value will allow a substantial compression of the original data. In summary, here is the algorithm used :

- We obtain:

$$f_p = \sum_{q=0}^{2^k-1} f_{p,q+1}\phi_{p,q}$$

- by the preceding formulas:

$$\begin{cases} f_{p-1,j} &= \dfrac{f_{p,2j} + f_{p,2j+1}}{2} \\ d_{p-1,j} &= \dfrac{f_{p,2j} - f_{p,2j+1}}{2} \end{cases}$$

- We calculate:

$$f_p = f_{0,0}\phi + \sum_{k=0}^{p-1}\sum_{q=0}^{2^{k-1}-1} d_{k,q}\psi_{k,q}$$

- putting:

$$\widetilde{d_{k,q}} = \begin{cases} d_{k,q} & \text{if } |d_{k,q}| \geq \varepsilon \\ 0 & \text{otherwise} \end{cases}$$

- And we display:

$$\widetilde{f_p} = f_{0,0}\phi + \sum_{k=0}^{p-1}\sum_{q=0}^{2^{k-1}} \widetilde{d_{k,q}}\psi_{k,q}$$

The corresponding program is just the preceding program, slightly modified:

```
:ond1(l,e)                              data in list l, precision e
:Func
:Local ls,ld,k,j,r,p
:dim(l)→p                               p : dimension of l
:ln(p)/(ln(2))→p                        p : power of 2
:{}→r                                   r : result list
:For j,p-1,0,-1                         calculation of sums and differences
:seq((l[2*k+1]+l[2*k+2])/2,k,0,2^j -1)→ls
:seq((l[2*k+1]-l[2*k+2])/2,k,0,2^j-1)→ld
:ls→l
:augment(ld,r)→r
:EndFor
:augment(l,r)→r                         result
:For j,1,dim(r)
:If abs(r[j])<e Then                    truncate to precision e
:0→r[j]
:EndIf
:EndFor
:r                                      final result
:EndFunc
```

Let's look at what our program gives for the following function:

$$f : x \mapsto -\frac{1}{10}\left(x - \frac{1}{2}\right)\sin(40x) - \frac{5}{2}\left(x - \frac{3}{2}\right)\left(x - \frac{1}{4}\right)$$

First, the graph of the function f, then, the definition of the two lists of data for this Plot:

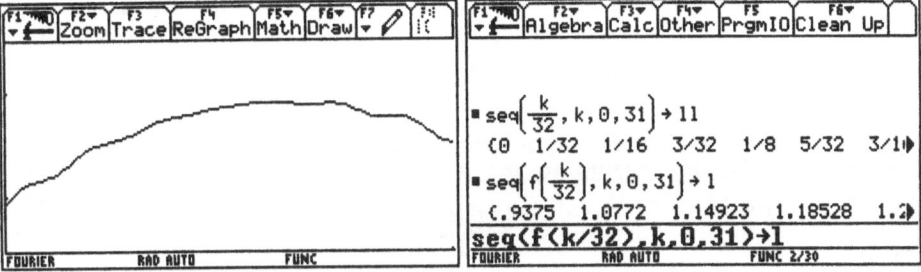

and then the definition of the statistical layout and the graph itself:

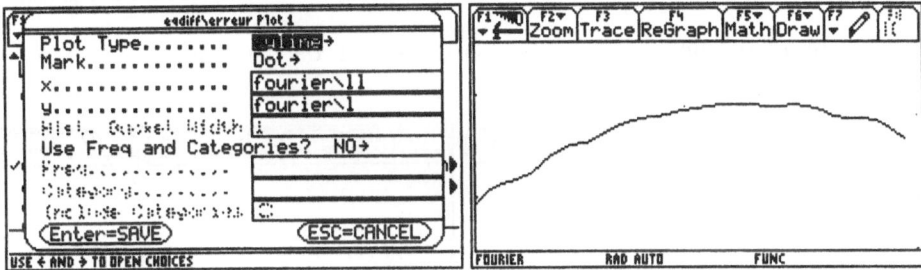

Shown below is the output for the truncated wavelets at a precision of 0.01, and the subsequent definition of the reconstituted function h. To define this function, we used the function Ψ, saved in y1 (on the screen at right).

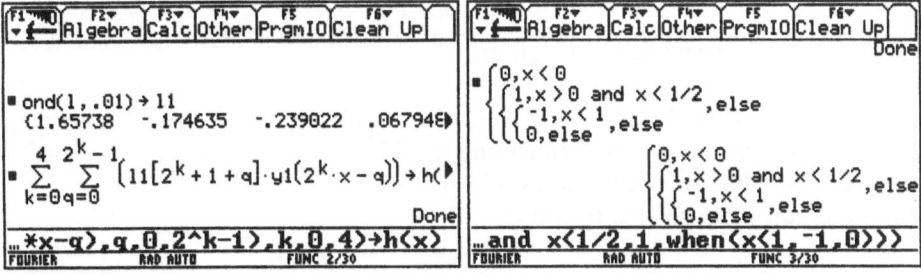

The graph of the initial data (crosses) and the function h defined above. The recovery of the data is pretty good considering the value ε of the precision and the small number of data (32)

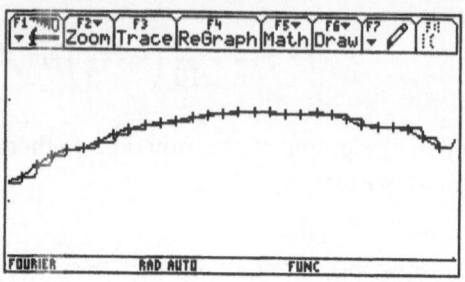

Let's perturb the function f by adding the function $x \mapsto e^{-10x} \sin(100x)$.

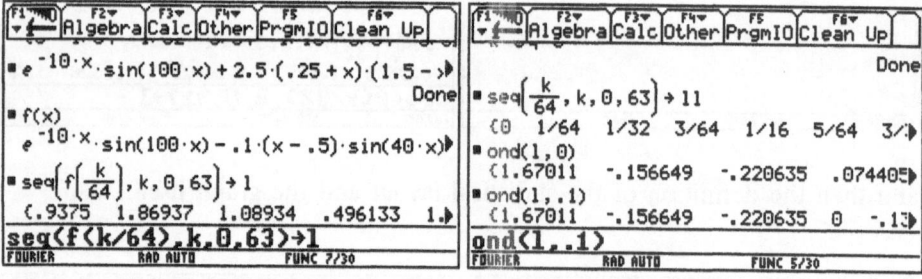

The graph of the new function f is on the screen on the right.

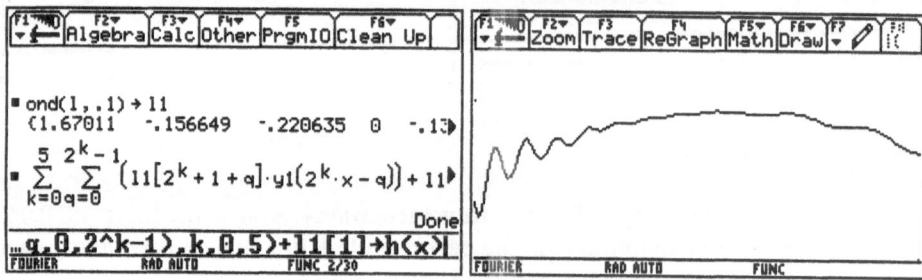

The graph of wavelet computations based on 64 sample points and the truncated function h defined above. The recovery of the original function is here clearly less successful, however, the precision is much less ($\varepsilon = 0.1$).

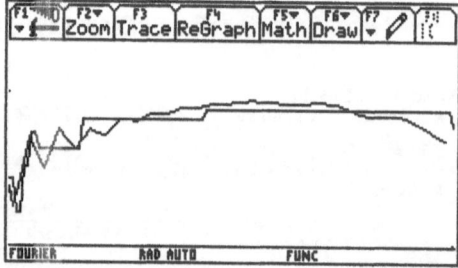

4

Interpolation and approximation

Interpolation is an important and rich mathematical theme. If only a sample of the values of a numerical function f are known, say at n distinct points $(x_1, x_2, \ldots, x_n)$, one may approximate f by a simple function P – the simplest being a polynomial function – which takes the same values as does f at these points. For all x in the interval defined by the points $(x_1, x_2, \ldots, x_n)$, one thus replaces the unknown values $f(x)$ by the values defined by $P(x)$: this is the principle of interpolation.

But is this "interpolation polynomial" in some sense "approximately equal" to the function f? Is interpolation always synonymous with "good approximation"? We will reply to this question in this chapter after we study different several interpolation methods.

1. Interpolating polynomials

We consider a family $\mathcal{F} = (A_1, A_2, \ldots, A_n)$ of n pairs $(x_k, y_k)_{(1 \leq k \leq n)}$, complex or real valued, with the x-values $(x_1, x_2, \ldots, x_n)$ being pairwise distinct.
There is a unique polynomial $P_{\mathcal{F}}$, of degree less than or equal to $(n-1)$, such that, for all $k \in \{1, \ldots, n\}$, $P_{\mathcal{F}}(x_k) = y_k$. This polynomial is called an *interpolating polynomial* of the family $\mathcal{F}$.
Here we will successively investigate several ways to determine $P_{\mathcal{F}}$. The proof of the existence be will made during each of the different constructions. Let's show the uniqueness once and for all.

Suppose on the contrary that there exist two polynomials P and Q of degree less than or equal to $(n-1)$ such that for all $i \in \{1, 2, \ldots, n\}, P(x_i) = Q(x_i)$. The polynomial $P - Q$ is of degree less than or equal to $(n-1)$ and has n roots $(x_1, x_2, \ldots, x_n)$. It is therefore identical to the zero polynomial by a well-known theorem of algebra, and thus the respective coefficients of P and Q are equal.

1. 1 Lagrange form of the interpolating polynomial

The first method leads to what one calls the Lagrange interpolating polynomial. For any k between 1 and n, we define a polynomial L_k by:

$$L_k(x) = \prod_{j=1, j \neq k}^{n} \frac{x - x_j}{x_k - x_j}$$

It is easy to show that each polynomial L_k is of degree $(n-1)$. In the same manner, we immediately verify the following equations, valid for all k and $j \in \{1, 2, \ldots, n\}$:

$$L_k(x_j) = \delta_{k.j} = \begin{cases} 1 & \text{if } k = j \\ 0 & \text{if } k \neq j \end{cases}$$

We then observe that the polynomial $P(x) = \sum_{k=1}^{n} y_k L_k(x)$ satisfies the requirements of the problem. Indeed, P is of degree less than or equal to $(n-1)$, and it satisfies the following equations, for all $1 \leq j \leq n$:

$$P(x_j) = \sum_{k=1}^{n} y_k L_k(x_j) = \sum_{k=1}^{n} y_k \delta_{k.j} = y_j$$

The following calculator function constructs this interpolating polynomial by using Lagrange form for the list of abscissas $X = \{x_1, x_2, \ldots, x_n\}$ and the list of y coordinates $Y = \{y_1, y_2, \ldots, y_n\}$. We have added a variable w, which contains the name of the variable in the final polynomial expression.

```
:lagr(x,y,w)                x,y: list of data; w: variable
:Func
:Local k,r,p
:0→p                        the polynomial is saved in p
:For k,1,dim(x)
:(w-x)/(x[k]-x)→r           the quotients
:1→r[k]                     for k = j, 1 instead
:p+y[k]*product(r)→p        we multiply the quotients by yₖ
:EndFor
:EndFunc
```

Here is an example using the function lagr. We obtain the interpolating polynomial in its expanded form. We may then check that the result is correct.

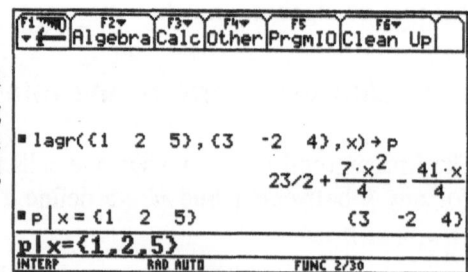

This shows use of the function lagr with complex points. The polynomial obtained is given with its real and its imaginary parts.

```
F1    F2   F3   F4   F5    F6
     Algebra Calc Other PrgmIO Clean Up
■ lagr({1 + i   2·i   -5 - i}, {3   -2 + 2·i  ▶
  i·(-37·x²  -  131·x  -  587 ) +  853·x  +  113·
       340        340      170      340      17(
■ p│x = {1 + i   2·i   -5 - i}
                      {3   -2 + 2·i   4}
 p│x={1+i,2*i,-5-i}
INTERP          RAD AUTO          FUNC 2/30
```

One may use the function lagr with formal data.

```
F1    F2   F3   F4   F5    F6
     Algebra Calc Other PrgmIO Clean Up
■ lagr({a   b   c}, {α   8   γ}, t) → p
  -((γ·b² - 8·c²)·a + ((α - 8)·c + (8 - γ)·a - (₁
■ expand(p, t)
  -(a·(8 - γ) - b·(α - γ) + c·(α - 8))·t²  , (a²,
        (a - c)·(a - b)·(b - c)
 expand(p,t)
INTERP          RAD AUTO          FUNC 4/30
```

Below are the definition of the graph of two lists xx and yy and the graph of the interpolating polynomial.

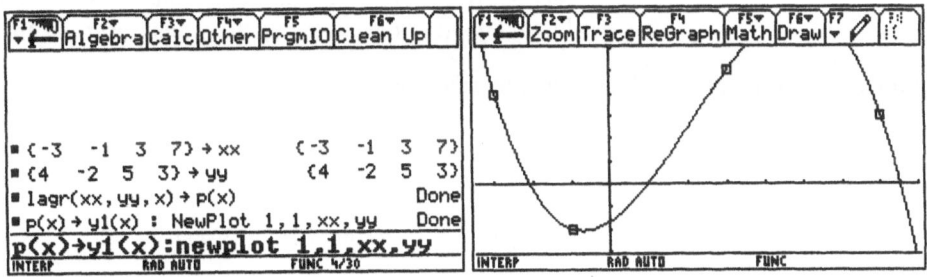

```
F1    F2   F3   F4   F5    F6
     Algebra Calc Other PrgmIO Clean Up



■ {-3   -1   3   7} → xx      {-3   -1   3   7}
■ {4   -2   5   3} → yy       {4   -2   5   3}
■ lagr(xx, yy, x) → p(x)              Done
■ p(x) → y1(x) : NewPlot 1, 1, xx, yy  Done
 p(x)→y1(x):newplot 1,1,xx,yy
INTERP          RAD AUTO          FUNC 4/30
```

1. 2 The Vandermonde form

If we denote the interpolating polynomial by $P(x) = \sum_{k=0}^{n-1} a_k x^k$, then to determine its coefficients is the same as to solve the system of equations:

$$\begin{cases} a_0 + a_1 x_1 + a_2 x_1^2 + \ldots + a_n x_1^{n-1} &= y_1 \\ a_0 + a_1 x_2 + a_2 x_2^2 + \ldots + a_n x_2^{n-1} &= y_2 \\ &\vdots &= \vdots \\ a_0 + a_1 x_n + a_2 x_n^2 + \ldots + a_n x_n^{n-1} &= y_n \end{cases}$$

which may be written in the following matrix form:

$$
\begin{pmatrix} y_1 \\ y_2 \\ \vdots \\ y_{n-1} \\ y_n \end{pmatrix} = \begin{pmatrix} 1 & x_1 & x_1^2 & \cdots & x_1^{n-1} \\ 1 & x_2 & x_2^2 & \cdots & x_2^{n-1} \\ \vdots & & \ddots & & \vdots \\ \vdots & & & \ddots & \vdots \\ 1 & x_n & x_n^2 & \cdots & x_n^{n-1} \end{pmatrix} \begin{pmatrix} a_1 \\ a_2 \\ \vdots \\ a_{n-1} \\ a_n \end{pmatrix}
$$

The matrix of this system is a square invertible matrix called Vandermonde's matrix.

Proposition 1: *The determinant of Vandermonde's matrix is equal to:*

$$
\prod_{1 \le i < j \le n} (x_j - x_i)
$$

Proof: If there exists $1 \le i \ne j \le n$ such that $x_i = x_j$, Vandermonde's determinant and the product $\prod_{1 \le i < j \le n} (x_j - x_i)$ are both equal to zero. We may therefore suppose that the (x_i) are distinct. Denote the Vandermonde determinant on the n points $(x_1, x_2, \ldots, x_n)$ by $\Delta(x_1, x_2, \ldots, x_n)$. Let's replace x_n by X. By expanding this determinant along the last row, we obtain a polynomial of degree $n - 1$ in X with leading coefficient $\Delta(x_1, x_2, \ldots, x_{n-1})$. Moreover, by replacing X successively by $x_i, 1 \le i \le n - 1$, we see that two rows of the determinant are equal, so the determinant vanishes for each $x_i, 1 \le i \le n - 1$. The $(n - 1)$ scalars $(x_1, x_2, \ldots, x_{n-1})$ are thus roots of the polynomial that can be factored into this form: $\Delta(x_1, x_2, \ldots, x_{n-1}) \prod_{i=1}^{n-1} (X - x_i)$.

By replacing X by x_n, this becomes:

$$
\Delta(x_1, x_2, \ldots, x_n) = \prod_{i=1}^{n-1} (x_n - x_i) \Delta(x_1, x_2, \ldots, x_{n-1})
$$

It remains only to finish the proof by recursion.
Thus, Vandermonde's matrix is invertible if and only if the scalars $(x_1, x_2, \ldots, x_n)$ are distinct.

The next program uses the expression that we just obtained.

```
:interpol(x,y,w)                                         x,y : list of data
:Func                                                    w : variable
:Local i,j,n,m
:dim(x)→n
:seq(seq(x[i]^j,j,n-1,0,-1),i,1,n)→m                     Vandermonde matrix
:polyEval(mat▸list(simult(m,list▸mat(y,1))),w)           interpolation
:EndFunc
```

Calculation of the interpolating polynomial by Vandermonde's method. On 3 points, it is a polynomial of degree 2, except when $a = -3$, in which case the three points lie on a line. It is then a polynomial of degree 1.

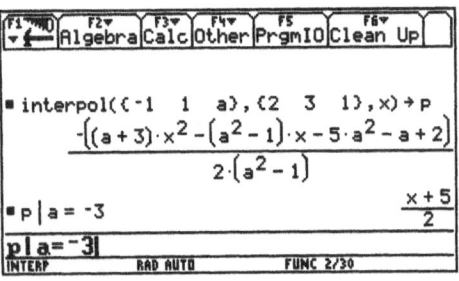

1.3 Newton's interpolating polynomial

The Lagrange form of the interpolating polynomial is only of theoretical interest. In practice, it is rarely used: we may reject it for requiring too many operations. Also, all calculations must be done again if we decides to add a new point.

The interpolating polynomial on n points may in fact be written in a form which is much more economical in terms of time of calculation and which asks only slight modifications each time that a new point has to be added. This is the Newton form which uses the principle of "divided differences".

Let $X = (x_1, x_2, \ldots, x_n)$ be a sequence of n scalars and $Y = (y_1, y_2, \ldots, y_n)$ a second sequence of n scalars which are associated with X.

We define the divided difference operator, denoted by [ ], in the following manner:

- if $n = 1$, we put $[x_1] = y_1$.
- if $n > 1$ we put:

$$[x_1, \ldots, x_n] = \frac{[x_2, x_3, \ldots, x_n] - [x_1, x_1, \ldots, x_{n-1}]}{x_n - x_1}$$

Thus, for example:

$$[x_1, x_2] = \frac{y_2 - y_1}{x_2 - x_1}$$

Remark: This notation makes implicit reference to the sequence Y without that sequence appearing; it is always necessary to keep in mind that Y is, in fact, being used.

We see that this recursive definition allows us to calculate the divided differences of two sequences X and Y of equal length n, step by step.

```
:dd(x,y)
:Func
:Local n
:dim(x)→n
:If n=1
:Return y[1]
:(dd(mid(x,2,n-1),mid(y,2,n-1))-dd(mid(x,1,n-1),
  mid(y,1,n-1)))/(x[n]-x[1])
:EndFunc
```

Here is an example using the recursive divided difference function in a symbolic calculation. Of course, the longer the lists x, y are, the more important will be the time of calculation.

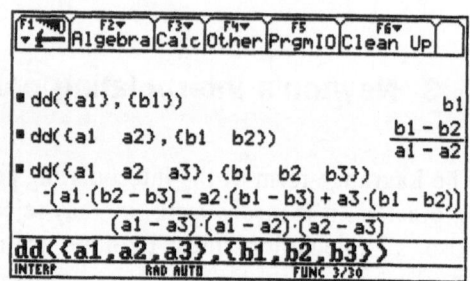

Any recursive function may be made "unrecursive". There is thus an iterative method to calculate the respective divided differences of the n scalar $(x_1, x_2, \ldots, x_n)$ and $(y_1, y_2, \ldots, y_n)$. Let's illustrate this with an example.

Suppose $n = 4$, and denote by $d_{i,k}$ the divided difference of the sequences $(x_i, \ldots, x_k)$ and $(y_i, \ldots, y_k)$, for all pairs (i, k), with $1 \leq i \leq k \leq 4$.

The divided difference calculation $d_{1,k}$, for $1 \leq k \leq 4$, may be done with the following table. Operations are done from the left column to the right, and on each column starting from the bottom to the top:

$$d_{1,1} = y_1$$

$$d_{1,2} = \frac{d_{2,2} - d_{1,1}}{x_2 - x_1}$$

$$d_{2,2} = y_2$$

$$d_{1,3} = \frac{d_{2,3} - d_{1,2}}{x_3 - x_1}$$

$$d_{2,3} = \frac{d_{3,3} - d_{2,2}}{x_3 - x_2}$$

$$d_{1,4} = \frac{d_{2,4} - d_{1,3}}{x_4 - x_1}$$

$$d_{3,3} = y_3$$

$$d_{2,4} = \frac{d_{3,4} - d_{2,3}}{x_4 - x_2}$$

$$d_{3,4} = \frac{d_{4,4} - d_{3,3}}{x_4 - x_3}$$

$$d_{4,4} = y_4$$

Indeed, it is possible to start with the list (y_1, y_2, y_3, y_4), and to replace it gradually by the list of divided differences $(d_{1,1}, d_{1,2}, d_{1,3}, d_{1,4})$. In the preceding calculation, each new divided difference replaces its left neighbor in the table.

We also see that if we wish to add a new point (x_0, y_0) to the list of existing points, it suffices to complete the given table by adding an upper diagonal containing successively the divided differences $(d_{0.0}, d_{0.1}, d_{0.2}, d_{0.3}, d_{0.4})$. The following calculator function, called ddi, successively calculates the divided differences $d_{1.k}$ of the sequences $X = (x_1, x_2, \ldots, x_n)$ and $Y = (y_1, y_2, \ldots, y_n)$, using the preceding notation.

The result is therefore the sequence $(d_{1.1}, d_{1.2}, \ldots, d_{1.n})$.

```
:ddi(x,y)
:Func
:Local n,j,k
:dim(x)→n
:For k,1,n-1
:For j,n,k+1,-1
:(y[j]-y[j-1])/(x[j]-x[j-k])→y[j]
:EndFor
:EndFor
:y
:EndFunc
```

Here are several tests of the function ddi. In the last example, we added a fourth point to a sequence of 3 numbers. We may compare times of calculation of the recursive version versus those of the iterative version.

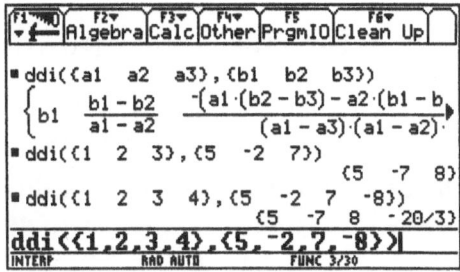

Divided differences are at the center of the Newton form of the interpolating polynomial. Recall that we know that there exists a unique polynomial P_n, of degree less than or equal to $(n-1)$, interpolating the sequences $(x_1, x_2, \ldots, x_n)$ and $(y_1, y_2, \ldots, y_n)$, where we suppose that the (x_i) are distinct.

The Lagrange form of P_n consists of expressing our polynomial in terms of the basis of $\mathbb{R}_n[X]$ formed of Lagrange polynomials $(L_k)_{1 \le k \le n}$.

The expression may be also be linked to the Vandermonde determinant and P_n may be expressed directly in terms of the canonical basis $(1, X, X^2, \ldots, X^{n-1})$ of $\mathbb{R}_{n-1}[X]$.

A third possibility uses the basis of polynomials $(B_k)_{0 \le k \le n}$, defined by:

$$B_0 = 1, \quad B_1 = X - x_1, \quad B_2 = (X - x_1)(X - x_2), \ldots, B_{n-1} = \prod_{j=1}^{n-1} (X - x_j)$$

Let's show first a general formula that we will use again in what follows.

Theorem 1:Newton's Identity. Let f be a function defined on $\mathbb{R}$, and let $X = (x_1, x_2, \ldots, x_n)$ be a sequence of n distinct points. Let $y_k = f(x_k)$, $1 \leq k \leq n$. For all real x, we may write:

$$f(x) = f(x_1) + [x_1, x_2](x - x_1) + [x_1, x_2, x_3](x - x_1)(x - x_2) + \ldots +$$
$$+ [x_1, x_2, \ldots, x_n](x - x_1)(x - x_2) \ldots (x - x_{n-1})$$
$$+ [x, x_1, x_2, \ldots, x_n](x - x_1)(x - x_2) \ldots (x - x_n)$$

Proof: The relationship is satisfied if x is equal to one of the x_i, so we will suppose that x is distinct from $(x_i)_i$. The demonstration is then finished by recursion on n:

• It is evident when $n = 1$.
• Suppose the proposition is proved for all sequences $(x, x_1, x_2, \ldots, x_{n-1})$, and let $X = (x_1, x_2, \ldots, x_n)$.
We know that:

$$[x, x_1, x_2, \ldots, x_n] = \frac{[x_1, \ldots, x_n] - [x, x_1, x_2, \ldots, x_{n-1}]}{x_n - x}$$

Therefore:

$$[x, x_1, x_2, \ldots, x_{n-1}] = [x_1, \ldots, x_n] - [x, x_1, x_2, \ldots, x_n](x - x_n)$$

It remains only to place this last expression in the form of f valid on $(x, x_1, \ldots, x_{n-1})$ to obtain the desired result.

Thus, for all real x, $f(x)$ may be written as:

$$f(x) = d_1 B_0 + d_2 B_1 + d_3 B_2 + \ldots + d_n B_{n-1} + R(x) = P(x) + R(x)$$

with:

$$R(x) = [x, x_1, x_2, \ldots, x_n](x - x_1)(x - x_2) \ldots (x - x_n)$$

But, for all $1 \leq k \leq n$, $R(x_k) = 0$. Thus, for all $1 \leq k \leq n$, $y_k = f(x_k) = P(x_k)$. The polynomial P is therefore the interpolating polynomial of $(x_1, x_2, \ldots, x_n)$ and $(y_1, y_2, \ldots, y_n)$.

Remark: The coefficient d_n of B_{n-1} is the leading coefficient of the interpolating polynomial. This polynomial does not depend on the order in which the points $(x_1, x_2, \ldots, x_n)$ are given, so we conclude that the divided difference of X and Y is independent of the order of elements in these sequences.

The Newton form of the interpolating polynomial has an added interest: it may be easily used with the Horner method of evaluating polynomials. Indeed, the next expression:

$$P_n(X) = d_1 + d_2(X - x_1) + \ldots + d_n(X - x_1)(X - x_2) \ldots (X - x_{n-1})$$

may be written in the form:

$$P_n(X) = d_1 + (X - x_1)\left((d_2 + (X - x_2)(d_3 + \ldots (d_{n-1} + d_n(X - x_{n-1}))\ldots)\right)$$

In other words, to calculate the value of the polynomial P_n at a point a, it is sufficient to construct the sequence $b_{n+1}, b_n, b_{n-1}, \ldots, b_1$ defined by the relationships:

$$b_{n+1} = 0, \text{ and } \forall\, k \in \{n, \ldots, 1\} \; : \; b_k = d_k + (a - x_k)b_{k+1}$$

We then have $b_1 = P(a)$. The function newt uses this method to form the expression of the interpolating polynomial P (or the value at a point of the former) for two sequences X and Y.

```
:newt(x,y,w)
:Func
:Local d,n,b,k
:ddi(x,y)→d           call to ddi
:dim(x)→n
:0→b
:For k,n,1,-1
:d[k]+(w-x[k])*b→b    Horner diagram
:EndFor
:EndFunc
```

This is an example of the function newt. The interpolating polynomial produced is not ordered by decreasing powers of the variable.

F1▼▼	F2▼ Algebra	F3▼ Calc	F4▼ Other	F5 PrgmIO	F6▼ Clean Up

$\blacksquare$ newt({1 4 5 6},{3 1 2 7},x)

$$\frac{193 \cdot x}{30} + \frac{19 \cdot x^3}{60} - \frac{11 \cdot x^2}{4} - 1$$

$\blacksquare$ expand$\left(\dfrac{193 \cdot x}{30} + \dfrac{19 \cdot x^3}{60} - \dfrac{11 \cdot x^2}{4} - 1\right)$

$$\frac{19 \cdot x^3}{60} - \frac{11 \cdot x^2}{4} + \frac{193 \cdot x}{30} - 1$$

expand(193*x/30+19*x^3/60−11*...

INTERP   RAD AUTO   FUNC 7/30

The Newton form of the interpolating polynomial may also be used to add a new point a to an interpolating sequence. We know that the polynomials P_n and P_{n+1} – both of degree less than or equal to n – take the same value y_k at each of the points x_k, $1 \leq k \leq n$. There exists therefore a constant λ such that $P_{n+1}(X) - P_n(X) = \lambda(X - x_1)\ldots(X - x_n)$.

The preceding investigation shows us that λ is equal to the divided difference $d_{1,n+1}$, but this is not important here; indeed, λ may be obtained by substituting x_{n+1} for the undetermined X in the preceding equation, and by taking into account the hypothesis $P_{n+1}(x_{n+1}) = y_{n+1}$, we obtain:

$$\lambda = \frac{y_{n+1} - P_n(x_{n+1})}{(x_{n+1} - x_1)(x_{n+1} - x_2)\dots(x_{n+1} - x_n)}$$

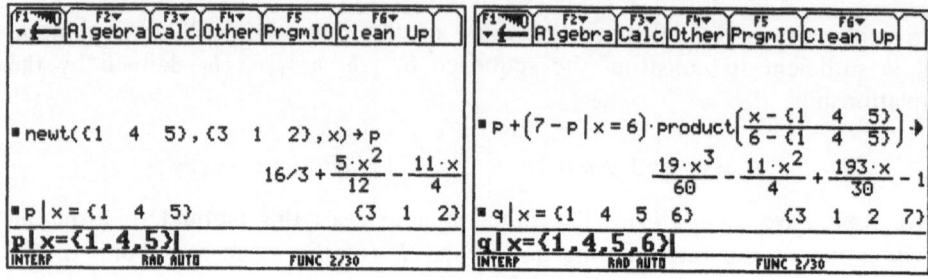

Here p is P_n and q is P_{n+1} in the previous equation.

1. 4 Neville's algorithm

The Neville algorithm is another method to find the interpolating polynomial P of two sequences X (abscissas) and Y (ordinates). This process is particularly useful to calculate other values of this polynomial. It reminds one a bit of the divided difference methods that we have seen previously.

Consider for example 3 points $A_1(x_1, y_1), A_2(x_2, y_2), A_3(x_3, y_3)$, with distinct x-values.

Let $P_{1,1}$ be the polynomial of degree 0 "interpolating" only the one point A_1: This is just the constant y_1. Similarly, let $P_{2,2}$ and $P_{3,3}$ be the constant polynomials interpolating respectively A_2 and A_3. By generalizing these notations, let $P_{1,2}$ be the (first degree) polynomial of degree less than or equal to 1 interpolating the two points A_1, A_2, and let $P_{2,3}$ interpolate A_2 and A_3. We easily verify that:

$$P_{1,2} = \frac{(X - x_2)P_{1,1} + (x_1 - X)P_{2,2}}{x_1 - x_2}, \quad P_{2,3} = \frac{(X - x_3)P_{2,2} + (x_2 - X)P_{3,3}}{x_2 - x_3}$$

Let $P_{1,2,3}$ be the (quadratic) polynomial of degree less than or equal to 2, interpolating the three points A_1, A_2, A_3. We observe (with a direct verification using the calculator) that:

$$P_{1,2,3} = \frac{(X - x_3)P_{1,2} + (x_1 - X)P_{2,3}}{x_1 - x_3}$$

More generally,

Proposition 2: Given a family of n points $A_k(x_k, y_k)_{1 \le k \le n}$ with distinct x-values, let $P_{1,2,\dots,n-1}$ be the interpolating polynomial of the $n-1$ first points $A_1, A_2, \dots, A_{n-1}$

and let $P_{2.3.....n}$ be the polynomial interpolating the $n-1$ last points $A_2, A_3, \ldots, A_n$. These two polynomials are of degree less than or equal to $n-2$. The polynomial:

$$P = \frac{(X - x_n)P_{1.2....,n-1} + (x_1 - X)P_{2,3.....n}}{x_1 - x_n}$$

of degree less than or equal to $n - 1$ is the interpolating polynomial of $A_1, A_2, \ldots, A_n$.

Proof: $P(x_1) = P_{1.2.....n-1}(x_1) = y_1$.

For all $2 \le k \le n$, $P(x_k) = \dfrac{(x_k - x_n)y_k + (x_1 - x_k)y_k}{x_1 - x_n} = y_k$,

and $P(x_n) = P_{2.3.....n-1}(x_n) = y_n$.

Starting from n points, we therefore apply the following algorithm:
• Write the n constant polynomials $P_{1.1}, P_{2.2}, \ldots, P_{n.n}$ "interpolating" each one of the given points.
•   Associate these constants pairwise to form the $n - 1$ polynomials $P_{1.2}, P_{2.3}, \ldots, P_{n-1.n}$ interpolating two consecutive points of the family.
• Associate these polynomials pairwise to form the $n-2$ polynomials $P_{1.2.3}, P_{2.3.4}$, etc., interpolating three consecutive points of the family.
• Continue this process until the polynomial $P = P_{1.2.....n}$ interpolating all n points is obtained.

Here is a function, called neville, that forms the interpolating polynomial for our family of n points.

```
:neville(x,y,w)
:Func
:Local n,j,k
:dim(x)→n
:For j,1,n-1
:For k,1,n-j
:((w-x[k+j])*y[k]+(x[k]-w)*y[k+1])/(x[k]-x[k+j])→y[k]
:EndFor
:EndFor
:y[1]
:EndFunc
```

We may make a comparison of the time for each of the three algorithms to calculate the interpolating polynomial of a given sequence of points. The Neville algorithm is particularly efficient for the calculation of values $P(a)$.

F1 F2▾ F3▾ F4▾ F5 F6▾
◆— Algebra Calc Other PrgmIO Clean Up
■ neville({-1 0 1 2 3},{-1 1 -7▶
2·x⁴ - 4·x³ - 7·x² + x + 1
■ newt({-1 0 1 2 3},{-1 1 -7 -▶
1 + x + 2·x⁴ - 4·x³ - 7·x²
■ lagr({-1 0 1 2 3},{-1 1 -7 -▶
1 + x + 2·x⁴ - 4·x³ - 7·x²
...0,1,2,3},{-1,1,-7,-25,-5},x)▶
INTERP   RAD AUTO   FUNC 3/30

$$P = \frac{(X - x_n)P_{1.2....,n-1} + (x_1 - X)P_{2,3.....n}}{x_1 - x_n}$$

To finish with the Neville method, let's use the following program, directly inspired by the previous function neville and whose role is to graph successively the partial interpolating polynomials constructed by the algorithm: first those of degree less than or equal to 1 interpolating 2 successive points, then these of degree less than or equal to 2 interpolating 3 successive points, etc.

```
:graphnev(a,b)
:Prgm
:Local n,j,k
:a→xx:b→yy
:dim(a)→n
:FnOff:PlotsOff
:NewPlot 1,1,xx,yy
:ZoomData
:For j,1,n-1
:ClrDraw
:For k,1,n-j
:((x-a[k+j])*b[k]+(a[k]-x)*b[k+1])/(a[k]-a[k+j])→b[k]
:DrawFunc b[k]
:EndFor
:Pause
:EndFor
:EndPrgm
```

Here we called the preceding program to display the polynomials interpolating $2 \leq n \leq 5$ points. From left to right: 4 lines interpolating 2 successive points, the 3 parabolas interpolating 3 points, the 2 cubics interpolating 4 successive points, and the quartic polynomial interpolating 5 points.

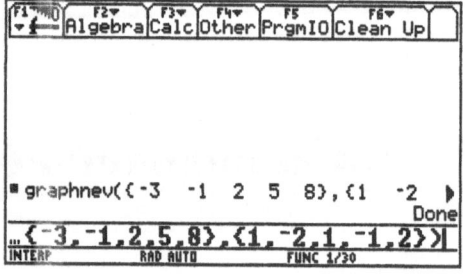

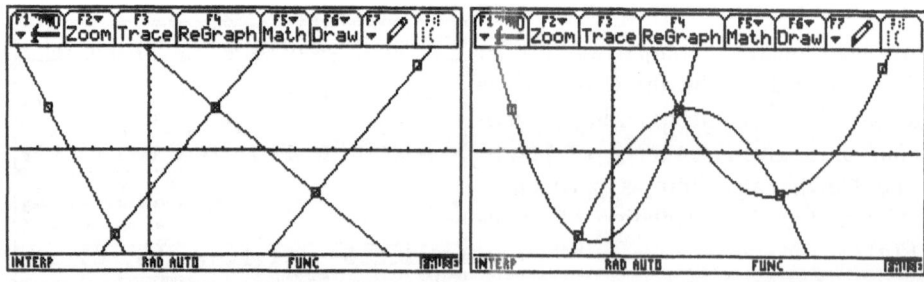

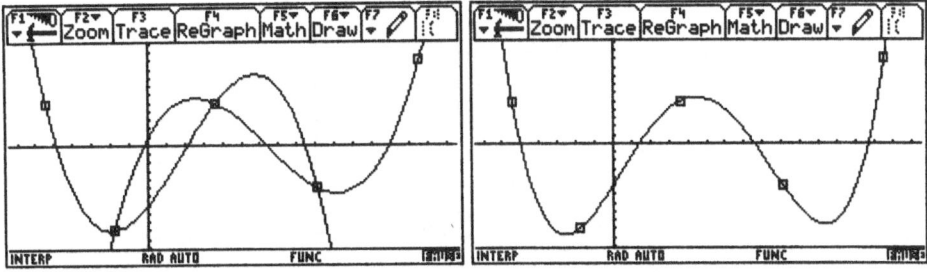

2. The Runge phenomenon

The usefulness of the interpolating polynomial of two sequences $X = (x_1, x_2, \ldots, x_n)$ and $Y = (y_1, y_2, \ldots, y_n)$ appears when $y_k = f(x_k)$, $1 \leq k \leq n$, where f is a function of the real variable x. Indeed, when a belongs to the interval formed by points x_k, we may hope to be able to approximate $f(a)$, by $P(a)$, where P is the interpolating polynomial of X and Y.

When we know the function f, we may judge how accurately the polynomial P approximates f. We also want to modify the values x_k – or to increase their number – to see if the approximation becomes more or less accurate.

We will therefore now study the accuracy of the approximation of a function f by the polynomial P that interpolates it at n points of an interval $[a, b]$ in its domain of definition.

We first present two small useful programs which create the graphs needed for our study. The first one shows the graph of f, of the interpolating polynomial and the interpolating points.

```
:plot1()
:Prgm
:FnOff:FnOn 1,2          the function f is saved in y1
:PlotsOff                the abscissas are in xx
:y1(xx)→yy               y ordinates f(x_k) in yy
:NewPlot 1,1,xx,yy       graph of the points X and Y
:DispG                   graph it!
:EndPrgm
```

Here is an example using the program plot1().

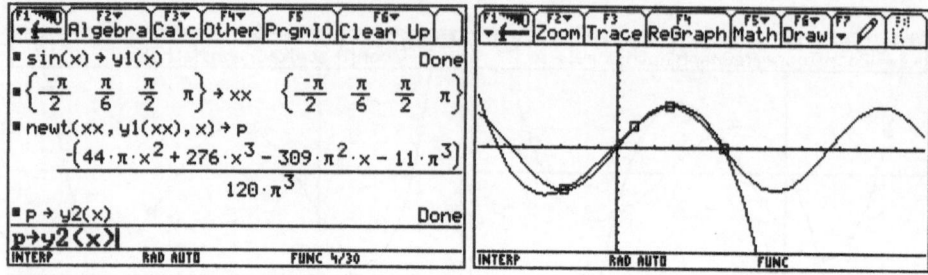

The program plot2 allows us to visualize more clearly the accuracy of the approximation by graphing the difference between f and its interpolating polynomial while limiting the window to the segment bounded by the extreme x-values.

```
:plot2()
:Prgm
:FnOff
:PlotsOff
:min(xx)→xmin:max(xx)→xmax     adjust the window
:y2(x)-y1(x)→y3(x)             difference between f and interpolating P
:ZoomFit                       zoom to fit
:EndPrgm
```

The graph plotted is that of the difference between f and its interpolating polynomial. We notice that even for a function as regular as $\sin x$, the approximation is not very good. (The error is of order 0.17). This program is rather slow.

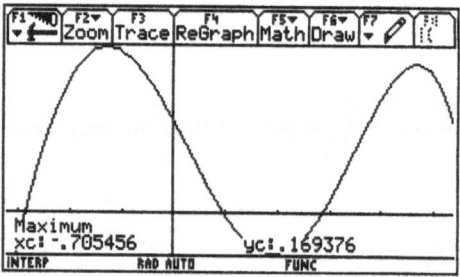

3. Interpolation by equally spaced x-values

In this paragraph, abscissas of interpolation will be equally distributed. The function xeq allows us to create a partition of n points of an interval $[a, b]$ with step size $h = \dfrac{b - a}{n}$.

```
:xeq(a,b,n)              interval [a,b], n:number of points
:Func
:Local h,x
:(b-a)/(n-1)→h           h : step
:seq(x,x,a,b+h/2,h)      creation of the partition
:EndFunc
```

Below we see approximation of $\cos x$ by its interpolating polynomials with 4, 5, 7 and 10 points equally spaced on $[0, 4\pi]$.

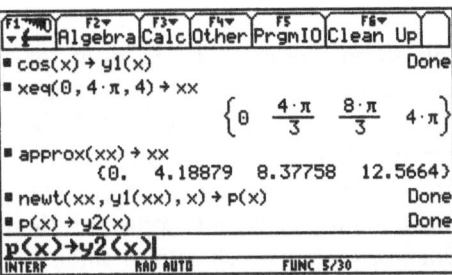

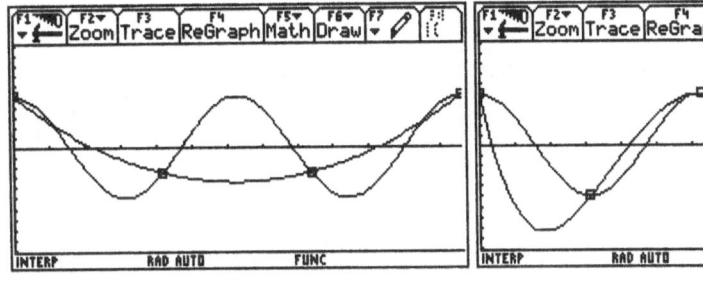

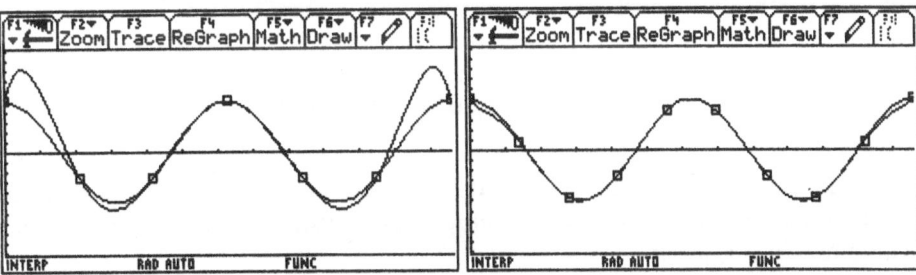

Note the graph of the error between f and its interpolating polynomial on 10 points. Although the approximation seems correct at the center of the interval, it is not so good on the edges.

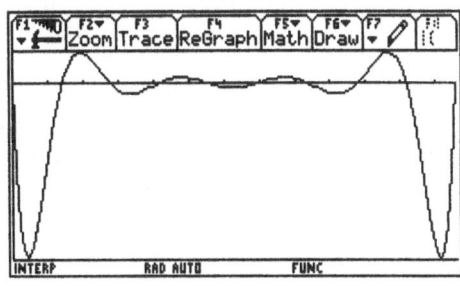

The German mathematician Carl Runge has shown clearly that the sequence of interpolating polynomials of a function f may not converge uniformly to f when one chooses equidistant abscissas, even when we increase the number of points of the partition. The example is the function $f : x \mapsto \frac{1}{1+x^2}$ on the interval $[-a, a]$. The behavior is reminiscent of the Gibbs phenomenon which is shown in the chapter on Fourier analysis.

The approximations of $\frac{1}{1+x^2}$ by its interpolating polynomials with $7, 10$ and 15 equally spaced points on $[-5, 5]$ are shown. Bumps near the edge of the interval do not go away, however. This phenomenon does not occur for the function $\cos(x)$ which is also smooth. Why?

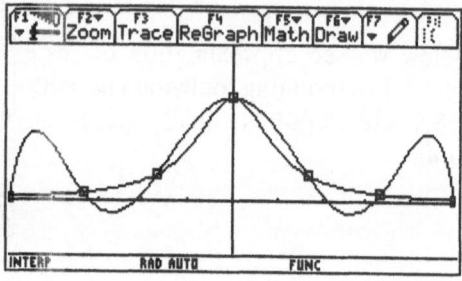

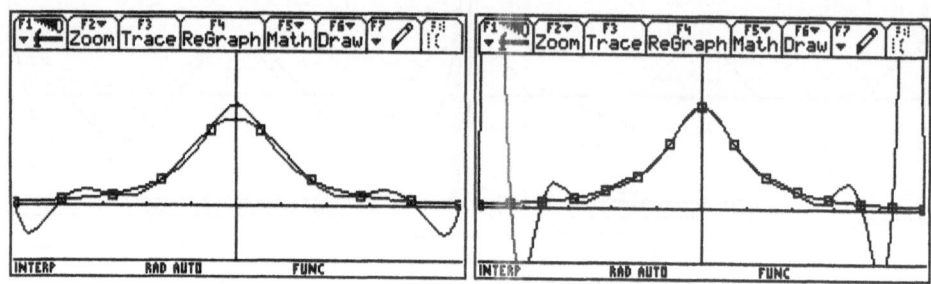

3. 1 Calculation of the interpolation error

Theorem 1: Let f be a function of class C^n on an interval $[a, b]$. Let P be its interpolating polynomial on a family S of n distinct points $(x_1, x_2, \ldots, x_n)$. Then, for all $x \in [a, b]$:

$$|f(x) - P(x)| \leq \frac{M_n}{n!}|(x - x_1)(x - x_2)\ldots(x - x_n)|$$

where $M_n = \sup_{x \in [a,b]} |f^{(n)}(x)|$.

Proof: Suppose x is distinct from the $(x_i)_i$. By the Newton identity, we know that:

$$f(x) - P(x) = [x, x_1, x_2, \ldots, x_n](x - x_1)(x - x_2)\ldots(x - x_n)$$

For all $t \in [a, b]$, let:

$$\Phi(t) = f(t) - P(t) - \lambda \prod_{k=1}^{n}(t - x_k)$$

with a real λ being chosen such that $\Phi(x) = 0$. The function Φ, of class C^n on $[a, b]$, vanishes $(n+1)$ times at $(x, x_1, \ldots, x_n)$. By Rolle's Theorem, there exists $\zeta(x) \in [a, b]$ (depending on x) such that $\Phi^{(n)}(\zeta(x)) = 0$, which, since the degree of P is less than or equal to $n - 1$, implies that:

$$\lambda = \frac{f^{(n)}(\zeta(x))}{n!}$$

By introducing this value of λ in the initial expression, we obtain the desired result. We likewise may show that there exists $\zeta(x) \in [a, b]$ such that:

$$[x, x_1, x_2, \ldots, x_n] = \frac{f^{(n)}(\zeta(x))}{n!}$$

Thus:

$$\sup_{x \in [a.b]} |f(x) - P(x)| \leq \frac{M_n}{n!} \sup_{x \in [a.b]} |(x - x_1)(x - x_2) \ldots (x - x_n)|$$

We see that to minimize the interpolation error, one may act on either:

- the real number $M_n = \sup_{x \in [a.b]} |f^{(n)(x)}|$. Hence, for $f : x \mapsto \dfrac{1}{1 + x^2}$, we have:

$$f(x) = \operatorname{Im}\left(\frac{1}{x - i}\right) \Rightarrow f^{(n)}(x) = -n! \operatorname{Im}\left(\frac{-1}{x - i}\right)^{n+1}$$

Therefore, $M_n = n!$ for even n and $M_n = n! \left(\cos \frac{\pi}{n+1}\right)^{n+1} \sim n!$ for odd n.

- or on the second factor, $\sup_{x \in [a.b]} |(x - x_1)(x - x_2) \ldots (x - x_n)| = \sup_{x \in [a.b]} |\Pi_S(x)|$.

This term is independent of f, and depends only on the choice of the family S of the n interpolating points.

Theorem 2: $\sup_{x \in [a.b]} |\Pi_S(x)| \geq \left(\dfrac{b - a}{2}\right)^n \dfrac{1}{2^{n-1}}$.

Proof: Suppose first that we are in the case where $[a, b] = [-1, 1]$.
For each $n \geq 0$, the function $T_n(x) = \cos(n \operatorname{Arccos}(x))$ is a polynomial of degree n with leading coefficient 2^{n-1}. Refer to the chapter on Orthogonality for properties of these polynomials, which are called the Chebishev polynomials. Obviously, for all $x \in [-1, 1]$ and for all $n \in \mathbb{N}, |T_n(x)| \leq 1$. It is easy to determine the zeros of T_n, since:

$$T_n(x) = 0 \Leftrightarrow \cos(n \operatorname{Arccos}(x)) = 0 \Leftrightarrow x_k = \cos\left(\frac{(2k + 1)\pi}{2n}\right), \quad (0 \leq k \leq n - 1)$$

Similarly:

$$T_n'(x) = \frac{n}{\sqrt{1-x^2}}\sin(n\operatorname{Arccos}(x)) = 0 \Leftrightarrow x = \alpha_k = \cos\left(\frac{k\pi}{n}\right),\ (1 \le k \le n-1)$$

The sequence $(\alpha_1, \alpha_2, \ldots, \alpha_{n-1})$ is an increasing sequence in $]-1, 1[$. Add to it $\alpha_0 = -1$ and $\alpha_n = 1$. We then obtain:

$$\forall\ k \in \{0, \ldots, n\}, T_n(\alpha_k) = \cos(k\pi) = (-1)^k$$

Let A_n be a polynomial of degree n with the same leading coefficient as that of T_n and such that for all $x \in [-1, 1], |A_n(x)| < 1$. The polynomial $A_n - T_n$ is of degree $(n-1)$ and:

$$\forall\ k \in \{0, \ldots, n\}, (A_n - T_n)(\alpha_k) = A_n(\alpha_k) - (-1)^k$$

This last expression alternates its sign. It follows that the polynomial $A_n - T_n$ vanishes at least once on each interval $]\alpha_{k+1}, \alpha_k[$, $(0 \le k \le n-1)$. It therefore has at least n zeroes, while its degree is strictly less than n. Therefore, it is identically zero. This is impossible, because for all $x \in [-1, 1], |A_n(x)| < 1$ and T_n alternately takes values -1 and 1.

Thus, among all polynomials P of degree n with leading coefficient 2^{n-1}, the polynomial T_n attains the minimum of $\sup_{x\in[-1,1]} |P(x)|$. This minimum is equal to 1. By multiplying the preceding result by $\frac{\lambda}{2^{n-1}}$, we return to the case of polynomials of degree n with leading coefficient λ. Thus:

$$\min_{P\ /P(x)=\lambda x^n+\ldots}\left(\sup_{x\in[-1,1]} |P(x)|\right) = \sup_{x\in[-1,1]} \frac{\lambda}{2^{n-1}}|T_n(x)| = \frac{|\lambda|}{2^{n-1}}$$

The change of variable $t = \frac{a+b}{2} + x\frac{b-a}{2}$ defines a one-to-one correspondence between functions g (of the variable x) defined on $[-1, 1]$ and functions f of the variable t) defined on $[a, b]$.

In particular, when $A(t)$ is a polynomial of degree n with leading coefficient 1 defined on $[a, b]$, the polynomial $B(x)$ which corresponds to it by the preceding correspondence is of degree n with leading coefficient $\left(\frac{b-a}{2}\right)^n$. Finally:

$$\sup_{t\in[a,b]} |A(t)| = \sup_{x\in[-1,1]} |B(x)| \ge \left(\frac{b-a}{2}\right)^n \frac{1}{2^{n-1}}$$

which proves the theorem. Morever, this inequality is an equality when B is proportional to the polynomial T_n, that is to say, when its roots are those of T_n. This accurs when roots of A are:

$$t_k = \frac{a+b}{2} + \frac{b-a}{2}\cos\left(\frac{(2k+1)\pi}{2n}\right), \ (1 \le k \le n)$$

Thus, when P is the interpolating polynomial of f at the preceding points, the inequality of the theorem becomes an equality, and:

$$\sup_{x \in [a,b]} |f(x) - P(x)| \le \frac{M_n}{n!} \left(\frac{b-a}{2}\right)^n \frac{1}{2^{n-1}}$$

Of course, for our example $(f(x) = \dfrac{1}{1+x^2}$ on $[-5,5])$, this only shows the uniform convergence of P to f on $[a,b]$ when $b - a < 4$. It is even possible to show that, whatever interpolation method you choose, there is a continuous function f such that the interpolating polynomial does not converge uniformly to f.

The next program calculates x-values for points defined by the Chebyshev polynomial of order n.

```
:xtcheb(a,b,n)
:Func
:approx((a+b)/2+(b-a)/2*seq(cos((2*k+1)*π/(2*n)),k,0,n-1))
:EndFunc
```

We show graphically that the sequence of interpolating polynomials of $f : x \mapsto \frac{1}{1+x^2}$ may converge to f on $[-5,5]$, although we have not demonstrated it formally. In the following screenshots, different interpolating polynomials at the Chebishev points are shown ($n = 7, 11$)

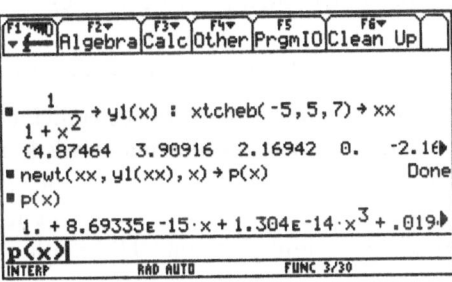

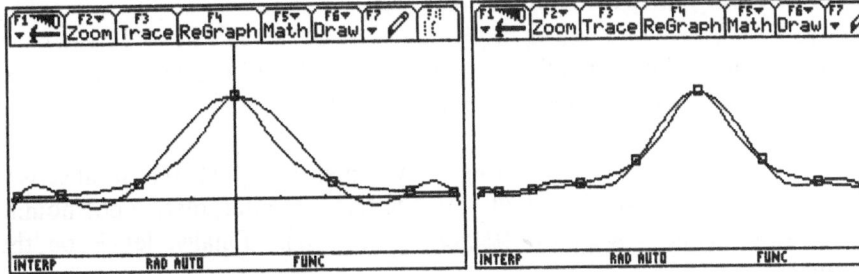

The x-values of the Chebishev points are more dense near the endpoints of the interval $[-5,5]$, which has suppressed the Runge phenomenon that appeared with equally spaced abscissas.

4. Hermite interpolation

Let f be a function which is differentiable on an interval $[a, b]$, and let n distinct points $x_1, x_2, \ldots, x_n$ of $[a, b]$ be given.

With Hermite interpolation we determine an interpolating polynomial P for f at the points $x_1, x_2, \ldots, x_n$ and which also satisfies the n equalities $P'(x_1) = f'(x_1), P'(x_2) = f'(x_2), \ldots P'(x_n) = f'(x_n)$.

The polynomial P, which must thus satisfy $2n$ conditions, is thus to be sought among polynomials of degree less than or equal to $2n - 1$, which indeed corresponds to $2n$ unknown coefficients.

Proposition 1: The problem just stated has a unique solution.

Proof: It suffices to consider the following map:

$$\Phi : \begin{array}{ccc} \mathbb{R}_{2n-1}[X] & \to & \mathbb{R}^{2n} \\ P & \to & (P(x_1), \ldots, P(x_n), P'(x_1) \ldots, P'(x_n)) \end{array}$$

The map Φ, which is clearly linear, is an isomorphism of vector spaces: indeed the dimensions of the vector spaces considered are the same; on the other hand, if P is in the kernel of Φ, then P has n distinct roots, all of multiplicity 2, so it has at least $2n$ roots. The hypothesis on the degree of P implies that P is the zero polynomial. In particular, there exists a unique polynomial P of degree less or equal to $2n - 1$ satisfying the equation: $\Phi(P) = (f(x_1), \ldots, f(x_n), f'(x_1), \ldots, f'(x_n))$, that is to say, which achieves the Hermite interpolation of f on the n abscissas $x_1, x_2, \ldots, x_n$.

There are several more or less efficient methods to determine P. The method of undetermined coefficients consists of proceeding as we have done with the function `interpol`: we write the polynomial P in its most general form (with $2n$ unknown coefficients) and we solves the system formed with the $2n$ conditions for Hermite interpolation. This method is quite difficult and we will not develop it here.

4. 1 From Lagrange to Hermite

Let f be a differentiable function on the interval $[a, b]$, and let $x_1, x_2, \ldots, x_n$ be n distinct points of this interval. For all $1 \leq k \leq n$, we let $y_k = f(x_k)$ and $z_k = f'(x_k)$.

Let L be the Lagrange interpolating polynomial for f at the n abscissas $x_1, x_2, \ldots, x_n$ ($\deg(L) \leq n-1$), and let H be the Hermite interpolating polynomial of f for these same n abscissas ($\deg(H) \leq 2n - 1$). Finally, let P be the polynomial:

$$P(X) = \prod_{k=1}^{n} (X - x_k)$$

The two polynomials L and H take the same value at each of the n abscissas $x_1, x_2, \ldots, x_n$: therefore there is a polynomial Q of degree less than or equal to $n-1$ such that $H - L = PQ$.

By differentiating and substituting one of the values x_k, we get: $z_k - L'(x_k) = P'(x_k)Q(x_k)$.

But for each integer k, $P'(x_k) = \prod_{j \neq k}(x_k - x_j)$. Knowing the polynomial L, we deduce the value of the $Q(x_k)$, which allows us to determine Q as an interpolating polynomial. Having thus obtained L and Q (by one of procedures that are already known), we concludes by writing $H = L + PQ$.

The following program, called Hermite, uses this method.

```
:hermite()
:Prgm
:Local y,i,p
:newt(xx,y1(xx),x)→i              xx: interpolating points
:product(x-xx)→p                  P(x)
:d(y1(x)-i,x)/(d(p,x))|x=xx→y
:i+p*newt(xx,y,x)→p(x)
:p(x)→y2(x)
:EndPrgm
```

Here is the Hermite interpolating polynomial P on the 6 points $(1, 1.5, 2, 2.5, 3)$ for the function $f : x \mapsto \ln x$. In the two next screenshots are the graph of $f - P$, then a zoom that allows us to evaluate the interpolation error.

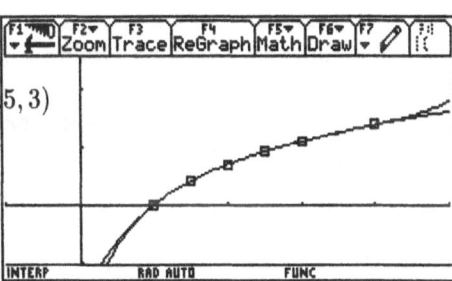

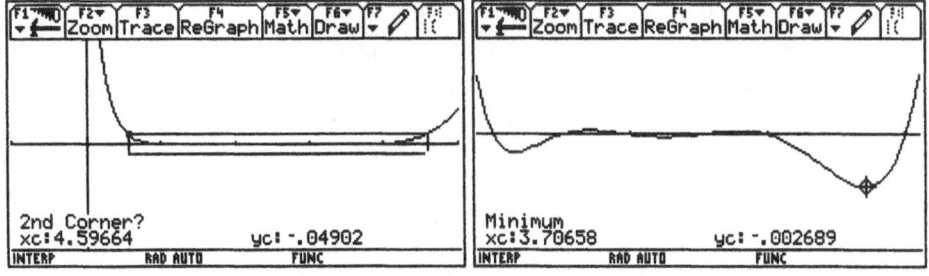

4. 2 Chebishev to the rescue of Hermite

As we did with the "classic" polynomial interpolation, we are now going to study the quality of the approximation of a function f by its Hermite interpolating polynomial at n points of an interval $[a, b]$ within its domain of definition. We will see that, in complete generality, the best choice of these n points is to take the abscissas of Chebishev which we saw previously. To demonstrate this, it is necessary to assume that the function f is $2n$ times continuously differentiable on the interval $[a, b]$.

Denote by S the family $(x_1, x_2, \ldots, x_n)$ of the n interpolating x-values, and by H_S the corresponding Hermite interpolating polynomial.

Proposition 2: For all x in $[a, b]$, there exists $\zeta(x) \in [a, b]$ such that:

$$f(x) - H_S(x) = \frac{f^{(2n)}(\zeta(x))}{(2n)!} \prod_{k=1}^{n} (x - x_k)$$

Proof: The equation is evident if x is one the x_k (since then $f(x_k) = H_S(x_k)$). We may therefore suppose that $x \notin (x_k)_k$ and define a function Φ, of class C^{2n} on $[a, b]$, like f, by:

$$\Phi(t) = f(t) - H_S(t) - \lambda \prod_{k=1}^{n} (t - x_k)^2$$

where the real λ is chosen such that $\Phi(x) = 0$. The function Φ thus vanishes at $n + 1$ distinct points which define n consecutive sub-intervals $I_1, I_2, \ldots, I_n$ of $[a, b]$.

By Rolle's Theorem, we deduce the existence of n x-values $a_1, a_2, \ldots, a_n$ such that $\Phi'(a_k) = 0$, with each a_k belonging to the corresponding sub-interval I_k. The $a_1, a_2, \ldots, a_n$ are therefore distinct from the n abscissas $x_1, x_2, \ldots, x_n$.

On the other hand, by construction, the derivative of Φ vanishes again at the points $x_1, x_2, \ldots, x_n$. This follows from the fact that f' and H' take the same value at each of these points, and that the derivative of $\prod_{k=1}^{n} (x - x_k)^2$ vanishes there.

Thus Φ is zero at $2n$ distinct points of $[a, b]$ that are $(a_1, \ldots, a_n, x_1, \ldots, x_n)$. By a repeated application of Rolle's Theorem, we conclude that Φ'' vanishes at $2n - 1$ distinct points, then that $\Phi^{(3)}$ vanishes at $2n - 2$ distinct points, etc. By recursion, we arrive at the existence of a point $\zeta(x) \in [a, b]$ such that $\Phi^{(2n)}(\zeta(x)) = 0$. Considering that the $2n$th derivative of H_S is zero (since its degree is less than or equal to $2n - 1$) and that of the polynomial $\prod_{k=1}^{n} (x - x_k)^2$ is equal to $(2n)!$, the condition $\Phi^{(2n)}(\zeta(x)) = 0$ implies that $f^{(2n)}(\zeta(x)) = \lambda(2n)!$

By introducing the resulting expression of λ in the equation $f(x) = 0$, one gets to the desired result.

Denote $M_{2n} = \sup\limits_{x \in [a,b]} |f^{(2n)}(x)|$. We then have, for all $x \in [a,b]$:

$$|f(x) - H_S(x)| \le \frac{M_{2n}}{(2n)!} \prod_{k=1}^{n} (x - x_k)^2 \le \frac{M_{2n}}{(2n)!} \sup_{x \in [a,b]} A_S^2(x)$$

by denoting $A_S(x) = \prod\limits_{k=1}^{n} (x - x_k)$.

We know that the quantity $\sup\limits_{x \in [a,b]} |A_S(x)|$ is optimal among all polynomials A of degree n with leading coefficient 1 when the roots of A are the n–Chebishev abscissas in $[a,b]$. Also, the minimum is equal to: $\left(\dfrac{b-a}{2}\right)^n \dfrac{1}{2^{n-1}}$. If we denote by T the family of the n–Chebishev values in the interval $[a,b]$, and by H_T the corresponding Hermite interpolating polynomial, then:

$$\sup_{x \in [a,b]} |f(x) - H_S(x)| \le \frac{M_{2n}}{(2n)!} \left(\frac{b-a}{2}\right)^{2n} \frac{1}{4^{n-1}}$$

As in our discussion of the Runge phenomenon, this does not constitute a proof of the uniform convergence of the polynomials H_S to f, but we may observe that the Chebishev abscissas make a very good choice. In order to do that, reconsider the example that we used to illustrate the Runge phenomenon.

Below, on the left, we graph of the Hermite interpolating polynomials for the function $x \mapsto \frac{1}{1+x^2}$ for 5, then 9 equally spaced points of $[-5,5]$. Screenshots on the right correspond to the graph of the difference $f - H_S$. The approximation is not very good.

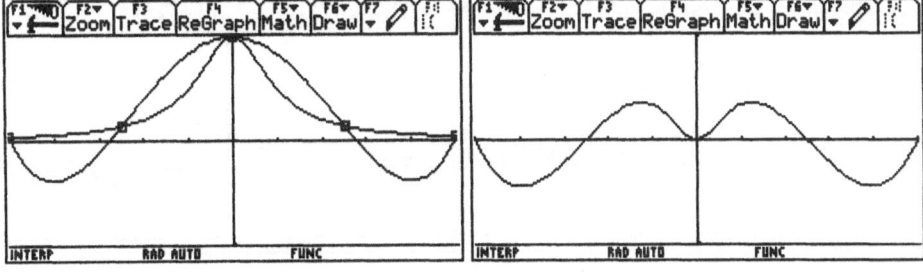

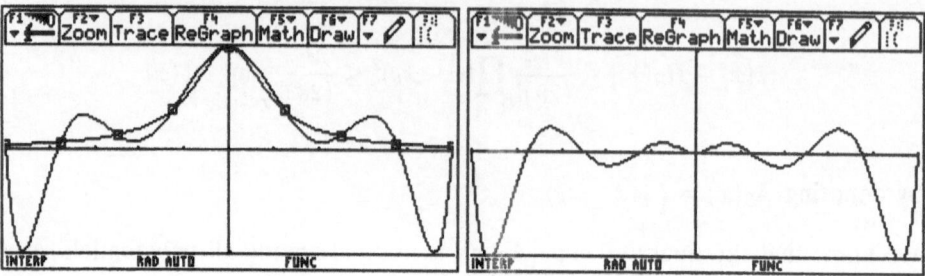

Here are the graphs of the Hermite interpolating polynomials for the same function for 5, 9, and then 13 Chebishev abscissas of $[-5, 5]$. Screenshots on the right show the difference $f - H_S$. The approximation is clearly much better!

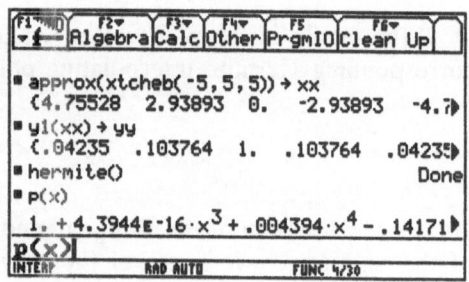

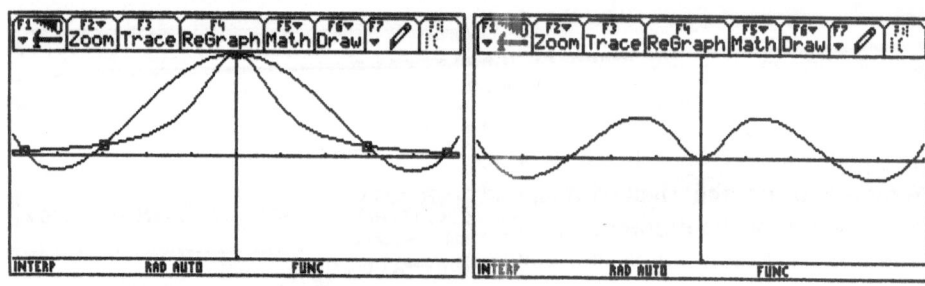

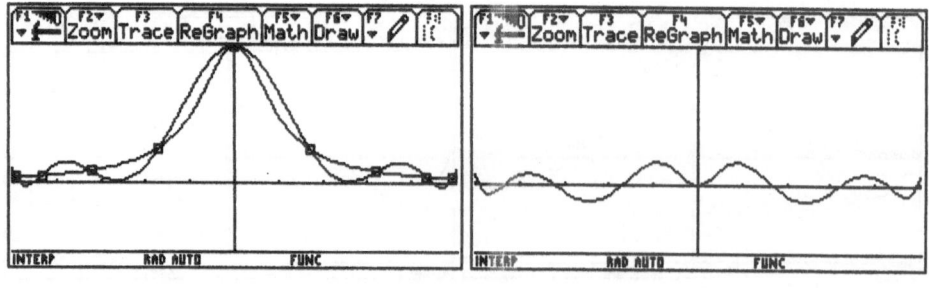

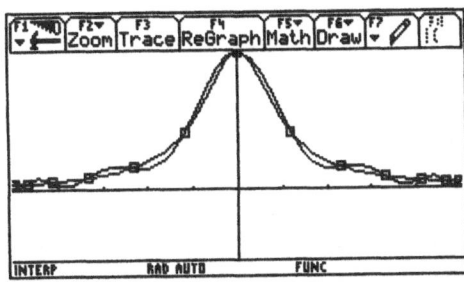
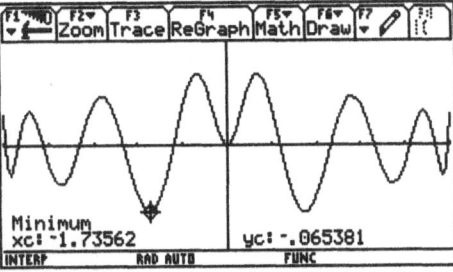

4. 3 Return to the divided differences

We know how to calculate the divided differences of $(x_1, x_2, \ldots, x_n)$ and $(y_1, y_2, \ldots, y_n)$. For example, the function ddi calculates the sequence of divided differences d_k corresponding to the k first elements of each of these two sequences. Let us see now how this notion may be extended to the case where the values $(x_1, x_2, \ldots, x_n)$ are not pairwise distinct.

In order to do that, we are going to examine the divided differences on very close x-values $(x_1, x_2, \ldots, x_n)$ with the values $(y_1, y_2, \ldots, y_n)$ being images of $(x_1, x_2, \ldots, x_n)$ under f, where f is a sufficiently differentiable application. We will then take to the limit as these x-values approach unique value a, and we will see how the limit value of these successive divided differences are expressed according to this common limit value.

Note the divided difference calculation and the passage to the limit when h tends to 0. Using an example, the limit seems to be $f(a)$, $f'(a)$, $f''(a)/2$, $f^{(3)}(a)/6$.

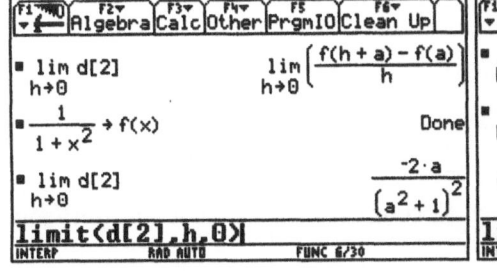

Proposition 3: Let f be a function of class C^n on a neighborhood of a point a. Then:

$$\lim_{h\to 0}[a, a+h, a+2h, \ldots, a+nh] = \frac{f^{(n)}(a)}{n!}$$

where $[a, a+h, a+2h, \ldots, a+nh]$ designates the sequence of divided differences of the points $(a+kh, f(a+kh))_{0\le k\le n}$.

Proof: We saw that, when f is a function of class C^n on an interval $[a, b]$, then for all $(x_1, x_2, \ldots, x_{n+1})$ of $[a, b]$, there exists a point ζ belonging to the interval determined by the points $(x_1, x_2, \ldots, x_{n+1})$ such that:

$$[x_1, x_2, \ldots, x_{n+1}] = \frac{f^{(n)}(\zeta)}{n!}$$

One ends the demonstration thanks to the continuity of $f^{(n)}$ when h tends to 0. One can therefore define the divided difference of $(a, a, \ldots, a)$ and $(f(a), f(a), \ldots, f(a$ (each sequence being length $n+1$): if f is of class C^n on a neighboorhood of a, one puts:

$$[a, a, \ldots, a] = \frac{f^{(n)}(a)}{n!}$$

It remains to modify the function ddi in order that it accepts a sequence of abscissas not necessarily distinct.

```
:ddi2(x)
:Func
:Local n,j,k,y
:dim(x)→n
:y1(x)→y
:For k,1,n-1
:For j,n,k+1,-1
:If when(x[j]=x[j-k],true,false,false) Then
:d(y1(x),x,k)/(k!)|x=x[j]→y[j]
:Else
:(y[j]-y[j-1])/(x[j]-x[j-k])→y[j]
:EndIf
:EndFor
:EndFor
:y
:EndFunc
```

The two screenshots following allow us to compare the two functions ddi and ddi2

```
┌─────────────────────────────────┐  ┌──────────────────────────────────────┐
│F1▼  F2▼  F3▼  F4▼  F5   F6▼      │  │F1▼  F2▼  F3▼  F4▼  F5   F6▼           │
│▼≠─│Algebra│Calc│Other│PrgmIO│Clean Up│ │▼≠─│Algebra│Calc│Other│PrgmIO│Clean Up│
├─────────────────────────────────┤  ├──────────────────────────────────────┤
│■ ln(x) → y1(x)            Done   │  │■ lim d → d1                           │
│■ {1  2-h  2  2+h  3} → xx        │  │  h→0                                  │
│        {1  2-h  2  h+2  3}       │  │ { 0  ln(2)  -(2·ln(2)-1)/2  8·ln(2)-5/8 ▶│
│■ ddi(xx, y1(xx)) → d             │  │■ ddi2({1  2  2  2  3}) → d2           │
│ {     -ln(-(h-2))   ln(-(h-2))+(h-1)·│ │ { 0  ln(2)  1/2-ln(2)  ln(2)-5/8  4·1 ▶│
│  0    ──────────   ──────────────── ▶│ │■ d1 - d2               {0 0 0 0 0}    │
│         h-1          h·(h-1)     │  │                                        │
├─────────────────────────────────┤  ├──────────────────────────────────────┤
│ddi(xx,y1(xx))→d                  │  │d1−d2                                  │
│INTERP    RAD AUTO    FUNC 3/30   │  │INTERP    RAD AUTO    FUNC 6/30        │
└─────────────────────────────────┘  └──────────────────────────────────────┘
```

4. 4 Polynomial interpolation "à la carte"

We resume the study in the preceding paragraph, starting with a function f and four abscissas x_1, x_2, x_3, x_4, all distinct at first. Then we consider what to do if the abscissas are not all distinct. We know that, in the case where these abscissas are distinct, the interpolating polynomial of f is written in its Newton form:

$$P(X) = d_{1,1} + d_{1,2}(X - x_1) + d_{1,3}(X - x_1)(X - x_2) + d_{1,4}(X - x_1)(X - x_2)(X - x_3)$$

It is interesting to note that this expression is valid when x_1, x_2, x_3, x_4 are not all distinct (by a passage to the limit).

Especially, if one resumes the three particular cases envisaged in the preceding paragraph, one obtains the following interpretations of the polynomial P.

• if $x_1 = x_2$ and $x_3 = x_4$, with $x_1 \neq x_3$, P has a contact of order 2 at x_1 and order 2 at x_3 with f, which means that $P(x_1) = f(x_1), P'(x_1) = f'(x_1)$, and $P(x_3) = f(x_3), P'(x_3) = f'(x_3)$. In others words, P is the Hermite interpolating polynomial of f for the abscissas x_1 and x_3.

• if $x_1 = x_2 = x_3$ and $x_1 \neq x_4$, P has a contact of order 3 at x_1 and order 1 in x_4 with f, which means that $P(x_1) = f(x_1), P'(x_1) = f'(x_1), P''(x_1) = f''(x_1)$, and $P(x_4) = f(x_4)$.

• if $x_1 = x_2 = x_3 = x_4$, P has a contact of order 4 at x_1 with f: $P(x_1) = f(x_1), P'(x_1) = f'(x_1), P''(x_1) = f''(x_1), P^{(3)}(x_1) = f^{(3)}(x_1)$. In this case, P is:

$$P(X) = f(x_1) + f'(x_1)(X - x_1) + \frac{f''(x_1)}{2!}(X - x_1)^2 + \frac{f^{(3)}(x_1)}{3!}(X - x_1)^3$$

In other words, P is the Taylor expansion of f at x_1 to the order 3.

We have in hand all the tools needed to write a function for calculation of the interpolating polynomial of a sufficiently regular function f, with any order of contact desired. In fact, it suffices to modify slightly the function newton.

```
:newt2(x,w)
:Func
:Local n,d
:dim(x)→n
:ddi2(x)→d
:∑(d[k]*∏(w-x[j],j,1,k-1),k,1,n)
:EndFunc
```

In this example, we interpolate $x \mapsto \cos x$ on $(0, \pi, \pi, 5\pi/2)$. The point π being repeated twice, we ask there a contact of order 2.

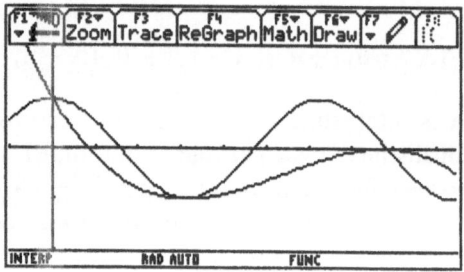

Here is the result, showing the graph of f and the interpolating polynomial, as well as the double contact at π.

The method that we just explained allows us to practice polynomial interpolation "à la carte", since for each abscissa one may choose the precision of the contact between the initial function f and the interpolating polynomial P. To do that, it suffices to repeat as many time as necessary each of the desired abscissas in order to obtain a higher order contact.

Recall that Hermite interpolation consists of requiring a contact of order 2 for each of the abscissas. In fact, one may generalize this idea – and it is the most widespread acceptance of the concept of Hermite interpolation – by fixing an integer m and by demanding that the following $n \times (m+1)$ conditions are realized:

$$\text{For all } 0 \le j \le m, \ \forall \ 1 \le k \le n, \ P^{(j)}(x_k) = f^{(j)}(x_k)$$

In other words, we ask for the polynomial P to take the same value that as the function f at each of the n points $x_1, \ldots, x_n$, as well as for their derivatives up to the m-th one.

As we have done for Lagrange interpolation and then for the Hermite form, we may show that the problem for precision m has an unique solution among polynomials of degree less than or equal to $n(m+1) - 1$.

```
:hermiteg(x,w,m)
:Func
:mat▸list((seq(x,k,0,m)))ᵀ→x
:newt2(x,w)
:EndFunc
```

Here are the definition of f and of the interpolating points. We have asked for contact of order 3 (function, first and second derivatives).

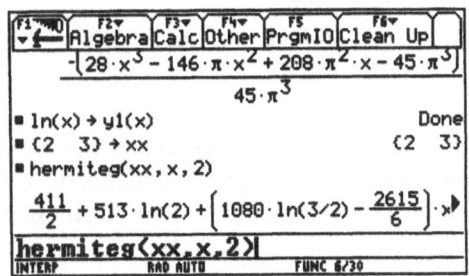

And this is the result, with the graph of f and the interpolating polynomial

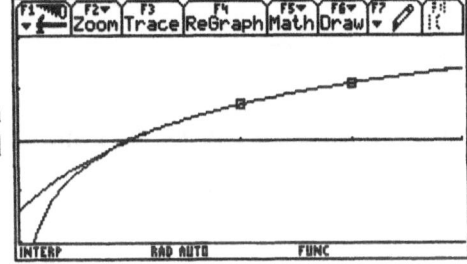

5. Bézier curves

Until now, we have tried to interpolate a function f on a interval $[a, b]$ by different polynomials in a more or less regular way. In most cases, the approximation of f by P was not satisfactory. We are therefore now going to search for a polynomial that approaches f on $[a, b]$ in a uniform manner, and we will draw upon a new idea for the interpolating problem. Let's begin with a very famous theorem.

Theorem 1:(The Weierstrass Approximation Theorem) The set $\mathbb{R}[X]$ of polynomials with real coefficients is dense in the vector space $C^0([a, b], \mathbb{R})$ of real valued continuous functions on $[a, b]$, using the norm of uniform convergence.

Another formulation of the theorem is:

Let f be an element of $C^0([a, b], \mathbb{R})$, and suppose $\varepsilon > 0$ is given. Then there exists a polynomial $P \in \mathbb{R}[X]$ such that:

$$\sup_{x \in [a,b]} |f(x) - P(x)| < \varepsilon$$

Proof: We may assume that $[a, b] = [0, 1]$. The demonstration of the general case follows from a bijection between $[0, 1]$ and $[a, b]$.

If $f \in C^0([0, 1], \mathbb{R})$, we put, for all $n \in \mathbb{N}$ and for all $x \in [0, 1]$:

$$B_n(f, x) = \sum_{k=0}^{n} \binom{n}{k} f\left(\frac{k}{n}\right) x^k (1 - x)^{n-k}$$

This sequence of polynomials, called the Bernstein polynomials, converges uniformly to f on $[0, 1]$. Indeed, if $x \in [0, 1]$ and if X is a random variable following a binomial law $\mathcal{B}(n, x)$, then:

- $\sum_{k=0}^{n} \binom{n}{k} x^k (1 - x)^{n-k} = 1$

- $\sum_{k=0}^{n} k \binom{n}{k} x^k (1 - x)^{n-k} = E(X) = nx$ (the expectation of X)

- $\sum_{k=0}^{n} (k - nx)^2 \binom{n}{k} x^k (1 - x)^{n-k} = V(X) = nx(1 - x)$ (the Variance of X)

Let $n \in \mathbb{N}, x \in [0, 1]$ and $\delta > 0$. Define

$$A_n = \left\{ k \in \mathbb{N}, | \, 0 \leq k \leq n, \left| \frac{k}{n} - x \right| \geq \delta \right\}$$

Then:

$$\sum_{k \in A_n} \binom{n}{k} x^k (1 - x)^{n-k} \leq \frac{1}{4n\delta^2}$$

It suffices to apply the Chebishev inequality because

$$\sum_{k \in A_n} \binom{n}{k} x^k (1 - x)^{n-k} = p(|X - E(X)| \geq n\delta) \leq \frac{V(X)}{n^2\delta^2} = \frac{x(1 - x)}{n\delta^2 2} \leq \frac{1}{4n\delta^2}$$

We finish the demonstration by using the uniform continuity of f on $[0, 1]$.

$$\forall \, \varepsilon > 0, \exists \, \delta > 0, \forall (x, x') \in [0, 1]^2, (|x - x'| < \delta \Rightarrow |f(x) - f(x')| < \varepsilon)$$

Finally, if $\mathcal{B}_n$ designates the complement of A_n in $\{1, 2, \ldots, n\}$:

$$|B_n(f,x) - f(x)| \leq \sum_{k=0}^{n} \binom{n}{k} \left| f\left(\frac{k}{n}\right) - f(x) \right| x^k (1-x)^{n-k}$$

$$= \sum_{k \in A_n} \binom{n}{k} \left| f\left(\frac{k}{n}\right) - f(x) \right| x^k (1-x)^{n-k}$$

$$+ \sum_{k \in B_n} \binom{n}{k} \left| f\left(\frac{k}{n}\right) - f(x) \right| x^k (1-x)^{n-k}$$

$$\leq 2 \sup_{x \in [0,1]} |f(x)| \sum_{k \in A_n} \binom{n}{k} x^k (1-x)^{n-k} + \varepsilon \sum_{k \in B_n} \binom{n}{k} x^k (1-x)^{n-k}$$

$$\leq 2M \frac{1}{4n\delta^2} + \varepsilon$$

This shows that the sequence of functions $(x \mapsto (B_n(f,x))_n$ converges uniformly on $[0,1]$ to f.

Here is the direct translation of the definition of the Bernstein polynomials.

```
:bernst(n)
:Func
:Local b
:∑(nCr(n,k)*f(k/n)*x^k*(1-x)^(n-k),k,0,n)→b(n,x)
:EndFunc
```

The definition of the function f and calculation of two Bernstein polynomials.

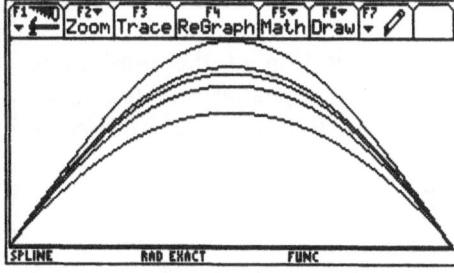

The graph of the function f and the Bernstein polynomials of order $3, 5, 7$ and 9 are shown. Although convergence of the polynomials $B_n(f)$ to f is uniform on $[0,1]$, it is extremely slow.

A *Bézier* curve of degree n defined by $(n+1)$ points $(a_0, a_1, \ldots, a_n)$ is just the Bernstein polynomial $\sum_{k=0}^{n} a_k \binom{n}{k} x^k (1-x)^{n-k}$. In fact, because they are sufficiently regular, in practice one only uses polynomials of degree 3, and therefore curves defined by two basis points (a_0, a_3) and two control points (a_1, a_2). Thus, the equation of a Bézier curve of the plane will be of the form:

$$\begin{cases} x(t) = a_0(1-t)^3 + 3a_1 t(1-t)^2 + 3a_2 t^2(1-t) + a_3 t^3 \\ y(t) = b_0(1-t)^3 + 3b_1 t(1-t)^2 + 3b_2 t^2(1-t) + b_3 t^3 \end{cases}$$

We will study some of their properties and their generalization, but we will also point out that these curves are both continuous (they are polynomials) and smooth. Here is a small program which graphs a Bézier curve for 4 points whose coordinates are stored in lists **xx** and **yy**.

```
:bezier()
:Prgm
:PlotsOff : FnOff
:NewPlot 1,2,xx,yy                      lists of the base points
:seq(nCr(3,k)*t^k*(1-t)^(3-k),k,0,3)→l(t)
:sum(yy*l(t))→yt1(t)                    parametric function
:sum(xx*l(t))→xt1(t)                    to be graphed
:DispG
:EndPrgm
```

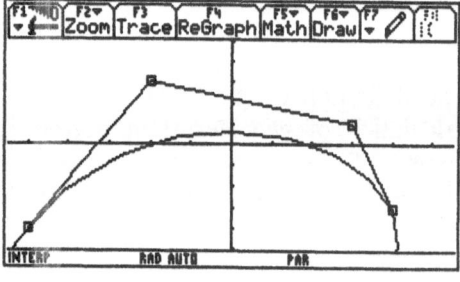

The Bézier curve is shown with the convex hull determined by the 4 points A, B, C, D. It is tangent to the lines (BA) and (CD) at A and D. Points B and C act as control points.

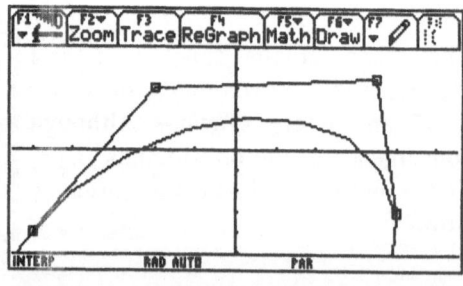

The modification of the third point changes the appearance of the curve. In practice, Bézier curves are used to smooth curves obtained from a family of points.

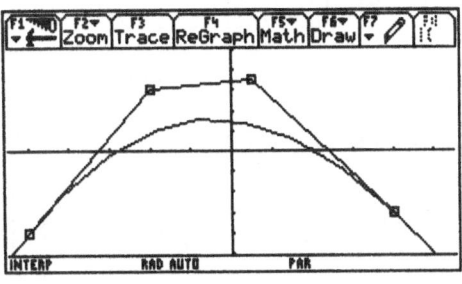

The third point has been moved again. The aspect of the curve is modified accordingly.

5. 1 The Casteljau algorithm

There is a recursive algorithm allowing us to calculate the Bézier curves (or Bernstein polynomials).

Proposition 1: If we puts for all $n \in \mathbb{N}$ and for all i such that $0 \le i \le n$, $B_{n,i}(x) = \binom{n}{i} x^i (1-x)^{n-i}$, then by supposing that $B_{n-1,-1} = B_{n-1,n} = 0$, for all $i \in \{0, 1, \ldots, n\}$, and for all real x, we have:

$$B_{n,i}(x) = (1-x)B_{n-1,i}(x) + x B_{n-1,i-1}(x)$$

Proof:
- For $i = 0$, to show that $B_{n,0}(x) = (1-x)B_{n-1,0}(x) + x B_{n-1,-1}(x)$, let:

$$(1-x)^n = (1-x)(1-x)^{n-1} + 0 \cdot x$$

- For $1 \le i \le n-1$, we must show that:

$$\binom{n}{i} x^i (1-x)^{n-i} = (1-x)\binom{n-1}{i} x^i (1-x)^{n-i} + x \binom{n-1}{i-1} x^{i-1}(1-x)^{n-i}$$

which is verified thanks to the relationship $\binom{n}{i} = \binom{n-1}{i} + \binom{n-1}{i-1}$.
- For $i = n$, we just show that $x^n = (1-x) \cdot 0 + x x^{n-1}$.

The following program implements this algorithm.

```
:castelj(l,x)                    lists of data, variable
:Func
:Local k,j,n
:dim(l)→n
:For k,2,n
:For j,1,n-k+1
:(1-x)*l[j]+x*l[j+1]→l[j]        recursive relation
:EndFor
:EndFor
:l[1]
:EndFunc
```

F1 Algebra Calc Other PrgmIO Clear a-z...

- f(x) $\sin(\pi \cdot x)$
- bernst(3) $\dfrac{-3 \cdot (x-1) \cdot x \cdot \sqrt{3}}{2}$
- castelj$\left(\text{seq}\left(f\left(\frac{k}{3}\right), k, 0, 3\right), x\right)$
                                          $\dfrac{-3 \cdot \sqrt{3} \cdot x \cdot (x-1)}{2}$

castelj(seq(f(k/3),k,0,3),x)
SPLINE   ∂ RAD EXACT   FUNC 3/30

F1 Algebra Calc Other PrgmIO Clear a-z...

- f(x) $\sin(\pi \cdot x)$
- castelj$\left(\text{seq}\left(f\left(\frac{k}{4}\right), k, 0, 4\right), x\right)$
  $-2 \cdot x \cdot (x-1) \cdot \left((2 \cdot \sqrt{2}-3) \cdot x^2 - (2 \cdot \sqrt{2}-3) \cdot x + \blacktriangleright\right.$
- expand(bernst(4), x)
  $-2 \cdot (x-1) \cdot x \cdot \left((2 \cdot \sqrt{2}-3) \cdot x^2 - (2 \cdot \sqrt{2}-3) \cdot x\right) - \blacktriangleright$

expand(bernst(4),x)
SPLINE   RAD EXACT   FUNC 3/30

6. Spline functions

Bézier curves furnish a new idea for interpolating a function f on an interval $[a, b]$. It consists of using several polynomials placed end to end, with one polynomial P_k for each interval $[x_k, x_{k+1}]$. It is also necessary that the function S that results from this piecewise definition is sufficiently regular.

6. 1 A first example

Let's begin with the classic example of the cubic spline. Let F be a family of n points $(x_1, y_1), (x_2, y_2), \ldots, (x_n, y_n)$, with $x_1 < x_2 < \ldots < x_n$. A cubic spline for the family F is a function S of class C^2 on $[x_1, x_n]$, that is, a polynomial of degree at most 3 on each interval $[x_k, x_{k+1}], (1 \leq k \leq n-1)$, and which satisfies $S(x_k) = y_k$ for all $1 \leq k \leq n$.

If f is a function defined on a interval $[a, b]$, one may interpolate f by a cubic spline function by choosing a partition $\sigma = (a = x_1, x_2, \ldots, x_n = b)$ of $[a, b]$: the ordinates are then obviously the $y_k = f(x_k)$, $1 \leq k \leq n$. The restriction of S to each of the $n-1$ intervals $[x_k, x_{k+1}]$ is a polynomial of degree at most 3, which therefore depends on four unknowns; the total number of unknown is thus $4(n-1)$.

The conditions $S(x_k) = y_k$ impose 2 linear constraints for each of the $n-1$ polynomials, which adds $2(n-1)$ constraints to the total.

The function S must be class of C^2 at each from points $x_2, \ldots, x_{n-1}$, which means that the second and first derivatives, at the left and at the right, have to coincide in each of these $n-2$ points. That translates into another $2(n-2)$ linear constraints.

Thus, the problem is: $4n-4$ unknowns and $4n-6$ linear equations. In principle, two additional conditions are necessary to determine the function S in a unique manner. These two supplementary conditions are most often chosen in one of these three ways:

- $CS_1 : S''(x_1) = S''(x_n) = 0$.

- $CS_2 : S'(x_1) = S'(x_n)$ and $S''(x_1) = S''(x_n)$.
- $CS_3 : S'(x_1) = f'(x_1)$ and $S'(x_n) = f'(x_n)$, when one seeks to interpolate a function f.

As an example, we plan to undertake the interpolation of the function $x \mapsto \cos(x)$ at the three points $(0, \pi/2, 3\pi/2)$ by means of a cubic spline. We will therefore need to find the two polynomials P_1 and P_2 of degree less or equal to 3, corresponding to this interpolation on the two intervals $[0, \pi/2]$ and $[\pi/2, 3\pi/2]$. The supplementary condition will be here CS_3.

We placed the function to be interpolated by a cubic spline in y1 (x). We get the 3 points, and the general form of the two polynomials to be determined. The following screenshots correspond to the equations obtained from the interpolation conditions.

```
F1 F2 F3 F4 F5 F6
   Algebra Calc Other PrgmIO Clean Up
 cos(x) → y1(x)                                    Done
 0 → x1 :  π/2 → x2 :  3·π/2 → x3           3·π/2
 a1·x³ + b1·x² + d1·x + e1 → p1
                      a1·x³ + b1·x² + d1·x + e1
 a2·x³ + b2·x² + d2·x + e2 → p2
                      a2·x³ + b2·x² + d2·x + e2
a2*x^3+b2*x^2+d2*x+e2→p2
INTERP          RAD AUTO              FUNC 4/30
```

```
F1 F2 F3 F4 F5 F6
   Algebra Calc Other PrgmIO Clean Up
 p1 = y1(x) | x = x1 → eq1                  e1 = 1
 d/dx(p1) = d/dx(y1(x)) | x = x1 → eq2       d1 = 0
 p1 = y1(x) | x = x2 → eq3
     a1·π³/8 + b1·π²/4 + d1·π/2 + e1 = 0
p1=y1(x)|x=x2→eq3
INTERP          RAD AUTO              FUNC 3/30
```

```
F1 F2 F3 F4 F5 F6
   Algebra Calc Other PrgmIO Clean Up
 p2 = y1(x) | x = x2 → eq4
     a2·π³/8 + b2·π²/4 + d2·π/2 + e2 = 0
 p2 = y1(x) | x = x3 → eq5
     27·a2·π³/8 + 9·b2·π²/4 + 3·d2·π/2 + e2 = 0
p2=y1(x)|x=x3→eq5
INTERP          RAD AUTO              FUNC 2/30
```

```
F1 F2 F3 F4 F5 F6
   Algebra Calc Other PrgmIO Clean Up
 d/dx(p2) = d/dx(y1(x)) | x = x3 → eq6
     27·a2·π²/4 + 3·b2·π + d2 = 1
 d/dx(p1) = d/dx(p2) | x = x2 → eq7
     3·a1·π²/4 + b1·π + d1 = 3·a2·π²/4 + b2·π + d2
d(p1,x)=d(p2,x)|x=x2→eq7
INTERP          RAD AUTO              FUNC 4/30
```

```
F1 F2 F3 F4 F5 F6
   Algebra Calc Other PrgmIO Clean Up
 d²/dx²(p1) = d²/dx²(p2) | x = x2 → eq8
     3·a1·π + 2·b1 = 3·a2·π + 2·b2
d(p1,x,2)=d(p2,x,2)|x=x2→eq8
INTERP          RAD AUTO              FUNC 1/30
```

The calculator solves the linear system of 8 equations with 8 unknowns. It remains only to graph the function and the cubic interpolating spline. The interpolation by splines does not give an excellent approximation, but it does preserve geometrical properties of the function such as curvature and convexity.

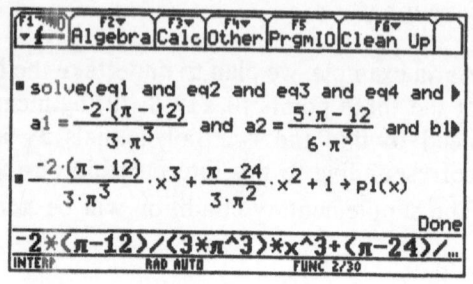

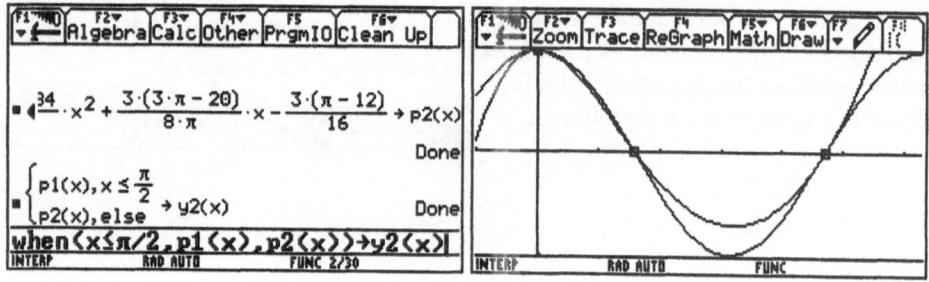

6. 2 Definitions and first properties for splines

Definition: Let $\sigma_n = (a = x_0 < x_1 < \ldots < x_n = b)$ be a partition of an interval $[a, b]$. Let ℓ a strictly positive integer. One calls spline of degree ℓ any application $s : [a, b] \to \mathbb{R}$ satisfying the following properties:

a) $s \in C_{\ell-1}[a, b]$

b) For all $0 \le k \le n - 1$, $s|_{[x_k, x_{k+1}[} \in R_\ell[X]$ (the restriction of s to $[x_k, x_{k+1}[$ is a polynomial function of degree less than or equal to ℓ).

We will note $S_\ell(\sigma_n)$ the set of spline functions of degree ℓ.

For example, define for all $0 \le k \le n - 1$, the function $q_{\ell,k}$ on $[a, b]$ by:

$$q_{\ell,k}(x) = (x - x_k)_+^\ell = \begin{cases} (x - x_k)^\ell & \text{if } x \ge x_k \\ 0 & \text{otherwise} \end{cases}$$

Each of these functions is a spline function of degree ℓ associated to the partition σ_n of $[a, b]$.

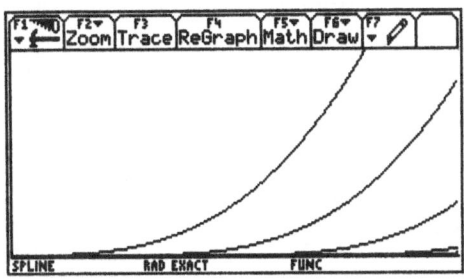

This is the graph of the spline of degree 3 defined above on the partition $(0, 1/4, 1/2, 3/4, 1)$ of $[0, 1]$.

This example is important. Indeed:

Theorem 1: The set $S_\ell(\sigma_n)$ is an $\mathbb{R}$-vector space of dimension $n + \ell$. A basis of this space is given by $(p_0, \ldots, p_\ell, q_{\ell,1}, \ldots, q_{\ell,n-1})$, where $p_k : x \to x^k, (0 \le k \le \ell)$.

Proof: We will show that any element $s \in S_\ell(\sigma_n)$ has a unique representation of the form:

$$s(x) = \sum_{k=0}^{\ell} a_k x^k + \sum_{j=0}^{n-1} b_j (x - x_j)_+^\ell, \qquad \forall x \in [a, b]$$

Let $s \in S_\ell(\sigma_n)$. On the first interval $[x_0, x_1]$, s is a polynomial, which is to say that on this interval $s(x) = \sum_{k=0}^{\ell} a_k x^k$. Our decomposition of s under the form

$$s(x) = \sum_{i=0}^{\ell} a_i x^i + \sum_{j=0}^{k-1} b_j (x - x_j)_+^\ell$$

is therefore valid on $I_k = [x_0, x_k]$, for $k = 1$.
Pose

$$\rho(x) = s(x) - \sum_{i=0}^{\ell} a_i x^i + \sum_{j=0}^{k-1} b_j (x - x_j)_+^\ell$$

The function ρ is of class $C^{\ell-1}$ on I_{k+1}, and $\rho(x) = 0$ for all $x \in I_k$. Morever, on the interval $[x_k, x_{k+1}[$, ρ is a polynomial function of degree less than or equal to ℓ and can therefore be considered as a solution of the differential equation $y^{(\ell+1)} = 0$ with the initial conditions $y(x_k) = y'(x_k) = \ldots = y^{(\ell-1)}(x_k) = 0$. Within a multiplicative constant, the solution of this equation is unique and may be written in the form: $\rho(x) = -b_k (x - x_k)_+^\ell$, for $x \ge x_k$.
We are going to show that our extension of s is valid for all $k \le n$. For $k = n$, this gives us the basis of $I_n = [a, b]$.

To end this paragraph, let's show that the splines of degree ℓ are not very "far away" from polynomials of degree less than or equal to $n + \ell - 1$.

Definition: A point $\zeta \in [x_k, x_{k+1}[\subset [a, b]$ is called a zero of $s \in S_\ell(\sigma_n)$, if $s(\zeta) = 0$, without s vanishing identically on the interval $[x_k, x_{k+1}[$.

Theorem 2: Any spline function $s \in S_\ell(\sigma_n)$ possesses at most $(n + \ell - 1)$ zeros on $[a, b]$, each zero being counted with its multiplicity.

Proof: let r be the number of zeros of s on $[a, b]$. By Rolle's Theorem, $s^{(\ell-1)} \in S_1(\sigma_n)$ has at least $r - \ell + 1$ zeroes. The function $s^{(\ell-1)}$ is continuous, piecewise linear; it can therefore have at most n zeros on $[a, b]$. Thus, $r - \ell + 1 \le n$ or $r \le n + \ell - 1$.

6. 3 Interpolation using spline functions

Spline functions are particularly interesting solutions to our problem. As opposed to the other interpolations already studied, they have the remarkable property of being able to interpolate on any number of values while retaining small degree (less than or equal to 3).

In what follows, we will only be interested in spline functions of odd degree.

The spline functions of degree 1 are particularly simple: they are continuous, piecewise linear functions on $[a, b]$. They are completely determined by the data of $(n + 1)$ points of interpolation.

If we suppose that $\ell = 2m - 1 \ (m \ge 2)$, so the dimension of $S_{2m-1}(\sigma_n)$ is equal to $2m + n - 1$, and if we want to interpolate at $(n + 1)$points $x_0, x_1, \ldots, x_n$, there remain $2m - 2$ degrees of freedom. There are chiefly three types of interpolation:

• Hermite interpolation. Let $f \in C^m([a, b])$. Determine $s \in S_{2m-1}(\sigma_n)$ such that:
a) for all $0 \le k \le n$, $s(x_k) = f(x_k)$
b) for all $1 \le j \le m - 1$, $s^{(j)}(a) = f^{(j)}(a)$ and $s^{(j)}(b) = f^{(j)}(b)$.

• Interpolation with natural boundary conditions. Let $f \in C^m([a, b])$ and $2 \le m \le n + 1$. Determine $s \in S_{2m-1}(\sigma_n)$ such that:
a) for all $0 \le k \le n$, $s(x_k) = f(x_k)$
b) for all $m \le j \le 2m - 2$, $s^{(j)}(a) = s^{(j)}(b) = 0$.

• Interpolation with periodic boundary conditions. Let $f \in C^m([a, b])$ such that for all $0 \le k \le m - 1, f^{(k)}(a) = f^{(k)}(b)$. Determine $s \in S_{2m-1}(\sigma_n)$ such that:
a) for all $0 \le k \le n$, $s(x_k) = f(x_k)$
b) for all $1 \le j \le 2m - 2$, $s^{(j)}(a) = s^{(j)}(b)$.

To solve these three problems, we first prove the next relationship, which is reminiscent of results on best approximation in the quadratic norm. (See the chapter on Orthogonality):

Proposition 1: Let $f \in C^m([a, b])$ and let $s \in S_{2m-1}(\sigma_n)$ be a spline interpolation function. Put $d = f - s$ and suppose that d satisfies the condition:

$$\sum_{k=0}^{m-2}(-1)^k s^{(m+k)}(a)d^{(m-k-1)}(a) = \sum_{k=0}^{m-2}(-1)^k s^{(m+k)}(b)d^{(m-k-1)}(b)$$

One has then:

$$\int_a^b (f^{(m)}(x))^2 dx = \int_a^b [f^{(m)}(x) - s^{(m)}(x)]^2 dx + \int_a^b (s^{(m)}(x))^2 dx$$

Remark: In the case of cubic splines ($m = 2$), our condition translates into $s''(a)d'(a) = s''(b)d'(b)$. This is satisfied in the three types of interpolation defined above.

Proof: Put

$$J = \int_a^b \left(f^{(m)}(x)s^{(m)}(x) - (s^{(m)}(x))^2 \right) dx$$

We must show that $J = 0$. But:

$$J = \int_a^b s^{(m)}(x)d^{(m)}(x)dx = \left[s^{(m)}(x)d^{(m-1)}(x) \right]_a^b - \int_a^b s^{(m+1)}(x)d^{(m-1)}(x)dx$$

By repeating with integration by parts, we get:

$$J = \sum_{k=0}^{m-3}(-1)^k \left[s^{(m+k)}(x)d^{(m-k-1)}(x) \right]_a^b + (-1)^{m-2} \int_a^b s^{(2m-2)}(x)d''(x)dx$$

Since $s \in C_{2m-2}([a, b])$, it is necessary to decompose our next integration by parts:

$$\int_a^b s^{(2m-2)}(x)d''(x)dx = \sum_{k=0}^{n-1} \left[s^{(2m-2)}(x)d'(x) - s^{(2m-1)}(x)d(x) \right]_{x_k}^{x_{k+1}}$$

$$+ \int_{x_k}^{x_{k+1}} s^{(2m)}(x)d(x)dx$$

$$= \left[s^{(2m-2)}(x)d'(x) \right]_a^b$$

because $s^{(2m)} = 0$ and $d(x_k) = 0$, for all $0 \le k \le n$. We therefore obtain:

$$J = \sum_{k=0}^{m-2}(-1)^k \left[s^{(m+k)}(x)d^{(m-k-1)}(x) \right]_a^b$$

Thus, using hypotheses of the proposition, $J = 0$.

We may next demonstrate our interpolation theorem.

Theorem 3: There is a unique solution to each of the three interpolation problems defined above.

Proof: By using the basis of $S_\ell(\sigma_n)$, $(p_0, \ldots, p_\ell, q_{\ell,1}, \ldots, q_{\ell,n-1})$, one can write:

$$s(x) = \sum_{k=0}^{\ell} a_k x^k + \sum_{j=1}^{n-1} b_j q_{\ell,j}$$

Our problem consists therefore in solve a system of $(n+\ell)$ equations of unknown $(a_0, \ldots, a_\ell, b_1, \ldots, b_{n-1})$.

In the three cases of interpolation, the associate homogeneous system corresponds to $f = 0$ whose interpolating spline is $s = 0$. It remains to demonstrate that the solution is unique. Our integral relationship entails $(f^{(m)} = 0 \Rightarrow s^{(m)} = 0)$ for all interpolating splines. Let's write:

$$s(x) = \sum_{k=0}^{2m-1} a_k \frac{x^k}{k!} + \sum_{j=1}^{n-1} b_j \frac{(x - x_j)_+^{2m-1}}{(2m-1)!}$$

In this case, $s^{(m)} \in S_{m-1}(\sigma_n)$ verifies for all $x \in [a, b]$:

$$s^{(m)}(x) = \sum_{k=m}^{2m-1} a_k \frac{x^{k-m}}{(k-m)!} + \sum_{j=1}^{n-1} b_j \frac{(x - x_j)_+^{m-1}}{(m-1)!} = 0$$

The independence of the present functions entails that all the coefficients are null, so that:

$$a_m = \ldots = a_{2m-1} = b_1 = \ldots = b_n = 0$$

and therefore:

$$s(x) = \sum_{k=0}^{m-1} a_k x^k$$

Each of the three interpolation conditions implies that these coefficients are all null. One may verify this immediately.

We are thus going to show that if $f = 0$, the unique interpolating spline of f is $s = 0$. The homogeneous equation system therefore has only the trivial solution.

The integral relationship that we have demonstrated puts the emphasis on another property of the interpolating splines. Indeed, if $f \in C^m([a, b])$ is given and $s \in S_{2m-1}(\sigma_n)$ is the unique interpolating spline of f, and if g is a function of $C^m([a, b])$ which satisfies the same interpolation conditions that s, it follows that:

$$||s^{(m)}||_2 \leq ||g^{(m)}||_2$$

In the case of cubic splines, this property is

$$\int_a^b (s''(x))^2 dx \leq \int_a^b (g''(x))^2 dx$$

and has the following geometrical interpretation:
with the local algebraic curvature of a function $x \mapsto g(x)$ defined as:

$$c(x) = \frac{g''(x)}{(1 + g'(x)^2)^{3/2}}$$

and supposing that $g'(x)$ is very small with respect to 1, the value of $\|c\|_2^2$ is approximately $\int_a^b (g''(x))^2 dx$. The interpolating spline s therefore minimizes the norm of the algebraic curvature in the class of functions $C^2([a, b])$ which satisfy to the interpolation conditions.

This is the geometrical property that has made the spline functions successful. For example, they are used for the creation of Postscript fonts because their geometrical properties allow them to be dilated without losing their approximation qualities.

6. 4 Convergence

Spline functions therefore allow good geometrical interpolation of a function f. But, when f is regular, the approximation itself is not bad at all!

Theorem 4: Let $f \in C^m([a, b])$, $(m \geq 2)$, and $s \in S_{2m-1}(\sigma_n)$ the interpolating spline found previously. We put $h = \max_k |x_{k+1} - x_k|$, when the (x_k) represent the partition σ_n. Then, for all $0 \leq j \leq m - 1$, we have:

$$\|f^{(j)} - s^{(j)}\|_\infty \leq \frac{m!}{\sqrt{m}} \frac{1}{j!} h^{m-j-1/2} \|f^{(m)}\|_2$$

Proof: Set $d = f - s$. This function is of class $C^m([a, b])$ and satisfies $d(x_k) = 0$ for $0 \leq k \leq n$. By Rolle's Theorem, for all $1 \leq j \leq m - 1$, $d^{(j)}$ has at least one zero in each interval $[x_k, x_{k+j}]$, with $k + j \leq n$.
Let ζ_j be such that $|d^{(j)}(\zeta_j)| = \|d^{(j)}\|_\infty$. The zero ξ_j closest to ζ_j satisfies $|\xi_j - \zeta_j| < (j + 1)h$.
For $j \leq m - 2$ (the inequality being roughly satisfied for $j = m - 1$), we have:

$$\|d^{(j)}\|_\infty = \left| \int_{\zeta_j}^{\xi_j} d^{(j+1)}(t) dt \right| \leq (j + 1)h \|d^{(j+1)}\|_\infty$$

$$\leq (j + 1)(j + 2)h^2 \|d^{(j+2)}\|_\infty \leq \cdots$$

$$\leq (j + 1)(j + 2)\ldots(m - 1)h^{m-j-1} \|d^{(m-1)}\|_\infty$$

$$= \frac{(m - 1)!}{j!} h^{m-j-1} \|d^{(m-1)}\|_\infty$$

By using the Schwartz inequality, we obtain:

$$||d^{(m-1)}||_\infty = \left| \int_{\zeta_{m-1}}^{\xi_{m-1}} d^{(m)}(t)dt \right|$$

$$\leq (mh)^{1/2} \left| \int_{\zeta_{m-1}}^{\xi_{m-1}} (d^{(m)}(t))^2 dt \right|^{1/2} \leq (mh)^{1/2}||d^{(m)}||_2$$

But we know that $||d^{(m)}||_2 \leq ||f^{(m)}||_2$. So we obtain:

$$||f^{(j)} - s^{(j)}||_\infty \leq \frac{m!}{\sqrt{m}} \frac{1}{j!} h^{m-j-1/2} ||f^{(m)}||_2$$

This theorem gives us the following result on uniform convergence:

Theorem 5: Let $\sigma_n([a,b])$ be a partition of the interval $[a,b]$ with step size h. Let $f \in C^m([a,b])$ and $s_n \in S_{2m-1}(\sigma_n)$ the interpolating splines of f. The sequence (s_n) converges uniformly to f as h tends to 0 when n tends to infinity. Moreover, if $m \geq 2$, for each $1 \leq j \leq m-1$, the sequence of derivatives $(s_n^{(j)})$ converges uniformly to $f^{(j)}$.

6. 5 Algorithm for calculation of cubic splines

We have seen an introductory example of the cubic spline interpolating a function f on a set of points $(x_0, x_2, \ldots, x_n)$. We will use here the interpolation with natural boundary conditions, (condition CS_1) which, let's recall, is expressed by:
Determine $s \in S_3(\sigma_n)$ with, for all $0 \leq k \leq n$, $s(x_k) = y_k$ and for which $s''(a) = s''(b) = 0$.
We will explain now an algorithm of calculation of the cubic interpolating spline of f on $(x_0, x_2, \ldots, x_n)$, and put, for all $1 \leq k \leq n, y_k = f(x_k)$.

• We denote $z_i = s''(x_i), 0 \leq i \leq n$, with $z_0 = z_n = 0$. If we suppose that we have calculated these points, then s is completely determined. Indeed, the restriction of s'' to the interval $[x_i, x_{i+1}]$ is a polynomial of degree 1, which we get by putting $h_i = x_{i+1} - x_i$:

$$s_i''(x) = \frac{z_{i+1}}{h_i}(x - x_i) + \frac{z_i}{h_i}(x_{i+1} - x)$$

By integrating twice and by using $s_i(x_i) = y_i$ and $s_i(x_{i+1}) = y_{i+1}$, we get:

$$s_i(x) = \frac{z_{i+1}}{6h_i}(x-x_i)^3 + \frac{z_i}{6h_i}(x_{i+1}-x)^2 + \left(\frac{y_{i+1}}{h_i} - \frac{h_i z_{i+1}}{6}\right)(x-x_i) + \left(\frac{y_i}{h_i} - \frac{h_i z_i}{6}\right)(x_{i+1}-x)$$

The spline s is then entirely determined by its restrictions $s_i, 0 \leq i \leq n-1$.

• We now show how to determine the coefficients (z_i).
One unused condition remains, that of the continuity of s'. By differentiating s and by using $s_{i-1}(x_i) = s'_i(x_i)$, it follows that:

$$s'_i(x_i) = -\frac{h_i}{6}z_{i+1} - \frac{h_i}{3}z_i + b_i$$

with

$$b_i = \frac{1}{h_i}(y_{i+1} - y_i)$$

In the same way:

$$s'_{i-1}(x_i) = \frac{h_{i-1}}{6}z_{i-1} + \frac{h_{i-1}}{3}z_i + b_{i-1}$$

so , for $1 \le i \le n-1$:

$$h_{i-1}z_{i-1} + 2(h_{i-1} + h_i)z_i + h_i z_{i+1} = 6(b_i - b_{i-1})$$

Put $u_i = 2(h_{i-1} + h_i)$ and $v_i = 6(b_i - b_{i-1})$. We then get a tridiagonal system:

$$\begin{pmatrix} 1 & 0 & & & & \\ h_0 & u_1 & h_1 & & & \\ & h_1 & u_2 & h_2 & & \\ & & \ddots & \ddots & \ddots & \\ & & & h_{n-2} & u_{n-1} & h_{n-1} \\ & & & & 0 & 1 \end{pmatrix} \begin{pmatrix} z_0 \\ z_1 \\ z_2 \\ \vdots \\ z_{n-1} \\ z_n \end{pmatrix} = \begin{pmatrix} 0 \\ v_1 \\ v_2 \\ \vdots \\ v_{n-1} \\ 0 \end{pmatrix}$$

It can be shown that this system has a solution, and it may be solved using the Pivot method or the **rref** function.

In summary, we get the following algorithm:

1. For $i = 0, 1, \ldots, n-1$ calculate:

$$\begin{cases} h_i = x_{i+1} - x_i \\ b_i = \frac{1}{h_i}(y_{i+1} - y_i) \end{cases}$$

2. Put

$$\begin{cases} u_1 = 2(h_0 + h_1) \\ v_1 = 6(b_1 - b_0) \end{cases}$$

then calculate for $i = 2, 3, \ldots, n-1$:

$$\begin{cases} u_i = 2(h_i + h_{i-1}) - \dfrac{h_{i-1}^2}{u_{i-1}} \\ v_i = 6(b_i - b_{i-1}) - \dfrac{h_{i-1}v_{i-1}}{u_{i-1}} \end{cases}$$

3. Put $z_n = z_0 = 0$ and calculate for $i = n-1, n-2, \ldots, 1$:

$$z_i = \frac{v_i - h_i z_{i+1}}{u_i}$$

We easily show by recursion that $u_i > h_i > 0$ and therefore that this algorithm is correct (stage 2). Indeed $u_1 > h_1 > 0$. If $u_{i-1} > h_{i-1}$, then:

$$u_i = 2(h_i + h_{i-1}) - \frac{h_{i-1}^2}{u_{i-1}} > 2(h_i + h_{i-1}) - h_{i-1} > h_i$$

Here are the programs which follow the process that we just developed.

Solution of the tridiagonal system:

```
:tridiag3(lx,ly)
:Func
:Local i,m,n,h,b,u,v,zz
:dim(lx)→n
:For i,1,n-1
:lx[i+1]-lx[i]→h[i]
:(ly[i+1]-ly[i])/(h[i])→b[i]
:EndFor
:2*(h[1]+h[2])→u[1]
:6*(b[2]-b[1])→v[1]
:For i,2,n-2
:2*(h[i]+h[i+1])-h[i-1]^2/(u[i-1])→u[i]
:6*(b[i]-b[i-1])-h[i]*[i-1]/(u[i-1])→[i]
:EndFor
:For i,1,n-1
:0→zz[i]
:EndFor
:For i,n-2,1,-1
:(v[i]-h[i]*zz[i+1])/(u[i])→zz[i]
:EndFor
:augment({0},zz)→zz
:EndFunc
```

Calculation of the cubic spline:

```
:spline3(lx,ly,lz,x)
:Func
:Local i,h,tmp,s,l
:{}→l
:For i,1,dim(lx)-1
:lx[i+1]-lx[i]→h
:lz[i]/2+(x-lx[i])*(lz[i+1]-lz[i])/(6*h)→tmp
:-h/6*(lz[i+1]+2*lz[i])+(ly[i+1]-ly[i])/h+(x-lx[i])*tmp→tmp
:ly[i]+(x-lx[i])*tmp→s
:augment(l,{s})→l
:EndFor
:l
:EndFunc
```

Here is the cubic spline interpolating the function $x \mapsto \cos(x)$ at the points $0, \pi/2, 3\pi/2$, with the boundary condition of CS_1. The approximation is not as good as with interpolation using condition CS_3. But calculations are here automated.

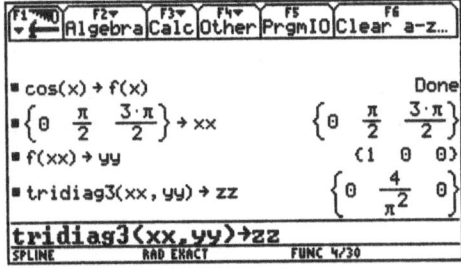

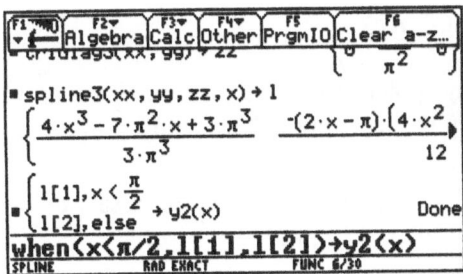

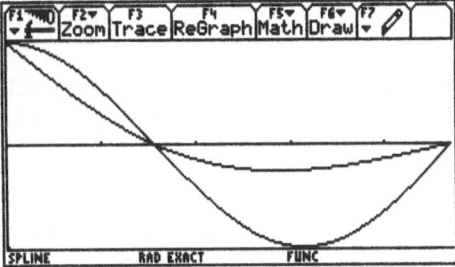

This is the cubic spline interpolating the function $f : x \mapsto 1/(1 + x^2)$ at the points **xx**, with the boundary condition CS_1.

Look at the graph of the spline and of the function f. The approximation does not seem too bad. For example, one may zoom on 0 to see the regular join of the polynomials.

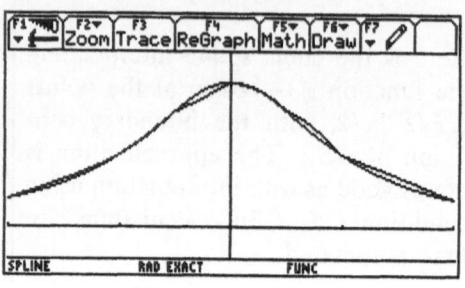

7. Interpolation by rational fractions

Instead of to seeking an interpolating polynomial, we are now going to concern ourselves with finding an interpolating "rational fraction", a quotient of polynomial.

Let n points $(x_1, y_1), (x_2, y_2), \ldots, (x_n, y_n)$ with pairwise distinct abscissas; determine a rational fraction R satisfying, for all $1 \leq k \leq n, R(x_k) = y_k$.

Suppose a priori that R is written in the form $R = \frac{A}{B}$, where A and B are two relatively prime polynomials of respective degree less than or equal to a and b respectively. The fraction R then depends on $a + b + 2$ arbitrary coefficients. Considering the fact that A and B are defined within a multiplicative coefficient (we may require that A is a monic polynomial, for example), the rational fraction R has $a + b + 1$ "degrees of freedom". Intuitively, one may hope to find a solution R when $a + b + 1 = n$, which offers several possibilities, the extreme possibility being with $a = n - 1$ and $b = 0$, which returns us to the problem of the polynomial interpolation.

It is logical to avoid high degrees which give birth to important roundoff errors, and therefore "to cut the pear in two", that is to say to choose a and b equal (or almost so). The equality $a + b = n - 1$ leads therefore to $a = b = (n-1)/2$ if the number from points n is odd and, if n is even to $a = n/2, b = a - 1 = n/2 - 1$ (for example, because one may make the inverse choice here).

In all cases, with this choice: $a = \lfloor (n/2) \rfloor$, and $b = \lfloor ((n-1)/2) \rfloor$. >From this, the rational fraction obtained has a total degree (the degree of the numerator minus the degree of the denominator) equal to 0 or to 1. In the case where the family of points is of the type $(x_k, f(x_k))_{1 \leq k \leq n}$, where f is an function, one thus obtains an interpolation of f by a rational fraction. Such an interpolation will certainly render better account of the behavior of f in the neighboorhood of one of its poles than can be done by a polynomial interpolation.

7. 1 A "Vandermonde style" method

We studied several methods of polynomial interpolation. One of them consisted of seeking the polynomial P in its most general form and then solving the system formed by the interpolation conditions. The matrix of this system was a Vandermonde matrix. We are going to see here an analogous method.

Suppose that the number n of interpolating points is odd $(n = 2m + 1)$. We then have to seek the rational fraction R in the form $R = \dfrac{A}{B}$, where A and B are two polynomials of degree m, the polynomial A with leading coefficient 1, for instance. The interpolation conditions $R(x_k) = y_k, 1 \le k \le n$ may be written $A(x_k) = y_k B(x_k)$, for all $1 \le k \le n$.

As an example, we consider the particular case where $n = 5$, and therefore $m = 2$. If we write $A(x) = x^2 + a_1 x + a_0$ and $B(x) = b_2 x^2 + b_1 x + b_0$, then the system formed by the 5 equalities $A(x_k) = y_k B(x_k), 1 \le k \le 5$ writes, in a matrix form:

$$
\begin{pmatrix}
-x_1 & -1 & y_1 x_1^2 & y_1 x_1 & y_1 \\
-x_2 & -1 & y_2 x_2^2 & y_2 x_2 & y_2 \\
-x_3 & -1 & y_3 x_3^2 & y_3 x_3 & y_3 \\
-x_4 & -1 & y_4 x_4^2 & y_4 x_4 & y_4 \\
-x_5 & -1 & y_5 x_5^2 & y_5 x_5 & y_5
\end{pmatrix}
\begin{pmatrix}
a_1 \\ a_2 \\ a_3 \\ a_4 \\ a_5
\end{pmatrix}
=
\begin{pmatrix}
x_1^2 \\ x_2^2 \\ x_3^2 \\ x_4^2 \\ x_5^2
\end{pmatrix}
$$

Remark: Contrary to polynomial interpolation, here there is no existence and uniqueness theorem for the rational interpolation when we fix the degree of the denominator and the numerator as we have done here; that comes partly from the fact that such rational fractions do not form a vector space. Tools that allow us to prove the existence of the interpolating polynomial (and that use the vector space language) are ineffectual here.

Obviously, the matrix above is not truly a Vandermonde matrix, and it is not simple to know whether it is really going to be invertible. Nevertheless, even if it is not as simple as a Vandermonde matrix, it is possible to construct it by a program. It clearly appears (look at the preceding example to convince yourself) that this matrix is formed of two juxtaposed blocks. It suffices to form these two sub-matrices separately, then to concatenate them.

```
:makemat(x,y)                                 x: abscissas, y: ordinates
:Func
:Local n,m,a1,a2
:dim(x)→n
:floor(n/2)-1→m
:-seq(seq(x[i]^j,j,m,0,-1),i,1,n)→a1          left part of M
:floor((n-1)/2)→m
:seq(seq(y[i]*x[i]^j,j,m,0,-1),i,1,n)→a2      right part of M
:augment(a1,a2)                               union of the two parts
:EndFunc
```

Now that we know how to construct the matrix of the system, it is easy enough to form the rational fraction interpolating a family of n points. One creates the matrix makemat and the second member of the linear system, that one solves with the function simult. One obtains then the vector composed of the coefficients of the numerator A and the denominator B (just add 1 to take account the leading coefficient of A). It suffices then to extract the coefficients of A, then those of B, and finally to form the rational fraction A/B.

```
:intrat1(x,y,w)                          abscissas, ordinates, variable
:Func
:Local a,m,s
:floor(dim(x)/2)→a                       degree of A
:makemat(x,y)→m                          matrix of the system
:list▶mat(x^a,1)→s                       free coefficients
:mat▶list(simult(m,s))→s                 resolution of the system
:augment(1,s)→s                          add 1 (leading coefficient)
:mid(s,1,a+1)→m                          A
:mid(s,a+2,dim(s))→s                     B
:polyEval(m,w)/(polyEal(s,w))            r = A/B
:EndFunc
```

Look at this example of a rational interpolation function. The leading coefficient of A is not equal to 1 because the calculator uses automatic simplifications.

This is the graph of the rational interpolation fraction in the preceding example.

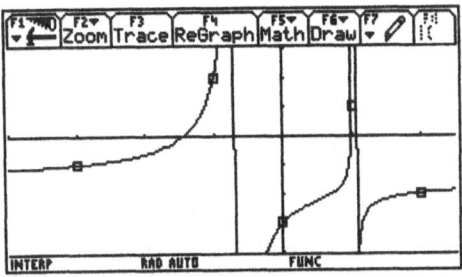

In the following example, we add the point $(3, -1)$ to the interpolation lists.

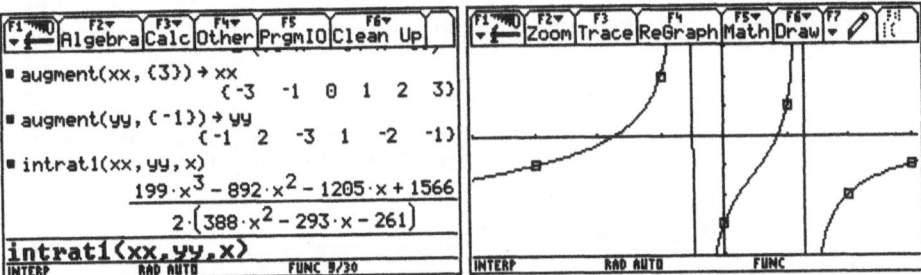

In general, we have $\deg(A)=\deg(B)$ for the degrees of the numerator A and the denominator B when the number n of points is odd, and $\deg(A)=\deg(B)+1$, when n is even. Of course, for n even, we may make the inverse choice. One may again write the two functions above, but if no ordinate is zero, as it is the case in the present example, here is how one may manage:

The screen speaks for itself...

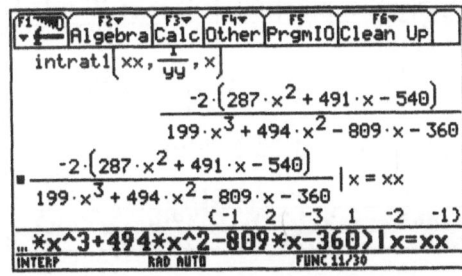

Here is the graph of the preceding function, a rational fraction which interpolates the same set of points. In general, there is not uniqueness of solutions to the problem of interpolation by rational fractions.

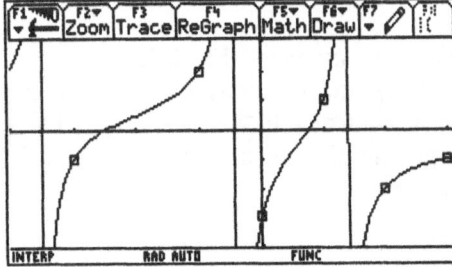

The problem of interpolation by rational fractions may have no solution. In this example, the point $(0, 1)$ is not interpolated. This is due to the fact that the matrix system that we solved was not equivalent to the problem of rational fraction interpolation.

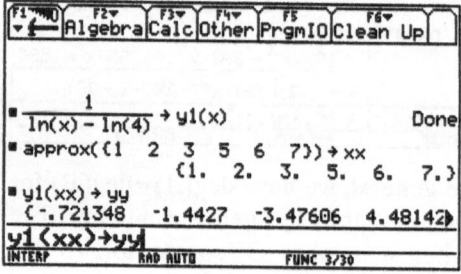

We are now going to study an example of interpolation of a function f which has a singularity. In such a case, comparison with polynomial interpolation is instructive.

The function $x \mapsto \frac{1}{\ln x - \ln 4}$ has a pole at $x = 4$. The interpolating rational fraction on the set $(1, 2, 3, 5, 6, 7)$ has a similar pole. On the graphic layout, the functions seem confounded.

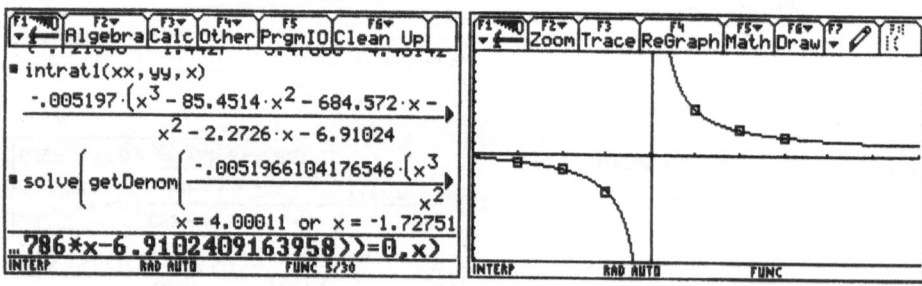

We have graphed the difference between f and its interpolating fraction on a window $-10^{-3} \le y \le 10^{-3}$. The approximation is not as good as we thought!

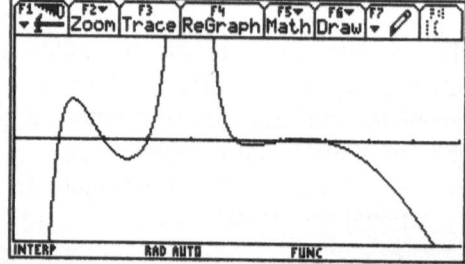

It is instructive to compare our result with polynomial interpolation. When f has a singularity, the interpolating polynomial, which must be regular, can not provide a good approximation.

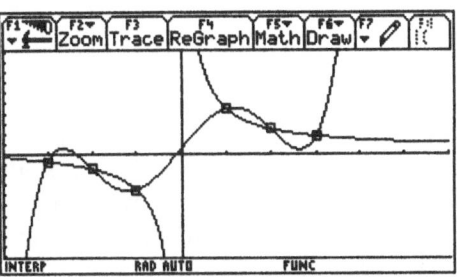

7. 2 Reciprocal divided differences

The preceding scheme seems effective, but it is nevertheless brutal enough; indeed, it consists of seeking the rational interpolation fraction R by an undetermined coefficient method, and therefore of solving a linear system which may be very voluminous. Several defects appear in this idea: the solution of the system can give birth to important roundoff errors (if it is ill conditioned), and especially, this method is not adaptable when we only need one or a few values of R. We propose now to look at another idea which looks similar to the divided difference method used for polynomial interpolation.

We define the *reciprocal divided difference* function in the following manner: Let $X = (x_1, x_2, \ldots, x_n)$ and $Y = (y_1, y_2, \ldots, y_n)$ be two finite sequences of the same length n. As in the case of divided differences, we will denote the reciprocal divided difference in the form $]x_1, x_2, \ldots, x_n[$.
If $n = 1$, we put: $]x_1[= y_1$.
If $n > 1$ we put :

$$\mathrm{ddr}(X, Y) =]x_1, x_2, \ldots, x_n[= \frac{x_{n-1} - x_n}{]x_1, \ldots, x_{n-1}[-]x_1, \ldots, x_{n-2}, x_n[}$$

For example:

$$]x_1, x_2[= \frac{x_1 - x_2}{y_1 - y_2}$$

Thus, the calculation of $]x_1, x_2[$ supposes that the ordinates y_1 and y_2 are distinct. Let's pause a bit on this kind of problem.
If we review the scheme of the preceding paragraph in the case where the number n of interpolating points is even, we look for the rational fraction R with the form $R = \frac{A}{B}$, with $\deg(A) = \frac{n}{2}$, and $\deg(B) = \frac{n}{2} - 1$, and with A having 1 as leading coefficient. In particular, if $n = 2$, we seek R in the form $R(x) = \frac{x + \alpha}{\beta}$.
If we denote (x_1, y_1) and (x_2, y_2) as the two interpolation points, the problem to be solved is:

$$\begin{cases} R(x_1) = y_1 \\ R(x_2) = y_2 \end{cases} \Leftrightarrow \begin{cases} x_1 + \alpha = \beta y_1 \\ x_2 + \alpha = \beta y_2 \end{cases} \Leftrightarrow \begin{cases} \beta =]x_1, x_2[\\ \alpha = \beta y_1 - x_1 \end{cases}$$

We may draw upon lesson of the preceding calculation. Our problem does not have a solution if $y_1 = y_2$. We presented this in one of the preceding screens. More generally, we may show that the rational interpolation, in the sense where it has been presented in the preceding paragraph (notably concerning questions of degrees of the numerator or at the denominator) may appear to be impossible in cases where several points of the data are in a "particular" configuration.

Such a particular situation for m points may arise when these m points can be interpolated by a polynomial of degree strictly less than $m - 1$. For example, when 2 points are on the same horizontal line, when 3 points are aligned, when 4 points are on the same parabola, etc. We could have therefore expected that rational interpolation fails in these particular cases.

But if we always consider the case of two interpolating points (x_1, y_1) and (x_2, y_2), the rational fraction we seed may be written:

$$R(x) = \frac{x + \alpha}{\beta} = y_1 + \frac{x - x_1}{\beta}, \quad \text{with } \beta =]x_1, x_2[$$

The preceding calculation is legal in the case where y_2 is different from y_1. If, on the contrary, $y_2 = y_1$, the expression of R is still valid, provided that we put $\frac{1}{\beta} = 0$, that is to say provided that we consider that β always exists but that it may be infinite. Indeed, we then find the rational fraction $R = Y_1$ which, as a constant, interpolates perfectly the two points (x_1, y_1) and $(x_2, y_2 = y_1)$.

The definition of a function ddr can therefore be secured by accepting the principle that a zero denominator gives birth to an "infinite" reciprocal divided difference. If the former appears later in the denominator of a new reciprocal divided difference, this last will be considered as zero (the calculator will facilitate us things because it automatically simplifies $1/\infty$ in 0). In this manner, the use that we are going to make of the reciprocal divided differences will generally be able to adapt to pathological cases by driving us to a rational fraction with degrees that are certainly less than the degrees we initially expected.

The iterative method of calculation of reciprocal divided differences that we propose is similar to that which was studied with polynomial interpolation (function ddi).

The function ddri is very much like the function ddi we saw previously.

```
:ddri(x,y)
:Func
:Local n,j,k,t,u
:dim(x)→n
:For k,1,n-1                column k
:For j,k+1,n               line j
:x[j]-x[k]→t               numerator t
:y[j]-y[k]→u               denominator u
:If u=undef:0→u            if infinity, put 0
:when(u=0,∞,t/u,t/u)→y[j]  if u = 0, put ∞ in line j
:EndFor
:EndFor
:EndFunc
```

Here are several examples using the reciprocal divided difference function, both numerical and formal.

7. 3 Use of reciprocal divided differences

As one may expect, reciprocal divided differences are going to be useful in calculating the rational fraction R interpolating a family of points (or only to calculate a value of R, without calculating all the coefficients).

Let $(x_1, x_2, \ldots, x_n)$ and $(y_1, y_2, \ldots, y_n)$ two finite sequences of same length n. Let $(d_1, d_2, \ldots, d_n)$ be the partial divided difference sequence (d_k is the reciprocal divided difference of the sequences $(x_1, \ldots, x_k)$ and $(y_1, \ldots, y_k)$). Let $R(x)$ be the rational fraction interpolating these n points. We define a sequence $R_n, R_{n-1}, \ldots, R_2, R_1$ of rational fractions (in this order) by putting:

$$R_n = d_n, \quad \forall\, k = n-1, n-2, \ldots, 1, \ R_k = d_k + \frac{x - x_k}{R_{k+1}(x)}$$

Theorem 1: The interpolating rational fraction R of the sequences $(x_1, x_2, \ldots, x_n)$ and $(y_1, y_2, \ldots, y_n)$ is equal to R_1.

Idea of the proof: Let $A = \dfrac{P_n}{Q_n}$ be the rational 'fraction that satisfies, for all $1 \leq k \leq n$: $A(x_k) = y_k$. We may write:

$$\frac{P_n(x)}{Q_n(x)} = y_1 + \frac{P_n(x)}{Q_n(x)} - \frac{P_n(x_1)}{Q_n(x_1)}$$

$$= y_1 + (x - x_1)\frac{P_{n-1}(x)}{Q_n(x)}$$

$$= y_1 + \frac{x - x_1}{Q_n(x)/P_{n-1}(x)}$$

But, for all $i > 1$:

$$\frac{Q_n(x_i)}{P_{n-1}(x_i)} = \frac{x_i - x_1}{y_i - y_1} =]x_1, x_i[$$

Therefore :

$$\frac{Q_n(x)}{P_{n-1}(x)} =]x_1, x_2[+ \frac{Q_n(x)}{P_{n-1}(x)} - \frac{Q_n(x_2)}{P_{n-1}(x_2)}$$

$$=]x_1, x_2[+ (x - x_2)\frac{Q_{n-1}(x)}{P_{n-1}(x)}$$

$$=]x_1, x_2[+ \frac{x - x_2}{P_{n-1}(x)/Q_{n-1}(x)}$$

and:

$$\frac{P_{n-1}(x_i)}{Q_{n-1}(x_i)} = \frac{x_i - x_2}{]x_2, x_i[-]x_1, x_2[}$$

One can then continue like that to get the required result. Thus, A will be represented as a "continued fraction":

$$y_1 + \cfrac{x - x_1}{(x_1, x_2) + \cfrac{x - x_2}{(x_1, x_2, x_3) + \cfrac{x - x_3}{\cdots}}}$$

```
:intrat2(x,y,w)
:Func
:Local k,r
:ddri(x,y)→y
:∞→r
:For k,dim(x),1,-1
:y[k]+(w-x[k])/r→r
:EndFor
:comDenom(r)
:EndFunc
```

Consider this calculation of rational interpolation with reciprocal divided differences. We be able to verify that the two functions intrat1 and intrat2 give the same result.

```
F1▾  F2▾    F3▾   F4▾   F5      F6▾
  Algebra Calc Other PrgmIO Clean Up
■ (-3  -1  0  1  2) → xx
                     (-3  -1  0  1  2)
■ (-1  2  -3  1  -2) → yy
                     (-1  2  -3  1  -2)
■ intrat2(xx, yy, x) → s
                -293·x² – 131·x + 414
                ─────────────────────
                 196·x² – 68·x – 138
intrat2(xx,yy,x)→s
INTERP        RAD AUTO         FUNC 3/30
```

When the points of interpolation are in a particular configuration, the interpolating rational fraction may not exist. In these cases, we must pay attention to the results rendered by our two functions.

```
F1▾  F2▾    F3▾   F4▾   F5      F6▾
  Algebra Calc Other PrgmIO Clean Up
■ (1  3  4  5) → xx : (1  1  3  -1) → yy
                     (1  1  3  -1)
■ intrat2(xx, yy, x)
■ intrat1(xx, yy, x) → r         undef
                -(2·x² – 19·x + 53)
                ───────────────────
                     11·x – 47
■ r | x = xx         (1  1  3  -1)
r|x=xx
INTERP        RAD AUTO         FUNC 4/30
```

And the results can be false!

```
F1▾  F2▾    F3▾   F4▾   F5      F6▾
  Algebra Calc Other PrgmIO Clean Up

■ (1  3  4  5) → xx : (1  3  1  -1) → yy
                     (1  3  1  -1)
■ intrat2(xx, yy, x)            9 – 2·x
■ intrat1(xx, yy, x) → r        -(2·x – 9)
■ r | x = xx         (7  3  1  -1)
r|x=xx
INTERP        RAD AUTO         FUNC 4/30
```

8. Trigonometric interpolation

Consider a complex or real valued function f, defined on the set of real numbers, which is periodic with period 2π. We suppose that we only know the values of f at n points of the interval $[0, 2\pi]$ (or any other interval of length 2π), and we wish to estimate the values of f at any point in this interval. In order to do that, we are going to look for a trigonometric function g (also 2π periodic) which is as simple as possible and which takes the same values as f at each of these n points. The function g will be then charged to interpolate f on the rest of the interval. The model is that of linear combinations of the functions $x \mapsto \exp(ikx)$, for $k \in \mathbb{Z}$. To minimize the frequencies, it is good to center values of the integers k around the origin. It is also logical to expect n unknown coefficients since we want to solve a problem of interpolation at n points. That leads us to the following function g (the interval of variation of the index k takes into account the parity of n; we denote $\lfloor m \rfloor$ as "the integer part of m).

$$g : (n, x) \mapsto \sum_{\lfloor -\frac{n}{2} \rfloor + 1}^{\lfloor \frac{n}{2} \rfloor} a_k \exp(ikx)$$

The function g is a sum of n terms. The calculator passes immediately into real mode and writes $g(n, x)$ as a combination of cosines and sines.

```
F1▼      F2▼   F3▼  F4▼   F5      F6▼
▼ ▪ Algebra Calc Other PrgmIO Clean Up

    floor(n/2)
  ■  Σ          (a[k]·e^(i·k·x)) → g(n, x)
   k=floor(-n/2)+1

                                    Done
  ■ g(3, x)
   (a[1] + a[-1])·cos(x) + a[0] + (a[1] − a[-1])▶
  g(3,x)
INTERP          RAD AUTO          FUNC 2/30
```

8. 1 Return to polynomial interpolation

We associate with g the polynomial $P : (n, X) \mapsto \sum_{k=0}^{n-1} a_{k+1+\lfloor -n/2 \rfloor} x^k$

We observe that for $m = \lfloor -n/2 \rfloor + 1$, by putting $\omega = \exp(ix)$:

$$g(n, x) = \sum_{k=m}^{\lfloor n/2 \rfloor} a_k \omega^k = \omega^m \sum_{k=m}^{\lfloor n/2 \rfloor} a_k \omega^{k-m} = \omega^m \sum_{k=0}^{n-1} a_{k+m} \omega^k = \omega^m P(n, \omega)$$

The problem of trigonometric interpolation on the n points $A_j = (x_j, y_j)$, for $1 \le j \le n$ is then defined as:
Determine n scalars a_k such that, for all $0 \le j \le n-1$, $g(n, x_j) = y_j$.

When we associate the polynomial function P with the trigonometric function g, we simply return to the problem of polynomial interpolation. Indeed, it suffices now to find the coefficients a_k of the polynomial P of degree less than or equal to $n-1$ such that, for all $0 \le j \le n-1$:

$$P(n, \omega_j) = \omega_j^m y_j, \text{ with } m = 1 + \lfloor -\frac{n}{2} \rfloor \text{ and } \omega_j = \exp(ix_j)$$

When the problem is solved—when the coefficients of the polynomial are found—the function g may be evaluated at any point x by writing:

$$g(n,x) = \omega_x^m P(n,\omega_x), \quad \omega_x = \exp(ix), \quad m = 1 + \lfloor -\frac{n}{2} \rfloor$$

```
:trigint1(x,y,w)
:Func
:Local m
:1+floor(-dim(x)/2)→m
:e^(i*x)→x          abscissas
:y*x^(-m)→y         ordinates
:e^(i*m*w)*newt(x,y,e^(i*w))  interpolation
:EndFunc
```

An example using trigonometric interpolation. The results can not be easily read. The calculator simplifies automatically and passes into real mode and writes $g(n,x)$ as a combination of sines and cosines.

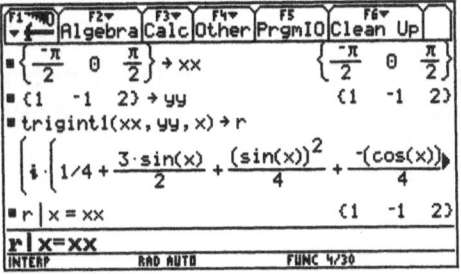

It is necessary to apply the function tcollect and/or texpand to simplify the result obtained. But it is futile to seek a symbolic trigonometric interpolation using our method with more than three points .

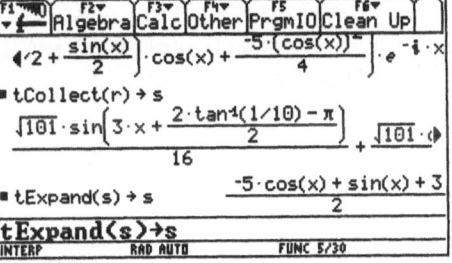

Here is the graph of the 3 points and the trigonometric interpolating polynomial.

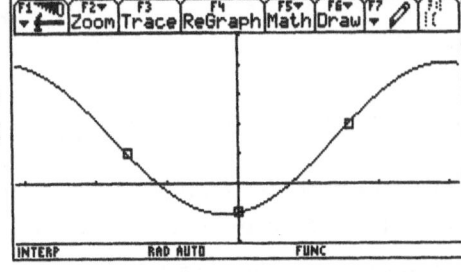

8. 2 Using undetermined coefficients

The problem of trigonometric interpolation is linear with respect to the coefficients (a_k) of the function g. It suffices therefore to just put the system in place, then to solve it. We denote, as we did before, $m = 1 + \lfloor -n/2 \rfloor$, then put $p = \lfloor n/2 \rfloor$.

With this notation, $g(n,x) = \sum_{k=m}^{p} a_k \exp(ikx)$. The system of the n conditions $g(n,x_k) = y_k, m \leq k \leq p$ is written:

$$\begin{pmatrix} e^{imx_1} & e^{i(m+1)x_1} & \cdots & e^{i(p-1)x_1} & e^{ipx_1} \\ e^{imx_2} & e^{i(m+1)x_2} & \cdots & e^{i(p-1)x_2} & e^{ipx_2} \\ \vdots & \vdots & & \vdots & \vdots \\ e^{imx_{n-1}} & e^{i(m+1)x_{n-1}} & \cdots & e^{i(p-1)x_{n-1}} & e^{ipx_{n-1}} \\ e^{imx_n} & e^{i(m+1)x_n} & \cdots & e^{i(p-1)x_n} & e^{ipx_n} \end{pmatrix} \begin{pmatrix} a_m \\ a_{m-1} \\ \vdots \\ a_{p-1} \\ a_p \end{pmatrix} = \begin{pmatrix} y_1 \\ y_2 \\ \vdots \\ y_{n-1} \\ y_n \end{pmatrix}$$

The next function uses this undetermined coefficient method to solve the problem of the trigonometric interpolation.

```
:trigint2(x,y,w)
:Func
:Local n,p,m,a
:dim(x)→n
:floor(n/2)→p
:p+1-n→m
:seq(seq(e^(i*j*x[i]),j,m,p),i,1,n)→a     creation of the matrix
:mat▸list(simult(a,list▸mat(y,1)))→a       solution of the system
:dotP(a,conj(seq(e^(i*k*w),k,m,p)))        g(n,x)
:EndFunc
```

This shows use of the "rapid" function trigint2 for the trigonometric interpolation.

8. 3 Equidistant abscissas

Let's return on the problem of trigonometric interpolation when the abscissas form a regular partition of the interval $[0, 2\pi]$, that is to say, when the partition of this interval is in the form $(x_0, x_1, \ldots, x_{n-1})$, with $x_k = \dfrac{2k\pi}{n}$, for all $0 \leq k \leq n-1$.

In this case, we don't need to solve a system of equations, since there is a formula allowing us to calculate directly the function $g(n, x)$.

Indeed, put $m = 1 + \lfloor -n/2 \rfloor$, and $p = \lfloor n/2 \rfloor$. The n equalities $g(n, x_h) = y_h$, $(0 \leq h \leq n - 1)$ may be written, for all $0 \leq h \leq n - 1$:

$$y_h = \sum_{k=m}^{p} a_k \exp\left(\frac{2ikh\pi}{n}\right)$$

Now put, for all $m \leq l \leq p$:

$$z_l = \sum_{h=0}^{n-1} y_h \exp\left(-\frac{2ilh\pi}{n}\right)$$

If we replace the (y_h) by their expressions as a function of the (a_k), we obtain:

$$z_l = \sum_{h=0}^{n-1} \left(\sum_{k=m}^{p} a_k \exp\left(\frac{2ikh\pi}{n}\right) \right) \exp\left(-\frac{2ihl\pi}{n}\right)$$

$$= \sum_{h=0}^{n-1} \left(\sum_{k=m}^{p} a_k \exp\left(\frac{2ih(k-l)\pi}{n}\right) \right)$$

$$= \sum_{k=m}^{p} a_k \left(\sum_{h=0}^{n-1} \exp\left(\frac{2i(k-l)\pi}{n}\right)^h \right) = \sum_{k=m}^{p} a_k \left(\sum_{h=0}^{n-1} (\omega^{k-l})^h \right)$$

with $\omega = \exp \dfrac{2i\pi}{n}$.

In the last expression of z_l, the inner sum is the sum of the n first terms of a geometrical sequence with ratio $q = \omega^{k-l}$. Both ω and q are nth roots of unity. This inner sum is therefore zero if q is different from 1, and equals n if $q = 1$. But the exponent $k - l$ of ω is between $-n + 1$ and $n - 1$ (k and l are indeed both in the same interval $[m, p]$ of length n): q is therefore equal to 1 only if this exponent is zero, that is to say, if $k = l$. We conclude that $z_l = na_l$. We have therefore obtained, for all $m \leq k \leq p$:

$$a_k = \frac{1}{n} z_k = \frac{1}{n} \sum_{h=0}^{n-1} y_h \exp\left(-\frac{2ihk\pi}{n}\right)$$

In fact, all that is quite standard. We deduce the (y_h) from the (a_k) by using the Discrete Fourier Transform. The Inverse Fourier Transform is given by the preceding equalities, and, as its name indicates, it allows us to find the (a_k) as a function of the (y_h). You may refer to the chapter devoted to Fourier series a study of this transformation and its rapid calculation algorithm.

We now have a formula for the calculation of the function g. If $m = 1 + \lfloor -n/2 \rfloor, p = \lfloor n/2 \rfloor$, and $\omega = \exp(\frac{2i\pi}{n})$, then, for all real x:

$$g(n,x) = \frac{1}{n}\sum_{k=m}^{p}\left(\sum_{h=0}^{n-1}y_h\omega^{-hk}\right)\exp(ikx)$$

```
:trigint3(y,w)
:Func
:Local n,ω,m,p
:dim(y)→n
:e^(2*i*π/n)→ω
:floor(n/2)→p
:p+1-n→m
:∑(e^(i*k*w)*∑(y[h+1]*ω^(-h*k),h,0,n-1),k,m,p)/n
:EndFunc
```

Here is an example of trigonometric interpolation on 6 points. The result is real and the interpolation is verified on the 6 points.

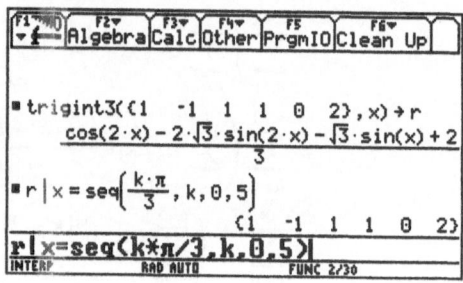

Here are trigonometric interpolation of the function $x \mapsto x(2\pi - x)$ on 4 equally distributed points and the graph of the function and its trigonometric interpolating polynomial.

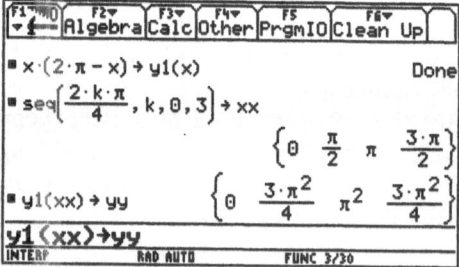

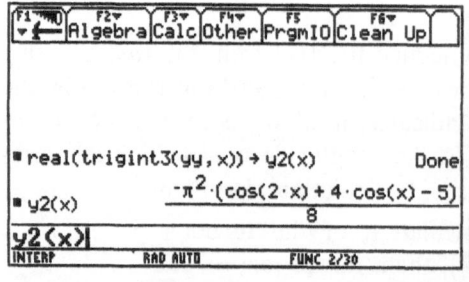

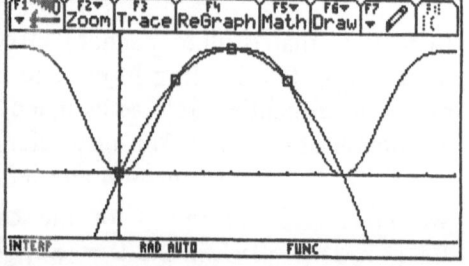

We doubled the number of interpolating points for the same function. The approximation does not seem too bad (at least graphically) on the interval $[0, 2\pi]$. We must not forget that the trigonometric interpolating polynomial can have an imaginary part which is not traced here.

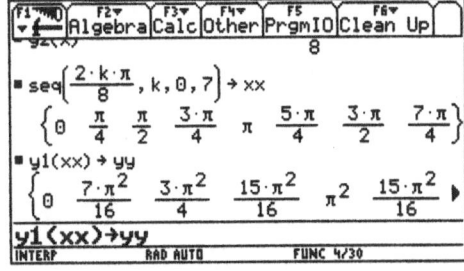

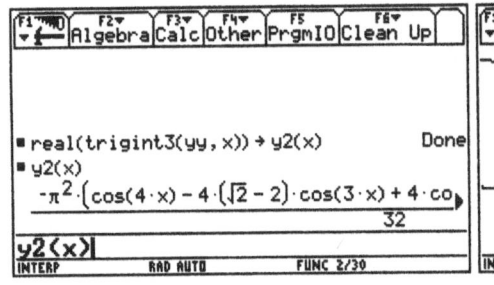

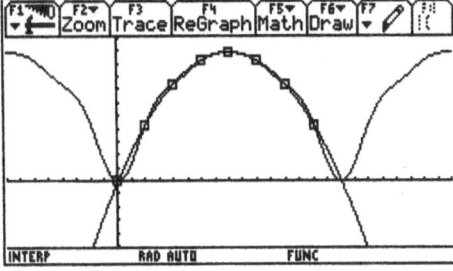

5

Orthogonality

1. PreHilbert or Inner Product Spaces

The word orthogonal is a mathematician's word for "perpendicular ". In order to address questions surrounding the notion of orthogonality, the theoretical setting is that of a preHilbert or an inner product vector space, which is equipped with an scalar product. This somewhat abstract notion allows us to address with the same approach situations in quite different settings, indeed some which are far from the "right angles " of classical Euclidean geometry.

In this first section we will recall some definitions.
In what follows, $\mathbb{K}$ designates $\mathbb{R}$ or $\mathbb{C}$, and E is a vector space over $\mathbb{K}$.

1. 1 Inner product

An scalar product over E is a map of $E \times E$ into $\mathbb{K}$, which we denote $(u, v) \mapsto\ < u, v >$ satisfying the following properties $(\forall (u, v, w) \in E^3, \forall (\alpha, \beta) \in \mathbb{K}^2)$:

- Semi-linearity from the left: $< \alpha u + \beta v, w > = \overline{\alpha} < u, w > + \overline{\beta} < v, w >$
- Right linearity: $< u, \alpha v + \beta w > = \alpha < u, v > + \beta < u, w >$
- Hermitian symmetry: $< v, u > = \overline{< u, v >}$
- Positive definiteness: $< u, u > \geq 0$ and $< u, u > = 0 \Leftrightarrow u = 0$

Sometimes the notation $u \cdot v$ or $(u|v)$ is used in place of (u, v) .

A map which satisfies all the preceding properties is known as an scalar product or as a scalar product or as a *positive definite sesquilinear form*.
When $\mathbb{K} = \mathbb{R}$, sesquilinearity is synonomous with linearity. An scalar product is then a *bilinear positive definite form*.
A vector space E equipped with an scalar product is called a preHilbert or an *inner product space*. If E is also finite dimensional, it is called Hermitian when $\mathbb{K} = \mathbb{C}$ or Euclidean when $\mathbb{K} = \mathbb{R}$.

An scalar product lets us define a norm (said to be a *Euclidean norm*), in the following way: $\forall u \in E, \|u\| = \sqrt{< u, u >}$. It may be verified that this satisfies the following properties $(\forall u, v, w \in E, \forall \lambda \in \mathbb{K})$:

$$\|u\| \geq 0; \|u\| = 0 \Leftrightarrow u = 0; \|\lambda u\| = |\lambda| \|u\| ; \|u + v\| \leq \|u\| + \|v\|$$

The last property (called the *triangle inequality*, or the *Minkowski inequality*), is an equality if and only if one of the two vectors u and v is the product of the other by a real positive number or zero, even if $\mathbb{K} = \mathbb{C}$.

The triangle inequality is a consequence of the following classical and very useful result, known as the *Cauchy-Schwarz inequality*:

$$\forall u, v \in E, |<u, v>| \leq \|u\| \, \|v\|$$

We note that this inequality is an equality if and only if u and v are linearly dependent.

Finally, the Euclidean norm satisfies the so-called *parallelogram identity*, which characterizes a norm derived from an scalar product:

$$\forall u, v \in E, \|u + v\|^2 + \|u - v\|^2 = 2 \left(\|u\|^2 + \|v\|^2 \right)$$

The Euclidean norm allows E to be equipped with a *distance* or a *metric*

$$\forall u, v \in E, d(u, v) = \|u - v\|$$

As for all metrics, this satisfies $(\forall u, v, w \in E)$:

$$d(u, v) = d(v, u); d(u, v) \geq 0; d(u, v) = 0 \Leftrightarrow u = v; d(u, w) \leq d(u, v) + d(v, w)$$

1. 2 Classical examples

The vector space K^n is equipped with an scalar product said to be the *canonical* or *natural* or *standard scalar product*:

$$\forall x = (x_1, x_2, \ldots, x_n), \forall y = (y_1, y_2, \ldots, y_n), <x, y> = \sum_{k=1}^{n} \overline{x_k} y_k$$

The associated norm is:

$$\forall x = (x_1, x_2, \ldots, x_n), \|x\| = \sqrt{<x, x>} = \sqrt{\sum_{k=1}^{n} |x_k|^2}$$

The instruction dotp of the TI-92+ calculates the scalar product. Its arguments may be two lists, two vectors, or two column-matrices.

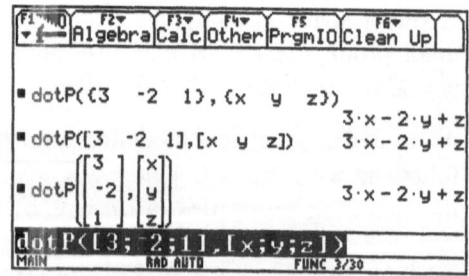

Contrary to our definition, the scalar product of the TI-92+, is semi-linear on the right. We see here how to define a function mysca1 adapted for our use, and a function mynorm accepting a vector, a list, or a column matrix.

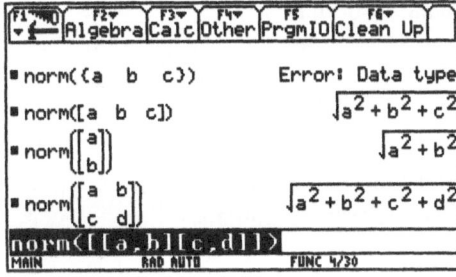

The TI-92+ has its own norm function, adapted to the cases of matrices or of vectors, but which doesn't accept lists.

Let I be an interval of $\mathbb{R}$ which is not empty and which is not a single point, and let $x \mapsto \omega(x)$ be a map (called a *weight function*) which is continuous on I, has positive values, and which vanishes only at isolated points of I.

Let E be the vector space of continuous functions f on I, with values in $\mathbb{K}$, and such that $|f|^2 \omega$ is integrable on I. We define an scalar product (and thus a norm) on E by setting:

$$\forall f, g \in E, < f, g > = \int_I \overline{f} g \omega; \quad \|f\| = \sqrt{\int_I |f|^2 \omega}$$

This very general definition covers a great many particular cases.

The program defsca1, as its name indicates, lets us define such an scalar product with the TI-92+. It displays a dialog box in which the user chooses the endpoints a and b of the interval, and the weight function ω.

The program defsca1 then creates the functions mysca1 and mynorm allowing calculation of the scalar product and the norm.

```
:defscal()
:Prgm
:DelVar x
:Dialog
:Text "<f,g>=∫(conj(f(x))g(x)w(x),x,a,b)"
:Request "Left endpoint a",θa
:Request "Right endpoint b",θb
:Request "Weight w(x)",θw
:EndDlog
:expr(θa)→θa:expr(θb)→θb:expr(θw)→θw
:∫(conj(θf)*θg*θw,x,θa,θb)→myscal(θf,θg)
:√(∫(abs(θf)^2*θw,x,θa,θb))→mynorm(θf)
:EndPrgm
```

This example shows how to define the scalar product

$$< f,g > = \int_{-1}^{1} \frac{\overline{f}g}{\sqrt{1-x^2}}.$$

This one will be used when we study Chebishev polynomials.

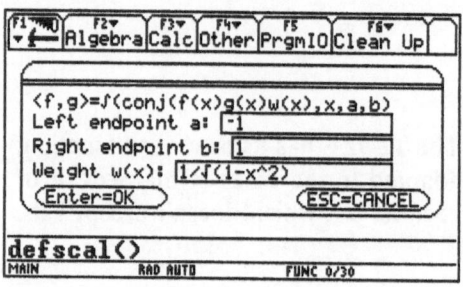

We may then express the scalar product or the norm of expressions in x, defined either symbolically or in explicit fashion. Here we calculated the norm of the polynomials 1, x, x^2. From the last example, we see that our choice of variable must be x.

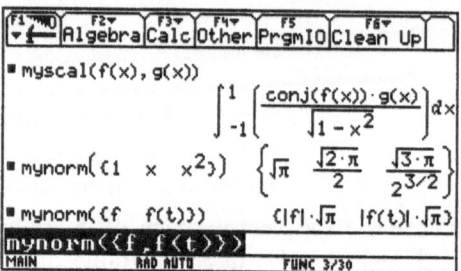

The definition of the scalar product which uses an integral may be extended to the case of piecewise continuous functions on I. But the map $(f,g) \mapsto < f,g >$ is then not properly called a scalar product. In fact, the hypothesis $\|f\| = 0$ no longer would imply $f \equiv 0$: we could only say that f is zero except possibly at isolated points of I.

1. 3 Orthonormal families

Two vectors u and v of an inner product space E are said to be orthogonal if $< u,v > = 0$. The vector u is a unit vector if $\|u\| = 1$.

If u and v are any two vectors of E, then:

$$\|u+v\|^2 = < u+v, u+v > = \|u\|^2 + 2\text{Re} < u,v > + \|v\|^2$$

If $\mathbb{K} = \mathbb{R}$, this equality reduces to $\|u+v\|^2 = \|u\|^2 + 2 <u,v> + \|v\|^2$.

In every case, $<u,v> = 0 \Rightarrow \|u+v\|^2 = \|u\|^2 + \|v\|^2$ (Pythagorean identity). This result may be extended to a finite family of pairwise orthogonal vectors, but the converse is only true if $\mathbb{K} = \mathbb{R}$, and then only in the case of two vectors.

We say that a family $(u_j)_{j \in J}$ of vectors in E is orthogonal if the vectors u_j are pairwise orthogonal. If they are not zero, such a family is linearly independent. In particular, this is the case if they are unit vectors: we then speak of an orthonormal set.

For example, suppose that E is of finite dimension $n \geq 1$. An orthonormal set is thus formed by at most n elements. If it has exactly n, it constitutes a basis, called an orthonormal basis, of E.

The simplest example of an orthonormal basis is furnished by the standard basis of $\mathbb{K}^n$, defined by the vectors:

$$e_1 = (1,0,0,\ldots,0)\,, e_2 = (0,1,0,\ldots,0)\,,\ldots, e_n = (0,0,\ldots,0,1)$$

The principal interest in orthonormal bases is that the scalar product and the norm may be expressed in very simple fashion with the aid of coordinates using such a basis.

In fact, if $(\varepsilon) = (\varepsilon_1, \varepsilon_2, \ldots, \varepsilon_n)$ is an orthonormal basis of E,

$$\forall x = \sum_{k=1}^{n} x_k \varepsilon_k, \quad \forall y = \sum_{k=1}^{n} y_k \varepsilon_k, \quad <x,y> = \sum_{k=1}^{n} \overline{x}_k y_k$$

Each vector x of E may thus be written in terms of this basis, $x = \sum_{k=1}^{n} <\varepsilon_k, x> \varepsilon_k$

We may also write this as a matrix product: if we denote $[u]_\varepsilon$ as the column matrix of coordinates of u in the basis (ε), then for all vectors $u = \sum_{k=1}^{n} x_k \varepsilon_k$ and $v = \sum_{k=1}^{n} y_k \varepsilon_k$,

$$<u,v> = \sum_{k=1}^{n} \overline{x}_k y_k = (\overline{x}_1 \quad \overline{x}_2 \quad \cdots \quad \overline{x}_n) \begin{pmatrix} y_1 \\ y_2 \\ \vdots \\ y_n \end{pmatrix} = \overline{[u]}_\varepsilon^\top [v]_\varepsilon$$

$\mathbb{K}^n$ (equipped with its standard basis) is the typical example of a finite dimensional inner product space.

Every inner product space E of dimension $n \geq 1$ equipped with an orthonormal basis (ε) may be identified with $\mathbb{K}^n$ by the isomorphism $\sum_{k=1}^{n} x_k \varepsilon_k \longmapsto (x_1, x_2, \ldots, x_n)$.

Orthonormal sets are likewise very useful in inner product spaces which are not finite dimensional, notably in function spaces using a scalar product defined by integrals.

The program defscal here lets us create the well known scalar product

$$< f, g > = \frac{1}{2\pi} \int_{-\pi}^{\pi} \bar{f} g.$$

which is at the center of the theory of Fourier series.

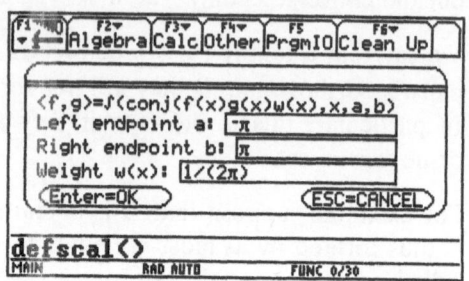

Here we verify that the maps $x \mapsto \exp(inx)$ form an orthonormal set for this scalar product. We see how to use the variable @n1, whose symbolic content is an integer to assure the orthogonality of these vectors.

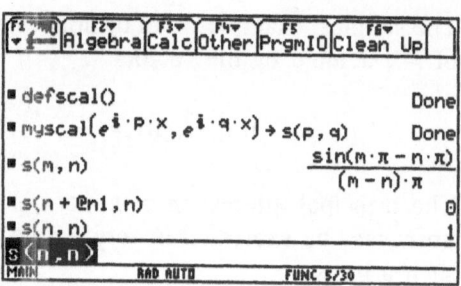

1. 4 Orthogonal complement of a subspace

Let X be any non-empty subset of an inner product space E.

We say that a vector u of E is orthogonal to X if u is orthogonal to every vector x of X.

The orthogonal complement of X, denoted $X^{\perp}$, is the set of vectors of E which are orthogonal to X.

We may easily verify the following properties:

- If $X \subset Y$, then $Y^{\perp} \subset X^{\perp}$.
- $X^{\perp}$ is a sub-space of E (even if X is not).
- If X is a sub-space of E and if $(e_j)_{j \in J}$ is a generating family (and notably if it is a basis), then $X^{\perp} = \{e_j, j \in J\}^{\perp}$. In other words a vector u of E is orthogonal to X if and only if it is orthogonal to each of the vectors e_j.
- The intersection of X and of $X^{\perp}$ is just $\overrightarrow{0}$. In particular, this means that if X is a sub-space of E, the sum $X + X^{\perp}$ is direct.
- It is evident that $\{\overrightarrow{0}\}^{\perp} = E$ and that $E^{\perp} = \{\overrightarrow{0}\}$.

For every family $u_1, u_2, \ldots, u_n$ of E, $\mathrm{Span}(u_1, u_2, \ldots, u_n)$ designates the subspace generated - or spanned - by this family, that is, it is the set of vectors of E which may be written as linear combinations $u_1, u_2, \ldots, u_n$.

Let $e_1, e_2, \ldots, e_n$ be an orthonormal set of E, and $F = \text{Span}(e_1, e_2, \ldots, e_n)$ the sub-space of E which it generates ($\dim(F) = n$).
Let u be a vector of F, with coordinates $\lambda_1, \lambda_2, \ldots, \lambda_n$ using the basis (e) of F. We thus have the equation $u = \sum_{k=1}^{n} \lambda_k e_k$.
If we take a scalar product of this equation with one of the vectors e_j, we find:

$$< e_j, u > = \sum_{k=1}^{n} \lambda_k < e_j, e_k > = \lambda_j.$$

The decomposition of u in terms of the basis (e) of F is thus $u = \sum_{k=1}^{n} < e_k, u > e_k$.
Now, let u be a vector of E. We denote $p(u) = \sum_{k=1}^{n} < e_k, u > e_k$.
$p(u)$ is clearly a vector of F.

As we saw before, if u is in F, then $p(u) = u$.
More generally, we state that: $\forall j \in \{1, \ldots, n\}, < e_j, p(u) > = < e_j, u >$.
Thus $u - p(u)$ is orthonormal to $e_1, e_2, \ldots, e_n$: it is thus orthogonal to F.
The equality $u = p(u) + (u - p(u))$ shows that u may be written as the sum of a vector of F and a vector of $F^{\perp}$. We deduce that $E = F + F^{\perp}$. Since we know that this is a direct sum of vector spaces (only the zero vector is in both F and $F^{\perp}$), we finally may write $E = F \oplus F^{\perp}$.
The map p thus looks like a projection of E onto F, parallel to $F^{\perp}$. We say that p is the orthogonal projection of E onto F.

1. 5 Gram-Schmidt orthogonalization

As we have seen, we have $E = F \oplus F^{\perp}$ for all sub-space F generated by a finite orthonormal set (e). The Gram-Schmidt orthogonalization process will show us that every non-empty finite dimensional sub-space F of E may be equipped with an orthonormal basis.

More precisely, let $u_1, u_2, \ldots, u_n$ be an independent set of E: then there exists a unique orthonormal set $e_1, e_2, \ldots, e_n$ of E such that:

$\forall k \in \{1, \ldots, n\}, \text{Span}(e_1, e_2, \ldots, e_k) = \text{Span}(u_1, u_2, \ldots, u_k)$ and $< u_k, e_k > > 0$

The proof follows by induction on the integer n.
For $n = 1$, it suffices to normalize the vector u_1. We put $e_1 = u_1 / \|u_1\|$.
Suppose that the property has been shown up to rank $k - 1 \geq 1$.
Let $F_{k-1} = \text{Span}(e_1, e_2, \ldots, e_{k-1}) = \text{Span}(u_1, u_2, \ldots, u_{k-1})$ and p the orthogonal projection of E onto F_{k-1} (the result of the preceding paragraph).

We put $e'_k = u_k - p(u_k) = u_k - \sum_{j=1}^{k-1} < e_j, u > e_j$.
The vector e'_k is not zero, (otherwise u_k will be in F_{k-1} and thus dependent on $u_1, u_2, \ldots, u_{k-1}$), and it is orthogonal to F_{k-1}. Then we put $e_k = e'_k / \|e'_k\|$.

The vector e_k is unitary and the family $(e_1, e_2, \ldots, e_k)$ is orthonormal.
On the other hand: $p(u_k) \in \mathrm{Span}(u_1, u_2, \ldots, u_{k-1}) \Rightarrow e_k \in \mathrm{Span}(u_1, u_2, \ldots, u_k)$.
We conclude that $F_k = \mathrm{Span}(e_1, e_2, \ldots, e_k) \subset \mathrm{Span}(u_1, u_2, \ldots, u_k)$ (and equality by reason of the dimension).
Finally, $< u_k, e_k > = \dfrac{1}{\|e'_k\|} < e'_k + p(u_k), e'_k > = \|e'_k\| > 0.$

The vector e_k is unique with respect to the preceding properties. In fact, suppose that f_k also has them. Necessarily f_k may be written $x_k e_k$ (since it belongs to F_k and it is orthogonal to $e_1, e_2, \ldots, e_{k-1}$).
We deduce that $1 = \|f_k\| = |x_k|$ and $< u_k, f_k > = x_k < u_k, e_k > = x_k \|e'_k\| > 0$.
Necessarily $x_k = 1$, that is, $f_k = e_k$.

We have thus shown the property for rank k and, by induction, for all $n \geq 1$.

We conclude that any non-zero, finite dimensional sub-space F of a vector space E which is equipped with an inner product, may be given an orthonormal basis: it suffices to apply the Gram-Schmidt orthogonalization process to any basis of F.

The following figure shows how this procedure transforms an independent set u_1, u_2, u_3 into an orthonormal set e_1, e_2, e_3:

- The vector e_1 is normalized from u_1: $e_1 = u_1 / \|u_1\|$.
- We then form orthogonal projection $p(u_2) = < e_1, u_2 > e_1$ of u_2 onto a line generated by e_1 (by u_1). The non-zero vector $e'_2 = u_2 - p(u_2)$, orthogonal to this line, gives rise to a unit vector $e_2 = e'_2 / \|e'_2\|$.
- Finally, we form the orthogonal projection $q(u_3)$ of u_3 onto the plane generated by $\{e_1, e_2\}$ (by $\{u_1, u_2\}$). The vector $e'_3 = u_3 - q(u_3)$ is orthogonal to this plane and not zero: it leads to a unit vector $e_3 = e'_3 / \|e'_3\|$.

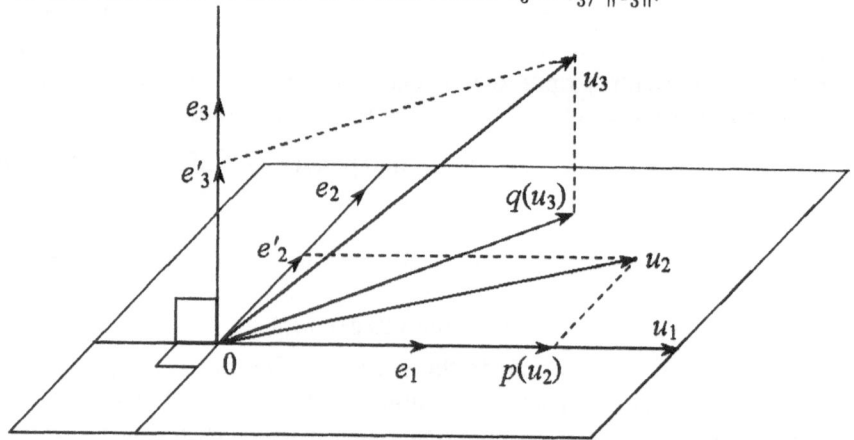

Orthonormalization of a set of three vectors u_1, u_2, u_3

It is very simple to program the Gram-Schmidt orthogonalization process with the calculator, as the following function shows.

The function schmidt takes a list as an argument (locally named u), consisting of vectors to be "treated ", which are progressively replaced by orthonormal vectors. (Each time through the For loop constructs a new unit vector, orthogonal to the preceding ones.)

The last list, representing the orthonormal set found, is returned as the output of the function schmidt.

This function was written to be used with the program defscal. It uses the functions myscal and mynorm to calculate the scalar product or the Euclidean norm.

The function schmidt won't work, for example, to normalize a set of vectors of $\mathbb{K}^n$, using the canonical scalar product. The calculator won't accept a list of vectors, and using a list of lists, which is automatically converted to a matrix, produces a syntax error at the level of the For loop.

```
:schmidt(u)
:Func
:Local j
:For j,1,dim(u)
:u[j]-Σ(myscal(u[k],u[j])*u[k],k,1,j-1)→u[j]
:u[j]/(mynorm(u[j]))→u[j]
:EndFor
:u
:EndFunc
```

With the classical integral scalar product, use of the Gram-Schmidt orthogonalization process gives interesting results, especially if we orthonormalize the family of polynomials $1, x, x^2, \ldots, x^n, \ldots$.

With the program defscal we put

$$< f, g > = \int_{-1}^{1} \overline{f} g$$

Thus, we define a scalar product on $\mathcal{C}([0,1], \mathbb{K})$, or more simply on $\mathbb{K}[X]$ (vector space of polynomials with coefficients in $\mathbb{K}$).

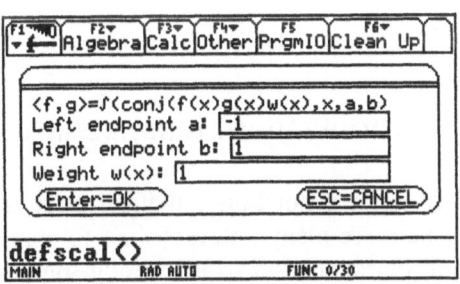

Then we call the function schmidt in order to orthonormalize $1, x, x^2$.
We verify with mynorm and myscal (which were created by defscal) that the polynomials obtained are indeed pairwise orthogonal and are unitary.

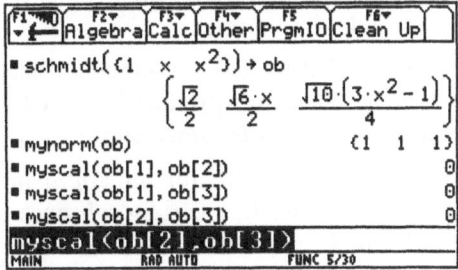

The Legendre polynomials are

$$P_n(x) = \frac{1}{2^n n!} \frac{d^n}{dx^n} \left((x^2 - 1)^n \right)$$

Applying the Gram-Schmidt orthogonalization process, to the family $(x^n)_{n \geq 0}$, leads to the polynomials $\sqrt{n + 1/2}\, P_n$.
We verify this for $0 \leq n \leq 2$.

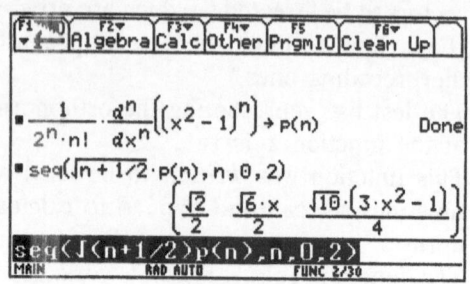

With defscal, we define the scalar product

$$< f, g > = \int_{-1}^{1} \frac{\overline{f} g}{\sqrt{1 - x^2}}.$$

then we orthonormalize $1, x, x^2, x^3$.

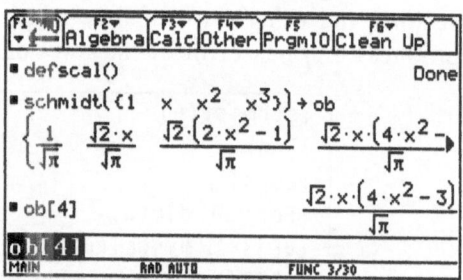

The Chebishev polynomials of the first space are defined by

$$P_n(x) = \cos \left(\mathrm{Arccos}(nx) \right).$$

The Gram-Schmidt orthogonalization process, applied to the family $(x^n)_{n \geq 0}$, leads to the family $\sqrt{\frac{2}{\pi}} P_n$.
We verify this here for $0 \leq n \leq 3$.

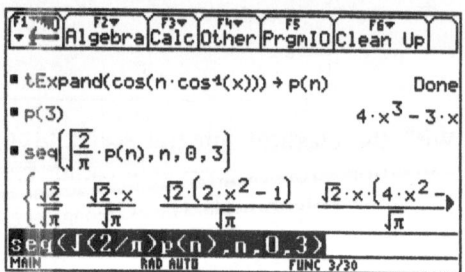

2. Problems of least squares

Consider a function f defined on an interval I with values in $\mathbb{K}$.
There are many different ways to approximate f on I: we must specify the type of approximation to use (using some measurement tool to make the decision), and we have to address the problem of deciding what is a good approximation of f and what is not.

Suppose that we use the inner product $< g, h > = \int_I \overline{g}(x) h(x) \, dx$. The distance given by this scalar product is a good way to measure the distance between two functions of E, functions whose squares are integrable on I.

In practice we look for an approximation f of a function of E by another function g from F, a certain subspace of E. Then the problem is the following: find g in F, which minimizes the quantity $\|f - g\|^2 = \int_I |f(x) - g(x)|^2 \, dx$.

If this problem has a solution g in F, we say that it is a best continuous approximation of f in the least squares sense.

If f is only known at $n+1$ points $x_0, \ldots, x_n$ of I, we will seek a function h (in a subspace F of E), which minimizes $\sum_{k=0}^{n} |f(x_k) - h(x_k)|^2$.

If this problem has a solution, we will say that h is a best discrete approximation of f in the least squares sense.

Now we are going to see how this problem may be formulated using the terminology of a preHilbert space.

2. 1 Distance to a subspace

The results obtained in **1.4** and **1.5** show that if F is a subspace of finite dimension in an inner product space E, then $E = F \oplus F^{\perp}$.

Every vector u of E may thus be written in a unique way: $u = u' + u''$, $u' \in F$, $u'' \in F^{\perp}$, and the theorem of Pythagoras gives $\|u\|^2 = \|u'\|^2 + \|u''\|^2 \geq \|u'\|^2$. The map p which associates with a vector u its component u' in F is called an orthogonal projection of E onto F.

For every v in F, we may write:
$$u - v = \big(u - p(u)\big) + \big(p(u) - v\big)$$
The two vectors $u - p(u)$ and $p(u) - v$ are orthogonal, so the Pythagorean theorem gives:

$$\|u - p(u)\| \leq \|u - v\|.$$

Thus, among the vectors of F $p(u)$ is the "nearest " to u.

We say that $d(u, F) = \|u - p(u)\|$ is the distance of u from F.

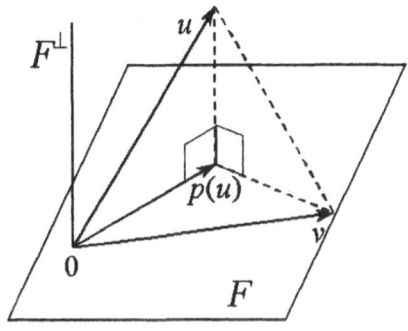

If $(e_1, e_2, \ldots, e_n)$ is a orthonormal basis of F (other than the trivial case where F reduces to $\{0\}$), we know that $p(u)$ may be written $p(u) = \sum_{k=1}^{n} <e_k, u> e_k$. The distance from a vector u to F satisfies:

$$d(u, F)^2 = \|u - p(u)\|^2 = \|u\|^2 - \|p(u)\|^2 = \|u\|^2 - \sum_{k=1}^{n} |<e_k, u>|^2$$

In particular, we note that $\sum_{k=1}^{n} |<e_k, u>|^2 \leq \|u\|^2$ (with equality if and only if $u \in F$).

Now, we suppose that E does not have a finite basis (E is not finite dimensional) but that there is an algebraic basis, a countably infinite set of vectors such that each vector of E is a finite linear combination of basis elements. The Gram-Schmidt orthogonalization process also shows here that there is an orthonormal sequence $(e_k)_{k \geq 0}$ in E.

If we apply the preceding arguments to the subspaces F_n generated by $(e_k)_{0 \le k \le n}$, we observe that the series with general term $|<e_k, u>|^2$ is convergent and that:

$$\sum_{k=0}^{\infty} |<e_k, u>|^2 \le \|u\|^2$$

This result is known as the Bessel Inequality. If equality holds, it is called the Parseval Identity: this applies in the case in the theory of Fourier series for piece-wise continuous functions.

The function distproj lets us simultaneously calculate $d(u, F)$ from a vector u to a sub-space F of finite dimension and the orthogonal projection $p(u)$ of u onto F.

The syntax is distproj(u,f), where f is a list of vectors forming a basis of F. This may be an arbitrary basis since it will be orthogonalized by the function schmidt).

The result is obtained in the form $\{d(u,F),p(u)\}$.

The function distproj uses the functions myscal and mynorm to calculate the scalar products and the norms: just like the function schmidt, it thus makes a sequence of calls to the function defscal.

```
:distproj(u,f)
:Func
:Local p,d
:schmidt(f)→ f
:Σ(myscal(f[j],u)*f[j],j,1,dim(f))→p
:√((mynorm(u))^2-(mynorm(p))^2)→d
:{d,p}
:EndFunc
```

Here is an example of using the function distproj: we want to know the distance from the function $x \mapsto \sqrt{x}$ to a subspace F of polynomials of degree less than or equal to 2, using the scalar product $<f, g> = \int_0^1 \overline{f} g$.

We start by calling the program defscal to define the scalar product. (In the dialogue box, we specify $a = 0$, $b = 1$, and $w(x) = 1$).

F is generated by $1, x, x^2$.
We thus call distproj with arguments $\sqrt{x}$ and $\{1, x, x^2\}$.
We see that the distance from the function $u(x) = \sqrt{x}$ to F is $\sqrt{2}/70$ and that its orthogonal projection onto F is:

$$p(u) = -\frac{4}{7}x^2 + \frac{48}{35}x + \frac{6}{35}$$

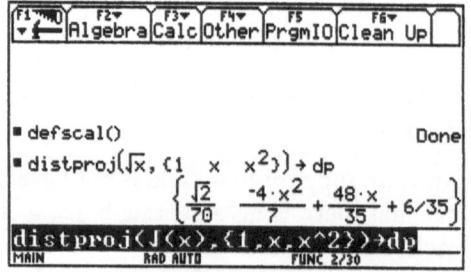

We may verify the preceding result by direct calculation:

We put in the variable d2 the square of the distance from $u(x) = \sqrt{x}$ to an arbitrary element of F, a polynomial of degree less than or equal to ≤ 2, with undetermined coefficients a, b, c. Use of expand will simplify somewhat the expressions which appear in the following calculations.

The quantity placed in d2 is a convex function of the variables a, b, c. It attains its minimum where the partial derivatives all vanish.

We store in eq1, eq2, eq3 the equations obtained by setting the partial derivatives equal to zero.
The we solve the system with solve:
We readily find the coefficients of the polynomial obtained by the function distproj.

We follow by using solve in a way specific to the TI-92 plus or the TI-89. In every case, we may make the calculations in the following way, which offers more control in the solution of the system:

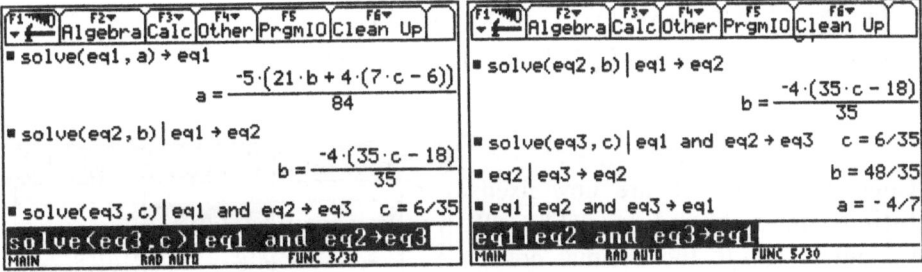

2. 2 "Continuous" or "discrete" least squares: a comparison

The calculations we are bow going to make will deliver the coefficients α, β, γ, which satisfy the following problem (here, with $\phi(x) = \sqrt{x}$):

(1) Find the minimum of $\int_0^1 \left(\phi(x) - (\alpha x^2 + \beta x + \gamma)\right)^2 dx.$

This does an approximation in the "continuous" least squares sense, as opposed to an approximation in the sense of "discrete least squares", which we describe as follows. If $S = \{x_0, x_1, \ldots, x_n\}$ is a family of n distinct points of the same interval $[0, 1]$:

(2) Find the minimum of $\sum_{k=0}^{n} \left(\phi(x_k) - (a_s x_k^2 + b_s x_k + c_s)\right)^2$.

For the TI calculators, this has a name: it may be done with an adjustment to the built-in function from statistics, "quadreg", which approximates a collection of points: $(x_0, \phi(x_0)), (x_1, \phi(x_1)), \ldots, (x_n, \phi(x_n))$, by a polynomial of degree less than or equal to 2.

We may also require that the solution (a_s, b_s, c_s) of problem (2), which depends on the family S, tends to a solution (α, β, γ) of problem (1), when the family S of points "best fills" the interval $[0, 1]$.
We will also try to respond to this question in this section, being content with only an empirical approximation since the problem is difficult.
As the family S we will take $S_n = \{x_0 = 0, \ldots, x_k = k/n, \ldots, x_n = 1\}$, formed of $n+1$ points equally spaced in the interval $[0, 1]$.

First, we resume the problem of approximating $\phi(x) = \sqrt{x}$ by a polynomial $P(x)$ of degree less than or equal to 2.
In the sense of problem (1), we have found $P(x) = \alpha x^2 + \beta x + \gamma$, with $\alpha = -4/7 \simeq -0.57142857$, $\beta = 48/35 \simeq 1.3714286$, $\gamma = 6/35 \simeq 0.17142857$.

We are going to use the statistical functions of the calculator to form the best approximation of type "QuadReg" to the collection of points $(x_k, \phi(x_k))_{0 \le k \le n}$.

Here is how to proceed:

- Use the table editor (press APPS,6,3). It does not matter what name we give the variable containing these "data".
- In the first row of column c1, put 5, for example.
- In case c2, put c2=seq(k/c1[1],k,0,c1[1]).
- In case c3, put c3=seq($\sqrt{}$(k/c1[1]),k,0,c1[1]).

Columns c2 and c3 are now fixed: their contents depend on c1[1]. If we put 5. (note the decimal point) instead of 5 in c1[1], the contents of column c2 and c3 are displayed in real format.

- Then pass to the screen for statistical calculations (press F5).
- Choose the type QuadReg, then put c2 and c3 in rows x and y.

- Confirm the table (press Enter twice). The calculator then displays the best approximation. The coefficient R^2 is a measure of the quality of the approximation: it is closer to 1 when the approximation is more accurate.

Here are the two principal steps:

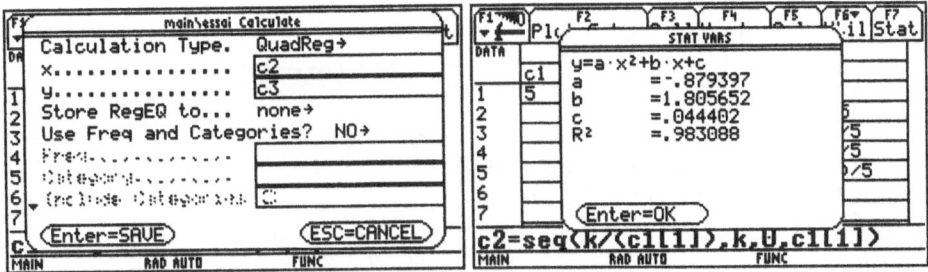

We note that with our 6 points $x_0, \ldots, x_5$ the coefficients a, b, c are values which are not particularly close to α, β, γ. The reason is that there are too few points. Thus, we repeat the same calculations, but with 101 points (thus c1[1]=100.), then 201 points (c1[1]=200.).

After confirming the contents of the last dialog box with Enter, we modify the contents of the cell c1[1] as indicated: columns c2 and c3 change as a consequence. Pressing F5, we return to the screen for statistics calculations which is confirmed by Enter since the first three "fields " are already filled.

Here is what we obtain with $n = 100$, then $n = 200$:

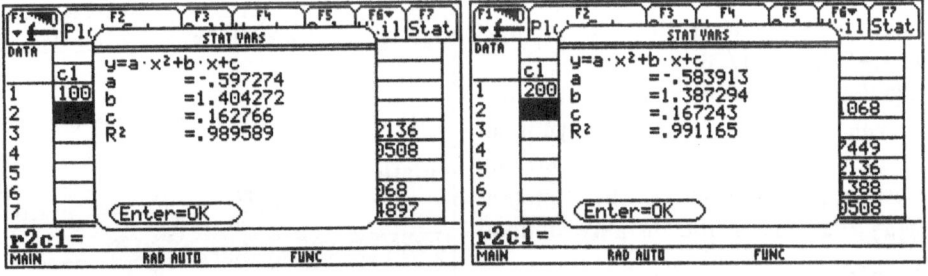

We observe that the coefficients a, b, c approach the coefficients α, β, γ. We will be content with this purely empirical observation.

Now we will make the same comparison and in the same context, but with the function $\phi(x) = x^3$, which will let us use a different method.

Recall that finding the best approximation in the sense of discrete least squares consists of minimizing $r = \sum_{k=0}^{n} f(x_k)$, where $f(x)$ designates $\left(\phi(x) - (ax^2 + \right.$

$bx + c))^2$, with $x_k = k/n$, as before.

We define $f(x)$, and we see that the calculator is capable of calculating the exact form of the sum r, which is a real accomplishment! (To be honest, this is made possible by our choice of a polynomial function for $\phi(x)$, here x^3).

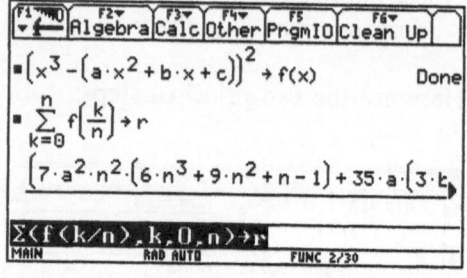

The minimum of the function f is attained for the values of a, b, c which cause the partial derivatives of r to vanish.
We calculate the three partial derivatives, then we form the system, which, all the same, is fairly monstrous.

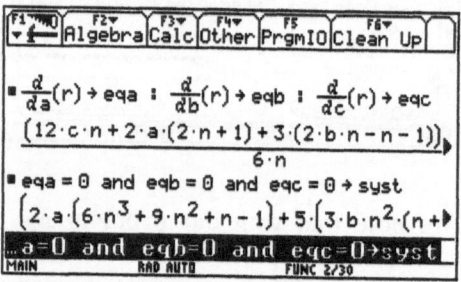

It only takes a few seconds for the calculator to solve this system exactly. We find three values a_n, b_n, c_n: $P_n(x) = a_n x^2 + b_n x + c_n$ is the best approximation of $\phi(x) = x^3$, in the sense of discrete least squares, for the $n+1$ points $x_k = k/n$, $0 \leq k \leq n$.

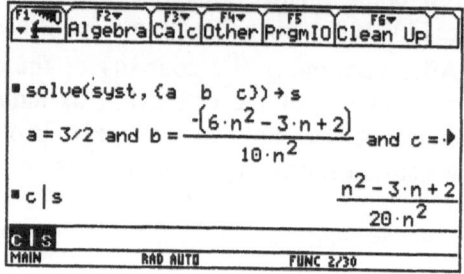

We see that when n tends to $+\infty$, the coefficients a_n, b_n, c_n tend to $a = 3/2$, $b = -3/5$, $c = 1/20$.
We finally have the pleasure of observing that the "limit" polynomial $P(x) = ax^2 + bx + cx$ is indeed the best approximation of the function $\phi(x) = x^3$, using the distance defined by $d(f, g) = \sqrt{\int_0^1 (f - g)^2}$.

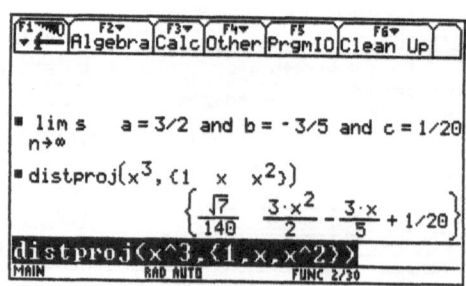

We end the comparison between the two modes of approximation by using a third method, showing that there are many possibilities to explore a given mathematical theme when using this calculator.
As in the first two examples, we use the interval $[0, 1]$, and we use the inner product defined by $< f, g > = \int_0^1 fg$.
At this time we are going to generalize the preceding problem by considering that the function ϕ which we want to approximate is arbitrary, while searching

for the best approximation of ϕ using polynomials $P(x) = ax + b$ of degree less than or equal to 1.

The problem of "discrete" approximation is then the same as finding a best fit of the "LinReg" type: it suffices to find the so-called "least squares line". (Readers who need a reminder about the formulas leading to the least squares line may wish to consult a statistics textbook.)

As before, we use the $n+1$ values $(x_k = k/n)_{0 \le k \le n}$.

If X_n is the sequence $x_0, \ldots, x_n$ and Y_n the sequence $\phi(x_0), \ldots, \phi(x_n)$, we find the least squares line of the set of pairs (X_n, Y_n) formed by $n+1$ points $(x_k, \phi(x_k))$.

We know that the equation of this line is $y = a_n x + b_n$, with:

$$a_n = \frac{\mathrm{cov}(X_n, Y_n)}{\mathrm{var}(X_n)}, \quad \text{et } b_n = \mathrm{E}(Y_n) - a_n \mathrm{E}(X_n)$$

- The expected value $\mathrm{E}(X_n)$ of the sequence X_n, may be written $\dfrac{1}{n+1} \displaystyle\sum_{k=0}^{n} \dfrac{k}{n}$

  and it has the value 1/2.
- The variance satisfies the Huyghens formula: $\mathrm{var}(X_n) = \mathrm{E}(X_n^2) - \mathrm{E}^2(X_n)$, with

$$\mathrm{E}(X_n^2) = \frac{1}{n+1} \sum_{k=0}^{n} \frac{k^2}{n^2} = \frac{2n+1}{6n}. \quad \text{Finally, we find: } \mathrm{var}(X_n) = \frac{n+2}{12n}.$$

Here is how the calculator gets the preceding results: first, the expected value of the sequence X_n and then of the sequence of its squares, X_n^2, then the variance of X_n.

F1	F2▾ Algebra	F3▾ Calc	F4▾ Other	F5 PrgmIO	F6▾ Clean Up

$$\blacksquare \frac{1}{n+1} \cdot \sum_{k=0}^{n}\left(\frac{k}{n}\right) \to ex \qquad\qquad 1/2$$

$$\blacksquare \frac{1}{n+1} \cdot \sum_{k=0}^{n}\left[\left(\frac{k}{n}\right)^2\right] \to ex2 \qquad \frac{2 \cdot n+1}{6 \cdot n}$$

$$\blacksquare ex2 - ex^2 \to vx \qquad\qquad \frac{n+2}{12 \cdot n}$$

`ex2-ex^2→vx`

MAIN   RAD AUTO   FUNC 3/30

- The expected value $\mathrm{E}(Y_n)$ of the sequence Y_n may be written $\dfrac{1}{n+1} \displaystyle\sum_{k=0}^{n} \phi\left(\dfrac{k}{n}\right)$

- The covariance of X_n and Y_n satisfies $\mathrm{cov}(X_n, Y_n) = \mathrm{E}(X_n Y_n) - \mathrm{E}(X_n)\mathrm{E}(Y_n)$,

  with $\mathrm{E}(X_n Y_n) = \dfrac{1}{n+1} \displaystyle\sum_{k=0}^{n} \dfrac{k}{n}\phi\left(\dfrac{k}{n}\right).$

The two quantities $\mathrm{E}(Y_n)$ and $\mathrm{E}(X_n Y_n)$ are clearly Riemann sums, and they have the obvious limits when n tends to ∞:

$$\lim_{n \to \infty} \mathrm{E}(Y_n) = \int_0^1 \phi(x)\,\mathrm{d}x, \quad \text{and} \quad \lim_{n \to \infty} \mathrm{E}(X_n Y_n) = \int_0^1 x\phi(x)\,\mathrm{d}x$$

Since the limit of $\mathrm{var}(X_n)$ is $1/12$, we deduce:

$$\lim_{n \to \infty} a_n = 12 \int_0^1 x\phi(x)\mathrm{d}x - 6 \int_0^1 \phi(x)\mathrm{d}x$$

$$\lim_{n \to \infty} b_n = \int_0^1 \phi(x)\mathrm{d}x - \frac{1}{2}\lim_{n \to \infty} a_n = 4 \int_0^1 \phi(x)\mathrm{d}x - 6 \int_0^1 x\phi(x)\mathrm{d}x$$

If we denote these two limits by a and b, it remains for us to verify that $P(x) = ax+b$ is indeed the best approximation of $\phi(x)$ in the sense of continuous least squares, or that it minimizes the quantity $I = \int_0^1 \left(\phi(x) - (ax+b)\right)^2 \mathrm{d}x$, considered as a function of a and b.

We store the preceding integral in the variable i and calculate its partial derivatives with respect to a and b, which we store in variables α and β.

We must now find the values of a and b which make the two partial derivatives vanish.
Wanting to immediately solve the system $\{\alpha = 0, \beta = 0\}$ is a little ambitious, as we see here.

It must be a trick! The quantities α and β are linear relative to a and b. To find the coefficients of a and b in these expressions, and the constant coefficients, the values of the partial derivatives must be evaluated at $a = b = 0$.

We may now solve the system $\{\alpha = 0, \beta = 0\}$ with respect to the variables a and b.
We get the same values of a and of b as by passing to the limit in the discrete least squares approximation above.

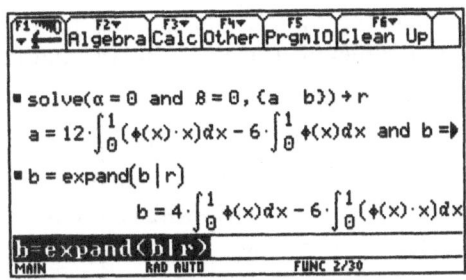

2. 3 "Discrete " least squares: a generalization

In the preceding paragraph, we compared two modes of least squares approximation: the continuous mode and the discrete mode.
The first allowed us to illustrate the notion of an orthogonal projection in an inner product space with an "integral " inner product. In the second, we took a more empirical approach, limiting ourselves to some simple cases. We are now going to make a more general presentation for the discrete case.

Let n be a natural number and let y be a vector in $\mathbb{K}^{n+1}$, equipped as usual with its canonical scalar product.
Let E be a $\mathbb{K}$-vector space of dimension $m + 1$, with $m \leq n$.
Let φ be an injective linear map from E to $\mathbb{K}^{n+1}$: φ thus defines an isomorphism of E onto $\operatorname{Im}\varphi$, a subspace of dimension $m + 1$ of $\mathbb{K}^{n+1}$.

The problem is the following: we seek among all elements of E, the one whose image under φ is the closest "approximation " to y.
It is clear that the problem has a unique solution a, the preimage in E by the map φ of the orthogonal projection $p(y)$ of y onto $\operatorname{Im}\varphi$. The following scheme illustrates this situation:

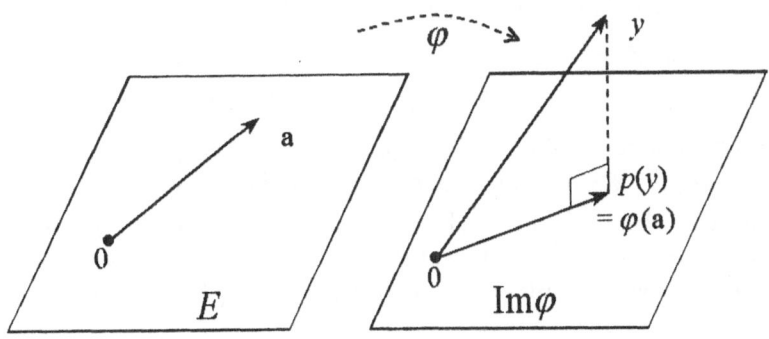

Now we will see how a search for the vector a leads to an invertible linear system.

Let (ε) be a basis of E, and let (e) be the standard basis of $\mathbb{K}^{n+1}$.
We denote by $[a]_\varepsilon$ and $[y]_e$ the column matrices respectively representing the coordinates of a in the basis (ε) and of y in the basis (e).
We are going to use the matrix representation of the scalar product in $\mathbb{K}^{n+1}$:
For each vector y and z of $\mathbb{K}^{n+1}$, $< y, z > = \overline{[y]_e}^\top [z]_e$ (see paragraph 1.3).
Let M be the transition matrix of φ between the bases (ε) of E and (e) of $\mathbb{K}^{n+1}$. This is a matrix with $n+1$ rows and $m+1$ columns): for each vector b of E, $[\varphi(b)]_e = M[b]_\varepsilon$.

The vector a is a solution of the preceding problem

$$\Leftrightarrow \varphi(a) \text{ is the orthogonal projection of } y \text{ onto } \operatorname{Im}\varphi$$

$$\Leftrightarrow \varphi(a) - y \in (\operatorname{Im}\varphi)^\perp \Leftrightarrow \forall z \in \operatorname{Im}\varphi, < z, \varphi(a) - y > = 0$$

$$\Leftrightarrow \forall b \in E, < \varphi(b), \varphi(a) - y > = 0 \Leftrightarrow \forall b \in E, \overline{[\varphi(b)]_e}^\top [\varphi(a) - y]_e = 0$$

$$\Leftrightarrow \forall b \in E, \overline{[b]_\varepsilon}^\top \overline{M}^\top \left(M[a]_\varepsilon - [y]_e \right) = 0 \Leftrightarrow \overline{M}^\top \left(M[a]_\varepsilon - [y]_e \right) = 0$$

$$\Leftrightarrow \overline{M}^\top M[a]_\varepsilon = \overline{M}^\top [y]_e$$

The matrix $\overline{M}^\top M$ is square of order $m+1$. The preceding equation thus is a system of $m+1$ equations in $m+1$ unknowns, the coordinates of a in the basis (ε) of E: we say that these are the *normal equations* of the problem.

This system has a unique solution. We may get it by observing that $\overline{M}^\top M$ is invertible since its "kernel " is just the zero vector.
In fact, for every column vector $[b]$ of $\mathcal{M}_{m+1,1}(\mathbb{K})$:

$$\overline{M}^\top M[b] = 0 \Rightarrow \overline{[b]}^\top \overline{M}^\top M[b] = 0 \Rightarrow \overline{M[b]}^\top M[b] = 0 \Rightarrow \|M[b]\|^2 = 0 \Rightarrow [b] = 0$$

The last implication results from the fact that the linear map φ associated as before with the matrix M is injective.

The matrix $N = \overline{M}^\top M$ is even more than just invertible:

- It likewise satisfies $\overline{N}^\top = N$: we say that N is *Hermitian*.
  In the "real " case ($\mathbb{K} = \mathbb{R}$), we say that N is *symmetric* ($N = N^\top$).

- For every non-zero column vector $[b]$, $\overline{[b]}^\top N[b] = \|M[b]\|^2 > 0$. We express this property by saying that N is *positive definite*.

One may show that N is diagonalizable over $\mathbb{K}$ (and even with an orthonormal basis) with strictly positive eigenvalues and that we may apply some specific algorithms to solve the system of normal equations. (We will revisit these facts later.)

The preceding presentation, being somewhat theoretical, seems far removed from the practical application of discrete least squares approximation which we untertook in the previous paragraph. We will see that this is not the case.

Here is how to pose the present problem in more concrete terms:

- Let f be a function defined on an interval I of $\mathbb{R}$, with values in $\mathbb{K}$.

- In I, we are given $n+1$ distinct points $x_0, x_1, \ldots, x_n$, and the vector $y = \big(f(x_0), \ldots, f(x_n)\big)$ (an element of $\mathbb{K}^{n+1}$) of values of f at these points.

- Finally, let $\varepsilon_0, \varepsilon_1, \ldots, \varepsilon_m$ be $m+1$ linearly independent maps of I into $\mathbb{K}$. Let E be the $\mathbb{K}$-vector space of dimension $m+1$ which they generate. The situation of interest is the one where the ε_k are the polynomial functions $x \mapsto x^k$, in which case $E = \mathbb{K}_m[x]$, the set of all polynomials with coefficients in $\mathbb{K}$.

- The problem is then as follows: among all functions a of E (linear combinations of $\varepsilon_0, \varepsilon_1, \ldots, \varepsilon_n$), we seek one which minimizes the quantity $\sum_{i=0}^{n} |f(x_i) - a(x_i)|^2$.

Let φ be the map which for each b of E associates the vector $\varphi(b) = \big(b(x_0), b(x_1), \ldots, b(x_n)\big)$: φ is linear from E into $\mathbb{K}^{n+1}$.

Thus, we are to find a in E whose image under φ is the "closest" to the vector y, that is, which minimizes $\|y - \varphi(a)\|^2 = \sum_{i=0}^{n} |f(x_i) - a(x_i)|^2$.

With this initial formulation of the problem there apparently remains one detail to verify: is the linear map φ injective? We will see that the response to this question can only be "almost always", since this depends on the choice of the $n+1$ points x_i.

Let $b = \sum_{j=0}^{m} \lambda_j \varepsilon_j$ be a function in E.
We know that the functions $\varepsilon_0, \varepsilon_1, \ldots, \varepsilon_m$ are linearly independent since they form a basis of E. This means that if b is identically zero on I, then the coefficients λ_j are zero.
But the problem of determining whether φ is injective may be expressed as follows: if b vanishes at $n+1$ points x_i, are the coefficients λ_j necessarily zero? (or again: is g necessarily zero on I ?).
If the family of $m+1$ functions $\varepsilon_0, \varepsilon_1, \ldots, \varepsilon_m$ is given, everything depends on the choice of the $n+1$ values $x_0, x_1, \ldots, x_n$.

For example, suppose that the ε_j are defined by $x \mapsto \varepsilon_j(x) = \sin(j+1)x$. If we choose, for all i of $\{1, \ldots, n\}$, $x_i = j\pi$, then we observe that all the functions ε_j (and thus their linear combinations) vanish at the points x_i. In this case, φ is identically zero and is thus far from being injective!

On the other hand, if the ε_j are defined by $x \mapsto \varepsilon_j(x) = x^j$, then φ is injective for any $n+1$ distinct points x_i! In fact, an element of E, that is, a polynomial of degree $\leq m$) which vanishes at these $n+1$ points, is then identically zero. This is a consequence of the fact that $n \geq m$).

In what follows, we suppose that the choice of $m+1$ functions $\varepsilon_0, \varepsilon_1, \ldots, \varepsilon_m$, or by default the choice of the $n+1$ points $x_0, x_1, \ldots, x_n$, is made "judiciously ", so that the map φ will be injective.

If we equip the space E with the basis (ε) and $\mathbb{K}^{n+1}$ with its standard basis, here is the matrix M of the linear transformation φ. This matrix has $m+1$ linearly independent columns and $n+1$ rows with $m \le n$.
We display also a particular classical case obtained when the functions ε_j are defined by $x \mapsto \varepsilon_j(x) = x^j$. M is then a so-called "Vandermonde " matrix:

$$M = \begin{pmatrix} \varepsilon_0(x_0) & \varepsilon_1(x_0) & \cdots & \varepsilon_m(x_0) \\ \varepsilon_0(x_1) & \varepsilon_1(x_1) & \cdots & \varepsilon_m(x_1) \\ \varepsilon_0(x_2) & \varepsilon_1(x_2) & \cdots & \varepsilon_m(x_2) \\ \vdots & \vdots & \ddots & \vdots \\ \varepsilon_0(x_n) & \varepsilon_1(x_n) & \cdots & \varepsilon_m(x_n) \end{pmatrix} \quad \text{Particular case } M = \begin{pmatrix} 1 & x_0 & \cdots & x_0^m \\ 1 & x_1 & \cdots & x_1^m \\ 1 & x_2 & \cdots & x_2^m \\ \vdots & \vdots & \ddots & \vdots \\ 1 & x_n & \cdots & x_n^m \end{pmatrix}$$

The solution $a = \sum_{j=0}^{m} a_j \varepsilon_j$ of the problem (approximation of the vector y with values of f at the points $x_0, \ldots, x_n$, by a function which is a linear combination of $\varepsilon_0, \ldots, \varepsilon_m$) may thus be obtained by solving the system: $\overline{M}^\top M[a] = {}^\top \overline{M}[y]$.

Note first that the "transpose " function of the calculator is in fact a "conjugate transpose ", as we see in this example: what the calculator designates by $M^\top$ is actually the transpose of the conjugate of M, that is, with our notation, $\overline{M}^\top$.

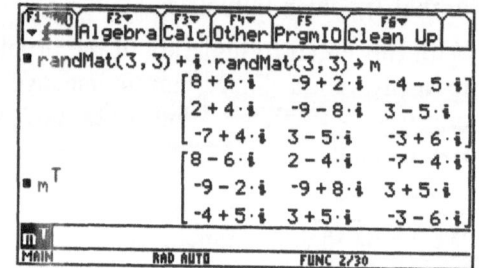

The function lsqrs gives the approximation in the sense of discrete least squares for a family of points given by the sequence $\{x_1, x_2, \ldots\}$ of these values (argument xx), the list $\{y_1, y_2, \ldots\}$ of the values (argument yy), and for a list $\{\varepsilon_1(x), \varepsilon_2(x), \ldots\}$ of "approximating " functions (argument ee):

```
:lsqrs(xx,yy,ee)
:Func
:Local n,m,a,i,j,r
:dim(xx)→n:dim(ee)→m
:If n≠dim(yy) or m>n:Return "Dim error"
:seq(seq(ee[j],j,1,m)|x=xx[i],i,1,n)→a
:list▷mat(yy,1)→y
:mat▷list(simult(aᵀ*a,aᵀ*yy))→r
:Σ(r[j]*ee[j],j,1,m)
:EndFunc
```

Remark: The function lsqrs, written in a "clear " manner and resorting to the instruction simult to solve the normal equations, is reserved for simple situations. Numerical methods which are applicable in all cases will be studied later.

In the example below, we have found the best approximation among the polynomials of degree ≤ 5 (these form a vector space of dimension 6) on the family of 7 points $(-2,2),(-1,-1),(3,1),(4,0),(6,-1),(1,3),(2,-2)$.

Then we graphed jointly the set of these points (using the instruction newplot 1,1,xx,yy) and the curve representing this polynomial:

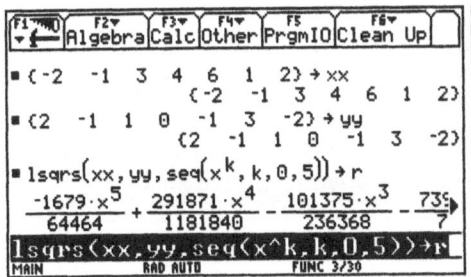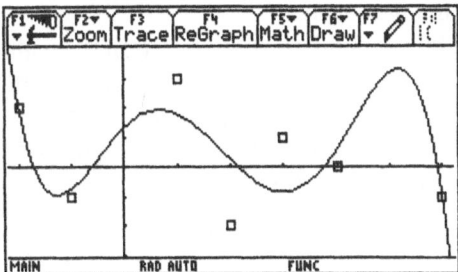

We mustn't forget that the calculator also knows how to do approximations in the sense of discrete least squares using polynomials of degree less than or equal than n, with $n = 1$ (instruction LinReg), $n = 2$ (QuadReg), $n = 3$ (CubicReg), and $n = 4$ (QuartReg).

In this example, we have approximated a set of 5 points by a polynomial function of degree ≤ 3, in two different ways:
• with the instruction CubicReg.
• with the function lsqrs.
The two results are identical.

```
F1▾  F2▾   F3▾   F4▾   F5    F6▾
▾◣─Algebra Calc Other PrgmIO Clean Up
■ {1  2  3  4  6}→xx   {1  2  3  4  6}
■ {2  1  4  0  1}→yy   {2  1  4  0  1}
■ CubicReg xx,yy : regeq(x)
        .1447·x³ − 1.561·x² + 4.564·x − 1.547
■ lsqrs(xx,yy,{x³  x²  x  1.})
        .1447·x³ − 1.561·x² + 4.564·x − 1.547
lsqrs(xx,yy,{x^3,x^2,x,1.})
MAIN          RAD AUTO        FUNC 4/30
```

With the same points, we have calculated in two ways the approximation by a function of type $a + b \ln x$.

Finally, we may very well imagine other "approximating " functions, such as the linear combinations of $a + bx + c \exp x$ here.

```
F1▾  F2▾   F3▾   F4▾   F5    F6▾
▾◣─Algebra Calc Other PrgmIO Clean Up
■ LnReg xx,yy : regeq(x)
                    2.167 − .5705·ln(x)
■ lsqrs(xx,yy,{1.  ln(x)})
                    2.167 − .5705·ln(x)
■ lsqrs(xx,yy,{1.  x  e^x})
                    −.0001·e^x − .233·x + 2.358
lsqrs(xx,yy,{1.,x,ε^(x)})
MAIN          RAD AUTO        FUNC 3/30
```

In this example, we see that the function lsqrs allows symbolic calculations. We have thus formed the equation of the least squares line for the three points $(1, -1)$, $(2, 3)$, and $(\lambda, 1)$, which isn't allowed by the instruction LinReg.

Here is a last example using the function lsqrs:
Here we find an approximation through the points $(1, 1+i)$, $(i, 1)$, $(0, 1)$ and $(1 - i, -i)$ by a polynomial of degree 2 with complex coefficients (Again, we could not use the instruction QuadReg).

3. Orthogonal polynomials

Let I be an interval of $\mathbb{R}$, and let w be a continuous function on I with positive values (possibly vanishing at isolated points of I).
We know that the equation $< f, g > = \int_I f(x)g(x)\omega(x)\mathrm{d}x$ defines a scalar product on the vector space E of continuous functions f on I, with values in $\mathbb{R}$, and which are such that $f^2\omega$ is integrable on I.

In all the classical particular cases based on this generic definition, E contains the vector space $\mathbb{R}[X]$ of polynomials with coefficients in $\mathbb{R}$. In particular, we could consider the restriction of the scalar product to $\mathbb{R}[X]$.

In this section, we will consider sequences $(P_n)_{n\geq 0}$ such that
• For every integer n, the polynomial P_n is of degree n.
• For all distinct integers m and n, P_n and P_m are orthogonal.

Such a family (P_n) will be called a sequence of orthogonal polynomials. We know that the Gram-Schmidt orthogonalization process, applied to the family $(x^k)_{k\geq 0}$, permits the creation of such a sequence (which moreover is constituted of monic polynomials).

3. 1 First properties

Let $(P_k)_{k\geq 0}$ be a sequence of orthogonal polynomials.

Proposition 1: For every natural number n, the polynomials $P_0, P_1, \ldots, P_n$ form a basis of $\mathbb{R}_n[X]$. More generally, the family $(P_k)_{k \geq 0}$ is a basis of $\mathbb{R}[X]$.

In fact, the polynomials P_k are of "consecutive " degree: thus, they form a linearly independent family (a classic result which is easily proved).
On the other hand, for each integer n, the $n + 1$ polynomials $P_0, P_1, \ldots, P_n$ appear in the vector space $\mathbb{R}_n[X]$ of polynomials of degree less than or equal to n, which is of dimension $n + 1$: they form a basis.

Let P be a polynomial of $\mathbb{R}_n[X]$. Write it as $P = \sum_{k=0}^{n} \lambda_k P_k$.
By scalar multiplication of this equation by P_k and using the orthogonality of P_k with the other polynomials of the sequence, we find:

$$\forall k \in \{0, \ldots, n\}, \ \lambda_k = \frac{< P_k, P >}{\|P_k\|^2}$$

Every polynomial P of $\mathbb{R}[X]$ may thus be written: $P = \sum_{k \geq 0} \frac{< P_k, P >}{\|P_k\|^2} P_k$

By the Pythagorean theorem: $\|P\|^2 = \sum_{k \geq 0} \left\| \frac{< P_k, P >}{\|P_k\|^2} P_k \right\|^2 = \sum_{k \geq 0} \frac{1}{\|P_k\|^2} |< P_k, P >|$

NB: The preceding sums are finite and bounded above by $k = \deg P$.

Proposition 2: For every natural number n, and for every polynomial P of degree strictly less than n, the scalar product $< P_n, P >$ is zero.

This is evident if $n = 0$. Otherwise, this results from P_n being orthogonal to $P_0, \ldots, P_{n-1}$ which generates the vector space $\mathbb{R}_{n-1}[X]$.

Proposition 3: Let $(Q_k)_{k \geq 0}$ be another orthogonal sequence of $\mathbb{R}[X]$.
Then, for every n of $\mathbb{N}$, there exists a non-zero coefficient λ_n such that $Q_n = \lambda_n P_n$.

Proof: let a_n and b_n be the non-zero coefficients of x^n in P_n and Q_n. If we put $\lambda_n = a_n/b_n$, and $R_n = P_n - \lambda_n Q_n$, then $\deg(R_n) \leq n - 1$. The polynomial R_n is thus orthogonal to P_n and to Q_n, thus to itself: it is the zero polynomial. Thus $P_n = \lambda_n Q_n$.

The preceding result means that, up to a given sequence of multiplicative non-zero coefficients, there only exists one orthogonal sequence. Thus, one supplementary condition (in general, a linear one) suffices to convince us of the existence and of the uniqueness of a given orthogonal family.

For example, there exists only one orthogonal sequence of monic polynomials (that is, with the coefficient of the highest degree term equal to one). Similarly the Legendre polynomials $(P_n)_{n \geq 0}$ form the only orthogonal family (for the scalar product $< P, Q > = \int_0^1 PQ$) satisfying: $\forall n \in \mathbb{N}, \ P_n(1) = 0$.

Proposition 4: *For each integer* $n \geq 1$, *a polynomial* P_n *of degree* n *has* n *distinct roots, which are all real, and which all appear in the interval* I.

Proof: Let m be the number of distinct roots of P_n, which are of odd multiplicity and which appear in I. Since P_n is of degree n, we know that we must show the equality $m = n$. Denote these roots by $x_1, x_2, \ldots, x_m$.
We set $Q_m = (x - x_1) \cdots (x - x_m)$ (if $m = 0$, we put $Q_0 = 1$). By definition of x_k, all the roots of $P_n Q_m$ which could appear in I are now of even multiplicity. The continuous map $x \mapsto P_n(x) Q_m(x) \omega(x)$ thus has a constant sign on the interval I. (Recall that ω has positive values.)
We reason by contradiction: suppose that $m < n$, that is, $\deg Q_m < \deg P_n$. Under these conditions P_n and Q_m are orthogonal.

We may thus write: $0 = <P_n, Q_m> = \int_I P_n(x) Q_m(x) \omega(x) \, dx$.

This equation shows that the function $P_n Q_m \omega$ (continuous and with constant sign) is in fact identically zero on I. The map ω is zero only possibly at isolated points of I, so the polynomial $P_n Q_m$ is necessarily identically zero on I, which is impossible since it is of degree $m + n \geq 1$.
The property is thus demonstrated: all the roots of P_n are real, distinct, and appear in the interval I.

Proposition 5: *For each integer* $n \geq 1$, *and for each polynomial* P *of degree strictly less than* $n - 1$, *the scalar product* $<xP_n, P>$ *is zero.*

This is a consequence of the first proposition, since the scalar product $<xP_n, P>$ may be written $<P_n, xP>$, and $\deg(xP) < n$.

Proposition 6: *There exist three real sequences* (a_n), (b_n), (c_n) *such that:* $\forall n \geq 1$, $xP_n = a_n P_{n-1} + b_n P_n + c_n P_{n+1}$.

Proof: Let n be an integer greater than or equal to 1. The polynomial xP_n, of degree $n + 1$, may be decomposed in the (orthogonal) basis $P_0, \ldots, P_n, P_{n+1}$ of $\mathbb{R}_{n+1}[X]$. But we know that it is orthogonal to every polynomial of degree less than $n - 1$, thus to $P_0, \ldots, P_{n-2}$. xP_n may thus be decomposed uniquely in terms of the polynomials P_{n-1}, P_n and P_{n+1}, which establishes the result.

Remark: To extend the preceding result and to define b_0 and c_0, we will write $xP_0 = b_0 P_0 + c_0 P_1$, expressing xP_0 in terms of the basis $\{P_0, P_1\}$ of $\mathbb{K}_1[X]$.

In the preceding relation, the coefficient c_n is non-zero (for reasons of degree). We could thus express P_{n+1} as a function of P_n and of P_{n-1}.
The relation then takes the following form: $P_{n+1} = (\alpha_n x + \beta_n) P_n + \gamma_n P_{n-1}$.
Such an equation lets us calculate the polynomials P_n step by step.

We may say nothing however about the sequences (a_n), (b_n), (c_n), (or of the sequences (α_n), (β_n) and (γ_n)) without having a supplementary hypothesis on the family of polynomials (P_n). (In fact, recall that they are only defined up to a sequence of multiplicative factors.) But we may state the following result:

Proposition 7: Let (P_n) be a sequence of orthogonal polynomials, of same norm beginning with $n = 1$. There exist two sequences (a_n) and (b_n) such that: $\forall n \geq 1, \; xP_n = a_n P_{n-1} + b_n P_n + a_{n+1} P_{n+1}$.

Proof: We know the relation: $\forall n \geq 1, \; xP_n = a_n P_{n-1} + b_n P_n + c_n P_{n+1}$.
If we take the product by P_{n+1} then by P_{n-1}, we find:
$\forall n \geq 1, \; < xP_n, P_{n+1} > = c_n \|P_{n+1}\|^2, \quad < xP_n, P_{n-1} > = a_n \|P_{n-1}\|^2$.
This last equation may be written: $\forall n \geq 0, \; < xP_{n+1}, P_n > = a_{n+1} \|P_n\|^2$.
Since $< xP_{n+1}, P_n > = < xP_n, P_{n+1} >$, and with the hypothesis on the norm of P_k, we find: $\forall n \geq 1, \; c_n = a_{n+1}$, which is the desired result.

Proposition 8: With the preceding notation (and still supposing that the P_k have the same norm if $k \geq 1$), the $n + 1$ roots $x_0, x_1, \ldots, x_n$ of P_{n+1} are the eigenvalues of the following matrix of order $n + 1$:

$$M_{n+1} = \begin{pmatrix} b_0 & c_0 & 0 & 0 & \cdots & 0 \\ a_1 & b_1 & a_2 & 0 & \cdots & 0 \\ 0 & a_2 & b_2 & a_3 & \cdots & 0 \\ \cdots & \cdots & \cdots & \cdots & \cdots & \cdots \\ 0 & \cdots & 0 & a_{n-1} & b_{n-1} & a_n \\ 0 & \cdots & \cdots & 0 & a_n & b_n \end{pmatrix}$$

Proof: For every real number x, let $U_{n+1}(x) = (P_0(x), P_1(x), \ldots, P_n(x))$, which we identify with the corresponding column vector. Because of its first component, this vector is never zero.

We are going to evaluate the product $M_{n+1} U_{n+1}(x)$ by using the recurrence relation satisfied by the polynomials P_k.

$$M_{n+1} U_{n+1}(x) = M_{n+1} \begin{pmatrix} P_0(x) \\ P_1(x) \\ \vdots \\ P_{n-1}(x) \\ P_n(x) \end{pmatrix} = \begin{pmatrix} b_0 P_0(x) + c_0 P_1(x) \\ a_1 P_0(x) + b_1 P_1(x) + a_2 P_2(x) \\ \vdots \\ a_{n-1} P_{n-2}(x) + b_{n-1} P_{n-1}(x) + a_n P_n(x) \\ a_n P_{n-1}(x) + b_n P_n(x) \end{pmatrix}$$

$$= \begin{pmatrix} x P_0(x) \\ x P_1(x) \\ \vdots \\ x P_{n-1}(x) \\ x P_n(x) - a_{n+1} P_{n+1}(x) \end{pmatrix} = x \begin{pmatrix} P_0(x) \\ P_1(x) \\ \vdots \\ P_{n-1}(x) \\ P_n(x) \end{pmatrix} - \begin{pmatrix} 0 \\ 0 \\ \vdots \\ 0 \\ a_{n+1} P_{n+1}(x) \end{pmatrix}$$

In particular, for each of the $n + 1$ (distinct) roots $x_0, \ldots, x_n$ of P_{n+1}, we see that $U_{n+1}(x_k)$ is an eigenvector of M_{n+1} for the eigenvalue x_k.
We thus have to find all the eigenvalues of the matrix M_{n+1}.

$$M_{n+1} \begin{pmatrix} P_0(x_k) \\ P_1(x_k) \\ \vdots \\ P_{n-1}(x_k) \\ P_n(x_k) \end{pmatrix} = x_k \begin{pmatrix} P_0(x_k) \\ P_1(x_k) \\ \vdots \\ P_{n-1}(x_k) \\ P_n(x_k) \end{pmatrix}$$

The preceding result proves that the characteristic polynomial of M_{n+1} is (up to a multiplicative constant) equal to the polynomial P_{n+1}.

We are now going to look at two classical sequences of orthogonal polynomials: Chebishev and Legendre polynomials.

3. 2 Chebishev polynomials

We consider here the scalar product $<f, g> = \int_{-1}^{1} \dfrac{f(x)g(x)}{\sqrt{1 - x^2}}\, dx$

Proposition 9: The equations $\cos n\theta = T_n(\cos\theta)$ $(n \in \mathbb{N}, \theta \in \mathbb{R})$ define a sequence $(T_n)_{n \geq 0}$ of orthogonal polynomials.

Proof: For every integer n, and for every real number θ:

$$\cos n\theta = \operatorname{Re} e^{in\theta} = \operatorname{Re}\left(\cos\theta + i\sin\theta\right)^n = \operatorname{Re}\left(\sum_{k=0}^{n}\binom{n}{k} i^k \sin^k\theta \cos^{n-k}\theta\right)$$

$$= \sum_{k=0}^{[n/2]}\binom{n}{2k}(-1)^k \sin^{2k}\theta\cos^{n-2k}\theta = \sum_{k=0}^{[n/2]}\binom{n}{2k}(-1)^k(1 - \cos^2\theta)^k\cos^{n-2k}\theta$$

which defines the polynomial $T_n(x) = \displaystyle\sum_{k=0}^{[n/2]}\binom{n}{2k}(x^2 - 1)^k x^{n-2k}$.

It is clear that T_n is of degree n, with dominant coefficient $\displaystyle\sum_{k=0}^{[n/2]}\binom{n}{2k} = 2^{n-1}$

The polynomial T_n is even if n is even, and odd if n is odd.
We immediately verify that: $T_0(x) = 1$, $T_1(x) = x$, $T_2(x) = 2x^2 - 1$ (without using the preceding formula, since we indeed know that $\cos 2\theta = 2\cos^2\theta - 1$).

It remains to verify the orthogonality of the sequence T_n.
Let m and n be two distinct natural numbers. In the integral giving the $<T_m, T_n>$, we effect the change of variable $x = \cos\theta$, with $0 \leq \theta \leq \pi$:

$$<T_m, T_n> = \int_{-1}^{1}\dfrac{T_m(x)T_n(x)}{\sqrt{1 - x^2}}\, dx = \int_{\pi}^{0}\dfrac{T_m(\cos\theta)T_n(\cos\theta)}{\sin\theta}(-\sin\theta)\, d\theta$$

$$= \int_{0}^{\pi}\cos m\theta\cos n\theta\, d\theta = \left[\dfrac{\sin(m-n)\theta}{2(m-n)} + \dfrac{\sin(m+n)\theta}{2(m+n)}\right]_{0}^{\pi} = 0$$

We have $\|T_0\| = \sqrt{\pi}$, But all the polynomials T_n, with $n \geq 1$, are of norm $\sqrt{\dfrac{\pi}{2}}$:

$$\forall n \geq 1, \|T_n\|^2 = \int_{0}^{\pi}\cos^2 n\theta d\theta = \dfrac{1}{2}\int_{0}^{\pi}(1 + \cos 2n\theta)d\theta = \dfrac{\pi}{2}$$

For every natural number n, and for every real number x of $[-1, 1]$, we may write (using the equality $x = \cos(\text{Arccos } x)$): $T_n(x) = \cos(n \, \text{Arccos } x)$.

To define the scalar product and the norm, we may use the program defscal, or do it directly, as here.
Here is a first definition of the polynomials T_n: we ask the calculator to expand $\cos(n \, \text{Arccos } x)$.

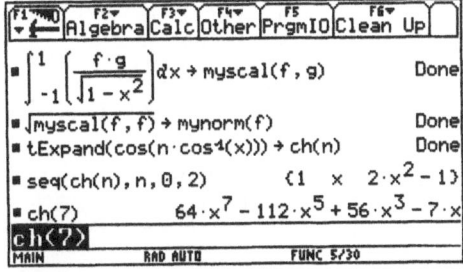

We may likewise use the explicit formula which gives the polynomial T_n. Even if this is not a very elegant formulation, the result is indeed obtained more rapidly than with the preceding method.

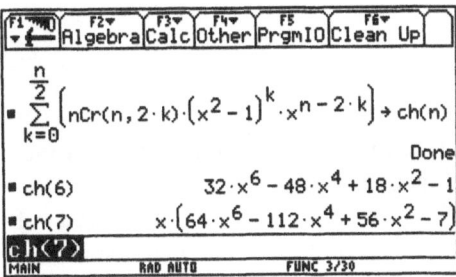

The polynomials T_n satisfy a differential equation:

Proposition 10: $\forall n \geq 0, (1 - x^2)T_n''(x) - xT_n'(x) + n^2 T_n(x) = 0$

Proof: We know that, for every real θ, $T_n(\cos \theta) = \cos n\theta$.
If we differentiate this equation twice, we find: $-\sin \theta \, T_n'(\cos \theta) = -n \sin n\theta$
then: $-\cos \theta \, T_n'(\cos \theta) + \sin^2(\theta) \, T_n''(\cos \theta) = -n^2 T_n(\cos \theta)$
If we put $x = \cos \theta$, we have thus found:
$\forall x \in [-1, 1], (1 - x^2)T_n''(x) - xT_n'(x) + n^2 T_n(x) = 0$
The polynomial in the first member is zero on $[-1, 1]$, thus on $\mathbb{R}$ for each integer, which establishes the result.

The T_n also possess some properties of orthogonal sequences which were studied in the preceding section.

In particular, the sequence of polynomials T_n must obey a recurrence relation linking three successive polynomials T_{n+1}, T_n and T_{n-1}. This relation stems from the equality $\cos(n+1)\theta + \cos(n-1)\theta = 2 \cos \theta \cos n\theta$ and if may be written:

$$\forall n \geq 1, T_{n+1}(x) = 2xT_n(x) - T_{n-1}(x) \text{ or again} : xT_n(x) = \frac{1}{2}\left(T_{n-1}(x) + T_{n+1}(x)\right)$$

The function Lch uses this recurrence formula to form the list of coefficients of the Chebishev polynomial of index n (ordered according to decreasing powers). This is a very rapid method, and we retrieve the algebraic form of the polynomial T_n for a given integer n by evaluating polyEval(Lch(n))).

```
:Lch(n)
:Func
:Local a,b,c,k
:{1} →b:If n=0:Return
:{1,0} →a
:For k,2,n
:  2*augment(a,{0})-augment({0,0},b)→c
:  a→b:c→a
:EndFor
:EndFunc
```

Here are some examples for the function Lch.

To give an idea of the speed of this function, the coefficients of the polynomial T_{20} have been obtained in two seconds.

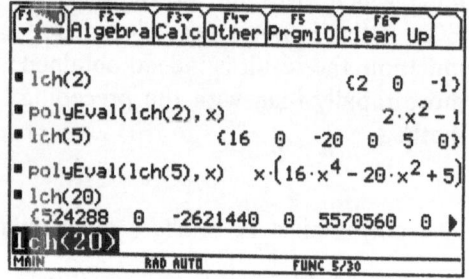

We know that the roots of the polynomial T_n (with $n \geq 1$) must be real, distinct, and must appear in the interval $I =]-1, 1[$. It is easy to confirm these properties since we may calculate them.

In fact, if we limit ourselves to the interval $[-1, 1]$:

$$T_n(x) = 0 \Leftrightarrow \cos(n \operatorname{Arccos} x) = 0 \Leftrightarrow n \operatorname{Arccos} x = \frac{\pi}{2} + k\pi, \ 0 \leq k \leq n-1$$

$$\Leftrightarrow x = x_{n,k} = \cos \theta_{n,k}, \ \text{with} \ \theta_{n,k} = \frac{\pi}{2n} + \frac{k\pi}{n}, \ 0 \leq k \leq n-1 \quad .$$

The zeros of T_n are thus the n real numbers $x_{n,k}$, which form a strictly decreasing sequence in $]-1, 1[$: $-1 < x_{n,n-1} < x_{n,n-2} < \cdots < x_{n,1} < x_{n,0} < 1$.

This sequence is symmetric with respect to the origin, an evident consequence of whether T_n is even or odd. The following figure illustrates, for example, the distribution of roots of the polynomial T_6:

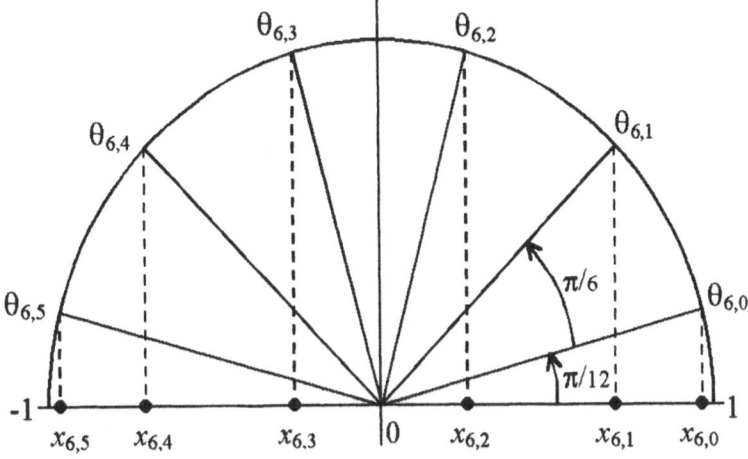

The function zch gives the roots of the polynomial T_n.

```
:zch(n):seq(cos((k+1/2)*π/n),k,0,n-1)
```

Here we form the list of zeros of T_4 with the function zch, then we confirm this in two different ways with the functions ch and Lch:
- By solving $T_4(x) = 0$.
- By evaluating T_4 at the points returned by the function zch.

We remark that, up to $n = 6$, the calculator is capable of symbolic solution of the equation $T_n(x) = 0$.
Further, the solution is done in the "approx " mode.
The function Lch evidently gives the roots quite rapidly.

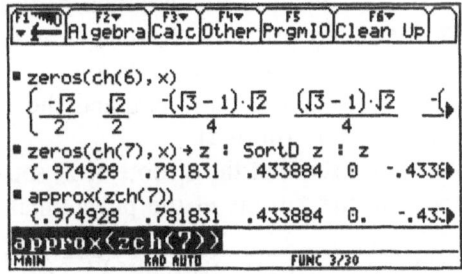

Proposition 11: $\forall n \geq 1$, The n roots of T_n separate the $(n+1)$ roots of T_{n+1}.

Proof: It suffices to show that the polynomial T_n vanishes between two successive zeros of T_{n+1}. In fact, for every integer k of $\{0, \cdots, n\}$:

$$T_n(x_{n+1,k}) = T_n(\cos\theta_{n+1,k}) = \cos(n\theta_{n+1,k}) = \cos((n+1)\theta_{n+1,k} - \theta_{n+1,k})$$

$$= \cos(\frac{\pi}{2} + k\pi - \theta_{n+1,k}) = (-1)^k \sin(\theta_{n+1,k})$$

Thus, the sign of $T_n(x_{n+1,k})$ is that of $(-1)^k$, since $0 < \theta_{n+1,k} < \pi$. The polynomial T_n thus vanishes between $x_{n+1,k+1}$ and $x_{n+1,k}$.

Even if the result is only anecdotal, the recurrence relation satisfied by the T_k and the results of the preceding paragraph show that the roots $x_{n+1,0}, x_{n+1,1}, \ldots, x_{n+1,n}$ of T_{n+1} are the eigenvalues of the matrix of order $n+1$ shown here:

$$M_{n+1} = \frac{1}{2} \begin{pmatrix} 0 & 2 & 0 & \cdots & 0 \\ 1 & 0 & 1 & \ddots & \vdots \\ 0 & \ddots & \ddots & \ddots & 0 \\ \vdots & \ddots & 1 & 0 & 1 \\ 0 & \cdots & 0 & 1 & 0 \end{pmatrix}$$

Now we pass to the global study of the polynomials T_n.

For every x of $[0,1]$, $|T_n(x)| = |\cos(n \operatorname{Arccos} x)| \leq 1$. More precisely:

$$|T_n(x)| = 1 \Leftrightarrow n \operatorname{Arccos} x = k\pi, \ 0 \leq k \leq n \Leftrightarrow x = x'_{n,k} = \cos\left(\frac{k\pi}{n}\right), \ 0 \leq k \leq n$$

The $x'_{n,k}$ form a strictly decreasing sequence of $n + 1$ points of $[-1,1]$ at which T_n alternate between the values -1 and 1. (To be precise, $T_n(x'_{n,k}) = \cos k\pi = (-1)^k$). They are separated by the n roots of T_n:

$$-1 = x'_{n,n} < x_{n,n-1} < x'_{n,n-1} < x_{n,n-2} < \cdots < x_{n,1} < x'_{n,1} < x_{n,0} < x'_{n,0} = 1$$

We likewise denote by $x'_{n,1}, \ldots, x'_{n,n-1}$ (all elements of $]-1,1[$) are roots of the derivative polynomial of T_n, which is of degree $n - 1$ and which may be written, on $]-1,1[$:

$$T'_n(x) = -\frac{n}{\sqrt{1-x^2}} \sin(n \operatorname{Arccos} x)$$

The calculator allows verification (or at least suggests) certain of the properties being established.

First, we looked at the variation of the polynomials T_n.
We began by graphing T_5 on $[-1,1]$ (with the same interval on the Oy axis): indeed, we see that the roots and the extrema alternate.
We then simultaneously graphed the three polynomials T_5 (normal trace), T_6 (bold) and T_7 (dotted).
Again, we see that the 5 roots of T_5 separate the 6 roots of T_6, which themselves separate the 7 roots of the polynomial T_7.

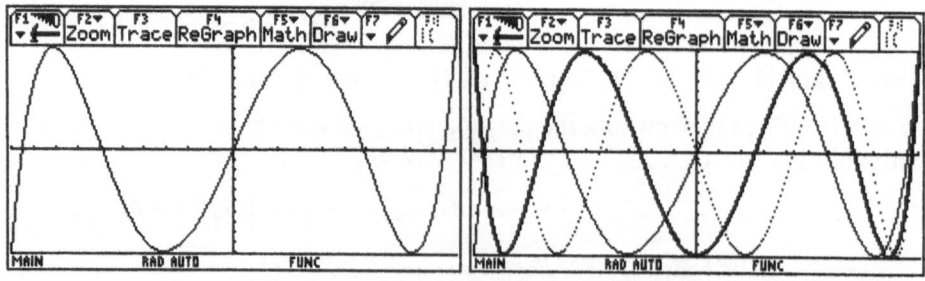

With a little recalculation, we indeed see the global behavior of the T_n: oscillations on $[-1, 1]$ between the values -1 and 1, contrasting with the more "extreme" oscillations outside of $[-1, 1]$.
We here graphed T_5, which is odd, and T_6, which is even (bold).

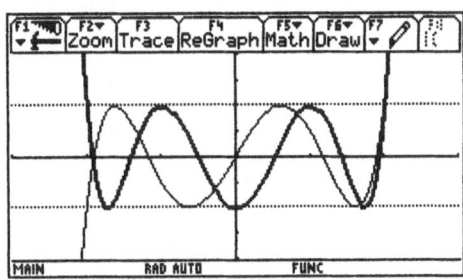

Proposition 12: For every monic polynomial P of degree $n \geq 1$,

$$\sup_{x \in [-1,1]} |P(x)| \geq \frac{1}{2^{n-1}}. \text{ The polynomial } P = \frac{1}{2^{n-1}} T_n \text{ satisfies the equality.}$$

Proof: Actually, $\frac{1}{2^{n-1}} T_n$ is monic and $\sup_{x \in [-1,1]} \left| \frac{1}{2^{n-1}} T_n(x) \right| = \frac{1}{2^{n-1}}$.

By contradiction, we suppose that A_n is monic and of degree n, such that

$$\sup_{x \in [-1,1]} |A_n(x)| < \frac{1}{2^{n-1}}.$$

By construction, $B_n = A_n - \frac{1}{2^{n-1}} T_n$ is non-zero and $\deg B_n \leq n - 1$.

On the other hand, $\forall k \in \{0, \dots, n\}, B_n(x'_{n,k}) = A_n(x'_{n,k}) - (-1)^k \frac{1}{2^{n-1}}$.

This quantity is alternately negative and positive (since $|A_n(x'_{n,k})| < \frac{1}{2^{n-1}}$).

The polynomial B_n of degree less than or equal to $n - 1$ thus has at least n distinct roots (one in each interval $]x'_{n,k+1}, x'_{n,k}[, 0 \leq k \leq n - 1)$.
This implies that B_n is the zero polynomial, which is absurd.

We have an analogous result on an arbitrary segment $[a, b]$, with the change of variable $t = \frac{a+b}{2} + x\frac{b-a}{2}$. In fact, to every monic polynomial function $A(t)$ defined on $[a, b]$ of degree n, there corresponds a polynomial function $B(x)$ on $[-1, 1]$ of degree n, whose dominant term is $\left(\frac{b-a}{2}\right)^n$.

The preceding proposition indicates that $\sup_{x \in [-1,1]} |B(x)| \geq \frac{1}{2^{n-1}} \left(\frac{b-a}{2}\right)^n$.

We know that the preceding inequality is an equality when B is proportional to T_n, that is when its roots are the $x_{n,k} = \cos\left(\frac{\pi}{2n} + \frac{k\pi}{n}\right)$. This happens when the roots of $A(t)$ are the coefficients $t_{n,k}$, with:

$$\forall k \in \{0, \dots, n-1\}, t_{n,k} = \frac{a+b}{2} + \frac{b-a}{2} x_{n,k} = \frac{a+b}{2} + \frac{b-a}{2} \cos\left(\frac{\pi}{2n} + \frac{k\pi}{n}\right)$$

We say that the $t_{n,k}$ are the Chebishev values (or sample points) of index n on the segment $[a, b]$.

We may thus state the following result for all real numbers $\alpha_1, \alpha_2, \ldots, \alpha_n$:

$$\sup_{a \leq t \leq b} \left| (t - \alpha_1)(t - \alpha_2) \cdots (t - \alpha_n) \right| \geq \frac{1}{2^{n-1}} \left(\frac{b-a}{2} \right)^n$$

(The equality is realized if the α_k are the Chebishev values of index n on $[a, b]$.)

3. 3 Chebishev polynomials and discrete least squares

Let f be a function defined on the interval $I =]-1, 1[$, with real values.

We want to calculate the best approximation of f at $n+1$ points of I by a polynomial function of degree less than or equal to m ($0 \leq m \leq n$), that is, by an element of the vector space $E = \mathbb{R}_m[X]$, which is of dimension $m + 1$.

A basis for a (ε) of E is formed by the Chebishev polynomials $T_0, T_1, \ldots, T_m$.

Suppose that we have the choice of $n+1$ values $x_0, \ldots, x_n$ in $[-1, 1]$. In particular, we could choose the $n+1$ roots of T_{n+1}:

$$\forall k \in \{0, \ldots, n\}, x_k = \cos \theta_k, \text{ with } \theta_k = \frac{(2k+1)\pi}{2(n+1)}$$

Proposition 13: $\forall p \in \{-2n, \ldots, 2n\}, \sum_{k=0}^{n} \cos(p\theta_k) = \begin{cases} n+1 \text{ if } p = 0 \\ 0 \text{ otherwise} \end{cases}$

Proof: This is evident if $p = 0$. We suppose thus $-2n \leq p \leq 2n$ and $p \neq 0$.

To simplify the notation, we put $\omega = \exp i\theta_0 = \exp \dfrac{i\pi}{2(n+1)}$.

We have: $\sum_{k=0}^{n} \cos(p\theta_k) = \mathrm{Re} \sum_{k=0}^{n} \exp(ip\theta_k) = \mathrm{Re} \sum_{k=0}^{n} \omega^{(2k+1)p} = \mathrm{Re} \left(\omega^p \sum_{k=0}^{n} (\omega^{2p})^k \right)$

ω is a $4(n+1)$-th root of unity: $\omega^{2(n+1)} = -1$, $\omega^{4(n+1)} = 1$.

But $-4n \leq 2p \leq 4n$ and $2p \neq 0$. We have deduced that $\omega^{2p} \neq 1$.
It follows then that:

$$\sum_{k=0}^{n} \cos(p\theta_k) = \mathrm{Re} \left(\omega^p \frac{\omega^{2p(n+1)} - 1}{\omega^{2p} - 1} \right) = \mathrm{Re} \left(\omega^p \frac{(-1)^p - 1}{\omega^{2p} - 1} \right) = \mathrm{Re} \left(\frac{(-1)^p - 1}{\omega^p - \omega^{-p}} \right)$$

However, $\omega^p - \omega^{-p} = 2i \sin \dfrac{p\pi}{2(n+1)}$ is pure imaginary.

We have deduced that $\sum_{k=0}^{n} \cos(p\theta_k) = 0$, which establishes the proposition.

Here is how the calculator allows us to confirm (or to intuit) the preceding result.

We have here calculated $\sum_{k=0}^{n} \cos(p\theta_k)$

- with $n = 4$ and $0 \leq p \leq 8$
- with $n = 5$ and $0 \leq p \leq 10$

We know (cf paragraph **2.3**) that the best approximation $P = \sum_{j=0}^{m} a_j T_j$ of the vector $y = \big(f(x_0), \ldots, f(x_n)\big)$ of values of f is obtained by solving the system $M^\top M[P] = M^\top[y]$ (here we are in the real case), with:

$$M = \begin{pmatrix} T_0(x_0) & T_1(x_0) & \cdots & T_m(x_0) \\ T_0(x_1) & T_1(x_1) & \cdots & T_m(x_1) \\ T_0(x_2) & T_1(x_2) & \cdots & T_m(x_2) \\ \vdots & \vdots & \ddots & \vdots \\ T_0(x_n) & T_1(x_n) & \cdots & T_m(x_n) \end{pmatrix}$$

We will observe that the matrix $M^\top M$ (square and of order $m+1$) is simply diagonal. This results from the choice of the basis $T_0, \ldots, T_m$ of $E = \mathbb{R}_m[X]$ and of the $n+1$ sample values $x_0, \ldots, x_n$ (the roots of T_{n+1}). In fact, for all indices i and j between 0 and m, the term of index (i, j) of $M^\top M$ (numbering from index 0) is:

$$\big(M^\top M\big)_{i,j} = \sum_{k=0}^{n} T_i(x_k)T_j(x_k) = \sum_{k=0}^{n} \cos(i\theta_k)\cos(j\theta_k)$$

$$= \frac{1}{2}\left(\sum_{k=0}^{n} \cos\big((i+j)\theta_k\big) + \sum_{k=0}^{n} \cos\big((i-j)\theta_k\big)\right)$$

We know that $0 \leq i, j \leq m \leq n$. We may conclude that $0 \leq i+j \leq 2n$ and $-n \leq i-j \leq n$.

According to the preceding proposition, the first of the two sums is $n+1$ if $i = j = 0$ and 0 otherwise. In the second, it is $n+1$ if $i = j$ and 0 otherwise.

We have deduced that $M^\top M$ is diagonal, its diagonal coefficients having value $n+1$ for the first oneand $\frac{(n+1)}{2}$ for the subsequent ones .

The function mch allows calculation of M.
The syntax is mch(n,m), where n, m have the previous meaning.
We use here the function zch (zeros of T_k) and Lch (coefficients of T_k)

```
:mch(n,m)
:Func:Local z
:If n<m:Return "Dim Error":zch(n+1)→z
:(seq(polyEval(Lch(k),z),k,0,m))ᵀ
:EndFunc
```

Here are some examples of use of the function mch in symbolic mode. In each case, we have verified that the matrix M^TM is diagonal.

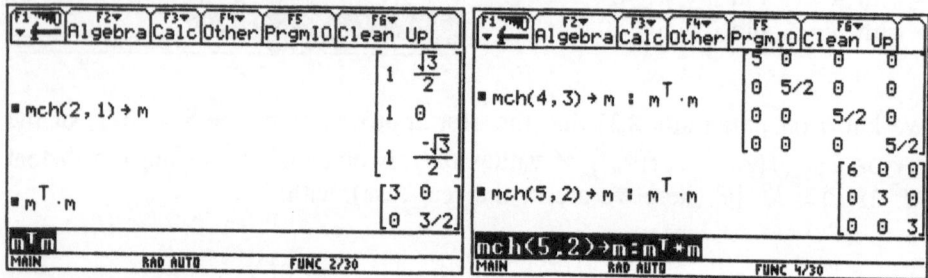

Here is an example in "approx" mode:

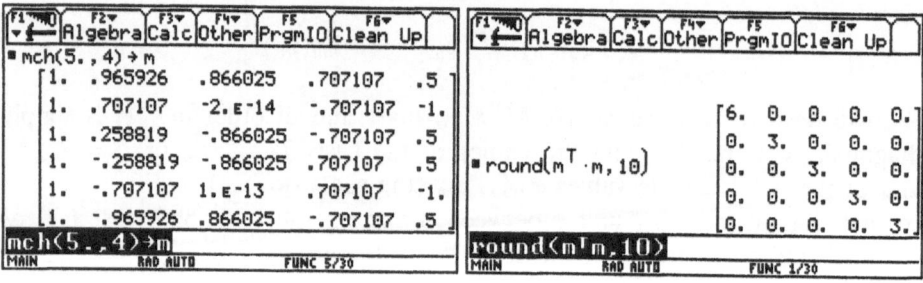

If $P = \sum_{j=0}^{m} a_j T_j$ is the best approximation in the sense of discrete least squares of a vector $y = (f(x_0), \ldots, f(x_n))$, then (denoting by $[P]$ and $[y]$ the column matrices of P and of y, respectively in the basis $T_0, \ldots, T_m$ of $E = \mathbb{R}_m[X]$ and in the standard basis of $\mathbb{R}^{n+1}$), then the column vector $[P]$ may be written $[P] = (M^TM)^{-1}M^T[y]$, that is:

$$[P] = \frac{1}{(n+1)} \begin{pmatrix} 1 & 0 & \cdots & 0 \\ 0 & 2 & \ddots & \vdots \\ \vdots & \ddots & \ddots & 0 \\ 0 & \cdots & 0 & 2 \end{pmatrix} \begin{pmatrix} T_0(x_0) & T_0(x_1) & \cdots & T_0(x_n) \\ T_1(x_0) & T_1(x_1) & \cdots & T_1(x_n) \\ \cdots & \cdots & \cdots & \cdots \\ T_m(x_0) & T_m(x_1) & \cdots & T_m(x_n) \end{pmatrix} \begin{pmatrix} f(x_0) \\ f(x_1) \\ \vdots \\ f(x_n) \end{pmatrix}$$

Stated otherwise, $P = \sum_{j=0}^{m} a_j T_j$, with (recall that $T_0 = 1$):

$$a_0 = \frac{1}{n+1} \sum_{k=0}^{n} f(x_k), \text{ and } \forall j \in \{1, \ldots, m\}, \ a_j = \frac{2}{(n+1)} \sum_{k=0}^{n} T_j(x_k) f(x_k)$$

The best approximation $P(x)$ of $f(x)$, in the sense of least squares, for the $n+1$ sample values $(x_0, \ldots, x_n)$, while the polynomials of degree $\leq m$, are thus:

$$P(x) = \frac{1}{n+1}\left(\sum_{k=0}^{n} f(x_k) + 2\sum_{j=1}^{m}\left(\sum_{k=0}^{n} T_j(x_k)f(x_k)\right)T_j(x)\right)$$

The function lsqch calculates the polynomial P (it accepts as an entry the list y of ordinates, followed by the degree m). It is independent of the functions which have been defined in this and previous paragraphs.

```
:lsqch(y,m)
:Local j,n,p,s,t,u,xx
:dim(y)-1→n
:If m>n:Return "Dim Error"
:seq(cos((k+1/2)*π/(n+1)),k,0,n)→xx
:sum(y)→p:{1} →s:{1,0} →t
:For j,1,m
:   p+2*dotP(polyEval(t,xx),y)*polyEval(t,x)→p
:   2*augment(t,{0})-augment({0,0},s)→u
:   t→s:u→t
:EndFor
:p/(n+1)
:EndFunc
```

Here is an example using the function lsqch.

First, we prepare the "scatter diagram" of points to be "fitted": we place in the variable xx the list of Chebishev values of index 5 (the roots of T_5) and in the variable yy the list of ordinates.

The instruction in the NewPlot program graphs this family of points.

The instruction lsqch(yy,3) calculates the best approximation, by a function polynomial of degree less than or equal to 3, of this set of points. The result is placed in the variable y1 to be graphed.

We then pass to the Graph screen: the family of points and the approximation polynomial are then graphed jointly. (The user decides the graph parameters: we may also employ ZoomData).

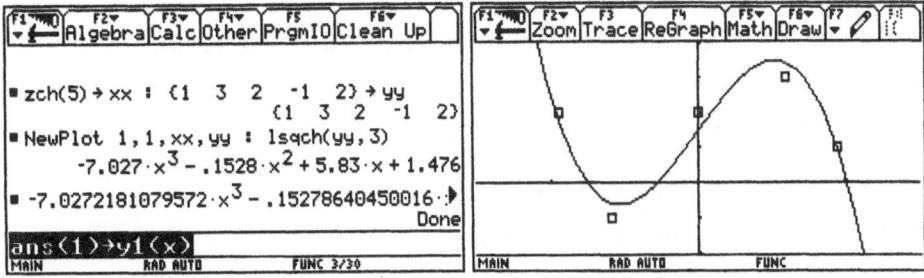

Up to now, we have approximated a function f defined on $I = [-1, 1]$.

If f is defined on the interval $J = [a, b]$, we remain in the preceding case with the changement of variable $t = \frac{a+b}{2} + x\frac{b-a}{2}$ (when x runs through I, t runs through J).

The program democh illustrates approximation of a function f defined on $J = [a, b]$ by a polynomial function of degree $\leq m$, based on the n Chebishev abscissas of the segment J. (We must have $m < n$).

```
:democh()
:Prgm
:Local n,m,a,b
:DelVar x,t:ClrIO:Disp
:Dialog
:Request "Function f(x)",θf
:Request "Number of points n",n
:Request "Degree m (m<n)",m
:Request "Left end a",a
:Request "Right end b",b
:EndDlog
:expr(θf)→ θf:expr(n)→n:expr(m)→m
:expr(a)→a:expr(b)→b
:(b-a)*approx(zch(n))/2+(b+a)/2→ θx
:FnOff :PlotsOff:expr(string(θf))→y1(x)
:θf|x=θx→ θy:lsqch(θy,m)→ θp
:expr(string(θp))|x=(2*t-a-b)/(b-a)→ θp
:θp|t=x→ θp
:NewPlot 1,1,θx,θy:expr(string(θp))→y2(x)
:ZoomData
:EndPrgm
```

Here is an example of use of the program democh.

The dialogue box lets us choose

- the function to be approximated (here, $f(x) = \sin(x)$).
- the number of Chebyshev values x_k to be considered (here, 5).
- the maximum degree of the polynomial approximating P (here, $m = 4$).
- The interval on which we are working (here, $[0, 6]$).

Then the program democh graphs on the same screen:

- the "scatter diagram " of points $(x_k, f(x_k))$.
- the function f, which is placed in the variable y1.
- the approximating polynomial P, placed in the variable y2.

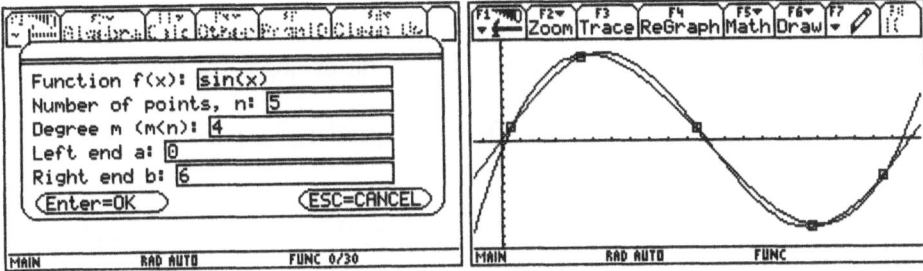

To extend the preceding example, we verify that:
- the list of values is in θx
- the polynomial P is in y2

We have calculated here:
- The values of f at the values x_k
- The values of P at the values x_k

For the quality of such an approximation, we refer to the earlier *Interpolation* chapter.

3. 4 Legendre polynomials

In this section, we consider the following scalar product (and the norm which is associated with it) defined on the vector space E of continuous functions on the interval $I = [-1, 1]$ with real values, and whose squares are integrable on I. (E contains the polynomial functions in particular):

$$< f, g > = \int_{-1}^{1} f(x)g(x)\, \mathrm{d}x, \qquad \|f\| = \sqrt{\int_{-1}^{1} f(x)^2\, \mathrm{d}x}$$

For this scalar product as for others, we know that there exist sequences $(P_n)_{n \geq 0}$ of orthogonal polynomials (in the sense defined in this section: $\forall m, n \in \mathbb{N}, \deg(P_n) = n$ and, if $m \neq n$, $< P_m, P_n > = 0$).

We also know that, up to a sequence of multiplicative factors, there is only one such sequence of orthogonal polynomials.

Thus, for this scalar product, there is a unique sequence $(L_n)_{n \geq 0}$ of orthogonal polynomials such that: $\forall n \in \mathbb{N}, L_n(1) = 1$.

We call them the *Legendre polynomials*. We are going to begin by establishing some of their properties. The first is Rodrigues' Formula, which lets us calculate directly any polynomial L_n.

Proposition 14: (Rodrigues' Formula) $\forall n \geq 0$, $L_n(x) = \dfrac{1}{n!2^n} \dfrac{d^n}{dx^n}(x^2 - 1)^n$.

Proof: We put $U_n(x) = (x^2 - 1)^n$ for each natural number n.
It is clear that U_n is an even polynomial function of degree $2n$, and that it has 1 and -1 as roots with multiplicity n.
If f is of class C^n on $[-1, 1]$, then after n integrations by parts:

$$\int_{-1}^{1} U_n(x) f^{(n)}(x)\, dx = \left[\sum_{k=0}^{n-1} (-1)^k U_n^{(k)}(x) f^{(n-k-1)}(x)\right]_{-1}^{1} + (-1)^n \int_{-1}^{1} U_n^{(n)}(x) f(x)\, dx$$

Since 1 and -1 are roots of U_n with the multiplicity n, the sucessive derivatives $U_n^{(k)}$ (pour $0 \leq k \leq n - 1$) vanish there.

Thus, the preceding equation holds and the term in square brackets is zero, reducing to the equation: $\displaystyle\int_{-1}^{1} U_n(x) f^{(n)}(x)\, dx = (-1)^n \int_{-1}^{1} U_n^{(n)}(x) f(x)\, dx$,

That is, we have the identity: $< U_n, f^{(n)} > = (-1)^n < U_n^{(n)}, f >$.

We conclude: $\forall P \in \mathbb{R}_{n-1}[X]$, $< U_n^{(n)}, P > = 0$ (since $P^{(n)} = 0$).

Thus, $P_n = U_n^{(n)}$, a polynomial of degree n, is orthogonal to all the polynomials of degree less than or equal to $n - 1$.
The family $(P_n)_{n \geq 0}$ thus constitutes a sequence of orthogonal polynomials.
We know that this implies the existence of a sequence $(\lambda_n)_{n \geq 0}$ of non-zero scalars such that: $\forall n \in \mathbb{N}, P_n = \lambda_n L_n$.

But $L_n(1) = 1$ implies that $P_n(1) = \lambda_n$. Hence, the Leibniz Formula gives us:

$$\forall n \in \mathbb{N}, P_n(x) = \left((x-1)^n (x+1)^n\right)^{(n)} = \sum_{k=0}^{n} \binom{n}{k} \left((x-1)^n\right)^{(k)} \left((x+1)^n\right)^{(n-k)}$$

In this sum, all the terms vanish at $x = 1$ except when $k = n$.
We may deduce that: $\forall n \in \mathbb{N}, P_n(1) = \left((x-1)^n\right)^{(n)} \left((x+1)^n\right)^{(0)} \Big|_{x=1} = n!2^n$

Stated otherwise: $\forall n \in \mathbb{N}, L_n = \dfrac{1}{n!2^n} P_n = \dfrac{1}{n!2^n} U_n^{(n)}$, which was to be shown.

Since the polynomial U_n is even, its n-th derivative (and thus the polynomial L_n) has the same parity as n.
The preceding result gives the coefficients of the polynomial L_n:

$$\forall n \in, L_n = \dfrac{1}{n!2^n} \dfrac{d^n}{dx^n} \sum_{k=0}^{n} \binom{n}{k} (-1)^k x^{2n-2k} = \dfrac{1}{n!2^n} \sum_{k=0}^{n/2} (-1)^k \binom{n}{k} \dfrac{(2n-2k)!}{(n-2k)!} x^{n-2k}$$

$$= \dfrac{1}{2^n} \sum_{k=0}^{n/2} (-1)^k \binom{n}{k} \binom{2(n-k)}{n} x^{n-2k}$$

The coefficient of the term of degree n is thus $a_n = \dfrac{1}{2^n}\dbinom{2n}{n} = \dfrac{1}{2^n}\dfrac{(2n)!}{n!^2}$

Rodrigues' Formula immediately gives:

$$L_0(x) = 1, \quad L_1(x) = \frac{1}{2}(x^2 - 1)' = x, \quad L_2(x) = \frac{1}{8}(x^4 - 2x^2 + 1)'' = \frac{1}{2}(3x^2 - 1)$$

Here are two methods of forming the polynomials L_n.
The first uses Rodrigues' Formula: it has the advantage of simplicity.
The second applies the result described above: it has the advantage of speed.

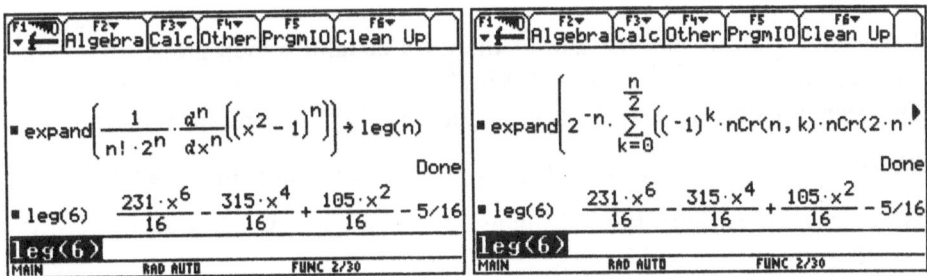

We know that every sequence of orthogonal polynomials satisfies a linear recurrence relation linking three successive polynomials.

Here is that relation which is satisfied for the Legendre polynomials.

Proposition 15: $\forall n \geq 1,\ (n+1)L_{n+1}(x) = (2n+1)xL_n(x) - nL_{n-1}(x)$ (E_1)

Proof: We know from **3.1** that there are three sequences $(\alpha_n)_{n\geq 1}$, $(\beta_n)_{n\geq 1}$ and $(\gamma_n)_{n\geq 1}$ such that, for every $n \geq 1$: $L_{n+1}(x) = (\alpha_n x + \beta_n)L_n(x) + \gamma_n L_{n-1}(x)$.
The polynomials L_{n+1}, xL_n, and L_{n-1} have the same parity (that of $n+1$), which is the parity opposite that of L_n.
The two polynomials $L_{n+1} - \alpha_n xL_n - \gamma_n L_{n-1}$ and $\beta_n L_n$, equal and of opposite parity, are thus zero. We may deduce that $\beta_n = 0$.
If we identify the terms of higher degree, we find:

$$\frac{1}{2^{n+1}}\frac{(2n+2)!}{(n+1)!^2} = \alpha_n \frac{1}{2^n}\frac{(2n)!}{n!^2} \quad \text{and thus} \quad \alpha_n = \frac{2n+1}{n+1}$$

Finally, putting $x = 1$ (and knowing that $\forall k,\ L_k(1) = 1$) we find:

$$1 = \alpha_n + \gamma_n \quad \text{puis} \quad \gamma_n = 1 - \frac{2n+1}{n+1} = -\frac{n}{n+1}$$

This establishes the recurrence relation between L_{n+1}, L_n and L_{n-1}.

Now we come to the calculation of the norm of the polynomial L_n.

Proposition 16: *For every integer* n, $\|L_n\| = \sqrt{\dfrac{2}{2n+1}}$

Proof: We remark that the relation (E_1) is still valid if $n = 0$, without knowing the value of L_{-1}, since $L_0 = 1$ and $L_1 = x$.

If we form the scalar product of the preceding relation by L_{n+1} and then by L_{n-1}, we obtain (taking into the account the orthogonality of the polynomials L_k):

$\forall n \geq 0,\ (n+1)\|L_{n+1}\|^2 = (2n+1) < xL_n, L_{n+1} > = (2n+1) < L_n, xL_{n+1} >$
$\forall n \geq 0,\ (2n+1) < L_{n-1}, xL_n > = n\|L_{n-1}\|^2$

This last equality may also be written:

$\forall n \geq -1,\ (2n+3) < L_n, xL_{n+1} > = (n+1)\|L_n\|^2$

The comparison of the first and of the third equalities then gives:

$\forall n \geq 0, (2n+3)\|L_{n+1}\|^2 = (2n+1)\|L_n\|^2$.

This signifies that the sequence $u_n = (2n+1)\|L_n\|^2$ is constant for $n \geq 0$.

We conclude: $\forall n \geq 0, (2n+1)\|L_n\|^2 = \|L_0\|^2 = 2$, which is the result we sought.

We have graphed here the polynomial L_5, then L_6. (We have also shown the dotted lines $y = -1$ and $y = 1$).

The graph window is $[-1.2, 1.2] \times [-1.2, 1.2]$ in each direction.

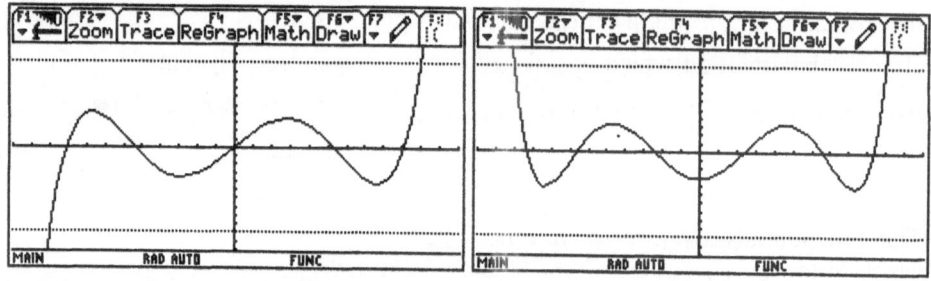

Just as was the case with the Chebishev polynomials, the Legendre polynomials satisfy a linear differential equation of order 2.

Proposition 17: $\forall n \geq 0,\ (1 - x^2)L_n''(x) - 2xL_n'(x) + n(n+1)L_n(x) = 0\ (E_2)$

Proof: If we again put $U_n = (x^2 - 1)^n$, we see that: $(1 - x^2)U_n' + 2nxU_n = 0$.
If we differentiate this equality $n+1$ times, we find (using the Leibniz Formula):

$$(1 - x^2)U_n^{(n+2)} - 2(n+1)xU_n^{(n+1)} - n(n+1)U_n^{(n)} + 2nxU_n^{(n+1)} + 2n(n+1)U_n^{(n)} = 0$$

That is,

$$(1 - x^2)U_n^{(n+2)} - 2xU_n^{(n+1)} + n(n+1)U_n^{(n)} = 0$$

and it suffices to divide this equation by $n!2^n$ to obtain the result.

Here are other equalities satisfied by the Legendre polynomials.

Proposition 18: For every integer $n \geq 1$, $\begin{cases} L'_{n+1} = xL'_n + (n+1)L_n & (E_3) \\ nL_n = xL'_n - L'_{n-1} & (E_4) \\ (x^2-1)L'_n = n(xL_n - L_{n-1}) & (E_5) \end{cases}$

Proof: Still with $U_{n+1} = (x^2-1)^{n+1}$, we see that: $U'_{n+1} = 2(n+1)xU_n$.
We differentiate this equation $n+1$ times: $U_{n+1}^{(n+2)} = 2(n+1)\left(xU_n^{(n+1)} + (n+1)U_n^{(n)}\right)$.
After division by $2^{n+1}(n+1)! = 2(n+1)(2^n n!)$, let: $L'_{n+1} = xL'_n + (n+1)L_n$.
We have thus shown (E_3).
For (E_4), we differentiate (E_1) then we use the expression for L'_{n+1} given by (E_4):

$$(2n+1)\left(xL'_n + L_n\right) - nL'_{n-1} = (n+1)L'_{n+1} = (n+1)\left(xL'_n + (n+1)L_n\right)$$

We deduce: $n^2 L_n = nxL'_n - nL'_{n-1}$, then the equation (E_4) on simplifying by n.
It remains to prove (E_5), both members of which vanish at $x = 1$. It thus suffices to verify that their derivatives are the same.
Hence, $n\left(xL_n - L_{n-1}\right)' = n(L_n + xL'_n - L'_{n-1}) = n(n+1)L_n$ (using (E_4)).
Similarly, $\left((x^2-1)L'_n\right)' = (x^2-1)L''_n + 2xL'_n = n(n+1)L_n$ (using (E_2)).
We have thus proved equation (E_5).

Proposition 19: $\forall n \geq 2$, The $n-1$ roots of L_{n-1} separate the n roots of L_n

Proof: The roots $x_1, x_2, \ldots, x_n$ of L_n are real and distinct and lie in the interval $]-1, 1[$ (general properties of orthogonal polynomials).
For every k, at $x = x_k$, the equality (E_5) gives: $(1 - x_k^2)L'_n(x_k) = nL_{n-1}(x_k)$.
Since $-1 < x_k < 1$, we observe that $L_{n-1}(x_k)$ has the sign of $L'_n(x_k)$.
From the fact that the roots x_k of L_n are of multiplicity 1, the quantities $L'_n(x_k)$ are alternately positive and negative. It is the same for the quantities $L_{n-1}(x_k)$, which proves that L_{n-1} changes sign between two roots of L_n.
Thus, L_{n-1} has a root on each open interval bounded by two successive roots of L_n, which was to be shown.

The function zleg calculates the roots of Legendre polynomials.
Contrary to the case of Chebyshev polynomials, these zeros don't satisfy a general formula.
The calculator gives the roots of L_n symbolically up to $n = 5$.

It may be useful to lay out all the roots of L_n (or of any family of orthogonal polynomials) at once, without having to calculate each time. The program zpol lets us do this.

Previously, we have named the function allowing the calculation of any Legendre function leg. If we execute the instruction zpol("leg",5), then the program zpol places in the variables zleg1, zleg2, ..., zleg5 the list of approximate values of the roots of the polynomials L_1, L_2, ..., L_5.

If, after this, we evaluate zpol("leg",7), then zpol is only going to calculate the list of roots of of L_6 and L_7 (which it places in zleg6 and zleg7).

In the same way, and since we have called ch, the function which calculates the Chebishev polynomials, the instruction zpol("ch",6) will place the list of roots of polynomials T_1 to T_6 in the variables zch1 to zch6.

The program zpol is a good opportunity to see how we may use the mechanism of indirection, using the character # in the assignment or the evaluation of names of variables on the calculator.

```
:zpol(pol,n)
:Prgm
:For k,1,n
:"z"&pol&string(k)→t
:If getType(#t)≠"LIST"
:approx(zeros(expr(pol&"(k)"),x))→#t
:EndFor
:EndPrgm
```

The instruction zpol("leg",7) calculates the roots of polynomials L_1 to L_6.

Here we display the list of roots of L_7, then the values of L_6 at different points.

We see that L_6 changes sign between two roots of L_7, which again shows the last instruction very well.

We are going to end this study of Legendre polynomials with two useful inequalities with upper bounds. We will only prove the first of these.

Proposition 20: For every integer n, and for every x of $[-1,1]$, $|L_n(x)| \leq 1$.

Proof: The result is evident if $n = 0$, so we suppose that $n \geq 1$. On the other hand, the function $x \mapsto |L_n(x)|$ is even, and it suffices to consider it only on $[0,1]$.

We define on $[0,1]$ the function $f : x \mapsto L_n(x)^2 + \dfrac{1-x^2}{n(n+1)} L_n'(x)^2$.

For every x of $[0,1]$, by using the differential equation (E_2):

$$f'(x) = \frac{2L_n'(x)}{n(n+1)}\Big(n(n+1)L_n(x) + (1-x^2)L_n''(x) - xL_n'(x)\Big) = \frac{2xL_n'(x)^2}{n(n+1)} \geq 0$$

Thus the function f is increasing on $[0,1]$. Hence, f majorizes L_n^2, that is, $f(x)$ is always greater than or equal to $L_n^2(x)$ on the interval of interest. We may conclude: $\forall x \in [0,1]$, $L_n(x)^2 \le f(x) \le f(1) = L_n(1)^2 = 1$.

Finally, we give the following result, which furnishes a least upper bound of $|L_n(x)|$ on the central part of the interval $[-1,1]$.

Proposition 21: For all $n \ge 1$, and for all x of $]-1,1[$, $|L_n(x)| \le \sqrt{\dfrac{2}{\pi n(1-x^2)}}$.

This bound shows that the sequence $(L_n)_{n \ge 0}$ converges uniformly to the zero function on every segment included in the open interval $]-1,1[$.

Here is the graph of the functions $x \mapsto \sqrt{\dfrac{2}{\pi n(1-x^2)}}$ and $x \mapsto |L_n(x)|$ when $n = 7$ and then when $n = 8$. We readily see the quality of the last upper bound. The graph window is $[-1,1] \times [0,1]$.

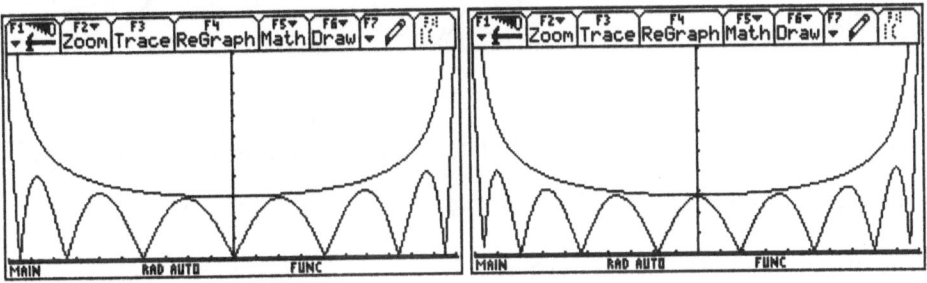

3. 5 Laguerre and Hermite polynomials

Very briefly, here are two other classical families of orthogonal polynomials.

• **Laguerre polynomials**

The interval of integration is $I = [0, \infty[$, and the weight function is $\omega(x) = e^{-x}$. The scalar product is thus defined by: $<f,g> = \displaystyle\int_0^{+\infty} f(x)g(x)e^{-x}\,dx$.

The polynomial of index n is $\mathcal{L}_n = \dfrac{1}{n!}e^x \dfrac{d^n}{dx^n}(x^n e^{-x})$. For every n of $\mathbb{N}$,

$$\mathcal{L}_n(x) = \sum_{k=0}^{n}(-1)^k \frac{1}{k!}\binom{n}{k}x^k = \frac{(-1)^n}{n!}x^n + \cdots$$

$$\|\mathcal{L}_n\|^2 = \int_0^{+\infty}\mathcal{L}_n^2(x)e^{-x}\,dx = 1$$

$$x\mathcal{L}_n''(x) + (1-x)\mathcal{L}_n'(x) + n\mathcal{L}_n(x) = 0$$

$$(n+1)\mathcal{L}_{n+1}(x) + (x-2n-1)\mathcal{L}_n(x) + n\mathcal{L}_{n-1}(x) = 0$$

$$x\mathcal{L}_n'(x) = n\big(\mathcal{L}_n(x) - \mathcal{L}_{n-1}(x)\big)$$

Here is a function lag allowing formation of the Laguerre polynomials.
We have thus calculated $\mathcal{L}_3$ and $\mathcal{L}_4$, then we have graphed jointly the bundle
of representative curves of $\mathcal{L}_1$ e $\mathcal{L}_6$.
Recall that by virtue of the general properties of sequences of orthogonal
polynomials, each $\mathcal{L}_n$ has exactly n roots, all real, distinct, and falling in the
interval $[0, +\infty[$. Moreover, the n roots of $\mathcal{L}_n$ "separate " the $n+1$ roots of
$\mathcal{L}_{n+1}$.
Contrary to the cases of Legendre or Chebishev polynomials, the Laguerre
polynomials are neither even nor odd.

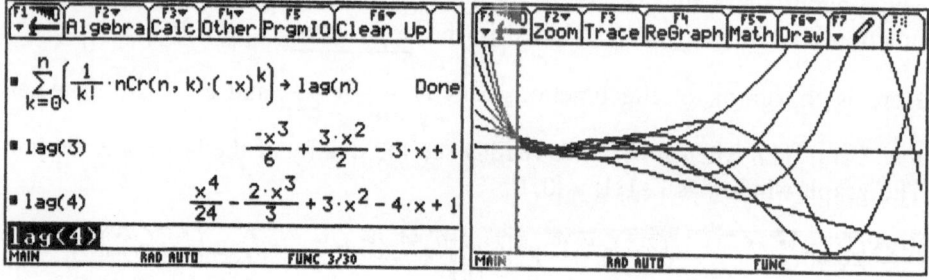

• Hermite polynomials

The interval of integration is $I = \mathbb{R}$. The weight function is $\omega(x) = \exp\left(-\frac{1}{2}x^2\right)$.

The scalar product is defined by: $<f, g> = \int_{-\infty}^{+\infty} f(x)g(x)\exp\left(-\frac{1}{2}x^2\right)dx$.

The orthogonal polynomial of degree n is $H_n = (-1)^n \exp\left(\frac{1}{2}x^2\right)\dfrac{d^n}{dx^n}\exp\left(-\frac{1}{2}x^2\right)$.

The polynomial H_n has the parity of n. For every natural number n,

$$\|H_n\|^2 = \int_{-\infty}^{+\infty} H_n^2(x)\exp\left(-\frac{1}{2}x^2\right)dx = n!\sqrt{2\pi}$$

$$H_n''(x) - xH_n'(x) + nH_n(x) = 0$$

$$H_{n+1}(x) - xH_n(x) + nH_{n-1}(x) = 0$$

$$H_n'(x) = nH_{n-1}(x)$$

The dominant coefficient of H_n is $\alpha_n = 1$. We see below how to calculate H_n.
We have formed H_5 and H_6, then we have graphed the polynomials $\frac{1}{n!}H_n$, with
$n = 1$ and $n = 6$. Each polynomial H_n has exactly n roots, all real and distinct
and which separate the $n+1$ roots of H_{n+1}.

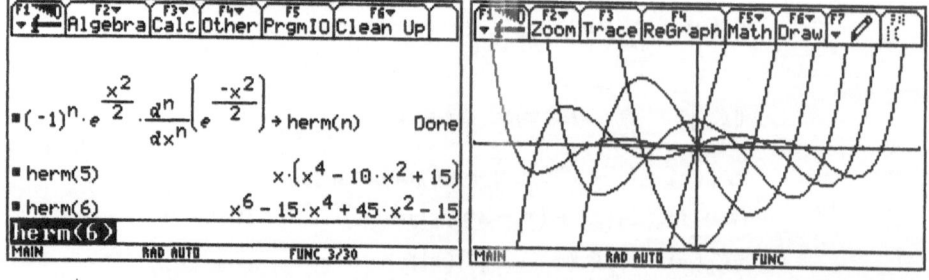

4. Gaussian Quadrature

In this section we will be concerned with the approximate calculation of definite integrals, which may seem a little removed from our previous, purely algebraic, preoccupation. However, this is one of the principal practical applications of orthogonal polynomials.

Let w be a continuous positive function, our weight function, which is positive on an interval I of $\mathbb{R}$, possibly vanishing at isolated points of I.

We suppose that for n in $\mathbb{N}$, map $x \mapsto x^n w(x)$ is integrable on I, and we equip $\mathbb{R}[X]$ with the scalar product $< P, Q > = \int_I P(x)Q(x)w(x)\mathrm{d}x$.

We know that there exists (up to a sequence of non-zero multiplicative coefficients) a unique family $(P_n)_{n \geq 0}$ of orthogonal polynomials for this pair, w and I.

We may thus speak of the roots of index n without risk of ambiguity. We know that these roots are all real and distinct and also that they appear in I.

4. 1 Introduction to the method

Let E be the vector space of functions $f : I \to \mathbb{R}$, such that fw is integrable on I. As noted, E contains $\mathbb{R}[X]$.

For every sequence $x_1, x_2, \ldots, x_n$ of n distinct points of I, and for every sequence $\lambda_1, \lambda_2, \ldots, \lambda_n$ of n real numbers, we consider the approximate integration formula

$$(F) \qquad \int_I f(t)w(t)\,\mathrm{d}t \approx \sum_{k=1}^n \lambda_k f(x_k)$$

Let m be a natural number. We will say that (F) is of order m if it gives equality for the polynomials 1, x, $\ldots$, x^m (and thus for all polynomials of degree less than or equal to m).

This signifies that the λ_k and the x_k satisfy the $m+1$ equalities

$$(E_r) \qquad \sum_{k=1}^n \lambda_k x_k^r = \int_I t^r w(t)\,\mathrm{d}t, \quad (0 \leq r \leq m)$$

There are $2n$ unknowns. We may thus hope that the system formed by the $2n$ equations (E_0), (E_1), $\ldots$, (E_{2n-1}) has a solution, or that the formula (F) is of order $2n - 1$.

We here set the problem with $I = [-1, 1]$, $\omega(x) = 1$, the Legendre setting explored earlier, and $n = 2$.
- The function e allows formation of the equations (E_0), (E_1), (E_2), (E_3).
- We place in the variable sys the system formed by these equations.
- We place in the variable r the solution of this system.

The result is a logical expression whose "root" is the operator or: this corresponds to two "symmetric" solutions $x_1 < x_2$ and $x_2 < x_1$.
We have here extracted the solution $x_1 < x_2$ using the part function. We find:

$$x_1 = -\tfrac{\sqrt{3}}{3}, \; x_2 = \tfrac{\sqrt{3}}{3}, \; \lambda_1 = \lambda_2 = 1.$$

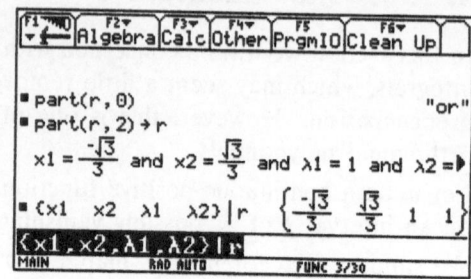

With the conditions of the preceding example, we see that the formula (F) is of order 3 when it is written:

$$\int_{-1}^{1} f(t)\,dt \approx f\left(-\tfrac{\sqrt{3}}{3}\right) + f\left(\tfrac{\sqrt{3}}{3}\right)$$

It may be found that $-\tfrac{\sqrt{3}}{3}$, and $\tfrac{\sqrt{3}}{3}$ are the roots of the orthogonal polynomial of index 2 for the scalar product $< P, Q > = \int_{-1}^{1} P(t)Q(t)\,dt$, that is, the Legendre polynomial $P_2 = \tfrac{1}{2}(3x^2 - 1)$.
This is not a random happenstance. We will soon see the intimate connection between approximate integration formulas and roots of orthogonal polynomials.

4. 2 The use of orthogonal polynomials

Proposition 1: The approximate integration formula (F) is of order $2n - 1$ if and only if the abscissas x_k are the roots of an orthogonal polynomial P_n of degree n. Under these conditions, the coefficients λ_k are strictly positive, and the formula (F) is not of order $2n$.

Preuve: We suppose that (F) is of order $2n - 1$. Let $A = \prod_{k=1}^{n} (x - x_k)$.
For every polynomial B of degree less than or equal to $n-1$, $\deg(AB) \le 2n-1$. We thus have the equation:

$$< A, B > = \int_{I} A(t)B(t)\omega(t)\,dt = \sum_{k=1}^{n} \lambda_k A(x_k)B(x_k) = 0$$

The polynomial A of degree n is orthogonal to $\mathbb{R}_{n-1}[X]$. It is thus proportional to the n-th orthogonal polynomial P_n for this scalar product. In particular, its roots $x_1, x_2, \ldots, x_n$ are those of P_n, which is what we had to show.

Conversely, suppose that $x_1, x_2, \ldots, x_n$ are the roots of P_n.
Let A be a polynomial of degree less than or equal to $2n-1$ and let $A = P_n Q + R$ be its quotient by P_n ($\deg Q \le n-1$ and $\deg R \le n-1$).
By definition of P_n: $\forall k \in \{1, \ldots, n\}, \; A(x_k) = R(x_k)$.

We may write: $< A,1 > = < P_n Q + R, 1 > = < P_n, Q > + < R, 1 > = < R, 1 >$

In other words: $\displaystyle\int_I A(t)\omega(t)\,dt = \int_I R(t)\omega(t)\,dt.$

For all k of $\{1,\ldots,n\}$, we put $\displaystyle B_k(t) = \prod_{1\le j\le n.\ j\ne k} \frac{t - x_j}{x_k - x_j}.$

Each B_k is of degree $n-1$ and satisfies: $B_k(x_k) = 1$ and, if $j \ne k$, $B_k(x_j) = 0$.

We have the equation: $\displaystyle R(t) = \sum_{k=1}^n R(x_k) B_k(t)$ (the Lagrange Interpolation

Formula), and we have deduced that:

$$\int_I A(t)\omega(t)\,dt = \sum_{k=1}^n R(x_k) \int_I B_k(t)\omega(t)\,dt = \sum_{k=1}^n \lambda_k A(x_k)$$

Here we have put $\displaystyle \lambda_k = \int_I B_k(t)\omega(t)\,dt = < B_k, 1 >$, for all k of $\{1,\ldots,n\}$. With
this choice of λ_k, the formula (F) is thus of order $2n-1$.

We remark that the polynomials B_k^2 are of degree $2n-2$.

We thus have the equations: $\displaystyle \forall j \in \{1,\ldots,n\}, \int_I B_j^2(t)\omega(t)\,dt = \sum_{k=1}^n \lambda_k B_j(x_k)^2 = \lambda_j,$

This proves that the coefficients $\lambda_1, \lambda_2, \ldots, \lambda_n$ are defined uniquely and that they
are strictly positive. (The function under the integral is continuous, positive,
and not identically zero.)

Finally, it remains to show that the formula (F) is not of order $2n$. This results
from the fact that the integral $\int_I P_n^2(t)\omega(t)\,dt$ is strictly positive, while the sum
$\sum_{k=1}^n \lambda_k P_n(x_k)^2$ is zero.

In what follows, we will suppose that the sample points x_k are roots of an
appropriate orthogonal polynomial and that the coefficients λ_k are defined as
above, and we will refer to the formula (F) as the Gaussian Quadrature Formula.
We will see that there is a lot of mathematics and math history surrounding
this formula for evaluating a definite integral.

4. 3 Precision of the method

The following result gives an idea of the precision of this formula. Here, we
denote by α_n the coefficient of the term of degree n in P_n.

Proposition 2: Let $f : I \mapsto \mathbb{R}$ be a map of class C^{2n}. Then there is an element
ξ of I such that: $\displaystyle \int_I f(t)\omega(t)\,dt - \sum_{k=1}^n \lambda_k f(x_k) = \frac{\|P_n\|^2}{(2n)!\,\alpha_n^2} f^{(2n)}(\xi)$

Proof: Consider the map ϕ of $\mathbb{R}_{2n-1}[X]$ in $\mathbb{R}^{2n}$, defined by:
$$\forall P \in \mathbb{R}_{2n-1}[X], \ \phi(P) = \big(P(x_1),\ldots,P(x_n), P'(x_1),\ldots,P'(x_n)\big)$$

ϕ is an isomorphism: its linearity is evident, the dimensions of $\mathbb{R}_{2n-1}[X]$ and of $\mathbb{R}^{2n}$ are the same, and ϕ is injective: $\phi(P) = 0 \Rightarrow x_1, x_2, \ldots, x_n$ are double roots of P. Hence $\deg P \le 2n - 1$. Thus, $P = 0$.

In particular, there is a unique polynomial H of degree less than or equal to $2n - 1$, such that: $\forall k \in \{1, \ldots, n\}$, $H(x_k) = f(x_k)$ and $H'(x_k) = f'(x_k)$. (H is the *Hermite Interpolation Formula* of f for the sample points $x_1, x_2, \ldots, x_n$).

(This was also considered in the chapter dedicated to interpolation.)

Under these conditions, and knowing that the Gaussian Quadrature Formula is of order $2n - 1$:

$$\sum_{k=1}^{n} \lambda_k f(x_k) = \sum_{k=1}^{n} \lambda_k H(x_k) = \int_I H(t)\omega(t)\,dt$$

We may deduce: $\displaystyle\int_I f(t)\omega(t)\,dt - \sum_{k=1}^{n} \lambda_k f(x_k) = \int_I \big(f(t) - H(t)\big)\omega(t)\,dt$

It remains to estimate the difference $f(t) - H(t)$ on the interval I.

Let t be an element of I, distinct from $x_1, x_2, \ldots, x_n$.

We consider the function g denined by: $\forall x \in I, g(x) = f(x) - H(x) - \lambda_t P_n^2(x)$, where λ_t is chosen such that $g(t) = 0$. (This is possible since $P_n(t) \ne 0$).

Observe that g (just as f) is of class C^{2n} on I, and also that it vanishes at the $n + 1$ distinct points $t, x_1, x_2, \ldots, x_n$.

Rolle's Theorem (applied to the n segments defined by these $n + 1$ points), shows that g' vanishes at n pairwise distinct points of I, but likewise distinct from $t, x_1, x_2, \ldots, x_n$. On the other hand it is clear that $g' = f' - H' - 2\lambda_t P_n P_n'$ again vanishes at $x_1, x_2, \ldots, x_n$. (See the definition of H).

Thus, g' vanishes at $2n$ distinct points of I. Rolle's Theorem shows that g'' vanishes at $2n - 1$ distinct points of I.

Repeated application of this same theorem shows that $g^{(2n-1)}$ vanishes at two distinct points of I, and finally that $g^{(2n)}$ vanishes at a point x_t of I.

Hence, $\deg H \le 2n - 1$ and the dominant term of $P_n(x)$ is $\alpha_n x^n$:

$$\forall x \in I, \ g^{(2n)}(x) = f^{(2n)}(x) - H^{(2n)}(x) - \lambda_t (P_n^2)^{(2n)}(x) = f^{(2n)}(x) - \lambda_t (2n)! \alpha_n^2$$

The definition of x_t then allows us to write: $\lambda_t = \dfrac{f^{(2n)}(x_t)}{(2n)!\alpha_n^2}$.

The initial definition of λ_t (the condition $g(t) = 0$) thus leads to the following equation:

$$f(t) - H(t) = \lambda_t P_n^2(t) = \frac{f^{(2n)}(x_t)}{(2n)!\alpha_n^2} P_n^2(t)$$

(This is true if t is distinct from the n points $x_1, x_2, \ldots, x_n$, but as is evident in the contrary case, x_t may then be arbitrary.) We may then write:

$$\int_I f(t)\omega(t)\,dt - \sum_{k=1}^n \lambda_k f(x_k) = \frac{1}{(2n)!\alpha_n^2} \int_I f^{(2n)}(x_t)P_n^2(t)\omega(t)\,dt$$

But the function $f^{(2n)}$ is continuous, and $P_n^2\omega$ remains positive. The Mean Value Theorem thus lets us affirm the existence of ξ in I such that:

$$\int_I f^{(2n)}(x_t)P_n^2(t)\omega(t)\,dt = f^{(2n)}(\xi)\int_I P_n^2(t)\omega(t)\,dt = f^{(2n)}(\xi)\|P_n\|^2,$$

This completes the proof of the proposition.

If we suppose $\left|f^{(2n)}\right| \le M_{2n}$, then $\left|\int_I f(t)\omega(t)\,dt - \sum_{k=1}^n \lambda_k f(x_k)\right| \le \dfrac{\|P_n\|^2}{(2n)!\alpha_n^2}M_{2n}.$

4. 4 The classic cases of Gaussian quadrature

The sample values $x_1, x_2, \ldots, x_n$ are the roots of the orthogonal polynomial P_n, and with the preceding notation we have the equations $\lambda_k = < B_k, 1 >$. For a given scalar product, these coefficients thus have values which may be determined once and which may then be placed in a table for subsequent use.

It is interesting, however, to see how to retrieve these values with the aid of our calculator for the cases already studied (Chebishev polynomials, Legendre polynomials, Laguerre, or Hermite polynomials).

The following result will allow us to calculate the coefficients λ_k (called the Christoffel coefficients).

We suppose that the polynomials $(P_k)_{k \ge 0}$ are of norm 1. We always denote $x_1, x_2, \ldots, x_n$ as the roots of the polynomial P_n, and α_k as the dominant coefficient of each P_k (that is, the coefficient of the term of degree k).

Proposition 3: For all k of $\{1, \ldots, n\}$, $\lambda_k = \dfrac{\alpha_n}{\alpha_{n-1}}\dfrac{1}{P_n'(x_k)P_{n-1}(x_k)}$

We are now going to apply the previous result to the most common cases: Chebishev, Lagrange, Laguerre, and Hermite polynomials.

• **Chebishev polynomials**

Here we utilize the notation of paragraph 3.3.

The n-th orthogonal polynomial of norm 1 is, for $n \ge 1$, $P_n = \sqrt{\dfrac{2}{\pi}}T_n$.

Its dominant coefficient is $\alpha_n = \sqrt{\dfrac{2}{\pi}}2^{n-1}$. Thus $\dfrac{\alpha_n}{\alpha_{n-1}} = 2$.

We may deduce $\lambda_k = \dfrac{\alpha_n}{\alpha_{n-1}}\dfrac{1}{P_n'(x_k)P_{n-1}(x_k)} = \dfrac{\pi}{T_n'(x_k)T_{n-1}(x_k)}.$

For all n of $\mathbb{N}$ and θ of $\mathbb{R}$: $T_n(\cos\theta) = \cos n\theta$ and $\sin\theta\, T_n'(\cos\theta) = n\sin n\theta$.

The roots of T_n may be written $x_k = \cos\theta_k$, with $\theta_k = \dfrac{2k-1}{2n}\pi$, $(1 \le k \le n)$.

With this notation, $\cos n\theta_k = 0$ and $\sin n\theta_k = (-1)^{k+1}$.

Thus, $T_{n-1}(x_k) = \cos(n-1)\theta_k = \cos n\theta_k \cos\theta_k + \sin n\theta_k \sin\theta_k = (-1)^{k+1}\sin\theta_k$.

Then $T_n'(x_k)T_{n-1}(x_k) = (-1)^{k+1}\sin\theta_k\, T_n'(\cos\theta_k) = (-1)^{k+1}n\sin n\theta_k = n$.

Thus, we have the very simple result: $\forall\lambda \in \{1,\ldots,n\}$, $\lambda_k = \dfrac{\pi}{n}$.

The Gaussian Quadrature Formula for Chebishev polynomials may thus be written:

$$\int_{-1}^{1} \frac{f(t)}{\sqrt{1-t^2}}\, dt \approx \frac{\pi}{n}\sum_{k=1}^{n} f(x_k), \quad \text{with} \quad x_k = \cos\left(\frac{2k-1}{2n}\pi\right)$$

The program gaussch allows us to test the quality of the preceding formula. Called with the syntax gaussch(n), where n is an integer, it places the list of zeros of T_n in the variable zchn and the second member of the preceding equation in the variable gchn. Then the function f must be stored in y1.

```
:gaussch(n)
:Prgm
:Local nx,np
:"zch"&string(n)→nx
:"gch"&string(n)→np
:seq(cos((k-.5)*π/n),k,1,n)→#nx
:DelVar y1
:π/n*Σ(y1(#nx[k]),k,1,n)→#np
:EndPrgm
```

Here is an example of using the program gaussch.

We tested the integration formula with n points ($n = 4$, then $n = 5$, then $n = 6$), and for the function $x \mapsto \exp x^2$. We indeed see that the quality of the approximation improves with n.

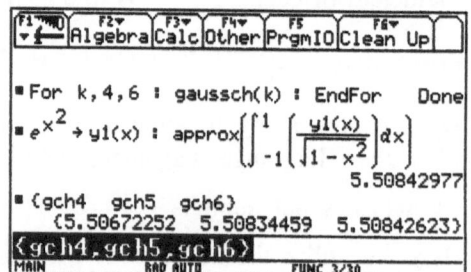

• **Legendre polynomials**

We use here the notation of paragraph **3.4**.

The n-th orthogonal polynomial of norm 1 is $P_n = \sqrt{\dfrac{2n+1}{2}}L_n$.

The dominant coefficient is $\alpha_n = \dfrac{(2n)!}{2^n n!^2}\sqrt{\dfrac{2n+1}{2}}$. Thus $\dfrac{\alpha_n}{\alpha_{n-1}} = \dfrac{1}{n}\sqrt{4n^2-1}$.

We may deduce that $\lambda_k = \dfrac{\alpha_n}{\alpha_{n-1}}\dfrac{1}{P_n'(x_k)P_{n-1}(x_k)} = \dfrac{2}{nL_n'(x_k)L_{n-1}(x_k)}$.

On the other hand, we know the equation $(1-x_k^2)L_n'(x_k) = nL_{n-1}(x_k)$. (See the proof of the proposition **6**, paragraph **3.4**).

We have deduced: $\forall k \in \{1,\dots,n\}$, $\lambda_k = \dfrac{2(1-x_k^2)}{n^2 L_{n-1}^2(x_k)}$.

For Legendre polynomials, the Gauss formula may thus be written:

$$\int_{-1}^{1} f(t)\,dt \approx \frac{2}{n^2}\sum_{k=1}^{n}\frac{1-x_k^2}{L_{n-1}^2(x_k)}f(x_k)$$

The program gaussleg allows us to test the quality of this formula. With the syntax gaussleg(n) (n an integer), it places the list of zeros of L_n in zleg$n$, the list of coefficients λ_k in λleg$n$ and the second member of the preceding equation in gleg$n$. The program gaussleg calls the program leg to form the polynomials L_{n-1} and L_n.
Then the function f must be stored in y1.

```
:gaussleg(n)
:Prgm
:Local nx,nλ,np
:DelVar x
:"zleg"&string(n)→nx
:"λleg"&string(n)→nλ
:"gleg"&string(n)→np
:approx(zeros(leg(n),x))→x:x→#nx
:2*(1-x^2)/(n*leg(n-1))^2→#nλ
:DelVar x,y1
:Σ(#nλ[k]*y1(#nx[k]),k,1,n)→#np
:EndPrgm
```

Here is an example using the program gaussleg.
We tested the integration formula with n points ($n = 4$, then $n = 5$, then $n = 6$) for the function $x \mapsto \exp x^2$.
As in preceding the case, the precision improves rapidly with the value of n.

```
F1    F2   F3   F4   F5      F6
   Algebra Calc Other PrgmIO Clean Up

■ For k,4,6 : gaussleg(k) : EndFor
                                  Done
■ e^x² →y1(x) : approx([∫₋₁¹ y1(x)dx])
                            2.92530349
■ {gleg4 gleg5 gleg6}
  {2.92454094  2.92526464  2.92530185}
{gleg4,gleg5,gleg6}
MAIN        RAD AUTO      FUNC 3/30
```

• Laguerre polynomials

The n-th polynomial of norm 1 is $\mathcal{L}_n$, with dominant coefficient $\alpha_n = \dfrac{(-1)^n}{n!}$.

We may deduce that $\dfrac{\alpha_n}{\alpha_{n-1}} = -\dfrac{1}{n}$, then $\lambda_k = -\dfrac{1}{n\mathcal{L}'_n(x_k)\mathcal{L}_{n-1}(x_k)}$.

The equation $x\mathcal{L}'_n(x) = n\big(\mathcal{L}_n(x) - \mathcal{L}_{n-1}(x)\big)$ implies that $x_k\mathcal{L}'_n(x_k) = -n\mathcal{L}_{n-1}(x_k)$.

Consequently: $\forall k \in \{1, \ldots, n\}$, $\lambda_k = \dfrac{x_k}{n^2 \mathcal{L}^2_{n-1}(x_k)}$.

Thus, the Gaussian Quadrature Formula in the "Laguerre " case may be written:

$$\int_0^{+\infty} f(t)\mathrm{e}^{-t}\,\mathrm{d}t \approx \frac{1}{n^2} \sum_{k=1}^n \frac{x_k}{\mathcal{L}^2_{n-1}(x_k)} f(x_k)$$

• Hermite polynomials

The n-th orthogonal polynomial of norm 1 is $P_n = \alpha_n H_n$, with $\alpha_n = \dfrac{1}{\sqrt{n!\sqrt{2\pi}}}$.
The dominant coefficient of P_n is α_n. On the other hand: $H'_n(x) = nH_{n-1}(x)$.

Thus, $\lambda_k = \dfrac{\alpha_n}{\alpha_{n-1}} \dfrac{1}{P'_n(x_k)P_{n-1}(x_k)} = \dfrac{1}{\alpha^2_{n-1}H'_n(x_k)H_{n-1}(x_k)} = \dfrac{n!\sqrt{2\pi}}{n^2 H^2_{n-1}(x_k)}$.

Thus, $\displaystyle\int_{-\infty}^{+\infty} f(t)\exp\left(-\tfrac{1}{2}t^2\right)\mathrm{d}t \approx \dfrac{n!\sqrt{2\pi}}{n^2} \sum_{k=1}^n \dfrac{1}{H^2_{n-1}(x_k)} f(x_k)$.

We leave to the reader the responsibility of writing the programs gausslag and gaussher, using for inspiration the functions lag and herm and the preceding examples. As a check of your work, they should produce the following results:

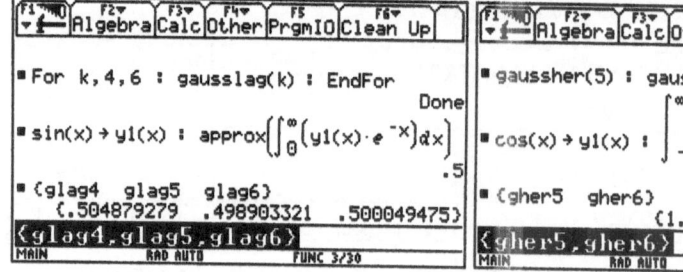

5. Orthogonal operators

After drawing many detailed computational formulas from our earlier foray into the theory of orthogonality, we now return to that general setting of vector spaces. Our explorations there will allow us to again wring some interesting and concrete mathematics from these abstract musings. Of course, our calculator will be our aide along the way.

In this section, we consider a Euclidean space E of dimension $n \geq 1$. The objects of interest will be linear transformations from E into E, that is, maps or functions which preserve linear combinations. Here we will call such a transformation an operator on E. They are also known as endomorphisms.

We denote as usual $< u, v >$ as the scalar product of two vectors and by $u \mapsto \|u\|$ the norm which it defines.

5. 1 General information about orthogonal operators

We say that an operator f on E is orthogonal, or again that f is a vector isometry, if for all vectors u and v of E, $< f(u), f(v) > = < u, v >$ (preservation of the scalar product).

Under these conditions, for all vectors u and v of E: $\|f(u)\| = \|u\|$ (preservation of the norm), and $\|f(u) - f(v)\| = \|u - v\|$ (preservation of distances).

Conversely, if f is an operator on E which preserves the norm, then f preserves the scalar product. This is a result of the "polarization identity" which holds for any norm which is defined by an inner product:

$$< u, v > = \frac{1}{2} \left(\|u + v\|^2 - \|u\|^2 - \|v\|^2 \right)$$

We denote by $O(E)$ the set of orthogonal operators on E.

We denote by $\mathcal{GL}(E)$ the group of the automorphisms of E (called the *linear group* of E).

Proposition 1: $O(E)$ is a subgroup of the group $\mathcal{GL}(E)$.

Proof: The map Id_E is evidently an element of $O(E)$.

On the other hand, if f is in $O(E)$, then f is injective (because the norm is preserved), thus it is also bijective (since we are in finite dimensions). Thus, $O(E) \subset \mathcal{GL}(E)$.

Finally, if f and g are in $O(E)$, then $g \circ f$ and f^{-1} are again orthogonal operators on E (an immediate verification).

We say that $O(E)$ is the *orthogonal group* of E.

Proposition 2: Let f be an operator on E, and let $(e) = (e_1, e_2, \ldots, e_n)$ be an orthonormal basis. The map f is a vector isometry if and only if the family $(f(e_1), \ldots, f(e_n))$ is an orthonormal basis of E. (f transforms every orthonormal basis of E into a orthonormal basis of E).

Proof: Let (e) be an orthonormal basis and f an orthogonal operator. For all i, j of $\{1, \ldots, n\}$: $< f(e_i), f(e_j) > = < e_i, e_j > = \delta_{i,j} = \begin{cases} 1 & \text{if } i = j \\ 0 & \text{if } i \neq j \end{cases}$

The family $\varepsilon_1 = f(e_1), \ldots, \varepsilon_n = f(e_n)$ is thus an orthonormal basis of E.

Conversely, suppose that the linear map f transforms an orthonormal basis (e) into another orthonormal basis (ε).

Then for all $u = \sum_{k=1}^{n} x_k e_k$: $f(u) = \sum_{k=1}^{n} x_k \varepsilon_k$ and $\|f(u)\|^2 = \sum_{k=1}^{n} x_k^2 = \|u\|^2$.

Now we will see that orthogonal operators are characterized by a property of their matrices in an arbitrary orthonormal basis.

Let (e) be an orthonormal basis of E, let f be an operator on E, and let M be the matrix of f in the basis (e), with general term $m_{i,j}$:

$$M = \begin{pmatrix} m_{1,1} & m_{1,2} & \cdots & m_{1,n} \\ m_{2,1} & m_{2,2} & \cdots & m_{2,n} \\ \vdots & \vdots & \ddots & \vdots \\ m_{n,1} & m_{n,2} & \cdots & m_{n,n} \end{pmatrix}$$

$$\forall j \in \{1,\ldots,n\}: f(e_j) = \sum_{i=1}^{n} m_{i,j} e_i.$$

The general term of $N = {}^{T}MM$ is $n_{i,j} = \sum_{k=1}^{n} m_{k,i} m_{k,j} = \; < f(e_i), f(e_j) >$.

Thus, $f \in O(E) \Leftrightarrow \bigl(f(e_1),\ldots,f(e_n)\bigr)$ is a orthonormal basis $\Leftrightarrow {}^{T}MM = I_n$.

We say that a matrix M of $\mathcal{M}_n(\mathbb{R})$, a set of real, square matrices, is orthogonal if ${}^{T}MM = I_n$, that is, if M is invertible and if $M^{-1} = {}^{T}M$.
We may thus affirm that:

Proposition 3: An operator f of E is orthogonal if and only if its matrix in an arbitrary orthonormal basis of E is an orthogonal matrix.

We denote by $O(n)$ the set of orthogonal matrices of order n. This is a group under the matrix product (isomorphic to $O(E)$ with the choice of an orthonormal basis).
Remark: A characterization of orthogonal matrices, equivalent to the preceding, is that they are the transition matrices between orthonormal bases. In particular, a matrix M of $\mathcal{M}_n(\mathbb{R})$ is orthogonal if and only if its column vectors (considered as elements of $\mathbb{R}^n$ expressed in the standard basis, orthonormal for the usual scalar product) are unit vectors and are pairwise orthogonal. (It is the same for row vectors since ${}^{T}M$ is also an orthogonal matrix).

Proposition 4: Let M be in $O(n)$. Then $\det M = \pm 1$. The real or complex eigenvalues of M are all of modulus 1. In particular, the only possible real eigenvalues of an orthogonal operator are 1 or -1.

Proof: We identify the elements of $\mathbb{C}^n$ and the column matrices of $\mathcal{M}_{n,1}(\mathbb{C})$. Let $X \in \mathbb{C}^n$ be an eigenvector of M for an eigenvalue λ of $\mathbb{C}$.

The equation $MX = \lambda X$ gives, by conservation of the norm: $\|X\| = \|\lambda X\|$, that is $|\lambda| = 1$.

The fact that the determinant of an orthogonal matrix M (and thus of an orthogonal operator) is equal to 1 or to -1 is an immediate consequence of the equation ${}^{T}MM = I_n$. The orthogonal matrices of determinant $+1$ are called positive or direct. They form a subgroup of $O(n)$ denoted $O^+(n)$ (or $SO(n)$) and called the special orthogonal group of index n.

The orthogonal matrices of determinant -1 are called negative or indirect. We denote by $O^-(n)$ the set of these matrices.

We likewise define the group $O^+(E)$ of vector isometries with determinant 1 (we call them vector rotations) and we define $O^-(E)$ as the set of isometries with determinant -1.

Proposition 5: Let f be an orthogonal operator on E. Let F be a subspace of E which is invariant under f. Then its orthogonal complement $G = F^\top$ is also stable under f, and the restrictions of f to F and G are vector isometries.

Proof: First, we remark that the hypothesis of invariance, which is expressed by $f(F) \subset F$, may in fact be written $f(F) = F$, since f is an automorphism of E (by preservation of the dimension).
Let u be a vector of $F^\top$. For every vector v of F, $< f(u), f(v) > = < u, v > = 0$.
The vector $f(u)$ is thus orthogonal to $f(F) = F$: it is in $F^\top$.
Thus, the subspace $F^\top$ is invariant under f.
Finally, the restrictions of f to F and $G = F^\top$ are evidently vector isometries. (We still have preservation of the scalar product for orthogonal operators.)

The preceding result has a kind of converse which may be stated in the following manner.

Proposition 6: Let $E_1, E_2, \ldots, E_m$, be m subspaces of E which are pairwise orthogonal and such that $E = E_1 \oplus E_2 \cdots \oplus E_m$. For every k of $\{1, \ldots, m\}$, let f_k be an orthogonal operator on E_k. Let u be an arbitrary vector of E and $\sum_{k=1}^{m} u_k$ its decomposition into a direct sum . Then the map f defined by $f(u) = \sum_{k=1}^{m} f_k(u_k)$ is an orthogonal operator of E.

Proof: The linearity of f is evident, and, with the notation as before:

$$\|f(u)\|^2 = \sum_{k=1}^{n} \|f_k(u_k)\|^2 = \sum_{k=1}^{n} \|u_k\|^2 = \|u\|^2$$

We may also express the preceding proposition in matrix form: if a square matrix M with real coefficients is block diagonal and if each of these blocks is an orthogonal matrix, then M is an orthogonal matrix.

Here is a last result, along the line of the last two.

Proposition 7: Let f be an orthogonal operator on E. The two subspaces $\mathrm{Inv}(f) = \mathrm{Ker}\,(f - Id_E)$ (invariant vectors) and $\mathrm{Opp}(f) = \mathrm{Ker}\,(f + Id_E)$ (vectors which transform into their opposite) are orthogonal.
If F is the orthogonal complement of $\mathrm{Inv}(f) \oplus \mathrm{Opp}(f)$ in E, the orthogonal direct sum $E = \mathrm{Inv}(f) \oplus \mathrm{Opp}(f) \oplus F$ is formed of three invariant subspaces of f, the restriction of f to F having no real eigenvalue.

Proof:

$\forall u \in \mathrm{Inv}(f), \ \forall v \in \mathrm{Opp}(f), \ <u,v> = <f(u), f(v)> = <u, -v> = - <u,v>$.
The orthogonality of $\mathrm{Inv}(f)$ and of $\mathrm{Opp}(f)$ follow. The rest of the proposition is an immediate consequence of the preceding results.

We may carry this property over in terms of matrices: let $M \in O(n)$.
Then, denoting $p = \dim \mathrm{Inv}(f)$ and $q = \dim \mathrm{Opp}(f)$, the matrix M is similar, with an orthogonal transition matrix, to a matrix written $\begin{pmatrix} I_p & 0 & 0 \\ 0 & -I_q & 0 \\ 0 & 0 & N \end{pmatrix}$, where N is an orthogonal matrix having no real eigenvalues (which are thus complex and of modulus 1). Here the integers p and q may be zero.

5. 2 Isometries of the plane

We know how to express all the orthogonal matrices of order 2:

- The matrices of $O^+(2)$ are the $R(\theta) = \begin{pmatrix} \cos\theta & -\sin\theta \\ \sin\theta & \cos\theta \end{pmatrix}$, $\theta \in \mathbb{R}$.

- The matrices of $O^-(2)$ are the $S(\theta) = \begin{pmatrix} \cos\theta & \sin\theta \\ \sin\theta & -\cos\theta \end{pmatrix}$, $\theta \in \mathbb{R}$.

We define here the function R, which allows us to observe that the group $O^+(2)$ is commutative.

More precisely:

- $R(\theta)R(\phi) = R(\theta + \phi) = R(\phi)R(\theta)$
- $R(0) = I_2$ and $R(\theta)^{-1} = R(-\theta)$

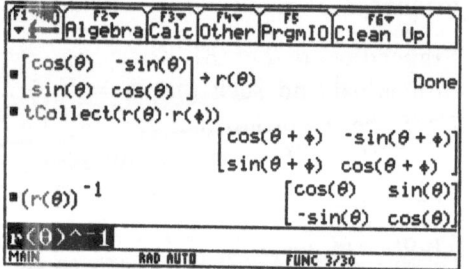

The set $O^-(2)$ is certainly not a subgroup of of $O(2)$. We verify here two classic properties:

- $S(\theta)S(\phi) = R(\theta - \phi)$
- $S(\theta)^{-1} = S(\theta)$, i.e. $S(\theta)^2 = I_2$

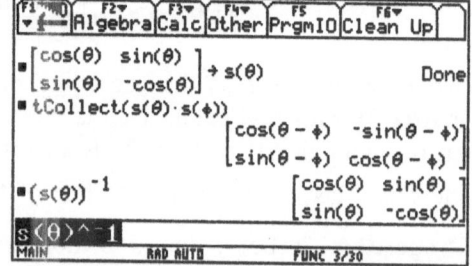

In the oriented plane $\mathbb{R}^2$, the elements of $O^+(2)$ are the transition matrices between orthonormal bases of the same orientation, and those of $O^-(2)$ are the transition matrices between bases with opposite orientations.

$R(\theta)$ and $S(\theta)$ represent respectively, in every positive orthogonal basis, the vector rotation r by an angle θ (mod 2π) and the orthogonal vector symmetry

s with respect to the line with polar angle $\theta/2 \pmod \pi$:

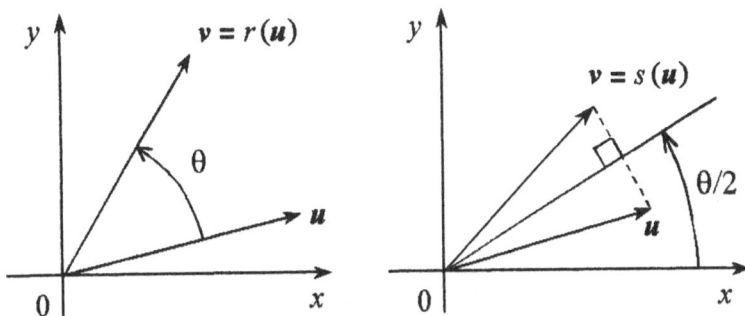

The fact that a rotation r of $\mathbb{R}^2$ by angle θ has the same matrix in every positive orthonormal basis is proved from $R(\phi)^{-1}R(\theta)R(\phi) = R(-\phi + \theta + \phi) = R(\theta)$. ($R(\phi)$ denotes the transition matrix to a positive orthonormal basis).

With the functions R and S defined above, the invariance of the vector rotation matrix is easy to verify. Likewise, $R(\phi)^{-1}S(\theta)R(\phi) = S(\theta)$, which assures that an orthogonal vector symmetry has the same matrix in every positive orthonormal basis.

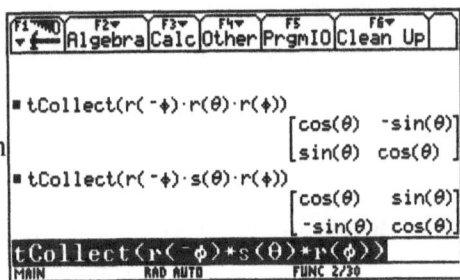

5. 3 Isometries of space

There isn't a simple form describing the orthogonal matrices of order 3. However, here are two examples:

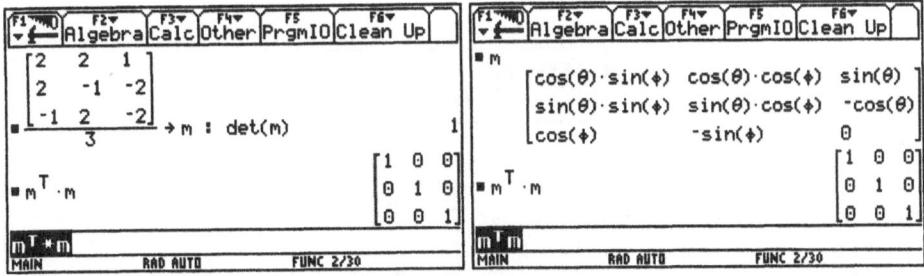

We know the classification of isometries of an oriented Euclidean vector space E of dimension 3. (The fundamental example, to which we will return, is $\mathbb{R}^3$ equipped with its usual scalar product and oriented by giving its standard basis considered as positive.)

Let f be an element of $O(E)$ and F be the subspace of invariant vectors of f.
- If $F = E$, f is the identity map Id_E. This is a positive isometry.
- If $\dim F = 2$, f is an orthogonal symmetry with respect to the plane F. This is a negative isometry.
- If $\dim F = 1$, f is a vector rotation about the line $D = F$ as an axis. This is a positive symmetry.
- If $F = \{0\}$, f may be written as $r \circ s = s \circ r$, where s is the orthogonal symmetry with respect to a plane P and r is a rotation (by angle $\theta \neq 0 \mod 2\pi$) about the axis $D = P^{\top}$.
  This is a negative isometry. If $\theta = \pi \mod 2\pi$, then $f = -Id_E$.

Let F be a vector plane of E and let u be a unit vector directing the line orthogonal to F.

The projection onto the line $D = F^{\perp}$ is defined by:

$$x \mapsto p(x) = < u, x > u$$

The symmetry with respect to the plane F is thus defined by:

$$x \mapsto s(x) = x - 2 < u, x > u$$

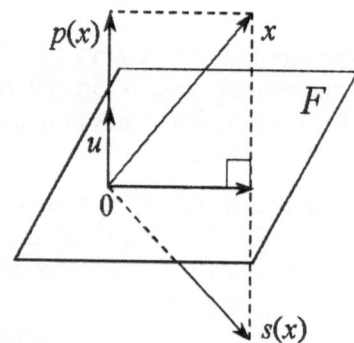

If we denote by $[v]$ the column matrix with coordinates of a vector v of $\mathbb{R}^3$ in the standard basis, then the equation $s(x) = x - 2 < u, x > u$ gives:

$$[s(x)] = [x] - 2^{\top}[u][x][u] = [x] - 2[u]^{\top}[u][x] = (I_3 - 2[u]^{\top}[u])[x]$$

The matrix of the orthogonal vector symmetry s is thus $S = I_3 - 2[u]^{\top}[u]$, or again $S = I_3 - \dfrac{2}{\|u^2\|}[u]^{\top}[u]$ if the vector u is not assumed to be a unit vector.

The function $\mathtt{sym}$ gives the matrix of the orthogonal symmetry with respect to the plane F with equation $ax + by + cz = 0$. (A vector orthogonal to F is $u = (a, b, c)$.)

```
:sym(u)
:1-2*uᵀu/dotP(u,u)
```

Here is the matrix of the orthogonal symmetry s with respect to the plane F with the equation $x + y + z = 0$. We verify that the vector $u = [1, 1, 1]$, orthogonal to F, is transformed into its opposite by the map s.

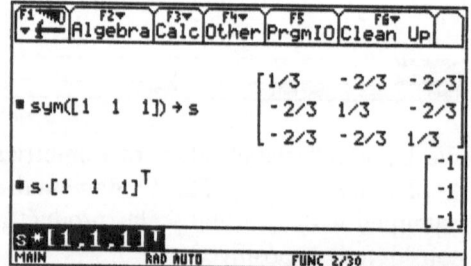

We represent here the image $r(x)$ of a vector x under the rotation r by angle θ about the unit vector u.

p and q are the projections of x onto the line generated by u and onto the plane P orthogonal to this line.

The vector v is equal to $u \times q = u \times x$ with the usual cross product.

The vector w is the projection of $r(x)$ onto P. This may be deduced from q by the rotation of P by the angle θ.

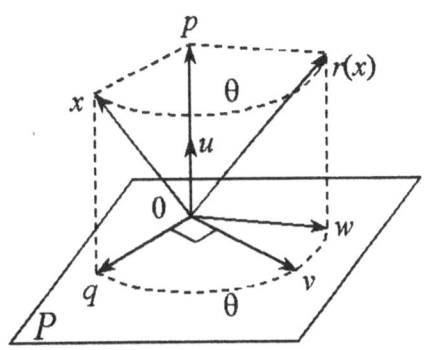

With the same notation, $r(x) = p + w = p + \cos\theta\ q + \sin\theta\ v$.

But $q = x - p$, $p = <u,x>\ u$, and $v = u \times x$.

Thus, $r(x) = \cos\theta\ x + (1 - \cos\theta)\ <u,x>\ u + \sin\theta\ (u \times x)$.

If $u = (a, b, c)$, the matrix of the map $x \mapsto u \times x$ is $A = \begin{pmatrix} 0 & -c & b \\ c & 0 & -a \\ -b & a & 0 \end{pmatrix}$.

In terms of matrices, the preceding definition of $r(x)$ may be written:

$$[r(x)] = \cos\theta\ [x] + (1 - \cos\theta)\ ^{\top}[u][x]\ [u] + \sin\theta\ A[x]$$
$$= \left(\cos\theta\ I_3 + (1 - \cos\theta)\ [u]^{\top}[u] + \sin\theta\ A\right)[x]$$

The matrix for the rotation r is thus: $M = \cos\theta\ I_3 + (1 - \cos\theta)\ [u]^{\top}[u] + \sin\theta\ A$.

The function rot calculates the matrix of the vector rotation r by an angle θ about the vector $u = [a, b, c]$. The syntax of the call is rot([a,b,c],θ).

NB: It is not necessary that the vector u be of norm 1 in this call since this normalization is automatically effected by the function rot.

```
:rot(u,θ)
:Func:Local i,a
:u/norm(u)→u:identity(3)→i
:seq(mat▷list(crossP(u,i[k])),k,1,3)→a
:cos(θ)*i+(1-cos(θ))*uᵀ*u+sin(θ)*aᵀ
:EndFunc
```

Here we calculate the matrix of a rotation by angle $\pi/2$ about $u = [1, 2, 2]$, then that of a rotation by angle θ about the vector $k = [0, 0, 1]$.

The matrix of a rotation may always remain in this last form with the convenient choice of the orthonormal basis i, j, k.

F1	F2▾	F3▾	F4▾	F5	F6▾
	Algebra	Calc	Other	PrgmIO	Clean Up

$\blacksquare\ \mathrm{rot}\!\left([1\ \ 2\ \ 2], \dfrac{\pi}{2}\right)$ $\begin{bmatrix} 1/9 & -4/9 & 8/9 \\ 8/9 & 4/9 & 1/9 \\ -4/9 & 7/9 & 4/9 \end{bmatrix}$

$\blacksquare\ \mathrm{rot}([0\ \ 0\ \ 1], \theta)$ $\begin{bmatrix} \cos(\theta) & -\sin(\theta) & 0 \\ \sin(\theta) & \cos(\theta) & 0 \\ 0 & 0 & 1 \end{bmatrix}$

rot([0,0,1],θ)

| MAIN | RAD AUTO | | FUNC 2/30 | |

Compose the rotation by angle $\pi/3$ about $u = [1, 1, 1]$ with the symmetry with respect to P which has equation $x + y + z = 0$, orthogonal to u.
The only invariant vector of the isometry f so obtained is $\overrightarrow{0}$:
$$f([x, y, z]) = [x, y, z]$$
$$\Leftrightarrow x = -y, \; y = -z, \; z = -x$$
$$\Leftrightarrow x = y = z = 0.$$

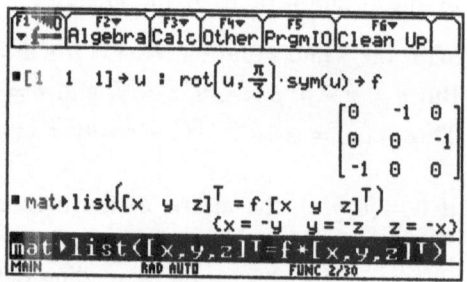

We thus know how to create the matrix of an arbitrary isometry of $\mathbb{R}^3$.

Conversely, it is interesting to be able to describe the characteristics of an isometry starting with its matrix in the standard basis.

This is exactly the role of the program isom, which takes as its argument a square matrix of order 3.

This matrix is assumed to be "exact": because of rounding errors, isom risks not identifying an orthogonal matrix which is in "real approximate" format.

```
:isom(m)
:Prgm
:Local r,a,i,t
:ClrIO:Disp:DelVar x,y,z:identity(3)→i
:Try
:   norm(mᵀ*m-i)→t:ref(m-i)→r
:   If when(t=0,false,true,true):Goto error
:   If r[1,1]=0 Then:Disp "Identity"
:   ElseIf r[2,2]=0 Then:r[1]→a:Goto sym
:   Else
:     (m-mᵀ)/2→a:[[a[3,2],a[1,3],a[2,1]]]→a:sin⁻¹(norm(a))→t
:     If when(t=0,false,true,true) Then
:       Disp "Rotation by angle "& string(t),"About ":Pause a
:     Else
:       If r[3,3]=1 Then:Disp "-Identity":Return
:       Else:Disp "Orth. symmetry wrt axis"
:         Disp "Oriented by the vector:":Pause crossP(r[1],r[2])
:       EndIf
:     EndIf
:     If r[3,3]=0:Return
:     Lbl sym:Disp "Orth. symmetry with respect to"
:     Disp "plane with equation:":Pause getNum(dotP(a,[[x,y,z]]))=0
:   EndIf
:Else
:   Lbl error:Disp "Not an isometry of R^3!"
:EndTry
:EndPrgm
```

Here is how the program isom, takes the matrix M to identify the corresponding isometry f.

First, verify that the matrix M is indeed orthogonal: ${}^{\top}MM = I_3$.

Next, apply Gauss reduction to $M - I_3$ by the instruction ref. Let R be the matrix obtained.

The number of its non-zero diagonal coefficients gives the rank of R, thus that of $M - I_3$, and thus (by complementing with 3) the dimension of the kernel of $M - I_3$, that is, the dimension of the subspace F of invariant vectors of f.

- If $R_{1,1} = 0$, that is, R is the zero matrix, so is $M = I_3$: f is the identity.

- Otherwise, if $R_{1,1} \neq 0$ and $R_{2,2} = 0$, the matrix R is of rank 1. f is thus the orthogonal symmetry with respect to the plane F of its invariant vectors.

Gauss reduction, which proceeds by row operations, does not modify the kernel of the matrix to which it is applied.

More concretely, there exists an invertible matrix Q such that $Q(M - I_3) = R$.

Thus, for every column vector X: $MX = X \Leftrightarrow (M - I_3)X = 0 \Leftrightarrow RX = 0$.

But R may be written $\begin{pmatrix} 1 & \alpha & \beta \\ 0 & 0 & 0 \\ 0 & 0 & 0 \end{pmatrix}$. Thus, $RX = 0 \Leftrightarrow x + \alpha y + \beta z = 0$ which

gives us the equation of the plane F for the symmetry f.

This explains the instruction `r[1]→a` leading to the jump to label `sym`.

• Otherwise, that is, if $R_{1.1} \neq 0$ and $R_{2.2} \neq 0$, the matrix R is of rank 2 or 3, and the subspace F of invariant vectors is thus of dimension 1 or 0.

Under these conditions, we know that f is a rotation r about a unit vector $u = (a, b, c)$, or the composition $r \circ s = s \circ r$ of such a rotation and of the symmetry with respect to the vector plane P orthogonal to u.

We are going to improve the notation of the scheme used to represent a vector rotation r of $\mathbb{R}^3$.

With this notation, $f(x) = \varepsilon p + w$, with $\varepsilon = 1$ if f is a rotation and $\varepsilon = -1$ if f is the composition of a rotation and of a symmetry.

The matrix of f may thus be written:

$$M = \cos\theta \; I_3 + (\varepsilon - \cos\theta) \; [u]^\top[u] + \sin\theta \; A, \text{ with } A = \begin{pmatrix} 0 & -c & b \\ c & 0 & -a \\ -b & a & 0 \end{pmatrix}$$

The matrices I_3 and $[u]^\top u]$ are symmetric, and A is antisymmetric.

We may deduce: $\dfrac{1}{2}(M - {}^\top M) = \sin\theta \; A$. (This is the "antisymmetric part " of M).

In `isom`, the instructions `(m-m`$^\top$`)/2→a` and `[[a[3,2],a[1,3],a[2,1]]]→a` place the matrix $\sin\theta \, A$, then the vector $\sin\theta \, u$, in the variable a. The instruction `sin`$^{-1}$`(norm(a))→t` next puts the angle θ in the variable t, determining it in $[0, \pi]$, since this lines up with the assumption that $\sin\theta \geq 0$.

If the content of t is non-zero, that is if $\theta \in]0, \pi[$, the program `isom` signals the rotation by the angle θ about the vector $a = \sin\theta \, u$.

Otherwise, when $\theta = 0 \mod \pi$, two cases may occur:

a) If the subspace of invariant vectors reduces to $\{\overrightarrow{0}\}$, which the program `isom` identifies from $R_{3.3} = 1$, then f is the map $-Id$. (This could have been computed more quickly.)

b) In the contrary case, f is the orthogonal vector symmetry with respect to the line generated by the vector u. The problem is that this vector is now inaccessible since $a = \sin\theta \, u$ is zero.

But $MX = X \Leftrightarrow RX = 0 \Leftrightarrow \begin{cases} x + \alpha y + \beta z = 0 \\ y + \gamma z = 0 \end{cases}$, since R is written $\begin{pmatrix} 1 & \alpha & \beta \\ 0 & 1 & \gamma \\ 0 & 0 & 0 \end{pmatrix}$.

This system defines the line sought as the intersection of two planes, orthogonal respectively to $v = (1, \alpha, \beta)$ and to $w = (0, 1, \gamma)$. A vector directing this line thus may be calculated by the vector product of v and w, which explains the instruction `Pause crossP(r[1],r[2])`.

Finally, it remains for isom to signal the symmetry with respect to a plane orthogonal to a. (In the case where f does not reduce to a rotation, that is, if $R_{3.3} \neq 0$.)

Now, here are some examples using the program isom. We used the functions rot and sym defined before to construct the matrices of isometries, then we had the program isom identify it.

• Symmetry with respect to the plane with equation $x + \cos\theta\, y + \sin\theta\, z = 0$.

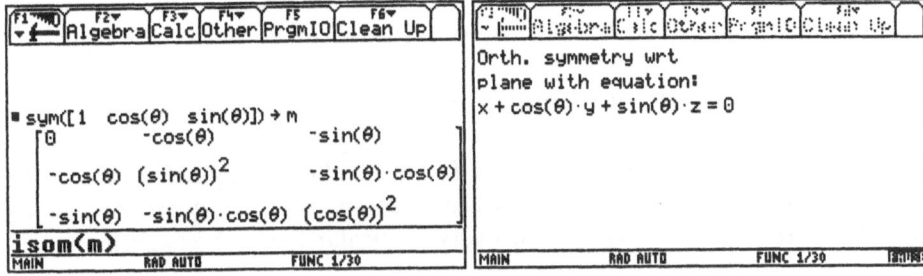

• Rotation by the angle $\pi/3$ about the vector $u = (1, 0, 1)$.

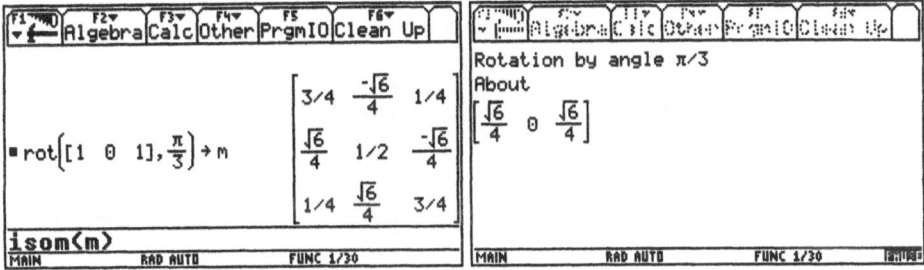

• Composition of the rotation by the angle $\pi/3$ about $u = (1, 1, 1)$ and of the symmetry with respect to the plane orthogonal to u.

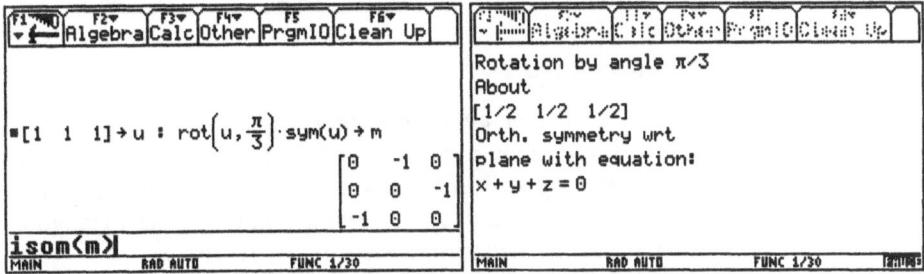

• Orthogonal symmetry with respect to the line directed by $u = (1, 2, 2)$.

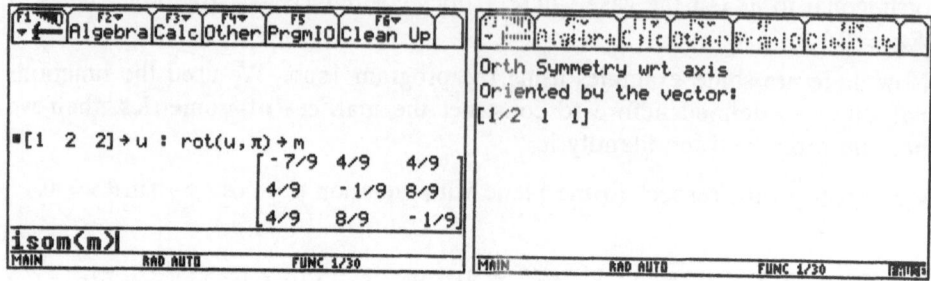

5. 4 Isometries in dimension n

So, in the preceding section, we have found all the orthogonal operators on $\mathbb{R}^3$. The passage to dimension n evidently complicates the question, and we will limit ourselves to looking at two very particular types of isometries.

• Symmetry with respect to a hyperplane

Let F be a subspace of dimension $n-1$ (a hyperplane) of $\mathbb{R}^n$.
The orthogonal complement of F is a line D.
Let u be a unit vector directing D.
The symmetry s with respect to the hyperplane F is defined by $x \mapsto s(x) = x - 2 <u, x> u$, supposing that we are in dimension 3.

Its matrix is $S = I_3 - 2[u]^\top[u]$ (or $S = I_3 - \dfrac{2}{\|u^2\|}[u]^\top[u]$ if u is not a unit vector).

The function sym, used for the study of isometries in dimension 3, does the work again.

Here, for example, is the matrix of the symmetry in $\mathbb{R}^6$, with respect to the plane (P) : $x_1 + 2x_2 - x_3 + 3x_4 + 2x_5 + x_6 = 0$, orthogonal to $u = (1, 2, -1, 3, 2, 1)$.

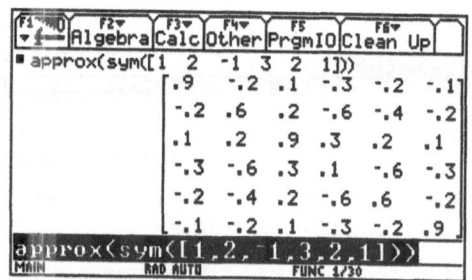

• Rotations in a coordinate plane

Let e_i, e_j be two distinct vectors of an orthonormal basis $(e) = (e_1, e_2, \ldots, e_n)$ of E, and let F be the plane generated and oriented by e_i and e_j (in this order). Let g be the rotation angle θ in the plane F. Consider the isometry f of E defined by g in the plane F, and by the identity map on the orthogonal complement of F, that is, by the equations: $\forall k \notin \{i, j\}, f(e_k) = e_k$.

We may thus say that f is the rotation by the angle θ in the plane oriented by e_i and e_j.

The function rotnijθ (so denoted because we must specify its arguments in that order) gives the matrix of this rotation in the basis (e).

```
:rotnijθ(n,i,j,θ)
:Func:Local m
:identity(n)→m
:cos(θ)→m[i,i]:-sin(θ)→m[i,j]
:sin(θ)→m[j,i]:cos(θ)→m[j,j]:m
:EndFunc
```

For example, here is the matrix of the rotation by the angle θ in the plane generated and oriented by the vectors e_2 and e_4.

NB: It is not necessary that the indices i and j satisfy $i < j$, as is the case here ($i = 2$, $j = 4$).

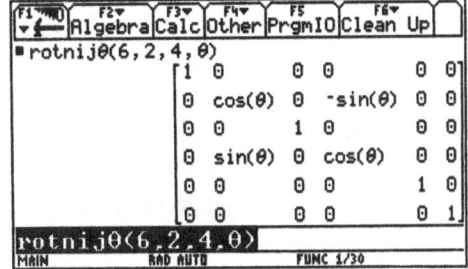

5. 5 Unitary matrices

In this section we have only considered matrices with real coefficients, but in $\mathcal{M}_n(\mathbb{C})$ the notion of a unitary matrix generalizes that of orthogonal matrix. Here is a very brief presentation.

Suppose that $\mathbb{C}^n$ is equipped with its usual scalar product.

A matrix M of $\mathcal{M}_n(\mathbb{C})$ is said to be unitary if $^\top\overline{M}M = I_n$, that is, if M is invertible and if $M^{-1} = {}^\top\overline{M}$ "the conjugate transpose of M ".

The set $U(n)$ of all the unitary matrices is a subgroup of the group $\mathcal{GL}(n, \mathbb{C})$ of invertible matrices of order n with complex coefficients.

- A matrix M of $\mathcal{M}_n(\mathbb{C})$ is unitary if and only if its column vectors form an orthonormal basis of $\mathbb{C}^n$. The unitary matrices are the transition matrices between orthonormal bases in $\mathbb{C}^n$.

- The elements of $U(n)$ are the matrices in the orthonormal bases of $\mathbb{C}^n$ of the unitary operators on $\mathbb{C}^n$, that is, those which preserve the scalar product (or the norm): $\forall u, v \in \mathbb{C}^n$, $< f(u), f(v) > = < u, v >$, $\|f(u)\| = \|u\|$.

- The determinant and the different eigenvalues of a unitary matrix M are of modulus 1: this follows from $^\top\overline{M}M = I_n$ (for the determinant) and from the conservation of the norm $\|MX\| = \|X\|$ (for the eigenvalues).

- Let M be a unitary matrix. Its eigenspaces are pairwise orthogonal, and their (direct) sum is $\mathbb{C}^n$.

6. QR factorization

In this section, we will be interested in one of the classic factorizations of a matrix: the "QR " decomposition.
This involves the orthonormalization of a family of vectors in an inner product space which has been carried over to the matrix setting.

Proposition 1: *Every invertible matrix A of $\mathcal{M}_n(\mathbb{C})$ may be written in a unique way in the form $A = QR$, where Q is a square unitary matrix and R is a square upper triangular matrix with diagonal coefficients which are strictly positive.*

Proof: Let $a_1, a_2, \ldots, a_n$ be the column vectors of A (considered as elements of $\mathbb{C}^n$). Since A is invertible, they form a basis of $\mathbb{C}^n$, denoted (a).
A is then the transition matrix $P_{e.a}$ from the standard basis (e) to the basis (a).
Let Q and R be two matrices of $\mathcal{M}_n(\mathbb{C})$, such that $A = QR$. These two matrices are necessarily invertible ($\det Q \det R = \det A \neq 0$).
The column vectors $q_1, q_2, \ldots, q_n$ of Q thus form a basis (q) of $\mathbb{C}^n$, and Q is the transition matrix $P_{e.q}$ from the basis (e) to the basis (q).
The equation $A = QR$ is then equivalent to $R = Q^{-1}A = P_{q.e}P_{e.a} = P_{q.a}$. Thus, R is the transition matrix from the basis (q) to the basis (a).
Under these conditions:

- To say that Q is unitary is to say that (q) is an orthonormal basis of $\mathbb{C}^n$.
- To say that R is upper triangular with positive diagonal coefficients is to say that, for every integer k between 1 and n, a_k is a linear combination of $q_1, q_2, \ldots, q_k$, its coordinate for q_k being strictly positive.

We recognize that the conditions describe the orthonormalization of the family $a_1, a_2, \ldots, a_n$. (See paragraph **1.5**). We know that these conditions define a family $q_1, q_2, \ldots, q_n$ and only one. Thus the matrix Q and the matrix R are unique. This establishes the result.

NB: Most often we want to form the factorization $A = QR$ for matrices with real coefficients. The matrix Q is then an orthogonal matrix.

The TI-92+ and the TI-89 have an instruction QR which does the factorization of the same name on a matrix with real or complex coefficients.
Here is an example of use of this instruction.
The calculation is here done in exact mode.
The matrix A has been chosen so that Q and R are particularly simple.

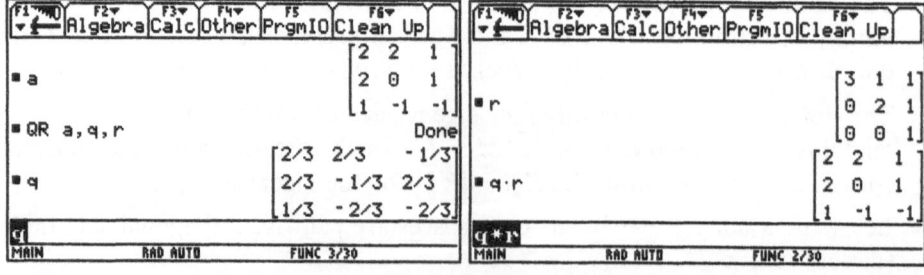

For an arbitrary matrix A, still in exact mode, Q and R are in general much more complicated, and the calculation time grows rapidly with the order of A.

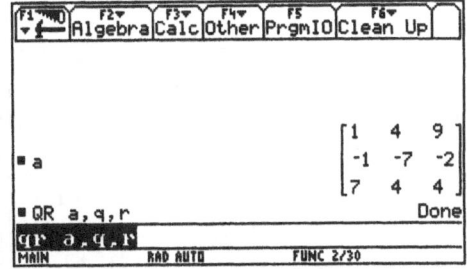

Here is a "reasonable " example with a matrix of order 3.

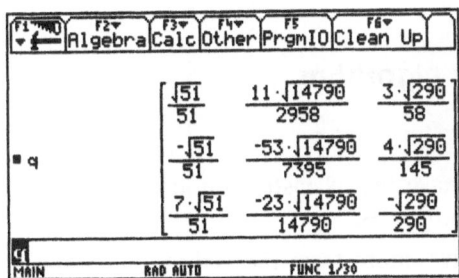

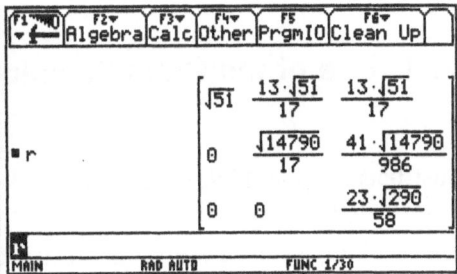

As soon as the matrix A contains at least one "decimal point ", the instruction QR works in "approx " mode and the calculations are then very rapid.

Here, for example, is the decomposition of a matrix of order 5 (with the display in "Float 4 ").

We may also factor "formal " matrices, or complex ones:

$$\bullet\ QR\begin{bmatrix} 1 & 1 \\ \lambda & 1 \end{bmatrix},q,r$$

Done

$$\bullet\ q$$

$$\begin{bmatrix} \dfrac{1}{\sqrt{\lambda^2+1}} & \dfrac{\lambda\cdot\mathrm{sign}(\lambda-1)}{\sqrt{\lambda^2+1}} \\[2ex] \dfrac{\lambda}{\sqrt{\lambda^2+1}} & \dfrac{-\mathrm{sign}(\lambda-1)}{\sqrt{\lambda^2+1}} \end{bmatrix}$$

$$\bullet\ QR\begin{bmatrix} 1+2\cdot i & 2-i \\ 1-i & 1+i \end{bmatrix},q,r$$

Done

$$\bullet\ q$$

$$\begin{bmatrix} \dfrac{\sqrt{7}}{7}+\dfrac{2\cdot\sqrt{7}}{7}\cdot i & \dfrac{2\cdot\sqrt{70}}{35}-\dfrac{\sqrt{70}}{35}\cdot i \\[2ex] \dfrac{\sqrt{7}}{7}-\dfrac{\sqrt{7}}{7}\cdot i & \dfrac{\sqrt{70}}{14}+\dfrac{\sqrt{70}}{14}\cdot i \end{bmatrix}$$

6. 1 Use of the Gram-Schmidt algorithm

Just because the TI-92+ and the TI-89 have the QR instruction doesn't keep us from programming this decomposition ourselves.

The techniques used are in fact instructive, more from the algorithmic point of view than from the mathematical point of view.

We begin by outlining the Gram-Schmidt algorithm, which offers an effective construction to orthonormalize a basis $(a) = a_1, a_2, \ldots, a_n$ of $\mathbb{C}^n$ (and thus also to construct the matrix Q which enters into the decomposition of the matrix A of these vectors in the standard basis of $\mathbb{C}^n$).

We know that the first column vector q_1 of Q is the normalized first column vector a_1 of A.

Let j be an index between 2 and the order n of the matrix A. We suppose that the vectors $q_1, \ldots, q_{j-1}$ are already constructed, and that we want to form q_j.

We know that q_j is the norm of $q'_j = a_j - \sum\limits_{k=1}^{j-1} < q_k, a_j > q_k$.

Denoting by $[u]$ the column of coordinates of a vector u of $\mathbb{C}^n$ gives:

$$[q'_j] = [a_j] - \sum_{k=1}^{j-1} {}^\top\overline{[q_k]}[a_j][q_k] = [a_j] - \sum_{k=1}^{j-1}[q_k]^\top\overline{[q_k]}[a_j] = \left(I_n - \sum_{k=1}^{j-1}[q_k]^\top\overline{[q_k]}\right)[a_j]$$

Thus, we obtain $[q'_j]$ by multiplying $[a_j]$ by $T_j = I_n - \sum_{k=1}^{j-1}[q_k]^\top\overline{[q_k]}$.

It is easy to form the sequence of matrices $T_2, \ldots, T_n$, starting with $T_1 = I_n$, as it is to calculate the new vectors q_k.

The program qrsch uses this method. The syntax is qrsch(a), and the matrices Q and R are placed in the global variables θq and θr.

```
:qrsch(a)
:Prgm:Local q,t,j
:(unitV(aᵀ[1]))ᵀ →q:q→ θq:1→t
:For j,2,colDim(a):t-q*qᵀ →t
: unitV(t*(aᵀ[j])ᵀ)→q:augment(θq,q)→ θq
:EndFor
:θqᵀ*a→ θr
:EndPrgm
```

Here is an example with the program qrsch using a matrix which has served to illustrate the instruction QR.
We represented the matrices Q and R, side by side.
Indeed, the result is the same as with QR, but it is obtained a little more rapidly.

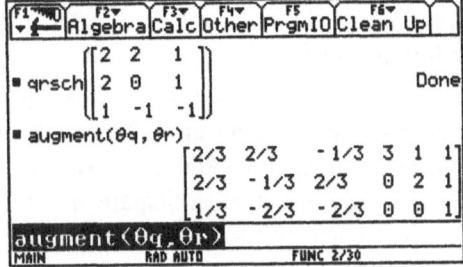

6. 2 Toward other methods

From now on, we assume that the matrices have real coefficients. We thus want to factor an invertible matrix A of $\mathcal{M}_n(\mathbb{R})$ in the form $A = QR$, where Q is orthogonal and R is upper triangular with strictly positive diagonal coefficients.

The two methods which we are going to see use the same principle:
We multiply A on the left by a succession of p orthogonal matrices $\Omega_1, \ldots, \Omega_p$ (and thus finally by the unique orthogonal matrix $\Omega = \Omega_p \times \cdots \times \Omega_1$), such that ΩA is an upper triangular matrix with positive diagonal coefficients R.
The matrix $Q = {}^T\Omega$ is orthogonal and satisfies the equation $A = QR$.

To calculate Ω, we observe that $\Omega = \Omega I_n = \Omega_p \cdots \Omega_1 I_n$. The matrix Ω may thus be deduced from I_n exactly in the same manner as R may be deduced from of A.

The idea is then to "border " A on the right by the identity matrix I_n, and by multiplying the tableau $(A|I_n)$ (of size $n \times 2n$) successively by $\Omega_1, \ldots, \Omega_p$. The final result is then the tableau $(R|\Omega)$ from which it is easy to extract R and $Q = {}^T\Omega$.

There are essentially two variants, which differ in the orthogonal matrices Ω_k. (In both cases, the corresponding isometries of $\mathbb{R}^n$ have been briefly described in paragraph **5.4**).

- *Givens' Method* uses rotations in the coordinate planes.
- *Householder's Method* uses orthogonal hyperplane symmetries.

Whatever the techniques used, we construct a seqence $A_0 = A$, $A_1 = \Omega_1 A_0$, $A_2 = \Omega_2 A_1 = \Omega_2 \Omega_1 A_0$ etc. of invertible matrices, which begins with the matrix A and lead to the triangular matrix R and the desired factorization.

Each matrix A_{k+1} must be "a little more triangular " than the preceding one to which it is related by the equation $A_{k+1} = \Omega_k A_k$. To accomplish this, we must "annihilate " - or change to zero - at least one of the "subdiagonal " coefficients of A_k, without changing those which are already zero. (This requires a judicious choice of the orthogonal matrices Ω_k).

The method of Givens consists of successively annihilating each subdiagonal term of A (in a column from top to bottom, and from the first to the last column). If the matrix A is of order n, we must thus anticipate $\frac{n(n-1)}{2}$ successive rotation matrices.

The method of Householder works in a little more "radical " fashion. Since each symmetry allows annihilation of all the subdiagonal coefficients of a given column at one blow, and since we must work from the first column to the next to last one, we must anticipate $n-1$ symmetry matrices. This is the method which is used by the TI-92+ to factor a matrix in "approx " mode.

6. 3   Givens' Method

Denote the standard basis of $\mathbb{R}^n$ by $(e) = (e_1, e_2, \ldots, e_n)$.

We consider a vector u of $\mathbb{R}^n$, written as $u = \sum_{k=1}^{n} a_k e_k$ in the basis (e).

Let i and j be two indices, with $1 \leq j < i \leq n$. We propose to annihilate the component a_i of u by means of a rotation r while rendering the component a_j positive and without affecting the coordinates with index $k \notin \{i, j\}$.

We must of course use a rotation in the plane $P_{j,i}$ generated (and oriented) by the vectors e_j and e_i.
Let θ be a measure of the angle of r.

If we denote $c = \cos\theta$ and $s = \sin\theta$, then:

- $r(e_j) = ce_j + se_i$
- $r(e_i) = -se_j + ce_i$

The image of u is $r(u) = \sum_{k=1}^{n} a'_k e_k$, with:

- $\forall k \notin \{i, j\}, a'_k = a_k$
- $a'_j = ca_j - sa_i$ and $a'_i = sa_j + ca_i$

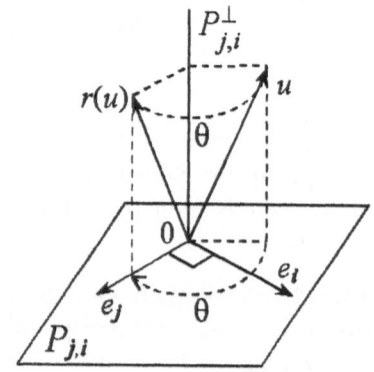

In order to annihilate the coefficient a'_i, the unit vector (c, s) must be orthogonal to the vector (a_i, a_j). We choose: $c = \dfrac{a_j}{\sqrt{a_i^2 + a_j^2}}$ and $s = \dfrac{-a_i}{\sqrt{a_i^2 + a_j^2}}$.

We observe then that $a'_j = \sqrt{a_i^2 + a_j^2}$ is strictly positive.

These formulas show that it is not necessary to calculate θ explicitly.

Of course it is assumed the coordinates a_j and a_i are not both zero, so that u is not in the orthogonal plane $P_{j,i}$. In the contrary case, the problem of annihilation of a_i evidently doesn't come up.

We imagine that u is the j-th column vector of a matrix M. We may thus annihilate the coefficient of M with index (i, j), (it is "subdiagonal"), by forming the product ΩM of M by an orthogonal matrix Ω (This represents a rotation of the plane oriented by e_j and e_i).

Beginning with an invertible matrix A, we may thus construct a sequence of matrices A_k by progressively annihilating the sub-diagonal terms. It is important to note that the coefficients which are already zeros remain equal to 0.

In fact, let A_k be one stage in this sequence which leads from A to a triangular matrix R. Let $A_1, A_2, \ldots, A_n$ be the successive column vectors of A_k. We suppose that the matrix A_k satisfies:

- All the subdiagonal coefficients of columns 1 though $j - 1$ are zero.
- The coefficients of column j, from row $j + 1$ to row $i - 1$, are zero.

At this step we thus annihilate the coefficient of index (i, j) of A_k, and render its j-th diagonal coefficient positive. Let r be the rotation (of the plane oriented by e_j and e_i) charged with the task, and let Ω_k be its matrix in the standard basis.

The preceding hypotheses imply that $u_1, u_2, \ldots, u_{j-1}$ are in the subspace generated by $e_1, e_2, \ldots, e_{j-1}$: they are thus invariant under r.

On the other hand the vector $r(u_j)$:

- conserves the zero components in $e_{j+1}, \ldots, e_{i-1}$ (invariant under r).
- has a positive component in e_j.
- has a zero component in e_i.

In terms of matrices, this signifies that the product $A_{k+1} = \Omega_k A_k$:

- does not modify the first $j - 1$ columns of A_k, which are thus still those of an upper triangular matrix with positive diagonal coefficients.
- does not modify the coefficients of rows $j + 1$ to $i - 1$ of the j-th column of A_k, which are still 0.
- makes positive the j-th diagonal coefficient of A_{k+1}.
- annihilates the coefficient with index (i, j) of A_{k+1}.

Indeed, we see that this method leads to the invertible matrix A with an upper triangular matrix R by successive annihilation of the subdiagonal coefficients of a given column, from the first to the next-to-last.

Remark 1: In this sequence of invertible matrices A_k which go from A to R, let R_j be the one which corresponds at the beginning of the "treatment" of column j. The first $j-1$ column vectors $u_1, u_2, \ldots, u_{j-1}$ of R_j are in $\text{Span}(e_1, e_2, \ldots, e_{j-1})$, but not u_j. At least one of the components a_i of u_j, with $i \geq j$, is thus not zero. This implies that at least once, during the treatment of the column j, we won't find the particular case $a_j = a_i = 0$ described before. At this point, the annihilation of a_i renders the j-th diagonal coefficient strictly positive, and it stays that way up to the end of the treatment of this column.

Remark 2: Let Ω_k be the rotation matrix of the subdiagonal coefficient of index (i, j) which permits the transition from A_k to $A_{k+1} = \Omega_k A_k$. With the notation already used, the product on the left by Ω_k summarizes the row operations: $L_j \leftarrow cL_j - sL_i$ and $L_i \leftarrow sL_i + cL_j$. We could thus easily program the products by Ω_k, rather than trusting the·calculator, which would be easier. We know that the products lead from the tableau $(A|I_n)$ to the tableau $(R|\Omega)$.

Remark 3: The last column of A is not treated because it has no subdiagonal coefficient. This signifies that the last diagonal coefficient of R is possibly negative. (It won't be zero since R, like A, is invertible; besides the sign of this coefficient is that of $\det R$, thus of $\det A$, since we have passed from A to R by rotation matrices whose determinant is always equal to 1). In this case, we apply to the tableau $(R|\Omega)$ symmetry with respect to the hyperplane orthogonal to e_n. (This follows by replacing the last row of this tableau by its opposite).

The program qrgivens applies the preceding method to decompose or factor an invertible matrix A with real coefficients into the form $A = QR$.

The matrices Q and R are stored in the global variables θq and θr.

```
:qrgivens(a):Prgm:Local n,j,i,t,u,s,c,m
:rowDim(a)→n:augment(a,identity(n))→ θr
:@ClrIO:Disp "(R|Qᵀ)=":Pause θr
:For j,1,n-1:For i,j+1,n
:  θr[j,j]^2+θr[i,j]^2→t
:  If when(t=0,true,false,false):Cycle
:  θr[j,j]/(√(t))→c:-θr[i,j]/(√(t))→s
:  c*θr[j]-s*θr[i]→t:s*θr[j]+c*θr[i]→u
:  For m,j,2*n:t[1,m]→θr[j,m]:u[1,m]→θr[i,m]:EndFor
:  @ClrIO:Disp "(R|Qᵀ)=":Pause θr
:EndFor:EndFor
:If when(θr[n,n]<0,true,false,false):mRow(-1,θr,n)→ θr
:(subMat(θr,1,n+1))ᵀ → θq:@ClrIO:Disp "Q=":Pause θq
:subMat(θr,1,1,n,n)→ θr:@ClrIO:Disp "R=":Pause θr
:EndPrgm
```

We have incorporated recognizable comments with the character @. If we remove these @ characters, the instructions which follow them are executed: these are messages allowing us to keep track of the progress of the method.

For example, here are four screens showing the progression in the QR factorization of the matrix A which serve to illustrate the program qrsch. Indeed, we see the regular formation of the triangular matrix R (starting with A) and of the orthogonal matrix Q (starting with I_3).

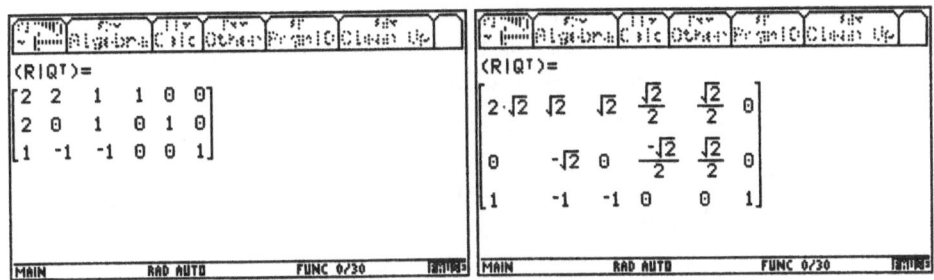

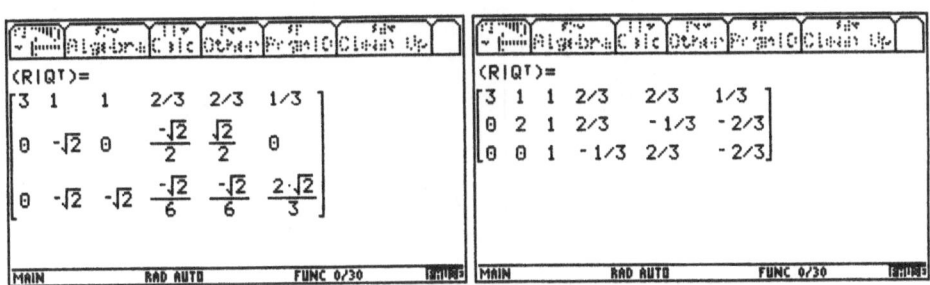

6. 4 Householder's method

We know that the orthogonal symmetry h of $\mathbb{R}^n$, with respect to the hyperplane orthogonal to a non-zero vector v, is given by:

$$\forall u \in \mathbb{R}^n, h(u) = u - \frac{2 <v, u>}{\|v^2\|} v$$

If we denote by $[w]$ the column matrix of a vector w in the standard basis, then the matrix H of h is defined by:

$$H[u] = [h(u)] = [u] - \frac{2}{\|v^2\|}{}^\top[v][u]\ [v] = \left([u] - \frac{2}{\|v^2\|}[v]^\top[v]\ [u]\right)$$

Thus, $H = I_n - \frac{2}{\|v^2\|}[v]^\top[v]$. We say that H is a Householder matrix.

We propose to transform the invertible matrix A of $\mathcal{M}_n(\mathbb{R})$ into an upper triangular matrix R with positive diagonal coefficients, by means of successive products by Householder matrices $H_1, H_2, \ldots, H_{n-1}$. Here, the matrix H_k is charged with "treating " the k-th column.

We will denote $A_0 = A$, $A_1 = H_1A$, $A_2 = H_2A_1 = H_2H_1A$, etc.

The passage from $A_0 = A$ to A_1 must be accompanied by the annihilation of all the subdiagonal coefficients of the first column, making the first diagonal coefficient positive (even if it already is).

If u_1 is the first column vector of A (identified as an element of $\mathbb{R}^n$), we must thus find a hyperplane symmetry h_1 such that $h_1(u_1)$ is collinear and of the same sense as the first vector e_1 of the standard basis. The only possibility taking into account the preservation of the norm is $h_1(u_1) = \|u_1\| e_1$.

Here we have represented the direct sum $\mathbb{R}^n = \mathbb{R}e_1 \oplus E$, where $E = \text{Vect}\{e_2, \ldots, e_n\}$.

We see that there only exist two hyperplane symmetries applying u_1 onto the line $\mathbb{R}e_1$. The two corresponding hyperplanes V_1 and W_1 are respectively orthogonal to vectors $v_1 = u_1 - \|u_1\| e_1$ and $w_1 = u_1 + \|u_1\| e_1$.

Only the symmetry h_1 with respect to V_1 works since it sends u_1 to $\|u_1\| e_1$.

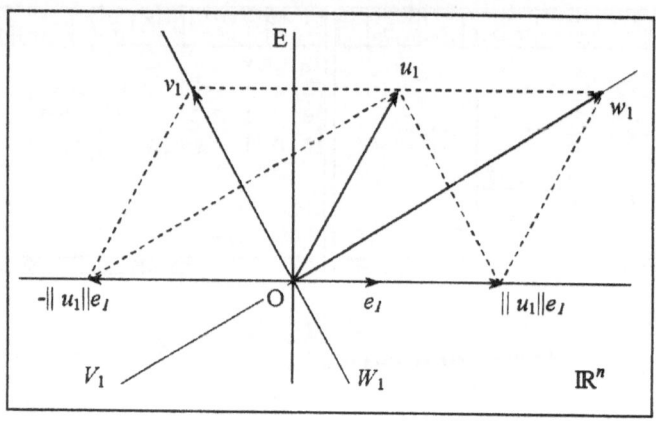

If we denote by H_1 the Householder matrix formed on the vector v_1, then the first column of $A_1 = H_1A$ only contains one non-zero coefficient, the diagonal coefficient, which is positive.

Suppose that we know the matrix A_{j-1}, with $j \leq n$: we want to form $A_j = H_jA_{j-1}$.

By hypothesis, the $j - 1$ first columns of A_{j-1} are those of an upper triangular matrix with positive diagonal coefficients.

Let $u_j = (a_1, a_2, \ldots, a_n)$ be the j-th column vector of A_{j-1}.

Let $u'_j = (a_1, \ldots, a_{j-1}, 0, \ldots, 0)$ be the projection of u_j on $E'_j = \text{Vect}\{e_1, \ldots, e_{j-1}\}$.

Let $u''_j = (0, \ldots, 0, a_j, \ldots, a_n)$ be the projection of u_j on $E''_j = \text{Vect}\{e_j, \ldots, e_n\}$.

Let $v_j = u''_j - \|u''_j\| e_j = (0, \ldots, 0, a_j - \|u''_j\|, \ldots, a_n)$ and h_j be the symmetry with respect to the hyperplane V_j orthogonal to v_j.

Since $e_1, e_2, \ldots, e_{j-1}$ are orthogonal to v_j, they are invariant under h_j. The restriction of h_j to E'_j is thus the identity. The restriction of h_j to E''_j is a hyperplane symmetry of E_j'' which sends the vector u_j'' onto $\|u_j''\| e_j$ (for the same reason as before).

Finally: $h_j(u_j) = h_j(u_j' + u_j'') = u_j' + \|u_j''\| e_j = (a_1, \ldots, a_{j-1}, \|u_j''\|, 0, \ldots, 0)$.

We thus observe that h_j sends the vector u_j to a linear combination of $e_1, \ldots, e_j$, the coefficient of e_j being positive. If H_j is the Householder matrix associated with v_j, the product $H_j A_{j-1}$ annihilates the subdiagonal coefficients of the j-th column (rendering the diagonal coefficient positive), and it does not modify the $j - 1$ first columns (as a result of the invariance of $e_1, \ldots, e_{j-1}$ under h_j): the work already done on these columns is thus preserved.

Thus, the successive products of A by the matrices $H_1, H_2, \ldots, H_{n-1}$ transform A into an upper triangular matrix R with positive diagonal coefficients (except possibly the last, since column n is not "treated", but we know from "Givens" how to fix this).

If we denote by H the orthogonal matrix $H_{n-1} \times \cdots \times H_1$, the equation $HA = R$ gives $A = QR$ with $Q = H^{-1} = {}^T H$.

The calculation of H could be done (as in Givens' Method) in parallel with that of R: it suffices to begin with $(A|I_n)$.

The program `qrholder` implements the QR decomposition using the preceding method. It takes the matrix to be factored as an argument and returns the matrices Q and R in the global variables θq and θr.

The listing which follows gives most of the details of the method, and it is thus easy to decipher. We simply explicate a few instructions:

- `(subMat(θr,j,j,n,j))`^{T} `→w:w[1,1]-norm(w)→w[1,1]`: using the notation above, we extract $w = (a_j, \ldots, a_n)$ from the j-th column of A_{j-1}, for the coordinates with indices from j to n, and we form $(a_j - \|w\|, \ldots, a_n)$

- `augment(newMat(1,j-1),w)→w`: we fill out to $(0, \ldots, 0, a_j - \|w\|, \ldots, a_n)$.

- `expand(1-2*w`^{T}`*w/(norm(w))^2)→h`: this is the Householder matrix.

We have included three comments in the program `qrholder`: if we suppress the @ characters which head them up, the program displays the steps of the decomposition, the matrices $A_0 = A$, H_1 (the first Householder matrix, charged with treating the first column of A), $A_1 = H_1 A_0$, H_2, $A_2 = H_2 A_1$, etc., then finally the matrices Q and R.

```
:qrholder(a)
:Prgm:Local n,j,w,h
:rowDim(a)→n:augment(a,identity(n))→θr
:For j,1,n-1
: @ClrIO:Disp "A"&string(j-1):Pause subMat(θr,1,1,n,n)
: (subMat(θr,j,j,n,j))ᵀ →w:w[1,1]-norm(w)→w[1,1]
: If j>1:augment(newMat(1,j-1),w)→w
: expand(1-2*wᵀ*w/(norm(w))^2)→h
: @ClrIO:Disp "H"&string(j):Pause h
: h*θr→ θr
:EndFor
:If when(θr[n,n]<0,true,false,false):mRow(-1,θr,n)→ θr
:(subMat(θr,1,n+1))ᵀ → θq:@ClrIO:Disp "Q=":Pause θq
:subMat(θr,1,1,n,n)→ θr:@ClrIO:Disp "R=":Pause θr
:EndPrgm
```

Here is the decomposition of a matrix of $\mathcal{M}_4(\mathbb{R})$. (Contrary to appearances, the matrix A was not chosen at random, but so the remaining calculations are the simplest possible).

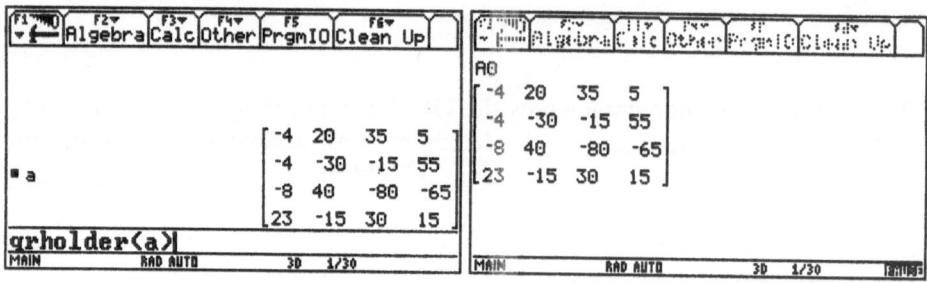

Calling the program   The matrix $A_0 = A$

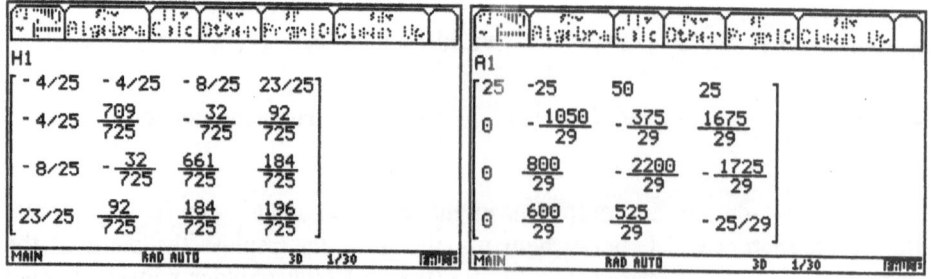

The Householder matrix H_1 The matrix $A_1 = H_1 A$

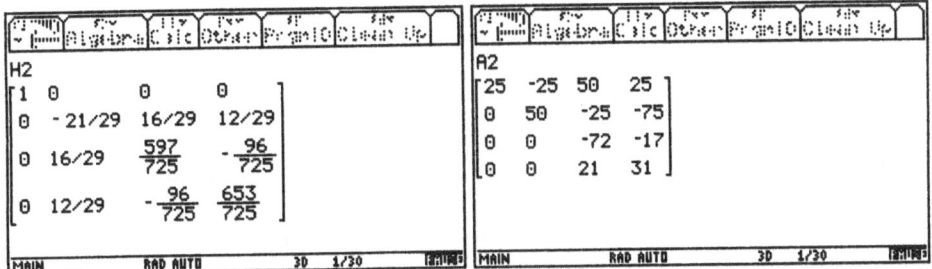

The Householder matrix H_2 $A_2 = H_2 A_1 = H_2 H_1 A$

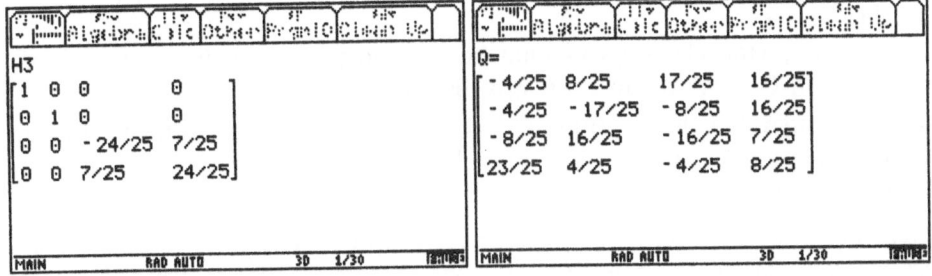

The Householder matrix H_3 The orthogonal matrix Q

Here is the triangular matrix R.
With the preceding notation, we have
$H_3 H_2 H_1 A = R$ and the matrix Q may
be written:
$$Q = (H_3 H_2 H_1)^{-1} = H_1^{-1} H_2^{-1} H_3^{-1}$$
$$= H_1 H_2 H_3.$$

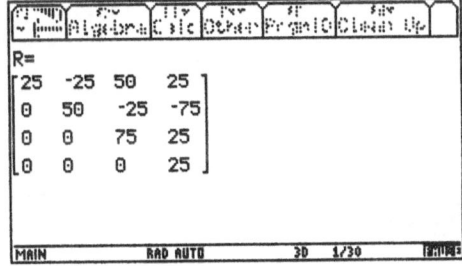

We see that the successive Householder matrices approach closer and closer
to I_n.   This follows from the fact that they are formed on the vectors
$v = (0, \ldots, 0, a_{k+1}, \ldots, a_n)$.   If we denote by H_v and H_w the Householder
matrices formed on the vectors v and $w = (a_{k+1}, \ldots, a_n)$, then:

$$H_v = I_n - \frac{2}{\|v\|^2}[v]^\top[v] = \begin{pmatrix} I_k & 0 \\ 0 & I_{n-k} \end{pmatrix} - \frac{2}{\|v\|^2}\begin{pmatrix} 0 \\ [w] \end{pmatrix}\begin{pmatrix} 0 & {}^\top[w] \end{pmatrix}$$

$$= \begin{pmatrix} I_k & 0 \\ 0 & I_{n-k} \end{pmatrix} - \frac{2}{\|w\|^2}\begin{pmatrix} 0 & 0 \\ 0 & [w]^\top[w] \end{pmatrix}$$

$$= \begin{pmatrix} I_k & 0 \\ 0 & I_{n-k} - \frac{2}{\|w\|^2}[w]^\top[w] \end{pmatrix} = \begin{pmatrix} I_k & 0 \\ 0 & H_w \end{pmatrix}$$

Thus, each Householder matrix H_j of our method is block diagonal: The first block is the identity matrix of order $j - 1$ and the second is a Householder matrix of order $n - j + 1$.

6. 5 Questions of precision

Up to now, we have decomposed exact matrices A. All the methods then give the same result - the decomposition is unique. But in general, we prefer the Gram-Schmidt process for ease of programming and the simple intermediate calculations. That is also the choice of the calculator.

It is otherwise when A contains approximate real numbers. The matrices Q and R are then necessarily contaminated with roundoff errors, due to the method itself or to the fact that the calculator truncates the real numbers to a certain number of decimals.

In this case, Householder's method has the advantage: it is not only more precise than Givens' method, but it is also faster.

We know that the decomposition $A = QR$ exists if A is invertible. But an approximate matrix A may be "almost" non-invertible. In this case, we may meet with serious errors, and it is the method which makes the difference.

Here we factor an "ordinary" matrix A, in "approx" format, with the program qrsch (the method of Gram-Schmidt). We observe that the matrix Q is almost orthogonal: $Q^{\mathsf{T}}Q$ is extremely close to of I_4.

The programs qrgivens and qrholder here will give precise results as well.

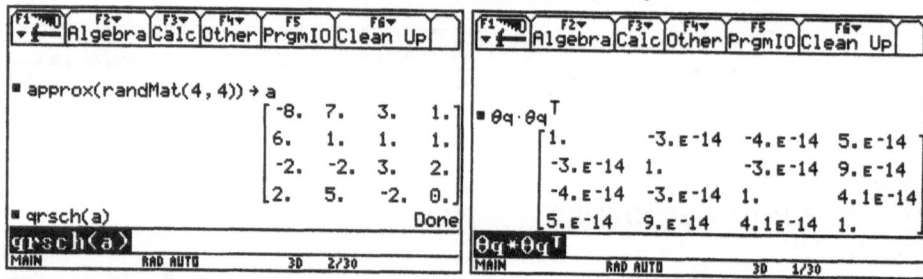

Now we construct a square matrix A of order 3 with a very small determinant, which we factor with the program qrsch.

We observe the matrix Q obtained is far from being orthogonal. (The product $Q^{\mathsf{T}}Q$ is in fact very far from the identity matrix).

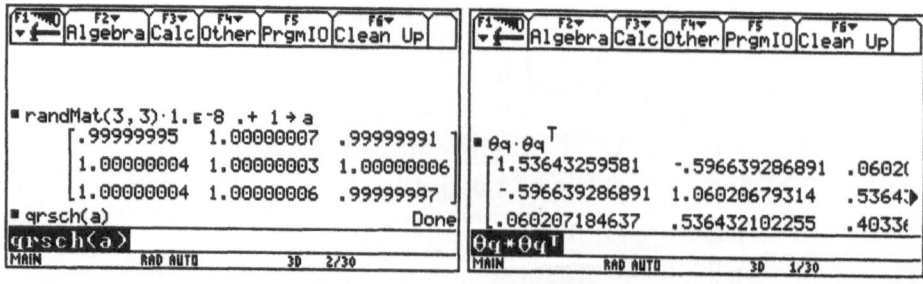

With the same matrix A, the programs qrgivens and qrholder are, to the contrary, very precise since $Q^\top Q$ is almost equal to the identity matrix.

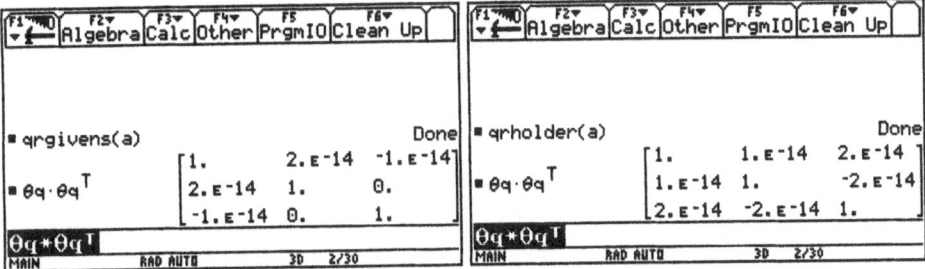

The bad behavior of the Gram-Schmidt method in pathological situations such as this may easily be explained.

In fact, we remember that in the transformation of a family $(A_1, A_2, \ldots, A_n)$ to an orthonormal basis $(e_1, e_2, \ldots, e_n)$, the vector e_k is obtained by normalizing the vector $e'_k = u_k - \sum_{j=1}^{k-1} < e_j, u_k > e_j$.

If the vector u_k is "almost " in the subspace E_{k-1} generated by the vectors $e_1, e_2, \ldots, e_{k-1}$ already constructed, then the vector e'_k is "almost " zero.

If we add that the scalar products $< e_j, u_k >$ which enter into the formation of e'_k are themselves only known with some imprecision, we doubt that the passage to $e_k = e'_k / \|e'_k\|$ may be made without important accompanying roundoff errors. The vector e_k thus may not be truly orthogonal to E_{k-1} and this property is often indispensable for the method to continue: the vectors remaining to be constructed are going to suffer from this imprecision.

An improvement of the Gram-Schmidt method consists of redoing the preceding notation, notably to "put right " the vector u_k leading to the vector e'_k, supposed to be orthogonal to E_{k-1}), by "putting right " the vector e'_k itself. We thus create the vector $e''_k = e'_k - \sum_{j=1}^{k-1} < e_j, e'_k > e_j$, which must, in principle, be "more " orthogonal to $e_1, e_2, \ldots, e_{k-1}$ than it was to e'_k. We next construct e_k by normalizing the vector e''_k.

This improvement is very simple to make in the program qrsch. It suffices to replace the instruction unitV(t*(a⊤[j])⊤)→q by the two instructions t*(a⊤[j])⊤)→q:unitV(t*q)→q.

We call the program thus modified qrsch2. The matrix A used here is the same as that used in the preceding comparative test.
We see that the precision is much better (comparable with that observed with Givens' method and that of Householder).

Householder's method itself may be improved, not quite to a level of precision to ward off certain pathological situations. To understand what this means, we refer to the figure which served to illustrate the principles of the method.

We show the diagram again with the two orthogonal hyperplane symmetries which send a vector u_1 onto the line generated by the first vector e_1 of the standard basis:

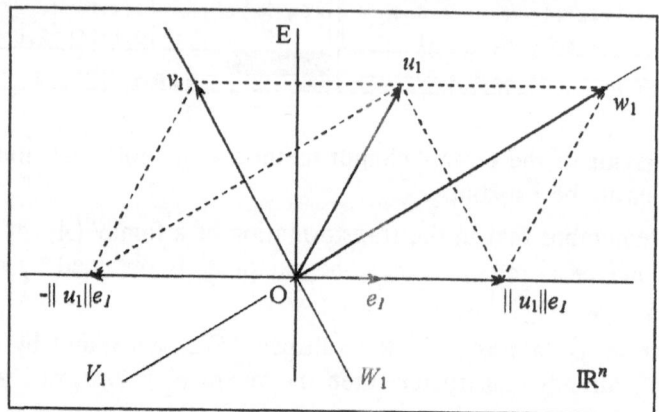

The two possible symmetries s_v and s_w with respect to the hyperplanes V_1 and W_1, are respectively orthogonal to vectors: $v_1 = u_1 - \|u_1\| \, e_1$, and $w_1 = u_1 + \|u_1\| \, e_1$.
To each of these is associated a Householder matrix:

$$H_v = I_n - \frac{2}{\|v_1\|^2} [v_1]^\top [v_1] \qquad \text{and} \qquad H_w = I_n - \frac{2}{\|w_1\|^2} [w_1]^\top [w_1]$$

In principle, we choose the symmetry s_v, which is the only one to send u_1 onto a vector in the same sense as e_1. A problem appears if the vector u_1 is already collinear in the same sense as e_1 (up to roundoff errors).
In this case the vector v_1 is zero or almost zero, and the calculation of the Householder matrix H_v will lead to a "division by zero" (because of the division by $\|v_1\|^2$) or to a large numerical imprecision.

We are thus interested in choosing between v_1 or w_1, the one which has the larger norm. The problem is that the diagonal coefficients of the matrix R created will no longer necessarily be positive. But this detail may be regulated with hindsight.

If we denote by a_1 the component of u_1 on e_1:

- $\|v_1\|^2 = \|u_1 - \|u_1\| e_1\|^2 = \|u_1\|^2 - 2\|u_1\| a_1 + \|u_1\|^2 = 2\|u_1\| \left(\|u_1\| - a_1 \right)$.
- $\|w_1\|^2 = \|u_1 + \|u_1\| e_1\|^2 = 2\|u_1\| \left(\|u_1\| + a_1 \right)$.

We see that we must choose the vector $w_1 = u_1 + \|u_1\| e_1$ if the component a_1 is positive, and the vector $v_1 = u_1 - \|u_1\| e_1$ if this component is negative.

In summary, we must choose $u_1 + \varepsilon \|u_1\| e_1$, where ε is the sign of a_1.

This remains valid throughout the method. It is thus easy to correct the program qrholder.

Here is the instruction to modify:

$$:\text{w[1,1]-norm(w)}\rightarrow\text{w[1,1]}$$

We replace it by:

$$:\text{when(w[1,1]>0,1,-1,-1)}\rightarrow\text{h:w[1,1]+h*norm(w)}\rightarrow\text{w[1,1]}$$

In the same way, the instruction in qrholder which assures the positivity of the last diagonal coefficient...

$$:\text{If when(}\theta r\text{[n,n]<0,true,false,false): mRow(-1,}\theta r\text{,n)}\rightarrow\theta r$$

...must here be modified in the following manner (since we are no longer certain of the sign of each of the diagonal coefficients of R):

$$:\text{seq(when(}\theta r\text{[j,j]<0,-1,1,1),j,1,n)}\rightarrow\text{h:diag(h)*}\theta r\rightarrow\theta r$$

If we take as qrhold2 the program so modified, below is an example of the decomposition of a matrix A, given that the sub-diagonal coefficients of the first column are zero.

We see that the program qrhold2 effects a very precise decomposition: we have calculated the norm of $Q^{\mathsf{T}}Q - I_4$ to verify that Q is "almost" orthogonal and the norm of $QR - A$ to check the precision of the equation $A = QR$. We see that the matrix R is, up to inevitable roundoff errors, upper triangular with positive diagonal coefficients.

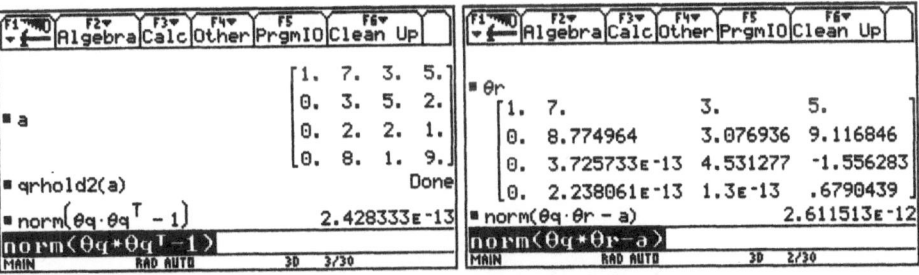

On the other hand, the program qrholder is interrupted by an error. The sub-diagonal coefficients of the first column are zero (and the diagonal coefficient is positive), so the vector v used to construct the Householder matrix is itself zero. This matrix is thus impossible to form.

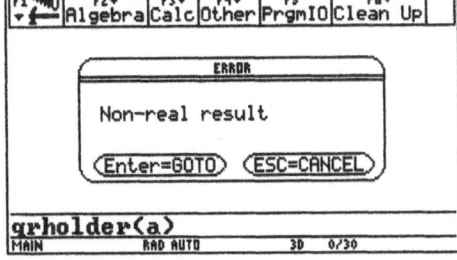

6

Eigenvalues and eigenvectors

The search for eigenvalues and eigenvectors of a linear map f, those scalars λ and the non-zero vectors u such that $f(u) = \lambda u$, is of considerable importance in linear algebra, as well as in the application of mathematics to economics, physics, and engineering.

An operator is a linear map from a vector space E into itself. To "reduce " an operator is to find subspaces, called eigenspaces, of E generated by eigenvectors on which f reduces to a dilation: for a vector in such a subspace, the image is just a multiple of that vector. Knowledge of eigenvalues and eigenvectors allows a better understanding of f and also is useful in a study of powers of f. Not only is the theory very rich, but the methods and the algorithms which allow us to find these values and vectors, generally only in approximate fashion, form an impressive part of the literature of mathematics.

In finite dimensions, operators are represented by matrices and that is what may be reduced to simpler form. There are general methods — not always very effective — and techniques which may be applied in particular cases — for symmetric matrices with real coefficients, for example.

After reviewing some theoretical ideas, we will see how the calculator permits one to better understand and to easily find eigenvalues and eigenvectors.

1. Review of Theory

Throughout this section, E is a vector space E over $\mathbb{K}$ ($\mathbb{K} = \mathbb{R}$ or $\mathbb{C}$). We denote by $\mathcal{L}(E)$ the set of operators on E. Thus, each member of $\mathcal{L}(E)$ is a linear transformation from E to E.

1. 1 First definitions

Let f be an operator on E, and let λ be a scalar (an element of $\mathbb{K}$).
We put $E_\lambda(f) = \mathrm{Ker}(f - \lambda I_n) = \{u \in E, f(u) = \lambda u\}$.
$E_\lambda(f)$ is a sub-space of E.
Particular cases: $E_0(f) = \mathrm{Ker}\, f$, $E_1(f) = \mathrm{Inv}(f)$ (vectors invariant under f) and $E_{-1}(f) = \mathrm{Opp}(f)$ (vectors mapped to their opposites).

Definition 1: We say that λ is an eigenvalue of f if there is a non-zero vector u such that $f(u) = \lambda u$, that is, if $E_\lambda(f)$ does not reduce to $\{\vec{0}\}$. We then say that u is an icopyeigenvector of f for the eigenvalue λ, and that $E_\lambda(f)$ is the eigenspace of f for λ. The *spectrum* of f, denoted $\mathrm{Sp}(f)$, is the set, possibly empty, of all the eigenvalues of f.

The following observations immediately result from the definition.

- An eigenspace E_λ never reduces to $\{\overrightarrow{0}\}$. More precisely, $E_\lambda(f)$ is the union of $\{\overrightarrow{0}\}$ and of the eigenvectors of f for all λ.
- For an eigenvalue λ, there is an infinity of eigenvectors. An eigenvector is never the only one for its eigenvalue.
- λ appears in the spectrum of $f \Leftrightarrow f - \lambda Id_E$ is not injective. In particular: 0 is an eigenvalue of $f \Leftrightarrow f$ is not injective. (We have the equation $E_0 = \mathrm{Ker}\, f$.)
- If $\dim E < \infty$, then: $\lambda \notin \mathrm{Sp}(f) \Leftrightarrow f - \lambda Id_E$ is an automorphism on E. In particular, $f \in \mathrm{Aut}(E) \Leftrightarrow 0 \notin \mathrm{Sp}(f)$.
- Let $\lambda \in \mathrm{Sp}(f)$. Then the restriction of f to E_λ is a dilation by λ.
- If f is the dilation $u \mapsto \lambda u$ of E, then $\mathrm{Sp}(f) = \{\lambda\}$ and $E_\lambda(f) = E$.
- Let F and G be two complementary subspaces of E, neither reducing to $\overrightarrow{0}$. Every vector of u from E may be written uniquely as $u = v + w$, $v \in F$, $w \in G$.
  Let p be the projection onto F, parallel to G: $u \mapsto p(u) = v$.
  Let s be a symmetry with respect to F, parallel to G: $u \mapsto s(u) = v - w$.
  Then, $\mathrm{Sp}(p) = \{0, 1\}$, $\mathrm{Sp}(s) = \{-1, 1\}$, $F = E_1(p) = E_1(s)$, $G = E_0(p) = E_{-1}(s)$.

Here are two extreme examples. In the first case the spectrum of f is all of $\mathbb{C}$, and in the second it is empty:

- If E is the $\mathbb{C}$-vector space of infinitely differentiable functions from $\mathbb{R}$ into $\mathbb{C}$, then the linear map $f : y \mapsto y'$ has every λ of $\mathbb{C}$ for an eigenvalue. In fact the differential equation $y' = \lambda y$ has for its general solution the vector line E_λ generated by the map $x \mapsto e^{\lambda x}$.
- If $E = \mathbb{C}[X]$, the linear map $f : P \mapsto XP$ has no eigenvalue (if $P \neq 0$ then $XP \neq \lambda P$, for reasons of degree).

Here are two other examples.

1. To solve the differential equation $y'' = \lambda y$, is the same as finding the eigenspaces of the operator $f : y \mapsto f(y) = y''$ on the vector space $C^\infty(\mathbb{R}, \mathbb{R})$.

We see how deSolve lets us find these eigenspaces: we have successively put $\lambda = \omega^2$, $\lambda = -\omega^2$ (with $\omega > 0$) and $\lambda = 0$.
All the reals are eigenvalues of f, and the eigenspaces are vector planes.

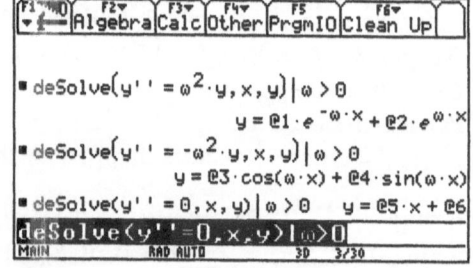

2. Now we consider $f \in \mathcal{L}(C^\infty(\mathbb{R}, \mathbb{R}))$, defined by $f : y \mapsto xy'$.

To find the eigenspaces of f, is to solve $xy' = \lambda y$, whose general solution on $I = \mathbb{R}^{+*}$ or $I = \mathbb{R}^{-*}$, is the line generated by $x \mapsto y_\lambda(x) = |x|^\lambda$.

The only y_λ which have C^∞ extensions to $\mathbb{R}$ are the $y_n : x \mapsto x^n$, where $n \in \mathbb{N}$ (if $\lambda \notin \mathbb{N}$, and if $p > \lambda$, $y_\lambda^{(p)}$ is not extendable to 0).

Sp(f) is thus equal to $\mathbb{N}$, and the eigenspace associated with n is the vector line generated by $x \mapsto x^n$.

```
F1┄┄  F2▾     F3▾   F4▾    F5      F6▾
▾┄▆▆▆ Algebra Calc Other PrgmIO Clean Up

• deSolve(x·y' = λ·y,x,y)          y = @1·x^λ
■ x^λ → y(x)                              Done
   d4
■ ───(y(x))     λ·(λ-3)·(λ-2)·(λ-1)·x^(λ-4)
   dx4
d(y(x),x,4)
MAIN              RAD AUTO        3D    3/30
```

The principal interest in eigenspaces of an operator f is that they are stable or invariant (the restriction of f to E_λ is just a dilation by λ) but also that they form a direct sum of E.

Proposition 1: Let f be an operator on a vector space E over $\mathbb{K}$. Let $\lambda_1, \lambda_2, \ldots, \lambda_p$ be p distinct eigenvalues of f, and let $E_1, \ldots, E_p$ be the corresponding eigenspaces. Then, the sum $E_1 + \cdots + E_p$ is direct.

Proof: This may be done by induction on $p \geq 2$.
Let u_1 be in E_1 and u_2 in E_2. Using $\lambda_1 \neq \lambda_2$, we find:

$$u_1 + u_2 = \overrightarrow{0} \Rightarrow \begin{cases} u_1 + u_2 = \overrightarrow{0} \\ f(u_1 + u_2) = \overrightarrow{0} \end{cases} \Rightarrow \begin{cases} u_1 + u_2 = \overrightarrow{0} \\ \lambda_1 u_1 + \lambda_2 u_2 = \overrightarrow{0} \end{cases} \Rightarrow u_1 = u_2 = \overrightarrow{0}$$

The two subspaces E_1 and E_2 are thus a direct sum.
Now we suppose that the property has been proved for rank $(p-1)$, and let $u_1, u_2, \ldots, u_p$ be in the subspaces associated with $\lambda_1, \lambda_2, \ldots, \lambda_p$.
We suppose that $\sum_{k=1}^{p} u_k = \overrightarrow{0}$. It must be shown that all the u_k are null.

$$\sum_{k=1}^{p} u_k = \overrightarrow{0} \Rightarrow \begin{cases} \sum_{k=1}^{p} u_k = \overrightarrow{0} \\ f\left(\sum_{k=1}^{p} u_k\right) = \overrightarrow{0} \end{cases} \Rightarrow \begin{cases} \lambda_p \sum_{k=1}^{p} u_k = \overrightarrow{0} \quad (1) \\ \sum_{k=1}^{p} \lambda_k u_k = \overrightarrow{0} \quad (2) \end{cases} \Rightarrow \sum_{k=1}^{p-1} (\lambda_p - \lambda_k) u_k = \overrightarrow{0}$$

By the recurrence hypothesis this implies: $\forall k \in \{1, \ldots, p-1\}$, $(\lambda_p - \lambda_k) u_k = \overrightarrow{0}$.
But the λ_j are pairwise distinct.
We may conclude that $u_1 = \cdots = u_{p-1} = \overrightarrow{0}$, then $u_p = \overrightarrow{0}$ by the initial equation.
The property is thus proved by induction.
This result may also be expressed in the following manner:

Let $(u_1, u_2, \ldots, u_p)$ be a family of p eigenvectors of f for the pairwise distinct eigenvalues $\lambda_1, \lambda_2, \ldots, \lambda_p$. Then the vectors $u_1, u_2, \ldots, u_p$ form a linearly independent set.

In particular, we deduce that if f is an operator on a vector space E over $\mathbb{K}$ with dimension n, then f has at least n distinct eigenvalues.

We note that the behavior of the map f is particularly simple on a direct sum $F = E_1 \oplus \cdots \oplus E_p$ of eigenspaces for the distinct eigenvalues $\lambda_1, \ldots, \lambda_p$.

In fact, if $u \in F$ decomposes into $u = \sum\limits_{k=1}^{p} u_k$ for this direct sum, then:

$$f(u) = \sum_{k=1}^{p} \lambda_k u_k \quad \text{and more generally:} \quad \forall m \in \mathbb{N}, \; f^m(u) = \sum_{k=1}^{p} \lambda_k^m u_k$$

We are now going to define the eigenvalues and the eigenvectors of a matrix with real or complex coefficients.

Definition 2: Let M be a square matrix of order n with coefficients in $\mathbb{C}$. We say that a complex number λ is a eigenvalue of M if there is a non-zero column vector X such that $MX = \lambda X$. As before, we define the spectrum of M and the eigenspace of M for the eigenvalue λ.

In fact, this second definition is a particular case of the first, with the operator on $\mathcal{M}_{n,1}(\mathbb{C})$ defined by $X \mapsto MX$.

This new definition also calls for some remarks.

If f is an operator on a $\mathbb{R}$-vector space, its eigenvalues must be real. But the eigenvalues of a matrix M with real coefficients may be real or complex. To erase all ambiguity, we denote by $\mathrm{Sp}_{\mathbb{R}}(M)$ the set of real eigenvalues of such a matrix.

Let E be a $\mathbb{C}$-vector space of dimension n with a basis (e). We denote by $[u]$ the column matrix with the coordinates of a vector u of E.

Let f be an operator on E, with matrix M in the basis (e).

For every scalar λ and every vector u: $f(u) = \lambda u \Leftrightarrow M[u] = \lambda [u]$.

In other words the eigenvalues of a square matrix M are those of every operator which may be represented by f, and the eigenvectors of M are the column matrices associated with the eigenvectors of f.

Two similar matrices M and N thus have the same spectra. More precisely, if P is an invertible matrix such that $N = P^{-1}MP$, then for every eigenvalue λ of M and N: $X \in E_\lambda(N) \Leftrightarrow PX \in E_\lambda(M)$ (the eigenspaces $E_\lambda(N)$ and $E_\lambda(M)$ are thus isomorphic by the map $X \mapsto PX$).

In fact:

$$X \in E_\lambda(N) \Leftrightarrow NX = \lambda X \Leftrightarrow P^{-1}MPX = \lambda X$$
$$\Leftrightarrow MPX = \lambda PX \Leftrightarrow PX \in E_\lambda(M)$$

Here we construct a somewhat pecu-
liar matrix M, square of order 3.

We observe that the vector $(1,1,1)$
(identified here with an element of
$\mathbb{R}^3$) is an eigenvector of M for the
eigenvalue $\lambda = 4$.

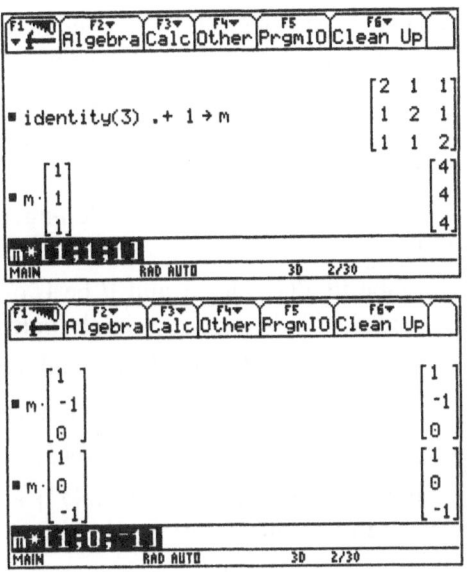

With the same matrix we see that
$(1,-1,0)$ and $(1,0,-1)$, which generate a plane, are eigenvectors for $\lambda = 1$.

Certainly the search for eigenvalues
and eigenspaces is not subject to
chance or to coincidence. There are
several methods which we are going
to describe.

1. 2 Polynomials of operators

In this paragraph, we recall – often without proof – some classical results.

Let f be an operator on the vector space E over $\mathbb{K}$.
We define integral powers of f by $f^0 = Id_E$, and $\forall p \in \mathbb{N}$, $f^{p+1} = f \circ f^p$.
Thus, $f^1 = f$, $f^2 = f \circ f$, etc.
If f is an automorphism of E, we extend this definition to negative exponents
by putting $f^{-p} = (f^{-1})^p$ for each natural number p.
Of course, we have the usual equations: $f^p \circ f^q = f^{p+q}$, $(f^p)^q = f^{pq}$.
Moreover, if f and g commute in $\mathcal{L}(E)$: $(f \circ g)^p = f^p \circ g^p$.

The *iterated kernels* of f are defined by $F_p = \operatorname{Ker} f^p$, for each integer $p \geq 0$.
These form a sequence, increasing by inclusion, of subspaces of E:
$$F_0 = \{\overrightarrow{0}\} \subset F_1 = \operatorname{Ker} f \subset F_2 = \operatorname{Ker} f^2 \subset \cdots$$
A standard exercise consists of showing that this sequence is either strictly
increasing or stationary. The second case is only possible if E is of finite
dimension.

The *iterated images* of f are the subspaces $G_p = \operatorname{Im} f^p$, $p \geq 0$.
They form a sequence ordered by inclusion, strictly decreasing or stationary:
$G_0 = E \supset G_1 = \operatorname{Im} f \supset G_2 = \operatorname{Im} f^2 \supset \cdots$.

Let $P = \displaystyle\sum_{k=0}^{m} a_k X^k$ be a polynomial with coefficients in $\mathbb{K}$. We put

$$P(f) = \sum_{k=0}^{m} a_k f^k.$$

Thus, we define the set $\mathbb{K}[f]$ of all the polynomials of an operator f.

We may verify that $\mathbb{K}[f]$ is a commutative subalgebra of $\mathcal{L}(E)$.

More precisely, for all polynomials P and Q, and for all scalars α and β:
$$\alpha P(f) + \beta Q(f) = (\alpha P + \beta Q)(f) \text{ and } P(f) \circ Q(f) = (PQ)(f) = (QP)(f)$$

If f and g are two operators of E which commute, then every element of $\mathbb{K}[f]$ commutes with every element of $\mathbb{K}[g]$.

If F is a subspace which is stable under f, it is invariant under each polynomial of f.

Let f be in $\mathcal{L}(E)$; let α and β be two scalars ($\alpha \neq 0$).

αf and $f - \beta Id_E$ are very simple polynomials of f, and the equivalences
$$f(u) = \lambda u \Leftrightarrow (\alpha f)(u) = (\alpha \lambda)u \text{ and } f(u) = \lambda u \Leftrightarrow (f - \beta Id_E)(u) = (\lambda - \beta)(u)$$
provide evidence for the following results:

- If $\alpha \neq 0$: $\operatorname{Sp}(\alpha f) = \{\alpha \lambda, \lambda \in \operatorname{Sp}(f)\}$ and $E_{\alpha\lambda}(\alpha f) = E_\lambda(f)$.
- $\operatorname{Sp}(f - \beta Id_E) = \{\lambda - \beta, \lambda \in \operatorname{Sp}(f)\}$ and $E_{\lambda-\beta}(f - \beta Id_E) = E_\lambda(f)$.
  (a useful result in certain algorithms for searching for eigenvalues).

Proposition 2: If u is an eigenvector of f for the eigenvalue λ, then:
- For each integer k, u is an eigenvector of f^k for λ^k.
- If f is an automorphism, u is an eigenvector de f^{-1} for $1/\lambda$.
- For every polynomial P, u is an eigenvector of $P(f)$ for $P(\lambda)$.

Proof: The equation $f(u) = \lambda u$ leads to $f^k(u) = \lambda^k u$ through an evident recursion.

Likewise, if f is bijective: $0 \notin \operatorname{Sp}(f)$, and $f(u) = \lambda u \Rightarrow f^{-1}(u) = \frac{1}{\lambda}u$.

More generally, if $P = \sum_{k=0}^m a_k X^k$:

$$P(f)(u) = \left(\sum_{k=0}^m a_k f^k\right)(u) = \sum_{k=0}^m a_k f^k(u) = \sum_{k=0}^m a_k \lambda^k u = \left(\sum_{k=0}^m a_k \lambda^k\right)u = P(\lambda)u$$

Proposition 3: (Theorem on decomposition of the kernels).

Let P and Q be two relatively prime polynomials, and let f be an operator on the vector space E. Then $\operatorname{Ker}(PQ)(f) = \operatorname{Ker} P(f) \oplus \operatorname{Ker} Q(f)$.

Proof: Since the polynomials are relatively prime, there are two polynomials A and B such that $AP + BQ = 1$ (Bezout's theorem). We conclude that $A(f) \circ P(f) + B(f) \circ Q(f) = 1(f) = Id_E$. Every vector u of E may thus be written $u = u_1 + u_2$, with $u_1 = A(f) \circ P(f)(u)$ and $u_2 = B(f) \circ Q(f)(u)$.

If u appears in $\operatorname{Ker} P(f) \cap \operatorname{Ker} Q(f)$, then $u_1 = \overrightarrow{0}$ and $u_2 = \overrightarrow{0}$. We conclude that the vector u is zero: the sum $\operatorname{Ker} P(f) + \operatorname{Ker} Q(f)$ is thus direct.

If $u \in \operatorname{Ker} PQ(f)$, $P(f)(u_2) = P(f) \circ A(f) \circ Q(f)(u) = A(f) \circ (PQ)(f)(u) = \overrightarrow{0}$. Thus, $u_2 \in \operatorname{Ker} P(f)$. Likewise, $u_1 \in \operatorname{Ker} Q(f)$, so $u \in \operatorname{Ker} P(f) \oplus \operatorname{Ker} Q(f)$, which proves the inclusion $\operatorname{Ker}(PQ)(f) \subset \operatorname{Ker} P(f) \oplus \operatorname{Ker} Q(f)$.

Finally: $u \in \operatorname{Ker} Q(f) \Rightarrow Q(f)(u) = \overrightarrow{0} \Rightarrow (PQ)(f) = \overrightarrow{0} \Rightarrow u \in \operatorname{Ker}(PQ)(f)$.

We conclude that $\operatorname{Ker} Q(f) \subset \operatorname{Ker}(PQ)(f)$. Likewise, $\operatorname{Ker} P(f) \subset \operatorname{Ker}(PQ)(f)$.

Thus, we have the inclusion $\operatorname{Ker} P(f) \oplus \operatorname{Ker} Q(f) \subset \operatorname{Ker}(PQ)(f)$, and finally the equation.

A finite recursion allows us to obtain a more general result:

Let f be an operator on a vector space E, and let $A_1, A_2, \ldots, A_p$ be pairwise relatively prime polynomials. Let $P = A_1 A_2 \ldots A_p$.

Then $\operatorname{Ker} P(f) = \operatorname{Ker} A_1(f) \oplus \operatorname{Ker} A_2(f) \oplus \cdots \oplus \operatorname{Ker} A_p(f)$.

This type of result lets us find, for example, the general solution of a homogeneous differential equation of arbitrary order.

In the vector space E of functions of class C^∞ from $\mathbb{R}$ into $\mathbb{R}$, we consider the differential equation with unknown function $x \mapsto y(x)$:

$$(\text{E}): \ a_m y^{(m)} + a_{m-1} y^{(m-1)} + \cdots + a_1 y' + a_0 y = 0$$

This may be written $\displaystyle\sum_{k=0}^{m} a_k y^{(k)} = 0$, that is, $P(D)(y) = 0$, where $P = \displaystyle\sum_{k=0}^{m} a_k X^k$

and where D is the differentiation operator $D : y \mapsto y'$.

Suppose that P may be factored over $\mathbb{K}$ into: $P = (X - \lambda_1)^{r_1} \cdots (X - \lambda_q)^{r_q}$.

The set of solutions of equation (E) is $S = \operatorname{Ker} P(D)$.

Thus, $S = \operatorname{Ker}(D - \lambda_1 Id)^{r_1} \oplus \cdots \oplus \operatorname{Ker}(D - \lambda_q Id)^{r_q}$.

It remains to determine $\operatorname{Ker}(D - \lambda Id)^r$, for any scalar λ and for every real r.

We denote by U the isomorphism of E defined: $y \mapsto U(y) = e^{-\lambda x} y$.

The inverse isomorphism is, of course, defined by: $y \mapsto U^{-1}(y) = e^{\lambda x} y$.

We observe that $D \circ U(y) = e^{-\lambda x}(y' - \lambda y)$. Thus, $U^{-1} \circ D \circ U = D - \lambda Id$.

More generally: $(D - \lambda Id)^r = U^{-1} \circ D^r \circ U$. We may thus deduce that:

$$(D - \lambda Id)^r(y) = 0 \Leftrightarrow (U^{-1} \circ D^r \circ U)(y) = 0 \Leftrightarrow D^r \circ U(y) = 0 \Leftrightarrow \big(U(y)\big)^{(r)} = 0$$

$$\Leftrightarrow U(y) \text{ is a polynomial } A_\lambda(x) \text{ of degree } \leq r - 1$$

$$\Leftrightarrow y \text{ may be written } y(x) = A_\lambda(x) e^{\lambda x}$$

For example, consider the equation: (E) $y^{(6)} - 3y^{(5)} + 6y^{(3)} - 3y'' - 3y' + 2y = 0$.

Here $P = X^6 - 3X^5 + 6X^3 - 3X^2 - 3X + 2 = (X - 1)^3 (X + 1)^2 (X - 2)$.

The general solution of equation (E) is thus formed from functions which may be written:

$$y(x) = (\alpha x^2 + \beta x + \gamma)e^x + (\lambda x + \mu)e^{-x} + \delta e^{2x}$$

1. 3 Polynomials of matrices

As we have defined polynomials of an operator f, we may define polynomials $P(M)$ of a square matrix M:

$$\text{If } P = \sum_{k=0}^{m} a_k X^k, \text{ we put } P(M) = \sum_{k=0}^{m} a_k M^k$$

This is a particular case of what preceded if we identify the matrix M with the operator on $\mathcal{M}_{n,1}(\mathbb{K})$ defined by $X \mapsto MX$. We thus retrieve properties analogous to those already stated above.

However, we point out the following results which will be easy for the reader to verify. P here designates an arbitrary polynomial.

- If M is the matrix of the operator f with respect to the basis (e) of E, then $P(M)$ is the matrix of $P(f)$ with the same basis.

- If M and N are two similar matrices, then $P(M)$ and $P(N)$ are similar. More precisely, for every invertible matrix Q: $Q^{-1}P(M)Q = P(Q^{-1}MQ)$.

- If M is an upper (resp. a lower) triangular matrix with diagonal coefficients $\lambda_1, \ldots, \lambda_n$, then $P(M)$ is upper (resp. lower) triangular with diagonal coefficients $P(\lambda_1), \ldots, P(\lambda_n)$.
  We have the same conclusion for diagonal matrices.

- More generally, if M is a block triangular matrix (the diagonal blocks being designated by $M_1, \ldots, M_r$) then $P(M)$ is a block triangular matrix, the diagonal blocks being $P(M_1), \ldots, P(M_r)$.
  We have the same conclusion for diagonal block matrices.

We see that it is easy to calculate a polynomial of matrices with the calculator. In this example, we have calculated the polynomial $P(M) = M^3 - 2M + I_3$ in two different ways. We have also calculated $M^3 - 3M^2 + 5M - 3I_3$ to see that a result which is the zero matrix.

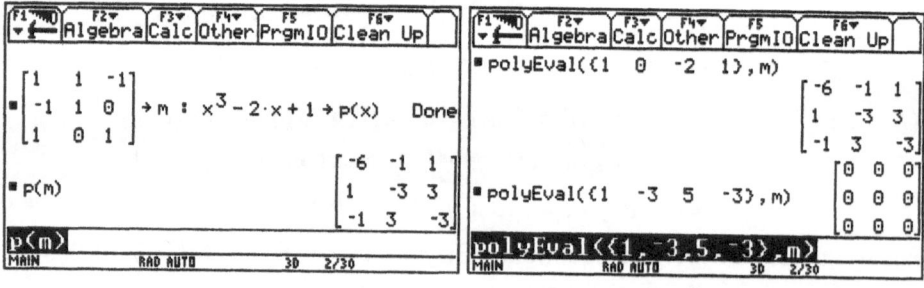

1. 4 Polynomial annihilators

We say that $P \in \mathbb{K}[X]$ is a *polynomial annihilator* of f if $P(f) = 0$.
If we denote by $\text{Ann}(f)$ the subset of these polynomials (which is non-empty, since it contains the zero polynomial), the following properties are evident:

- $\forall P \in \text{Ann}(f), \forall Q \in \mathbb{K}[X], \ PQ \in \text{Ann}(f)$.
- $\forall P, Q \in \text{Ann}(f), \forall \alpha, \beta \in \mathbb{K}, \alpha P + \beta Q \in \text{Ann}(f)$.

If $\dim E = n < \infty$, every operator on f has a non-zero annihilating polynomial. In fact the space $\mathcal{L}(E)$ is of dimension n^2: there is thus a non-trivial linear relation between the $Id_E, f, f^2, \ldots, f^{n^2}$. But we are quickly going to see that we may find polynomial annihilators of much less degree...

Let f be a rotation by the angle $\theta = 2\pi/n$ in the oriented Euclidean plane ($n \geq 3$).
Since $f^n = Id$, a annihilating polynomial of f is $X^n - 1$.

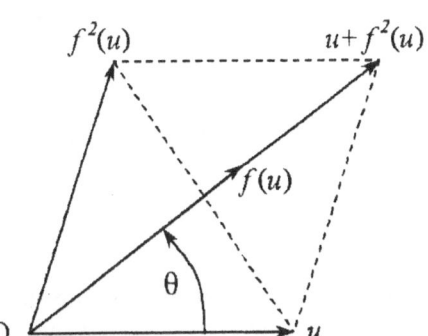

However, for every vector u
$$u + f^2(u) = 2\cos\theta \; f(u).$$
A annihilating polynomial of f is thus
$$X^2 - 2\cos\theta X + 1.$$

From Proposition 2, we immediately draw the following consequence:
If λ is an eigenvalue of f, and if P is a annihilating polynomial of f, then λ is a root of P.

In fact $P(\lambda)$ is an eigenvalue of $P(f)$. Hence $P(f) = 0$. Thus $P(\lambda) = 0$.

A particular case: If f is nilpotent ($P(f) = 0$ with $P = X^m$ for some "m".) then $\mathrm{Sp}(f) = \{0\}$.

Definition 3: If $\mathrm{Ann}(f)$ does not reduce to $\{0\}$, it contains a non-zero polynomial of minimum degree, and even one which is a monic polynomial - its leading coefficient is "1". We call this the minimal polynomial of f.

Proposition 4: Let A be the minimal polynomial of f. The annihilator polynomials of f are multiples of A: $\mathrm{Ann}(f) = \mathbb{K}[X]A = \{QA, Q \in \mathbb{K}[X]\}$.

Proof: $A(f) = 0 \Rightarrow$ for each multiple $P = QA$ of A, $P(f) = Q(f) \circ A(f) = 0$.
Conversely, let $P \in \mathrm{Ann}(f)$ and $P = QA + R$ be its quotient by A.
The polynomials P and A annihilate f: It is thus the same for R.
Hence, $\deg R < \deg A$. The definition of A implies that $R = 0$: P is a multiple of A.

Examples:

- If p is a projection distinct from 0 and Id, its minimal polynomial is $X^2 - X$.
- If s is a symmetry distinct from $\pm Id$, its minimal polynomial est $X^2 - 1$.
- f is nilpotent $\Leftrightarrow$ its minimal polynomial is a power of X.

Of course, we define the polynomial annihilators and the minimal polynomial of a square matrix M.

- Two similar matrices M and N ($N = P^{-1}MP$) have the same polynomial annihilators and thus the same minimal polynomial.

- For every f of $\mathcal{L}(E)$ (with $\dim E = n$), if M is the matrix of f with respect to the basis (e), then $\mathrm{Ann}(M) = \mathrm{Ann}(f)$: f and M have the same minimal polynomial.

We are now going to define the essential utility which permits localization of the eigenvalues of a matrix M of order n: the characteristic polynomial χ_f of f: it is of degree n, it annihilates M (this is the famous Cayley-Hamilton theorem), and its roots are the eigenvalues of M.

1. 5 The characteristic polynomial

Definition 4: Let $M \in \mathcal{M}_n(\mathbb{K})$. We call the character -istic polynomial of M the polynomial χ_M, defined by: $\chi_M(X) = \det(M - XI_n)$.

χ_M is effectively a polynomial, it is of degree n, and we have easily found three coefficients:

$$\chi_M = (-1)^n X^n + (-1)^{n-1}\mathrm{tr}(M)X^{n-1} + \cdots + \det M$$

$$\chi_M = \begin{vmatrix} a_{1.1} - X & a_{1.2} & \cdots & a_{1.n} \\ a_{2.1} & a_{2.2} - X & \cdots & a_{1.1} \\ \cdots & \cdots & \cdots & \cdots \\ a_{n.1} & a_{n.1} & \cdots & a_{n.n} - X \end{vmatrix}$$

A matrix and its transpose ${}^{T}M$ have the same character -istic polynomial:

$$\chi_{{}^{T}M} = \det({}^{T}M - XI_n) = \det({}^{T}(M - XI_n)) = \det(M - XI_n) = \chi_M$$

Two similar matrices have the same character -istic polynomial:

$$N = P^{-1}MP \Rightarrow \det(N - XI_n) = \det(P^{-1}MP - XI_n) = \det\left(P^{-1}(M - XI_n)P\right)$$
$$= \det(P^{-1})\det(M - XI_n)\det(P) = \det(M - XI_n)$$

We may thus define the character -istic polynomial χ_f of an operator f of a vector space of finite dimension n: it is that of the matrix of f in an arbitrary basis of E.

It is quite easy to calculate the characteristic polynomial of a matrix using the function charpol defined here.

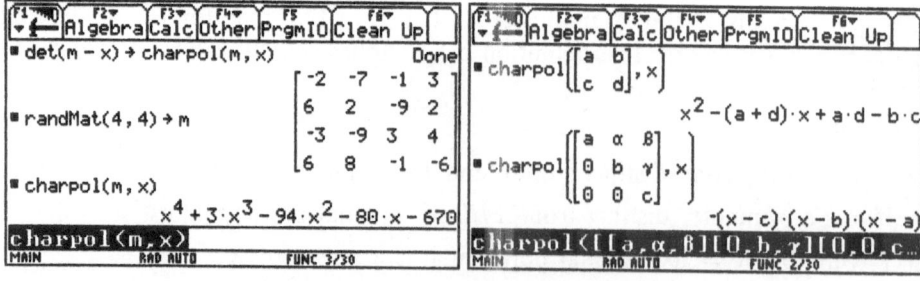

In the second example above, we indeed see that for a square matrix of order 2, $\chi_M(X) = X^2 - \mathrm{tr}(M)X + \det M$.

The third example illustrated the following property:

The character -istic polynomial of a triangular (or diagonal) matrix M with diagonal coefficients $\lambda_1, \cdots, \lambda_n$ is $\chi_M(X) = \prod_{k=1}^{n}(\lambda_k - X)$.

Here is a particular case: The characteristic polynomial of $M = \lambda I_n$ is $(\lambda - X)^n$.

The character -istic polynomial is, at least in theory, the best means of determining the eigenvalues of a matrix. In fact:

Proposition 5: The eigenvalues of a square matrix are the roots of its character -istic polynomial.

Proof: This is evident since $\lambda \in \mathrm{Sp}(M) \Leftrightarrow M - \lambda I_n$ is non-invertible $\Leftrightarrow \det(M - \lambda I_n) = 0 \Leftrightarrow \chi_M(\lambda) = 0$.

Recall that by default we assumed that the matrices have complex coefficients, and that we are considering their eigenvalues in $\mathbb{C}$.

We know that every polynomial of $\mathbb{C}[X]$ splits, that is, it factors into a product of factors of first degree. The regrouping of the possibly identical roots makes apparent the multiplicity of these roots.

Definition 5: We say that λ is an eigenvalue of multiplicity m of the square matrix M if λ is a root of multiplicity m of the polynomial χ_M.

We say that an eigenvalue is simple if $m = 1$, double if $m = 2$, etc.

If we denote by $\lambda_1, \ldots, \lambda_p$ the different eigenvalues of M, with the respective multiplicities $m_1, \cdots, m_p$, then the character -istic polynomial of M may be factored into:

$$\chi_M = (\lambda_1 - X)^{m_1} \cdots (\lambda_p - X)^{m_p} = \prod_{k=1}^{p}(\lambda_k - X)^{m_k}$$

Every square matrix of order n thus has exactly n eigenvalues, each counted as many times as its multiplicity.

If $M \in \mathcal{M}_n(\mathbb{R})$, $\lambda \in \mathrm{Sp}(M) \Leftrightarrow \overline{\lambda} \in \mathrm{Sp}(M)$ and the multiplicities are the same.

As with eigenspaces, $X \in E_\lambda(M) \Leftrightarrow \overline{X} \in E_{\overline{\lambda}}(M)$.

Comparison between the factored form and the expanded expression of χ_M immediately gives the following properties, where $M \in \mathcal{M}_n(\mathbb{C})$.

- The sum of the eigenvalues of M, each counted as many times as its multiplicity, is equal to the trace of M.

- The product of the eigenvalues of M, each counted as many times as its multiplicity, is equal to the determinant of M.

It is often difficult to "devine " the eigenvalues, except in the obvious case of a triangular matrix, where they are the diagonal coefficients (their multiplicity being the number of times they appear on the diagonal).

Here is another particular case: a nilpotent matrix M (of order n) only has the eigenvalue 0, with the multiplicity n. (Its characteristic polynomial is of degree n, having only the root 0: it is thus $(-1)^n X^n$).

We define the multiplicity of an eigenvalue λ of an operator of a vector space E of finite dimension n over $\mathbb{K}$ ($\mathbb{K} = \mathbb{R}$ or $\mathbb{C}$) as before, starting with the characteristic polynomial of f.

If $\mathbb{K} = \mathbb{R}$, the two preceding properties about the sum and product of eigenvalues are only true if χ_f splits over $\mathbb{R}$.

Here we calculate the eigenvalues of a matrix M by first solving the equation $\chi_M(x) = 0$, then by using the built-in function eigVl. In the latter case, the answer is approximate.

That 0 is an eigenvalue is indicated by the fact that M is not invertible.

Proposition 6: Let $f \in \mathcal{L}(E)$, with $\dim E = n \geq 1$. Let G be a subspace of E, distinct from $\{\vec{0}\}$ and stable under f, and let g be the restriction of f to G. Then the polynomial χ_g divides the polynomial χ_f.

Proof: Suppose that G is distinct from E, otherwise this is immediate.

Let $(e) = (e_1, e_2, \ldots, e_n)$ be a basis of E, obtained by extending a basis $(e') = (e_1, e_2, \ldots, e_p)$ of G. Let M be the matrix of f with respect to (e).

The matrix M may be written $\begin{pmatrix} N & P \\ 0 & Q \end{pmatrix}$ where N is the matrix of g with respect to (e').

We conclude that $\chi_f = \chi_M = \det(M - XI_n) = \det(N - XI_p)\det(Q - XI_{n-p}) = \chi_N \chi_Q$.

Thus the polynomial χ_N, that is to say χ_g, divides the polynomial χ_f.

This proposition implies an important result.

Proposition 7: Let $f \in \mathcal{L}(E)$, where $\dim E = n \geq 1$. Let λ be an eigenvalue of f, of multiplicity $m(\lambda)$. Let $d(\lambda) = \dim E_\lambda(f)$: then $d(\lambda) \leq m(\lambda)$.

Proof: We apply the preceding result to $G = E_\lambda(f)$. The restriction g of f to this subspace is the dilation $u \mapsto \lambda u$, whose characteristic polynomial is $\chi_g = (\lambda - X)^{d(\lambda)}$. The inequality $d(\lambda) \leq m(\lambda)$ derives from the fact that χ_g divides χ_f which has the same λ as a root of multiplicity $m(\lambda)$.

Consequence: The eigenspace with a simple eigenvalue ($m(\lambda) = 1$) is necessarily a vector line.

We know now that finding the eigenvalues of a matrix is the same as finding the roots of its characteristic polynomial. The methods of finding approximately the roots of a polynomial may thus be utilities for finding the spectrum of a matrix.

Conversely, we may ask whether an arbitrary polynomial P (apparently given in non-factored form, otherwise the problem would be too easy) is the character-istic polynomial of an appropriately chosen matrix M. The response is affirmative and uses the notion of companion matrix.

This will show that to find the roots of a polynomial we may consider using techniques of calculating eigenvalues approximately.

Definition:
Let P be a monic polynomial of degree n,
$P = X^n - a_{n-1}X^{n-1} - \cdots - a_1 X - a_0$.
The square matrix M of order n shown here, is called the companion matrix of P. (Sometimes the transpose of this matrix is taken as the definition.)

$$M = \begin{pmatrix} 0 & 0 & \cdots & 0 & a_0 \\ 1 & 0 & \cdots & 0 & a_1 \\ 0 & 1 & \ddots & \vdots & \vdots \\ \vdots & \ddots & \ddots & 0 & a_{n-2} \\ 0 & \cdots & 0 & 1 & a_{n-1} \end{pmatrix}$$

It may be verified that $\chi_M = (-1)^n P(X)$.

Proof: This is obvious is $n = 1$. We establish the result by induction on n.
If it is true for rank $n - 1$, we expand χ_M by its first row.
We obtain $\chi_M(X) = (-1)^{n+1}a_0 - X \chi_N(X)$, where N is the companion matrix of the polynomial $Q(X) = X^{n-1} - a_{n-1}X^{n-2} - \cdots - a_2 X - a_1$.
By hypothesis $\chi_N(X) = (-1)^{n-1}Q(X)$.
We conclude that $\chi_M(X) = (-1)^n XQ(X) + (-1)^{n+1}a_0 = (-1)^n P(X)$, which shows the property for rank n and completes the induction.

The function below, called $\mathtt{compmat}$ constructs the companion matrix of a polynomial $P = X^n - a_{n-1}X^{n-1} - \cdots - a_1 X - a_0$.
With this polynomial, the syntax is $\mathtt{compmat(\{a_0, a_1, \ldots, a_{n-1}\})}$.

```
:compmat(p):Func:Local n,t:dim(p)-1→n
:seq(seq(when(i=j,1,0),j,1,n),i,0,n)→t
:augment(t,list▷mat(p,1))
:EndFunc
```

Here is an example using the function $\mathtt{compmat}$, with the polynomial
$P = x^4 - a_3 X^3 + a_2 X^2 - a_1 X - a_0$.
In calculating the characteristic polynomial of the matrix M obtained, we retrieve $(-1)^4 P = P$.

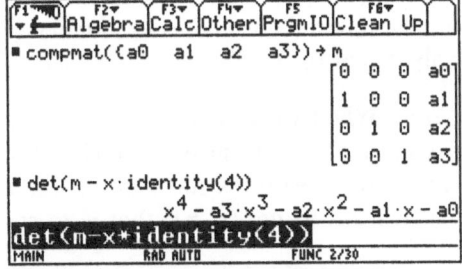

We now come to a famous result which indicates that the character-istic polynomial of a matrix M is a multiple of its minimal polynomial. This is not, however, of much practical use.

Proposition 8: (*The Cayley-Hamilton Theorem*) Let $M \in \mathcal{M}_n(\mathbb{K})$. The character-istic polynomial of M is a annihilating polynomial of M.

(Stated loosely: a matrix satisfies its character -istic polynomial.)

Proof: We must show that $\chi_M(M) = 0$.

If $f \in \mathcal{L}(E)$, where $\dim E = n \geq 1$, it amounts to the same to show that $\chi_f(f) = 0$.

Let u be an arbitrary element of E. It must be proved that $\chi_f(f)(u) = \overrightarrow{0}$.

The result is evident if u is zero, so we suppose $u \neq \overrightarrow{0}$.

Let p be the maximum index $(p \geq 1)$ such that $e_0 = u, e_1 = f(u), \ldots, e_{p-1} = f^{p-1}(u)$ are independent, and let G be the subspace of E which they generate.

From the definition of p, there are coefficients $a_1, a_2, \ldots, a_{p-1}$ such that:

$$f^p(u) = a_0 u + a_1 f(u) + \cdots + a_{p-1} f^{p-1}(u) = \sum_{k=0}^{p-1} a_k f^k(u).$$

The preceding inequality may be written:
$f(e_{p-1}) = a_0 e_0 + a_1 e_1 + \cdots + a_{p-1} e_{p-1}$.
Hence, $\forall k \in \{0, \ldots, p-2\}$, $f(e_k) = e_{k+1}$.
G is thus stable under f and the matrix, with respect to $e_1, e_2, \ldots, e_{p-1}$, of the restriction g of f to G is the companion matrix shown.

$$M = \begin{pmatrix} 0 & 0 & \cdots & 0 & a_0 \\ 1 & 0 & \cdots & 0 & a_1 \\ 0 & 1 & \ddots & \vdots & \vdots \\ \vdots & \ddots & \ddots & 0 & a_{p-2} \\ 0 & \cdots & 0 & 1 & a_{p-1} \end{pmatrix}$$

We know that χ_g divides χ_f: $\exists Q \in \mathbb{K}[X]$, $\chi_f = Q\chi_g$. Thus, $\chi_f(f) = Q(f) \circ \chi_g(f)$.

Hence $\chi_g = (-1)^p \left(X^p - \sum_{k=0}^{p-1} a_k X^k \right)$ and $\chi_g(f)(u) = (-1)^p \left(f^p(u) - \sum_{k=0}^{p-1} a_k f^k(u) \right) = \overrightarrow{0}$

We conclude that $\chi_f(f)(u) = \overrightarrow{0}$, which is what we had to prove.

In the example below, we formed a random matrix M of order 3, and the instruction det(m-x)→p(x) created a function p giving the character -istic polynomial of M.

Note: When we evaluate p(m), or similarly p(x)|x=m, we get 0, which is not a confirmation of the Cayley-Hamilton Theorem. In fact, in both cases the calculator substitutes the name m for the name x in the expression det(m-x), which leads to det(m-m), that is to say to det(0)=0.

A correct verification of the Cayley-Hamilton Theorem on the calculator is done by defining the function p by -(x^3+7x^2+30x+50)→p(x), then, as shown below, to store the expression of the polynomial in a variable q.

An actual verification of the theorem thus comes from the last instruction, with the result being the zero matrix of order 3.

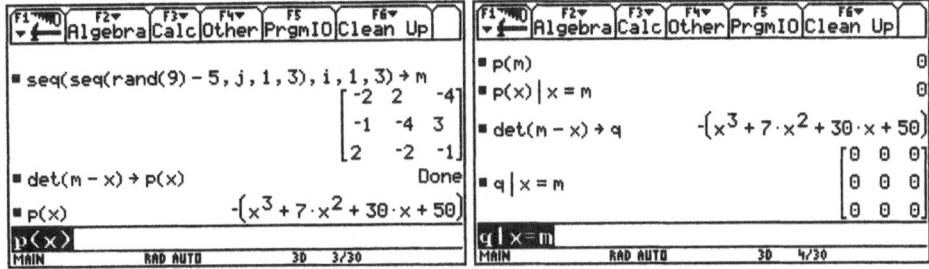

We end this section with a very simple observation.

We know that the minimal polynomial of a matrix M is a divisor of its character -istic polynomial (from the Cayley-Hamilton Theorem).

It is sometimes a strict divisor.

In this example we have formed a square matrix M of order 5.

Its character -istic polynomial is $P = -(X-6)(X-1)^4$ (6 is a simple eigenvalue and 1 is a quadruple eigenvalue).

Of course, $P(M) = 0$, which we don't verify here.

On the other hand, it is more interesting to note that the polynomial $(X-6)(X-1)$ likewise annihilates the matrix M, and then that it is only of degree 2: this is its minimal polynomial (neither $X-6$ nor $X-1$ annihilate M...).

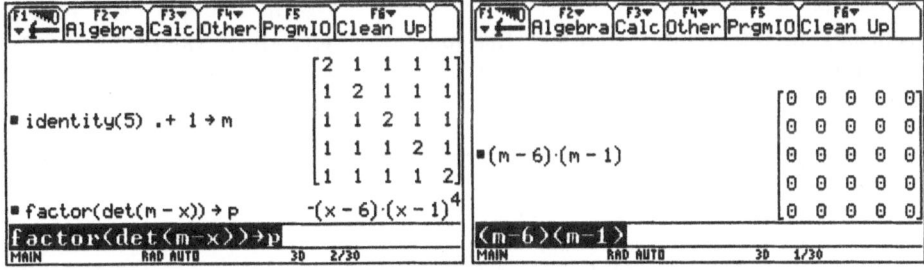

2. Reduction of operators

After that preliminary theory, we are now at the heart of the subject.

To reduce an operator f on E is to find a basis E adapted to f, that is, one for which the behavior of f is particularly simple. Knowing the eigenvectors is the most useful approach to do this.

To reduce a matrix M of $\mathcal{M}_n(\mathbb{K})$ is to find a matrix Δ, similar to M, which is as simple as possible, the ideal being that Δ is a diagonal matrix.

The two problems are equivalent: to reduce f is to reduce its matrix M in an arbitrary basis for E.

To attain these objectives, especially with our calculator, there are both symbolic and numeric methods. We are going to begin with the symbolic ones by creating a number of utilities.

- The first step in the reduction of M, is to find its eigenvalues and and their multiplicities, that is, we must do a complete factorization of the character -istic polynomial of M.

- The second consists of finding a basis of eigenvectors for each of the eigen-spaces of M. The system $(M - \lambda I_n)X = 0$ must be solved over $\mathbb{K}^n$, for each of the eigenvalues λ.

- If the family of eigenvectors obtained is not sufficient to form a basis for $\mathbb{K}^n$, it must be completed "as much as possible ".

2. 1 Eigenvalues and multiplicities

The program mzeros gives the list of real or complex roots of a polynomial P, each counted as many times as its multiplicity. If $P = a_n X^n + \cdots + a_1 X + a_0$, the syntax is mzeros(p(x),x).

The principle of the program is simple: cZeros gives the list of roots of P, then forms a list of successive derivatives of P. We then repeat each root z according to the number of times the successive derivatives vanish (if z nulls P' but not P'', it is a double root: it thus repeats only once, which leads to two copies of z in the final list).

```
:mzeros(p,x)
:Func:Local j,k,d,r,z
:cZeros(p,x)→z:{} →d
:While when(p=0,false,true,true)
:   d(p,x)→p:augment(d,{p})→d
:EndWhile:{} →r
:For j,1,dim(z):For k,1,dim(d)-1
:   augment(r,{z[j]})→r
:   If when(d[k]=0|x=z[j],false,true,true):Exit
:EndFor:EndFor:r
:EndFunc
```

Here are some examples using the program mzeros. The second screen shows for example that the eigenvalues of the matrix M are 0 (double), $2(\sqrt{21} + 4)$ and $-2(\sqrt{21} - 4)$ (both of multiplicity 1).

The following example of a square matrix of order 4 only has one eigenvalue, $\lambda = 1$, with multiplicity 4.

This matrix is however far from being triangular with diagonal elements all equal to 1...

2. 2 Finding eigenspaces

Let M be a matrix of $\mathcal{M}_n(\mathbb{K})$, identified (in the standard basis) with an operator on $\mathbb{K}^n$.

Let λ be an eigenvalue of M of multiplicity $m(\lambda)$ and let E_λ be the eigenspace associated with it. (It is a subspace of $\mathbb{K}^n$, not reducing to the zero vector, and of dimension less than or equal to $m(\lambda)$).

We want to find a basis of E_λ. For this we are going to use the instruction rref (reduced row echelon form) of the calculator.

Recall that rref transforms any matrix A of $\mathcal{M}_{n,p}(\mathbb{K})$ into a matrix B of the same size but of upper echelon form, by a succession of elementary row operations. Looking at B tells us the rank of A, the number of its non-zero pivots, and the dimension of its kernel.

Here we create a 3×6 matrix A and apply the instruction rref to it.

The result is a matrix B which has two non-zero pivots (equal to 1).

The matrix A is thus of rank 2. It represents a linear map from $\mathbb{K}^6$ to $\mathbb{K}^3$ whose kernel is of dimension $6 - 2 = 4$.

Each succession of elementary operations on rows conducted on a matrix A and leading to a matrix B, may be interpreted as the product $B = PA$ of A by an invertible matrix P.

To find P, it suffices to adjoin the identity matrix to A on the right and to apply rref to the tableau $(A|I)$: the result is the tableau $(B|P)$.

In the following example, with the same matrix A, we have placed in the variable bp the result of the instruction rref on the tableau $(A|I)$, then we have extracted the matrix P. We observe, of course, that the product PA returns the matrix B which had been obtained earlier.

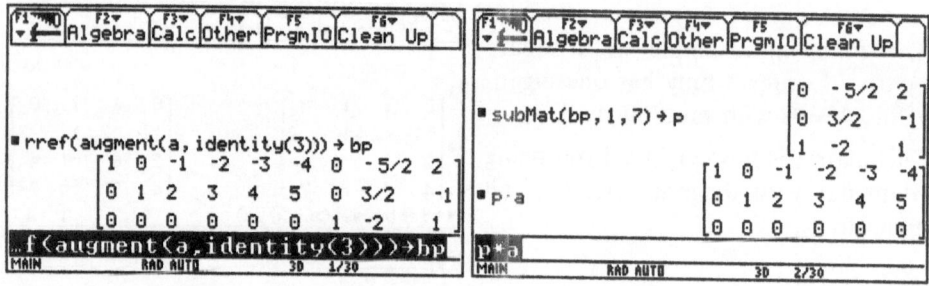

With the preceding notation, the equation $B = PA$ shows us that the matrices A and B have the same kernel.

In fact, for every column vector X, $BX = 0 \Leftrightarrow PAX = 0 \Leftrightarrow AX = 0$.

The matrix B is in echelon form due to rref, and it is very easy to find the reduced equations of its kernel.

For example, taking the example above, and denoting by x, y, z, t, u, v the coordinates in the standard basis of $\mathbb{K}^6$, a system of equations with the kernel of A is $\begin{cases} x - z - 2t - 3u - 4v = 0 \\ y + 2z + 3t + 4u + 5v = 0 \end{cases}$.

To find the eigenspace of a matrix M for an eigenvalue λ, we find the kernel of the matrix $A = M - \lambda I_n$.

The instruction rref allows us to find a reduced system of equations.

In this example, we see that the matrix M has two eigenvalues, 3 (double) and 4 (simple).

For $\lambda = 3$, a system of equations for the eigenspace is $\begin{cases} x + \frac{1}{2}y = 0 \\ z = 0 \end{cases}$.

We could also find the reduced equations of an eigenspace with solve, as the following example shows, with the same matrix M:

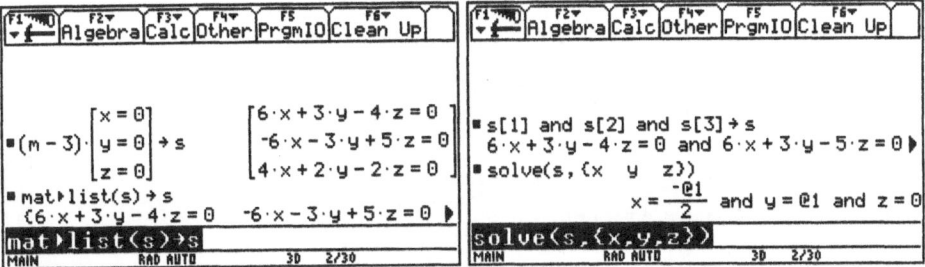

All the same, it is preferable to program a function to solve this type of work, since the preceding method is very tricky.

We propose the function eigeqns. The syntax is eigeqns(m, λ) (where λ must be an eigenvalue of M) and the result is a logical expression representing the equations defining the eigenspace.

The function eigeqns uses the variable names x_1, x_2, etc. to designate the coordinates in the standard basis.

```
:eigeqns(m,λ)
:Func
:Local k,n,v,eq
:rowDim(m)→n
:rref(m-λ*identity(n))→m
:seq(#("x_"&string(k)),k,1,n)→v
:m*list▷mat(v,1)→m
:true→eq
:For k,1,n
:   eq and solve(m[k,1]=0,v[k])→eq
:EndFor
:EndFunc
```

We revisit here the matrix of order 4 whose only eigenvalue is $\lambda = 1$ with multiplicity 4.

We see that the eigenspace has for its equations $\begin{cases} x_1 = x_3 - \frac{1}{2}x_4 \\ x_2 = -\frac{1}{2}x_4 \end{cases}$.

This is thus a plane in $\mathbb{R}^4$.

If we reconsider the preceding example, we obtain a basis of the eigenspace E_1 by giving to the variables x_3 and x_4 (here treated as parameters) the values (for example) $x_3 = 1, x_4 = 0$, then $x_3 = 0, x_4 = -2$.

We then find the vectors $v_1 = (1, 0, 1, 0)$ and $v_2 = (1, 1, 0, -2)$.
We see here how to form v_1 and v_2 starting from the preceding results, and how to verify that they are actually eigenvectors of M for the eigenvalue 1 (in this case, that they are invariants of M).

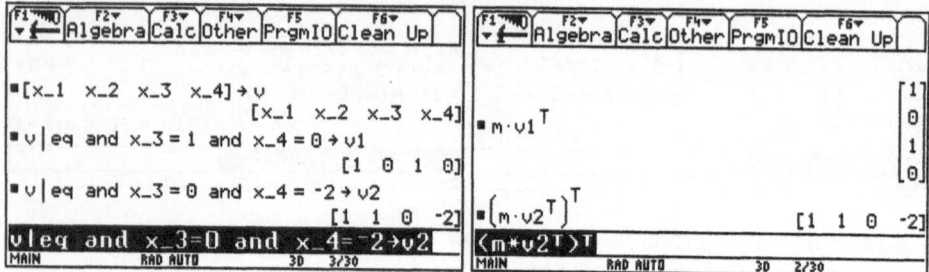

When reducing an operator (of a matrix), it is usually preferable to display the eigenspace, rather than the system of equations.
The preceding method is a little laborious, and we are going to relieve that a bit, again by applying the instruction rref.

We have seen earlier that a succession of elementary operations on the rows of a matrix A (and leading to a matrix B) are equivalent to the equation $B = PA$, where P is a square invertible matrix, and this results in the equality of the kernels of A and of B.

In particular, this is the case when one passes from A to the upper echelon matrix B with the instruction rref.

In the same manner, the passage from A to C by a succession of column operations leads to the equation $C = AQ$, where Q is a square invertible matrix, and it results from this equation that the images of A and C are identical.

Our calculator doesn't possess an instruction which does column operations on a matrix A or which is the analog of the instruction rref.
However, we may just as well work on columns by first transposing the matrix A (the rows of the transpose are the columns of A), doing the row operations, and then transposing the result.

We create here a square matrix A of order 3, and we see very well how the instruction $\left(\text{rowSwap}(a^\top, 1, 2)\right)^\top$ allows exchange of the first two columns of A, a little like having evaluated $\text{colSwap}(a, 1, 2)$.

A double transposition likewise allows us to do a column reduction to echelon form, as the example here shows.

The two matrices A and B here have the same image: the plane of $\mathbb{R}^3$ generated by the vectors $(1, 0, -1)$ and $(0, 1, 2)$.

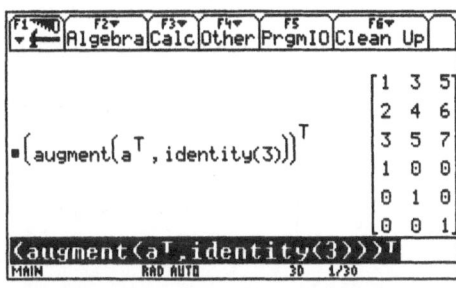

If we pass from a matrix A to a matrix B by a succession of column operations, it is interesting to be able to retrieve the invertible matrix Q such that $B = AQ$.

To do this, it suffices to apply the same operations to the tableau obtained by adjoining the identity matrix below A.

In fact, $\begin{pmatrix} A \\ I \end{pmatrix} Q = \begin{pmatrix} AQ \\ Q \end{pmatrix} = \begin{pmatrix} B \\ Q \end{pmatrix}$.

Here we superpose A and I_3. Then, as before, we do a column reduction using transposes.
From the result, stored in bq, we extract the matrix Q.
Finally, we verify that the product AQ actually returns B.

We examine the preceding example: the square matrix A of order 3 is (looking at the matrix B) of rank 2. Its kernel is thus of dimension $3 - 2 = 1$.

The third column of $B = AQ$ is all zeroes, which signifies that the third column vector of Q is in the kernel of A (and that it thus constitutes a basis since Ker A is of dimension 1).

This observation may easily be generalized. Suppose that the square matrix A of order n is not invertible and thus is of rank $r < n$ The dimension of its kernel is then $n - r$.

Column reduction produces a square matrix B of order n whose first r columns are linearly independent (they form a basis for the image of A) and whose last $n - r$ columns are zeroes.

This last point, added to the fact that there is an invertible matrix Q such that $B = AQ$, proves that the last $n - r$ columns of Q (which are independent) are in the kernel of A. Those columns thus form a basis for the kernel.

If λ is an eigenvalue of a square matrix M and if we apply the above argument to $A = M - \lambda I_n$, we may obtain a basis of the kernel of A, that is, of the eigenspace of M for the eigenvalue λ.

We present the function eigbasis (syntax eigbasis(M,λ), where λ must be an eigenvalue of M). The result is then a matrix whose column vectors form a basis of the eigenspace of M for λ.

```
:eigbasis(m,λ)
:Func
:Local n,r,t
:rowDim(m)→n:m-λ*identity(n)→m
:rref(augment(mᵀ,identity(n)))→t
:subMat(t,1,1,n,n)→m
:subMat(t,1,n+1)→t
:For r,n,1,-1
:   If when(norm(m[r])≠0,true,false,true)
:   Exit
:EndFor
:(subMat(t,r+1))ᵀ
:EndFunc
```

Here A is square of order 5 and has eigenvalues 1 (quadruple) and 6 (simple).
A basis of the eigenspace E_1 is formed by the vectors:
$u_1 = (1,0,0,0,-1)$, $u_2 = (0,1,0,0,-1)$
$u_3 = (0,0,1,0,-1)$, $u_4 = (0,0,0,1,-1)$
A basis of E_6 is formed by the vector
$u_5 = (1,1,1,1)$.

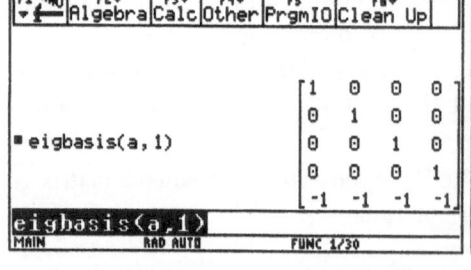

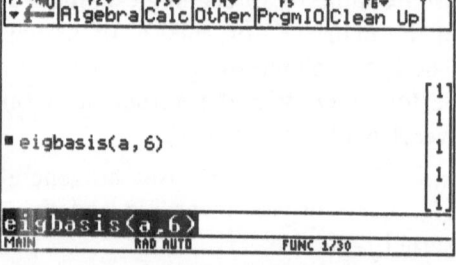

2. 3 Diagonalization

Let f be an operator on a vector space E of dimension n ($n \geq 1$) over $\mathbb{K}$ ($\mathbb{K} = \mathbb{R}$ or $\mathbb{C}$), let M be the matrix of f with respect to a basis (e) of E ($M \in \mathcal{M}_n(\mathbb{K})$), and let $\mathrm{Sp}(f)$ be the spectrum of f (which is likewise the spectrum $\mathrm{Sp}(M)$ of M in $\mathbb{K}$).

We denote by $m(\lambda)$ the multiplicity of an eigenvalue λ and by $d(\lambda)$ the dimension of the associated eigenspace E_λ. Recall that $d(\lambda) \leq m(\lambda)$.

One says that the character -istic polynomial of f (which is also that of M: $\chi(X) = \det(M - XI_n)$) splits over $\mathbb{K}$ if it factors entirely over $\mathbb{K}[X]$ into a product of polynomials of first degree.

This property always is satisfied if $\mathbb{K} = \mathbb{C}$ (The Fundamental Theorem of Algebra or d'Alembert's theorem), but it may be false if $\mathbb{K} = \mathbb{R}$. For example, the character -istic polynomial of $M = \begin{pmatrix} 0 & -1 \\ 1 & 0 \end{pmatrix}$ is $X^2 + 1$: it does not split over $\mathbb{R}$.

Remarks:

- When a polynomial splits over $\mathbb{K}$, the calculator arrives at a complete factorization (with `factor` over $\mathbb{R}$ and `cfactor` over $\mathbb{C}$).

- When we consider an operator f on a vector space E over $\mathbb{R}$, we don't want to split the character -istic polynomial of f over $\mathbb{C}$, since the only possible eigenvalues of f are real.
  To say that χ_f splits here thus signifies "splits over $\mathbb{R}$ ".

- On the other hand if $M \in \mathcal{M}_n(\mathbb{R})$, we could also consider M as an element of $\mathcal{M}_n(\mathbb{C})$. In this case, we must be precise about the field, $\mathbb{R}$ or $\mathbb{C}$, over which we consider that χ_M is split.

- To say that the character -istic polynomial of f (of M) splits over $\mathbb{K}$, is to say that the number of its roots (which are the eigenvalues of f and of M over $\mathbb{K}$) is equal to the dimension n of E (to the order n of M).
  If we regroup those which are the same, this is equivalent to saying that the sum of all the multiplicities of $m(\lambda)$ is equal to n.

- If χ_M (where χ_f) splits, then the product (and respectively, the sum) of all the eigenvalues (each counted as many times as its multiplicity) is equal to the determinant (respectively, the trace) of M (or of f).

Definition 1: We say that f is diagonalizable if any of the following equivalent conditions are satisfied:

- There is a basis of E in which the matrix of f is a diagonal matrix.
- There is a basis of E formed of eigenvectors of f.
- E is the direct sum of the different eigenspaces of f.
- $\chi(f)$ splits over $\mathbb{K}$, and $\forall \lambda \in \mathrm{Sp}(f)$, $d(\lambda) = m(\lambda)$.

Definition 2: We say that $M \in \mathcal{M}_n(\mathbb{K})$ is diagonalizable in $\mathbb{K}$ if it is similar to a diagonal matrix D of $\mathcal{M}_n(\mathbb{K})$, that is, if there exists an invertible matrix P of $\mathcal{M}_n(\mathbb{K})$ such that $D = P^{-1}MP$.

Remarks:

- It is clear that f is diagonalizable if and only if the matrix M, which may be represented in an arbitrary basis (e) of E, is itself diagonalizable.

- If M is diagonalizable, the reduced D has on its diagonal the different eigenvalues of M, each counted as many times as its multiplicity. The column vectors of P form a basis of eigenvectors for M in the order corresponding to that of the eigenvalues on the diagonal of D.

- If $M \in \mathcal{M}_n(\mathbb{R})$ is diagonalizable over $\mathbb{R}$ it is obviously so over $\mathbb{C}$ with the same reduction $D = P^{-1}MP$. The converse is false (only when $\chi(M)$ has non-real roots).

- If M has n distinct eigenvalues, then it is diagonalizable. (All the eigenspaces are vector lines). This condition is sufficient but not necessary.

- If $\mathrm{Sp}(M) = \{\lambda\}$ with multiplicity n, M is diagonalizable $\Leftrightarrow M = \lambda I_n$.

Proposition 1: The operator f of E is diagonalizable if and only if it is annihilated by a split polynomial with simple roots.

Proof: Suppose that f is diagonalizable, with eigenvalues $\lambda_1, \lambda_2, \ldots, \lambda_p$, and let $E_1, E_2, \ldots, E_p$ be the eigenspaces of f which correspond to them.
We know that $E = E_1 \oplus E_2 \cdots \oplus E_p$.
Any vector u of E may be decomposed into this direct sum: $u = \sum_{k=1}^{p} u_k$.
Let $P(X) = (X - \lambda_1)(X - \lambda_2) \cdots (X - \lambda_p)$.
It is clear that $P(f) = (f - \lambda_1 Id) \circ (f - \lambda_2 Id) \circ \cdots \circ (f - \lambda_p Id)$, all of whose factors commute, annihilating every vector u of E. (For example, the factor $(f - \lambda_k Id)$ annihilates the component u_k of u in E_k). Thus, $P(f) = 0$.
Conversely, if $P(f) = 0$, with $P = (X - \alpha_1)(X - \alpha_2) \cdots (X - \alpha_p)$ (the α_k are distinct), then $E = \mathrm{Ker}\, P(f) = \mathrm{Ker}(f - \alpha_1 Id) \oplus \mathrm{Ker}(f - \alpha_2 Id) \cdots \oplus \mathrm{Ker}(f - \alpha_p Id)$ (the theorem of decomposition of the kernel).
Certain of the $\mathrm{Ker}(f - \alpha_k Id)$ may reduce to $\overrightarrow{0}$. As for the others, they form a direct sum of eigenspaces covering all of E, which signifies that f is diagonalizable.

Remark: again, the preceding proposition signifies that f is diagonalizable if and only if its minimal polynomial splits with simple roots.

To diagonalize a matrix with the calculator when possible:

- Form the character -istic polynomial χ_M of M (over $\mathbb{C}$).

- Find the eigenvalues of M (the roots of χ_M) with their multiplicities. This work is the job of mzeros, but again we may have to use the function cSolve to succeed with this factorization.

- For each eigenvalue of multiplicity m, determine a basis and the dimension d of the eigenspace (with the function eigbasis). If even once $d < m$, then

M is not diagonalizable. Otherwise, the juxtaposition of column vectors thus obtained gives a basis of eigenvectors of M.

Here is the program di ago, whose role is (to try) to diagonalize a square matrix M (syntax di ago(M)).

The program di ago calls the functions mzeros and ei gbasis.

```
:diago(m)
:Prgm: Local e,i,j,n,s,w,λ
:DelVar x_:rowDim(m)→n:ClrIO
:mzeros(det(m-x_),x_)→w
:Disp "Eigenvalues:":Pause w
:If dim(w)<n:Goto err
:augment(w,{rand()})→w:newMat(n,1)→ θp:1→i
:While i≤n
: w[i]→ λ:string(λ)→s
: ClrIO:Disp "For λ = "&s:i→j
: Loop:i+1→i:If string(w[i])≠s:Exit:EndLoop
: i-j→j:Disp "Multiplicity = "&string(j)
: Disp "Eigenspace generated by :"
: eigbasis(m,λ)→e:colDim(e)→ θd
: For j,1,θd:Disp (conj(e))ᵀ[j]:EndFor
: Pause :augment(θp,e)→ θp
:EndWhile
:If colDim(θp)=n+1 Then
: subMat(θp,1,2,n,n+1)→ θp:diag(left(w,n))→ θd
: ClrIO:Disp "Transition matrix (in θp)":Pause θp
: ClrIO:Disp "Reduced diagonal (in θd)":Pause θd
:Else
: Lbl err:DelVar θp,θd
: Pause "Diagonalization impossible"
:EndIf
:EndPrgm
```

Here we are going to diagonalize a square matrix M of order 5 (and diagonalizable over $\mathbb{R}$, since all symmetric matrices have real coefficients).
The second screen shows that M has two eigenvalues: $\lambda = 1$ (of multiplicity 4) and $\lambda = 6$ (of multiplicity 1).

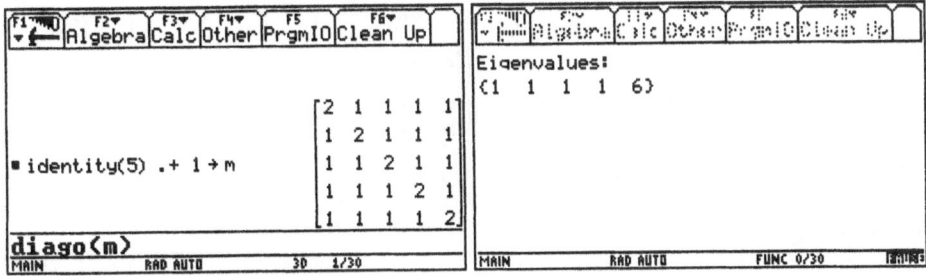

We now find a basis for each of the eigenspaces of M, of dimensions 4 and 1 respectively (which confirms that M is diagonalizable).

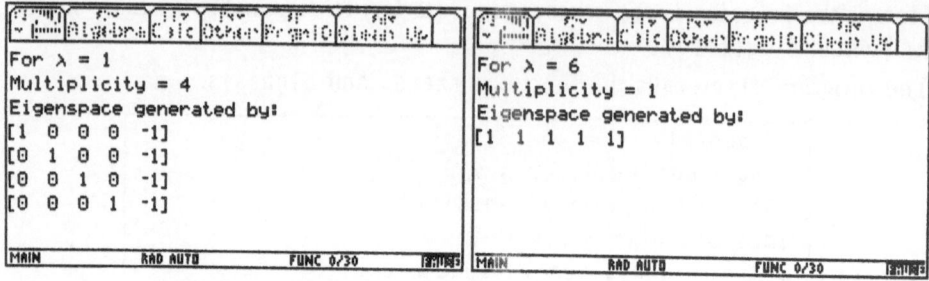

The program diago next displays a transition matrix P to a basis of eigenvectors of M and the reduced diagonal matrix D such that $D = P^{-1}MP$.

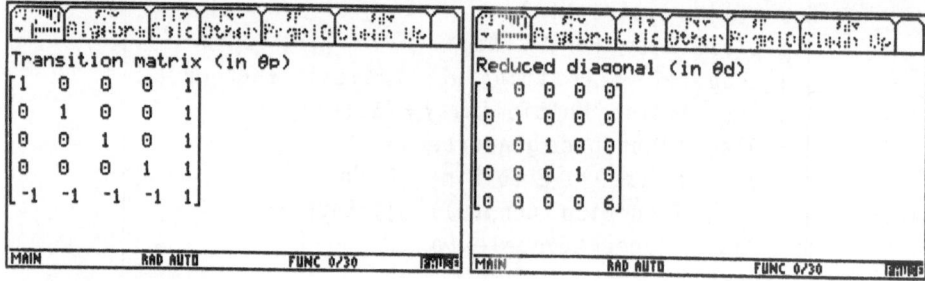

The program diago stores the matrices P and D in the global variables θp and θd.

Here is how to verify the result:
We form the product θp$\times\theta$d$\times\theta$p^{-1}, and we indeed retrieve the matrix M given at the outset.

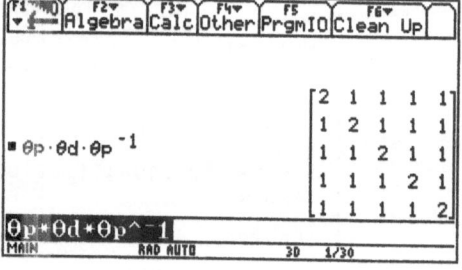

In the second example we reuse the matrix of order 4 which has a single eigenvalue ($\lambda = 1$, with multiplicity 4). We see that the eigenspace is a plane. The matrix M is thus not diagonalizable.

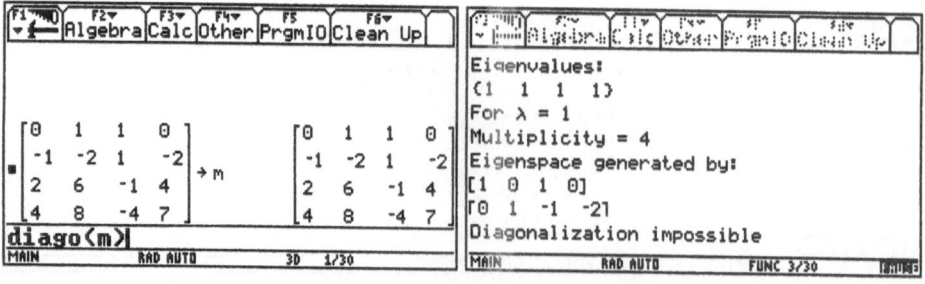

2. 4 Triangularization

As we have learned, there exist matrices M which are not diagonalizable (neither in $\mathbb{R}$, nor in $\mathbb{C}$).
We could always try to reduce such a matrix M to one with a form which approximates as closely as possible a diagonal matrix.
The first step consists of triangularization M, that is, to reduce it to a triangular form. We are going to see that this is always possible in $\mathbb{C}$.
We could then try to move to a more simple reduced triangular form: this will lead us to Jordan reduction.

Definition 3: We say that a matrix M of $\mathcal{M}_n(\mathbb{K})$ is triangularizable in $\mathbb{K}$ if it is similar to an upper triangular matrix T of $\mathcal{M}_n(\mathbb{K})$, that is, if there is an invertible matrix P of $\mathcal{M}_n(\mathbb{K})$ such that $T = P^{-1}MP$.

Remarks:

- If M is triangularizable, every matrix N similar to M is also triangularizable: if Q is an invertible matrix such that $N = QMQ^{-1}$, then $T = (QP)^{-1}N(QP)$.
- If M is triangularizable $(T = P^{-1}MP)$, the matrix T has the eigenvalues of M on its diagonal, each counted as many times as its multiplicity.

Definition 4: One says that an operator f of E (a vector space of dimension $n \geq 1$ over $\mathbb{K}$) is triangularizable if its matrix in an arbitrary basis of E is triangularizable.

Remarks:

- To say that f is triangularizable is to say that there is a basis $(e) = (e_1, e_2, \ldots, e_n)$ of E such that, for every k of $\{1, \ldots, n\}$, $f(e_k) \in \text{Span}(e_1, e_2, \ldots, e_k)$.
- With the preceding notation, the matrix of f in the basis (e) is upper triangular. But it is equivalent to say that the matrix of f in the basis $(e') = (e_n, e_{n-1}, \ldots, e_1)$ is lower triangular. The choice of upper triangular matrices thus does not constitute a restriction.

Proposition 2: A matrix M of $\mathcal{M}_n(\mathbb{K})$ is triangularizable in $\mathbb{K}$ if and only if its character -istic polynomial splits over $\mathbb{K}$. In particular, every square matrix with real or complex coefficients is triangularizable in $\mathbb{C}$.

Of course, we may state an analogous proposition for operators of a $\mathbb{K}$-vector space of dimension $n \geq 1$.

Proof: We remark that if a matrix is triangularizable, its character -istic polynomial obviously splits since it is that of the reduced triangular matrix T of M, which may be written $\chi_T = \prod_{k=1}^{n}(\alpha_k - X)$, where $\alpha_1, \alpha_2, \ldots, \alpha_n$ are the diagonal elements of T (the eigenvalues of M).

The converse may be shown by induction on n (the order of M and the dimension of E).

The property is evident if $n = 1$. Thus, we suppose that the result holds for order $n-1$ $(n \geq 2)$ and let f be an operator on E $(\dim E = n)$.

The character -istic polynomial of f splits over $\mathbb{K}$, so it has a root α at least in $\mathbb{K}$, which is thus a eigenvalue of f: let ε_1 be an associated eigenvector, and let F be the complement in E of the vector $\mathbb{K}\varepsilon_1$.

We extend ε_1 to a basis $(\varepsilon) = (\varepsilon_1, \varepsilon_2, \ldots, \varepsilon_n)$ of E.

The matrix of f in (ε) may be written $M = \begin{pmatrix} \alpha & L \\ 0 & N \end{pmatrix}$, where N is square and of order $n-1$ and L is a $1 \times (n-1)$ row matrix.

The character -istic polynomial of N divides that of M ($\chi_M = (X - \alpha)\chi_N$), and it thus splits. By the induction hypothesis, we conclude that N is triangularizable. There thus exist matrices Q (invertible of order $n-1$) and S (upper triangular with order $n-1$) such that $Q^{-1}NQ = S$.

We then define the invertible square matrix P of order n, by $P = \begin{pmatrix} 1 & 0 \\ 0 & Q \end{pmatrix}$.

Then $P^{-1}MP = \begin{pmatrix} 1 & 0 \\ 0 & Q^{-1} \end{pmatrix} \begin{pmatrix} \alpha & L \\ 0 & N \end{pmatrix} \begin{pmatrix} 1 & 0 \\ 0 & Q \end{pmatrix} = \begin{pmatrix} \alpha & L \\ 0 & Q^{-1}N \end{pmatrix} \begin{pmatrix} 1 & 0 \\ 0 & Q \end{pmatrix}$

$= \begin{pmatrix} \alpha & LQ \\ 0 & Q^{-1}NQ \end{pmatrix} = \begin{pmatrix} \alpha & LQ \\ 0 & S \end{pmatrix}$ is upper triangular.

Thus, the matrix M (and thus the operator f) are triangularizable, which proves the statement for rank n and finishes the induction argument.

Remark: The matrix $M = \begin{pmatrix} 0 & -1 \\ 1 & 0 \end{pmatrix}$, has the character -istic polynomial $X^2 + 1$ which doesn't split over $\mathbb{R}$, and which is not triangularizable over $\mathbb{R}$. It does split over $\mathbb{C}$ where M is even diagonalizable (two eigenvalues : i and $-i$).

2. 5 Jordan Matrices

When a matrix M is diagonalizable, there is not just one pair (D, P) (D diagonal, P invertible) such that $M = PDP^{-1}$. In fact, we have the choice of the order in which we list the eigenvalues, and of a basis in each of the eigenspaces E_λ (especially when $\dim E_\lambda > 1$).

For the triangularization, the choice is even greater, and we must find the simplest reduced triangular matrix.

The Jordan reduction of a matrix M (given that the character -istic polynomial splits) consists of finding a reduced triangular T having the following property: among the non-diagonal coefficients of T, the only ones which are not zero are those which are immediately on either side of the diagonal, and these coefficients have values 0 or 1.

The diagonal coefficients of T are the eigenvalues of M, each counted as many times as its multiplicity. Of course, if M is diagonalizable, this matrix T is the reduced diagonal matrix of M.

The form indicated for the matrix T is that of a block diagonal matrix, each block being a matrix $J(m, \lambda)$, in the sense of the following definition.

Definition 5: Let $m \in \mathbb{N}^*, \lambda \in \mathbb{C}.$ $J(m, \lambda) = \begin{pmatrix} \lambda & 1 & 0 & \cdots & 0 \\ 0 & \lambda & 1 & \ddots & \vdots \\ 0 & 0 & \lambda & \ddots & 0 \\ \vdots & \ddots & \ddots & \ddots & 1 \\ 0 & \cdots & 0 & 0 & \lambda \end{pmatrix} \in \mathcal{M}_m(\mathbb{C}).$

The matrices $J(m, \lambda)$ are called *elementary Jordan matrices*.

The function ejordan lets us create such a matrix.

```
:ejordan(n,λ)
:seq(seq(when(j=i+1,1,0),j,1,n),i,1,n)+λ
```

The matrix $J(n, \lambda)$ may obviously be written $J(n, \lambda) = \lambda I_n + T_n$, where T_n is a strictly triangular matrix which is thus nilpotent − some power equals the zero matrix. The successive powers T_n^k themselves have a very simple form, the "super-diagonal" moves up a position each time, finally with $T_n^{n-1} \neq 0$ and $T_n^n = 0$.

For example, with $T_4 = \begin{pmatrix} 0 & 1 & 0 & 0 \\ 0 & 0 & 1 & 0 \\ 0 & 0 & 0 & 1 \\ 0 & 0 & 0 & 0 \end{pmatrix}$:

$$T_4^2 = \begin{pmatrix} 0 & 0 & 1 & 0 \\ 0 & 0 & 0 & 1 \\ 0 & 0 & 0 & 0 \\ 0 & 0 & 0 & 0 \end{pmatrix}, \quad T_4^3 = \begin{pmatrix} 0 & 0 & 0 & 1 \\ 0 & 0 & 0 & 0 \\ 0 & 0 & 0 & 0 \\ 0 & 0 & 0 & 0 \end{pmatrix}, \quad T_4^4 = \begin{pmatrix} 0 & 0 & 0 & 0 \\ 0 & 0 & 0 & 0 \\ 0 & 0 & 0 & 0 \\ 0 & 0 & 0 & 0 \end{pmatrix}.$$

We deduce the powers of $J(n, \lambda)$: $\forall p \in \mathbb{N}, \ J(n, \lambda)^p = \sum_{k=0}^{n-1} \binom{p}{k} \lambda^{n-k} T^k.$

For example: $\forall p \in \mathbb{N}, \ T_4^p = \begin{pmatrix} \lambda^p & p\lambda^{p-1} & \frac{p(p-1)}{2}\lambda^{p-2} & \frac{p(p-1)(p-3)}{6}\lambda^{p-3} \\ 0 & \lambda^p & p\lambda^{p-1} & \frac{p(p-1)}{2}\lambda^{p-2} \\ 0 & 0 & \lambda^p & p\lambda^{p-1} \\ 0 & 0 & 0 & \lambda^p \end{pmatrix}.$

This formula is still valid for negative exponents if $J(n, \lambda)$ is invertible, that is, if $\lambda \neq 0$. We see below the formation of an elementary Jordan matrix J of order 3, the calculation of J^{10}, then that of J^{-1}.

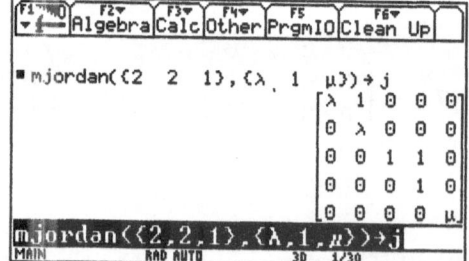

A Jordan matrix will be any square, block diagonal matrix with the blocks being themselves elementary Jordan matrices.

The function mjordan constructs a "Jordan block diagonal ".

The syntax is $mjordan(\{d_1, d_2, \ldots, d_p\}, \{\lambda_1, \lambda_2, \ldots, \lambda_p\})$, where the first list contains the successive dimensions of the diagonal blocks, and the second contains the parameters λ which correspond to these elementary Jordan matrices.

```
:mjordan(d,λ)
:Func:Local n,m,dd,i,j
:sum(d)→n:newMat(n,n)→m:augment({0},cumSum(d))→dd
:For i,1,dim(d)
:    For j,dd[i]+1,dd[i+1]:λ[i]→m[j,j]:EndFor
:    For j,dd[i]+1,dd[i+1]-1:1→m[j,j+1]:EndFor
:EndFor
:m
:EndFunc
```

Here we create a Jordan matrix J of order 5, formed of 3 diagonal blocks of orders respectively 2, 2, and 1.

The function pjordan calculates the n-th power of a Jordan matrix J. The syntax is pjordan(J,n). (The method, of course, consists of raising to the power n each of the elementary blocks which compose J).

Interest in this function resides in the fact that it may be used with a symbolic exponent n. (For explicitly defined integer exponents, it would evidently be better to use the power function of the calculator).

NB: The function pjordan does not verify that J is really a Jordan matrix. If that is not the case, the result has no significance.

Here is the n-th power of the Jordan matrix J formed in the preceding example .

If we try to evaluate j^n with the calculator, it will respond with "Error: Data type".

Here is the listing of the function pjordan

```
:pjordan(m,p)
:Func:Local n,i,ii,c,j,k,t
:rowDim(m)→n
:For i,1,n:i→ii:m[i,i]→c
:   While i<n
:      If m[i,i+1]=1 Then:i+1→i
:      Else:Exit:EndIf
:   EndWhile
:   For k,0,i-ii
:      nCr(p,k)*c^(p-k)→t
:      For j,ii,i-k:t→m[j,j+k]:EndFor
:   EndFor
:EndFor
:m
:EndFunc
```

2. 6 Jordan reduction

Now we are going to prove that every square matrix with real or complex coefficients may be "Jordanized " in $\mathbb{C}$. Because of this result we say that a square matrix may be put in Jordan canonical form.

We begin with some remarks based on the elementary Jordan matrices.
Let f be an operator on a $\mathbb{C}$-vector space E of dimension n.

We suppose that the matrix M of f in the basis $(e) = (e_1, e_2, \ldots, e_n)$ of E is an elementary Jordan matrix $J(n, \lambda)$.

This signifies that $f(e_1) = \lambda e_1$ and

$\forall k \in \{2, \ldots, n\}$, $f(e_k) = e_{k-1} + \lambda e_k$.

$$J(n,\lambda) = \begin{pmatrix} \lambda & 1 & 0 & \cdots & 0 \\ 0 & \lambda & 1 & \ddots & \vdots \\ 0 & 0 & \lambda & \ddots & 0 \\ \vdots & \ddots & \ddots & \ddots & 1 \\ 0 & \cdots & 0 & 0 & \lambda \end{pmatrix}$$

The preceding conditions may also be written:

- $\forall k \in \{1, \ldots, n-1\}, e_k = (f - \lambda Id)(e_{k+1}) = (f - \lambda Id)^{n-k}(e_n)$.
- $(f - \lambda Id)(e_1) = (f - \lambda Id)^n(e_n) = \overrightarrow{0}$.

Thus, the basis $(e) = e_n, e_{n-1} \ldots, e_1$ may be written $(e) = \big(e_n, g(e_n), \ldots, g^{n-1}(e_n)\big)$, where g designates the operator $f - \lambda Id$. This basis is thus constituted of iterates of a vector e_n of E which does not appear in $\operatorname{Ker} g^{n-1}$ (since its image under g^{n-1} is the non-zero vector e_1).

Just as the matrix $T_n = J(n, \lambda) - \lambda I_n$ satisfies $T^{n-1} \neq 0$ and $T^n = 0$, the linear map $g = f - \lambda Id$ satisfies $g^{n-1} \neq 0$ and $g^n = 0$. We say that the index of nilpotence of T is n.

The sequence of the iterated kernels of g (increasing under inclusion) satisfies:
$$\forall k \in \{1, \ldots, n\}, \ \operatorname{Ker} g^k = \operatorname{Span}(e_1, e_2, \ldots, e_k).$$

Conversely, if f is an operator on E such that $g = f - \lambda Id$ has a nilpotence index n ($g^{n-1} \neq 0$, $g^n = 0$), then for every u of E such that $g^{n-1}(u) \neq 0$, the family $(\varepsilon) = \big(g^{n-1}(u), \ldots, g(u), u\big)$ is a basis of E in which the matrix of g is T_n and thus in which the matrix of f is $J(n, \lambda)$.

More generally, let F be a sub-space of dimension m of E ($m \geq 1$), stable under f. We will say that F is cyclic for f if there is a scalar λ and a basis of F in which the matrix of the restriction of f is $J(m, \lambda)$.

As we will see, by always putting $g = f - \lambda Id$ this is equivalent to the existence of a vector u of F such that $g^{m-1}(u) \neq 0$, and $g^m(u) = 0$. Such a vector is called a "generalized eigenvector".

The family $\big(u, g(u), \ldots, g^{m-1}(u)\big)$ is then a basis of F.

We will denote by $F(m, u, \lambda)$ such a cyclic subspace for f.

- $F(m, u, \lambda) = \operatorname{Vect}\{u, (f - \lambda Id)(u), \ldots, (f - \lambda Id)^{m-1}(u)\}$
- $(f - \lambda Id)^{m-1}(u) \neq \overrightarrow{0}$, $(f - \lambda Id)^m(u) = \overrightarrow{0}$

Remark 1:

If $\lambda \neq 0$, the restriction of f to the subspace $F(m, u, \lambda)$ is an isomorphism (its determinant is λ^m). We conclude that $F(m, u, \lambda) = f\big(F(m, u, \lambda)\big)$.

Remark 2:

If $\lambda = 0$, then $F(m, u, 0)$ is generated by u and by the vectors $f(u), \ldots, f^{m-1}(u)$, the latter being elements of $f\big(F(m, u, 0)\big)$. We conclude that every v of $F(m, u, 0)$ may be written $v = \alpha u + f(w)$, where $\alpha \in \mathbb{K}$ and $w \in F(m, u, 0)$.

We have now come to the principal result of this section.
There are several proofs. The one which we develop here is quite simple and is quite recent (1996). It is by Israel and Seymour Goldberg of the University of Tel Aviv.

Proposition 3: Let $M \in \mathcal{M}_n(\mathbb{K})$, and suppose that its character-istic polynomial splits in $\mathbb{K}$. Then M is similar to a Jordan matrix J of $\mathcal{M}_n(\mathbb{K})$.

Recall that J is called "Jordan" if it may be written as a block diagonal matrix, each of whose blocks is an elementary Jordan matrix.
The preceding statement has an equivalent in term of linear maps.

Proposition 4: Let $f \in \mathcal{L}(E)$, where E is a $\mathbb{K}$-vector space of dimension $n \geq 1$. If the character -istic polynomial χ_f of f splits in $\mathbb{K}$, then E is the direct sum of cyclic subspaces for f.

Proof:
We begin by defining a practical neologism: we will say that a basis (e) of E "Jordanizes" f if the matrix of f in terms of (e) is a Jordan matrix, which is to say that (e) is obtained by juxtaposition of bases of cyclic subspaces for f.
We remark that for every scalar μ, if a basis (e) Jordanizes f then it Jordanizes $f - \mu Id$ (if J is a Jordan matrix, then $J - \mu I_n$ is again a Jordan matrix).
If we choose for μ an eigenvalue of f (which is possible since χ_f splits), we thus see that the problem reduces to "Jordanizing" the non-bijective operator $g = f - \mu Id$.
We may thus suppose without loss of generality that f is itself not bijective.

The proof proceeds by induction on the dimension n of E. If $n = 1$, the result is evident. We suppose that $n \geq 2$ and that the proposition is true for $n - 1$.

Let F be a subspace of E of dimension $n - 1$ containing $\mathrm{Im}\, f$ (Recall that f is not an isomorphism and thus that $\dim \mathrm{Im}\, f \leq n - 1$.)
The subspace F is invariant under f since $f(F) \subset \mathrm{Im}\, f \subset F$.
By the induction hypothesis, F is the direct sum of cyclic subspaces:

$$F = F(m_1, u_1, \lambda_1) \oplus F(m_2, u_2, \lambda_2) \oplus \cdots \oplus F(m_p, u_p, \lambda_p)$$

We may always suppose that the coefficients λ_k in this sum which are equal to zero (if there are any) appear at the front.
We designate by q the maximum index k for which $\lambda_k = 0$, and we put:

- $G = F(m_1, u_1, 0) \oplus \cdots \oplus F(m_q, u_q, 0)$
- $\forall k \in \{1, \ldots, q\}, G_k = F(m_1, u_1, 0) \oplus \cdots \oplus F(m_k, u_k, 0)$.
- $H = F(m_{q+1}, u_{q+1}, \lambda_{q+1}) \oplus \cdots \oplus F(m_p, u_p, \lambda_p)$

Of course, we have $F = G \oplus H$.
There are two particular cases: if $q = 0$ (none of the λ_k are zero), then $G = \{0\}$ and $H = F$, and if $q = p$ (all the λ_k are zero), then $G = F$ and $H = \{0\}$.

In the sum defining G, we could suppose that the dimensions $m_1, \ldots, m_q$ of the successive cyclic subspaces satisfy $m_1 \leq m_2 \leq \ldots \leq m_q$.

We know that the restriction of f à $F(m_k, u_k, 0)$ (for $1 \leq k \leq q$) is nilpotent of index m_k. The preceding convention thus implies that for every k of $\{1, \ldots, q\}$, the m_k-th power of f is identically zero on G_k.

Let a be a vector of E not appearing in F.

Since the vector $f(a)$ is in $\mathrm{Im}\, f$ and thus in F, remarks (1) and (2) show that it may be written $f(a) = \sum_{k=1}^{q} \alpha_k u_k + f(b)$, with $b \in F$.

Putting $c = a - b$, we thus have the equation $f(c) = \sum_{k=1}^{q} \alpha_k u_k$.
Since $a \notin F$ and $b \in F$, the vector c is not in F.

If it happens that $f(c) = \vec{0}$, then the vector line $\mathbb{K}c$ is itself a cyclic subspace for f (in fact $\mathbb{K}c = F(1, c, 0)$), and the equation $E = F \oplus \mathbb{K}h$ together with the decomposition of F shows that the vector space E is a direct sum of cyclic subspaces for f.

We thus suppose that $f(c) \neq \vec{0}$ and let r be the maximum index k such that $\alpha_k \neq 0$. With the notation introduced above, the vector $f(c)$ is in G_r. We may then write $f(c) = \sum_{k=1}^{r-1} \alpha_k u_k + \alpha_r u_r$.

Putting $d = \dfrac{1}{\alpha_r} c$ and $h = \dfrac{1}{\alpha_r} \sum_{k=1}^{r-1} \alpha_k u_k$, it follows that $f(d) = h + u_r$, with $h \in G_{r-1}$.

To simplify the notation, we put $m = m_r$.

We know that, by definition, $G_r = G_{r-1} \oplus F(m, u_r, 0)$.

We are going to show that $G_r = G_{r-1} \oplus F(m, f(d), 0)$.

The component of $f^k(f(d))$ in $F(m, u_r, 0)$ (relative to the direct sum $G_r = G_{r-1} \oplus F(m, u_r, 0)$) is $f^k(u_r)$ which is non-zero (just as $f^k(f(d))$) as long as $k \in \{1, \ldots, m-1\}$.

On the other hand, since $f(d)$ appears in G_r the vector $f^m(f(d))$ is zero.

We may conclude that $\text{Vect}(f(d), f^2(d), \ldots, f^m(d))$ is a cyclic subspace of f. More precisely, this works with $F(m, f(d), 0)$. It is of dimension m.

Since $f(d)$ is in G_r, all of its iterates under f are equal. We may conclude that $F(m, f(d), 0)$ and thus also $G_{r-1} + F(m, f(d), 0)$ are included in G_r.

Let v be an element of $G_{r-1} \cap F(m, f(d), 0)$. It may be written $v = \sum_{k=1}^{m} \beta_k f^k(d)$.

It may thus also be written $v = \sum_{k=1}^{m} \beta_k f^k(h) + \sum_{k=1}^{m} \beta_k f^{k-1}(u_r)$.

The first of these two sums is an element of G_{r-1}.
Because this is also the case for v, the second sum is likewise in G_{r-1}.
We may deduce that it is zero since it is in $G_{r-1} \cap G_r$.
All the β_k are thus zero ($u_r, f(u_r), \ldots, f^{m-1}(u_r)$ is a basis of G_r).
Thus, $v = \vec{0}$.

We have proved that the sum $G_{r-1} + F(m, f(d), 0)$ is direct.
However, it is included in G_r and has the same dimension.
We have thus proved: $G_r = G_{r-1} \oplus F(m, f(d), 0)$.

Finally, we remark that:
$$\mathbb{K}d + F(m, f(d), 0) = \mathbb{K}d + \text{Vect}(f(d), \ldots, f^m(d))$$
$$= \text{Vect}(d, f(d), \ldots, f^m(d)) = F(m+1, d, 0).$$

We conclude that:
$$E = F \oplus \mathbb{K}d = \oplus_{k \neq r} F(m_k, u_k, \lambda_k) \oplus F(m, f(d), 0) \oplus \mathbb{K}d$$
$$= \oplus_{k \neq r} F(m_k, u_k, \lambda_k) \oplus F(m+1, d, 0).$$

Thus, we have proved that E may be decomposed into a direct sum of cyclic subspaces for f. The property is thus proved for rank n, which completes the induction and finishes the proof.

We know now that every square matrix whose character -istic polynomial splits is "triangularizable ". In particular, every matrix with real or complex coefficients is triangularizable in $\mathbb{C}$.

The proof of this result is already established, but an attempt to transform it into an algorithm for "Jordanization " would be a little ambitious. We will thus be content to study some examples.

We will consider here a square matrix M of order 4 (already used several times) and which only has a single eigenvalue $\lambda = 1$ with multiplicity 4.

The matrix M is not diagonalizable (otherwise it would be similar to I_4, thus equal to I_4, which is not the case). But its characteristic polynomial $(1 - X)^4$ splits over $\mathbb{R}$. M is thus Jordanizable in $\mathbb{R}$.

The Cayley-Hamilton Theorem assures us that $(M - I_4)^4 = 0$.

The calculation below in fact shows that $(M - I_4)^3 = 0$ and $(M - I_4)^2 \neq 0$: the minimal polynomial of M is thus $(X - 1)^3$.

The matrix M represents in the standard basis, a certain operator f.

The first column of $(M - I_4)^2$ represents the image (clearly not zero) of the first vector $\varepsilon_0 = (1, 0, 0, 0)$ of this basis by the map $(f - Id)^2$.

As one may see, $(f - Id)^2(\varepsilon_0) \neq \overrightarrow{0}$ and $(f - Id)^3(\varepsilon_0) = \overrightarrow{0}$.

The subspace $F = F(3, \varepsilon_0, 1) = \text{Vect}\big(\varepsilon_2 = (f - Id)^2(\varepsilon_0), \varepsilon_1 = (f - Id)(\varepsilon_0), \varepsilon_0\big)$ is thus cyclic for the operator f.

It is stable under f and the matrix of the restriction of f to F is, in the basis $(\varepsilon_2, \varepsilon_1, \varepsilon_0)$, the elementary Jordan matrix $J(3, 1) = \begin{pmatrix} 1 & 1 & 0 \\ 0 & 1 & 1 \\ 0 & 0 & 1 \end{pmatrix}$.

We see here how to form the vectors ε_0, ε_1, ε_2, then the matrix P of coordinates of these three vectors in the standard basis (pay attention to the order: beginning with ε_2, then ε_1 , then ε_0).

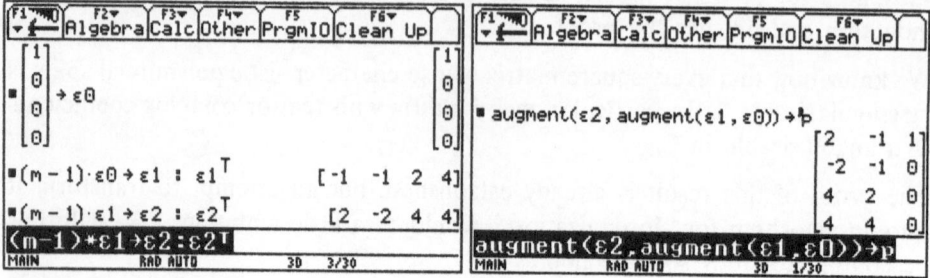

This family must now be completed by a fourth vector of $\mathbb{R}^4$.

We know that the kernel of the restriction of $f - Id$ to F is a vector line (generated by of the vector ε_2).

Stated another way, this signifies that the intersection of the eigenspace E_1 of f with F is a vector line.

However, the following calculation (using our function eigbasis), shows that E_1 is a plane. We are thus certain that we may find in E_1 a vector which is not in F and which thus extends ε_2, ε_1, ε_0 to a basis of $\mathbb{R}^4$.

We have constructed the matrix B of coordinates of two vectors for forming a basis of E_1. We see on the second screen that each of these two vectors is linearly independent from ε_2, ε_1, ε_0.

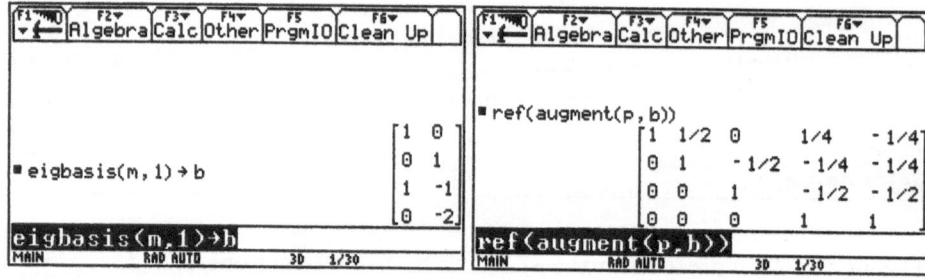

We thus complete P by one of the two column vectors of B. (Here we chose the first.) We obtain a square, invertible matrix P of order 4. P is indeed the transition matrix to a basis which Jordanizes f as the second screen shows.

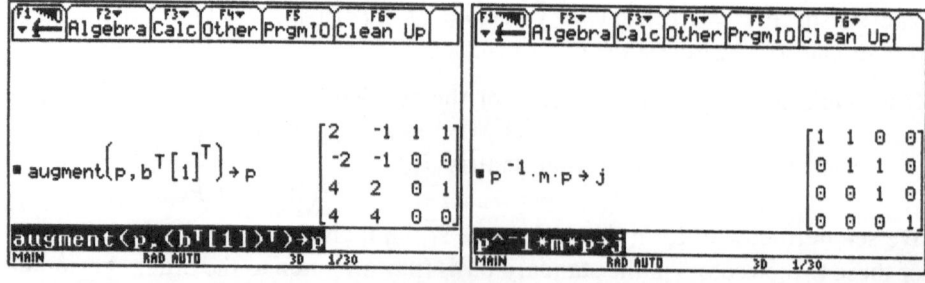

Our function pjordan permits computation of the n-th power of the reduced Jordan matrix J (which is here stored in the variable jn), then that of the matrix M (using the equation $M^n = PJ^n P^{-1}$).

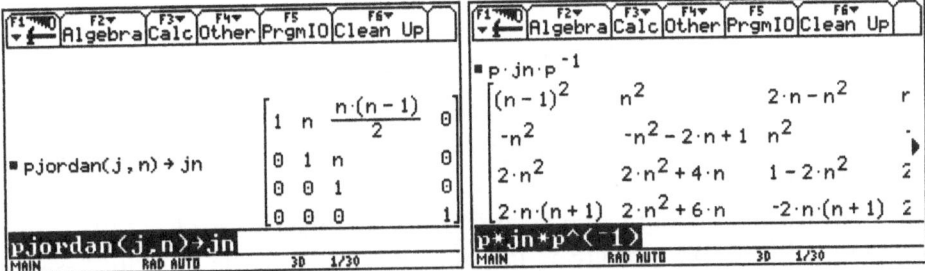

Now here is a square matrix M of order 5 which has two eigenvalues, -2 (simple) and 1 (quadruple). Let f be the linear map associated with this matrix in the standard basis.

The function eigbasis shows that the eigenspace E_1 reduces to a plane. The matrix M is thus not diagonalizable. But its character -istic polynomial $\chi_M = (1 - X)^4(-2 - X)$ splits: M is thus Jordanizable.

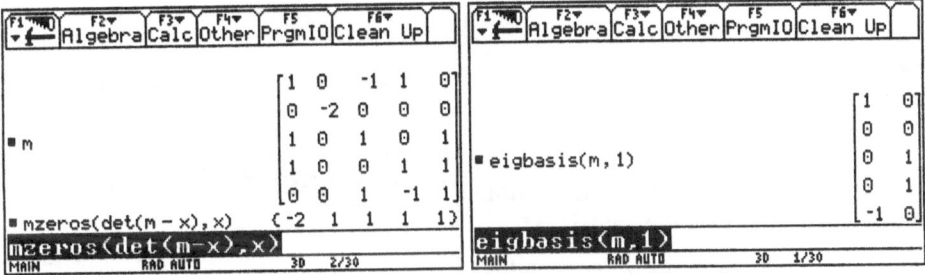

We observe directly from M that the eigenvector line E_{-2} is generated by the vector $\varepsilon_1 = (0, 1, 0, 0, 0)$.

The Cayley-Hamilton Theorem assures us that $(f + 2Id) \circ (f - Id)^4 = 0$. The following calculation shows in fact that $(f + 2Id) \circ (f - Id) \neq 0$ and $(f + 2Id) \circ (f - Id)^2 = 0$.

Thus, the minimal polynomial of M and of f is $(X + 2)(X - 1)^2$.

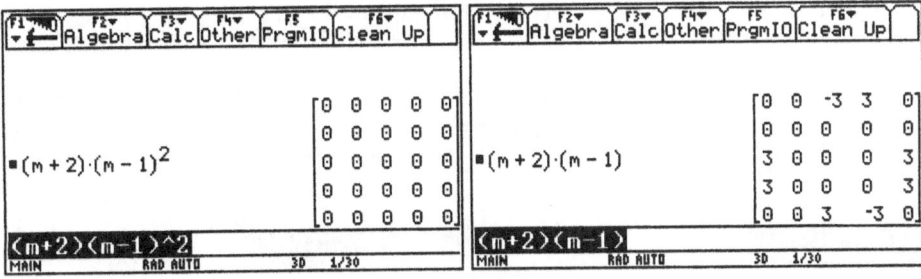

The result below shows that the vectors $e_3 = (0,0,1,0,0)$ and $e_5 = (0,0,0,0,1)$ are not annihilated by $(f + 2Id) \circ (f - Id)$.

A fortiori, neither are they annihilated by $(f + 2Id)$.

We consider then the vectors $\varepsilon_3 = (f + 2Id)(e_3)$ and $\varepsilon_5 = (f + 2Id)(e_5)$.

We observe that:

• $\varepsilon_2 = (f - Id)(\varepsilon_3) = (f - Id) \circ (f + 2Id)(e_3) \neq \vec{0}$ and $(f - Id)^2(\varepsilon_3) = \vec{0}$

• $\varepsilon_4 = (f - Id)(\varepsilon_5) = (f - Id) \circ (f + 2Id)(e_5) \neq \vec{0}$ and $(f - Id)^2(\varepsilon_5) = \vec{0}$

The subspaces $F = \text{Vect}(\varepsilon_2, \varepsilon_3)$ and $G = \text{Vect}(\varepsilon_4, \varepsilon_5)$ are thus two cyclic planes for f (for the eigenvalue 1).

We now see how to form the vectors ε_1, ε_3, ε_5, then $\varepsilon_4 = (f - Id)(\varepsilon_5)$ and $\varepsilon_2 = (f - Id)(\varepsilon_3)$. We next create the transition matrix in the standard basis to the basis $\varepsilon_1, \ldots, \varepsilon_5$. ($P$ is invertible: its determinant is 1.)

NB: Here we have used lists for simple reasons of presentation (because then everything took only two screens!).

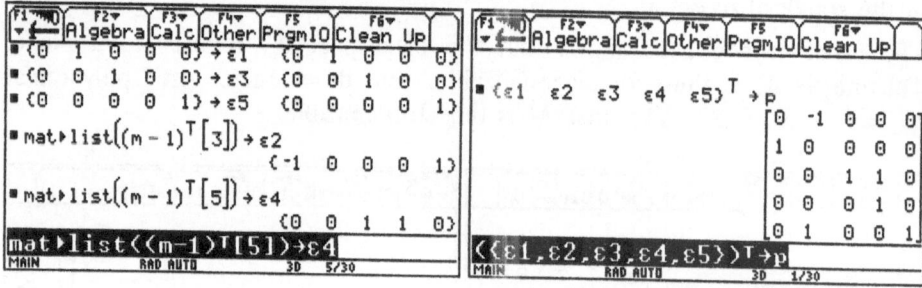

Indeed we observe that the transition matrix P Jordanizes the matrix M.

The difference with the first example lies in the fact that for $\lambda = 1$, there are two non-trivial elementary Jordan blocks here instead of just one.

As we did for the first example of Jordan reduction, here is how to calculate the n-th power of the reduced Jordan matrix J, then that of the initial matrix M.

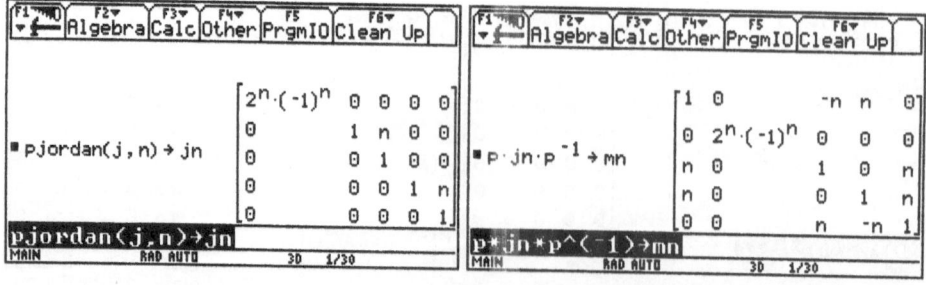

The formula giving M^n is equally valuable for negative exponents. (M is invertible since it does not have 0 as an eigenvalue).

We verify this by forming the product of mn (n having a certain value k) by mn (n having then the value $-k$): the result is the identity matrix.

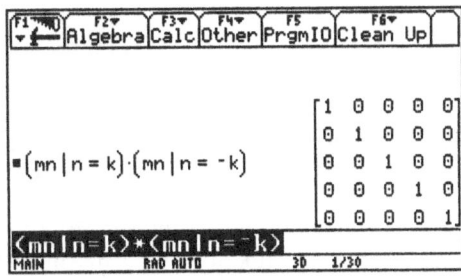

2. 7 Characteristic supspaces

Let f be an operator on a $\mathbb{K}$-vector space E of dimension $n \geq 1$. We suppose that the character -istic polynomial χ_f splits over $\mathbb{K}$ (which is automatically realized if $\mathbb{K} = \mathbb{C}$).

Let $\lambda_1, \lambda_2, \ldots, \lambda_p$ be the different eigenvalues of f of respective multiplicities $m_1, m_2, \ldots, m_p$.

For every eigenvalue λ_k and each integer r, we will denote $S_k(r) = \mathrm{Ker}(f - \lambda_k Id)^r$.

With this definition $S_k(0) = \{\vec{0}\}$, and $S_k(1) = \mathrm{Ker}(f - \lambda_k Id)$ is the eigenspace of f pour λ_k. It does not reduce to $\{\vec{0}\}$.

The sequence $S_k(r)_{r \geq 0}$ is thus the sequence of iterated kernels of $f - \lambda_k Id$. As every sequence of iterated kernels is in finite dimension, it increases under inclusion before becoming stationary.

The character -istic polynomial of f may be written $\chi_f = \prod_{k=1}^{p}(\lambda_k - X)^{m_k}$.

The Cayley-Hamitton theorem affirms that $\chi_f(f) = 0$. The theorem of decomposition of the kernel (Proposition **3** of paragraph **1.2**) thus gives:

$$E = \mathrm{Ker}\,\chi_f(f) = \bigoplus_{k=1}^{p} \mathrm{Ker}(f - \lambda_k Id)^{m_k} = \bigoplus_{k=1}^{p} S_k(m_k) \qquad (1)$$

Recall that the minimal polynomial of f, denoted here μ_f, is among all the polynomial annihilators of f the one which is of minimum degree (and we are assured of its uniqueness by requiring it to be monic). (See paragraph **1.4** of this chapter.)

We know that the polynomial annihilators of f (which are χ_f) are the multiples of μ_f.

We likewise know that each of the eigenvalues of f is a root of all of the polynomial annihilators of f and thus of μ_f.

We may conclude that $\mu_f = \prod_{k=1}^{p}(X - \lambda_k)^{s_k}$ where, $\forall k \in \{1, \ldots, p\},\ 1 \leq s_k \leq m_k$.

Since $\mu_f(f) = 0$, it follows:

$$E = \operatorname{Ker}\mu_f(f) = \bigoplus_{k=1}^{p} \operatorname{Ker}(f - \lambda_k Id)^{s_k} = \bigoplus_{k=1}^{p} S_k(s_k) \qquad (2)$$

The equation **(1)**,**(2)** and the fact that for every k of $\{1,\dots,p\}$ we have $S_k(s_k) \subset S_k(m_k)$ lead to the equations: $\forall k \in \{1,\dots,p\}, S_k(s_k) = S_k(m_k)$.

In other words, the sequence of iterated kernels $S_k(r) = \operatorname{Ker}(f - \lambda_k Id)^r$ is stationary with rank s_k (we easily show that up to this step it is strictly increasing by inclusion).

We may thus summarize the situation by writing, for every k of $\{1,\dots,p\}$:

$$\{\overrightarrow{0}\} \subset \operatorname{Ker}(f - \lambda_k Id) \subset \cdots \subset \operatorname{Ker}(f - \lambda_k Id)^{s_k} = \cdots = \operatorname{Ker}(f - \lambda_k Id)^{m_k}$$

The inclusions preceding the equal sign are strict inclusions.

Definition 6: The subspace $S_k = \operatorname{Ker}(f - \lambda_k Id)^{s_k}$ is called a characteristic, or spectral subspace of f for the eigenvalue λ_k.

We remark that S_k is invariant under f. (It is the kernel of $(f - \lambda_k Id)^{s_k}$, which commutes with f). We denote by f_k the restriction of f to S_k.

The preceding shows that $h_k = f_k - \lambda_k Id$ is a nilpotent operator. (To be precise, its index of nilpotence is s_k). The character -istic polynomial of f_k is thus $(\lambda_k - X)^{d_k}$, denoting by d_k the dimension of the subspace S_k.

Since $E = S_1 \oplus S_2 \oplus \cdots \oplus S_p$, and because each S_k is stable under f, the character -istic polynomial of f is the product of those of f_k: $\chi_f = \prod_{k=1}^{p} (\lambda_k - X)^{d_k}$.

This equation proves that the dimension d_k of each subspace S_k is equal to the multiplicity m_k of the eigenvalue λ_k.

2. 8 The Dunford decomposition and applications

Now f will always designate an operator on a $\mathbb{K}$-vector space E ($\dim E \geq 1$), with a character -istic polynomial which splits.

We form a basis (e) of E by juxtaposition of a basis $(e)_k$ of each of the spectral subspaces S_k of f.

The matrix B of f in (e) is block diagonal, each block B_k being the matrix in the basis $(e)_k$ of the restriction f_k of f.

$$B = \begin{pmatrix} B_1 & 0 & \cdots & 0 \\ 0 & B_2 & \ddots & \vdots \\ \vdots & \ddots & \ddots & 0 \\ 0 & \cdots & 0 & M_p \end{pmatrix}$$

The matrix B_k may itself be decomposed into $B_k = \lambda I_{m_k} + N_k$, where N_k is the matrix of $h_k = f_k - \lambda_k Id$ in the basis $(e)_k$.

This leads to the decomposition $B = D + N$, where:

$$D = \begin{pmatrix} \lambda_1 I_{m_1} & 0 & \cdots & 0 \\ 0 & \lambda_2 I_{m_2} & \ddots & \vdots \\ \vdots & \ddots & \ddots & 0 \\ 0 & \cdots & 0 & \lambda_p I_{m_p} \end{pmatrix} \quad \text{and} \quad N = \begin{pmatrix} N_1 & 0 & \cdots & 0 \\ 0 & N_2 & \ddots & \vdots \\ \vdots & \ddots & \ddots & 0 \\ 0 & \cdots & 0 & N_p \end{pmatrix}$$

The matrices D and N commute (since this is the case for the blocks which correspond pairwise). D is evidently diagonal, and N is nilpotent (since each of the blocks N_k is nilpotent). More precisely, $N^s = 0$ where s is the maximum exponent of an eigenvalue in the minimal polynomial (and in every case, $N^m = 0$, where m is the maximum multiplicity of an eigenvalue of f).

Let d and h be the endorphisms of E with matrices D and N in the basis (e). Just like D and N, they commute. h is nilpotent since N is nilpotent. d is diagonalizable since its matrix D in the basis (e) is diagonal.

We have thus shown the following result (the Dunford decomposition)

Proposition 5: Each operator f of a $\mathbb{K}$-vector space E $(\dim E \geq 1)$, with character -istic polynomial which splits, may be written as the sum $f = d + h$ of a diagonalizable operator d and nilpotent operator h. Moreover, the operators d and h commute.

We may show that this decomposition is unique.

In terms of matrices, this result is:

Proposition 6: Let $M \in \mathcal{M}_n(\mathbb{K})$, where character -istic polynomial splits. Then M may be written uniquely as $M = \Delta + H$ where Δ and H commute, D is diagonalizable and H is nilpotent.

Evidently, if M is itself diagonalizable, the matrix H of this decomposition is the zero matrix.

The principal interest in this type of result resides in the possibility of calculating the powers (and thus polynomials) of M, or even the "usual " transcendental functions of M (its exponential, for example).

In fact: $\forall r \in \mathbb{N}, M^r = \sum_{k=0}^{m} \binom{r}{k} H^k \Delta^{r-k}$, this sum being limited by the nilpotence of H (whose index is less than or equal to the maximum multiplicity m of the eigenvalues of M).

In the same manner, since H and Δ commute: $\exp M = \exp H \exp \Delta$.

- The matrix $\exp H$ may be calculated with the finite sum $\sum_{k=0}^{m} \frac{1}{k!} H^k$.

- If P is a diagonalizing transition matrix Δ $(\Delta = PDP^{-1})$, then $\exp \Delta = P \exp D P^{-1}$, and $\exp D$ may be calculated by taking the exponential of the diagonal coefficients of D.

We are going to see how to calculate and then to use the Dunford decomposition of this matrix M.

To begin we see that M has two eigenvalues, 1 (double) and 4 (triple). The character -istic polynomial of M splits.

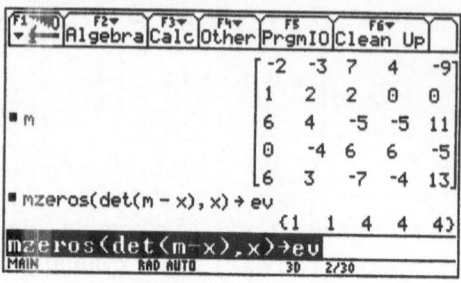

We see how the eigenspace of M for the eigenvalue double 1 is reduced to a vector line. The matrix M is thus not diagonalizable.

Next we calculate $\text{Ker}(M - I_5)^2$, which is a plane (this is normal: it is the spectral subspace of M for $\lambda = 1$).

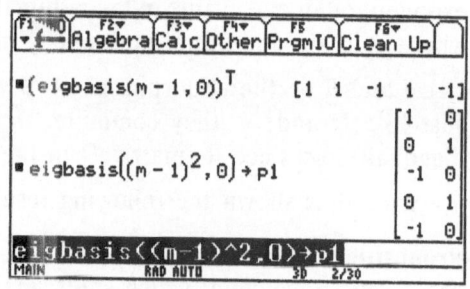

Now we see that $\text{Ker}(M - 4I_5)$, $\text{Ker}(M - 4I_5)^2$ and $\text{Ker}(M - 4I_5)^3$ are respectively of dimensions 1, 2 and 3.

We have thus found the characteristic sub-space of M for the eigenvalue 4.

The fact that it is of dimension 3 is normal (this must be the multiplicity of the eigenvalue).

The fact that we succeeded for the exponent 3 shows that $(X - 4)^3$ is a factor of minimal polynomial of M.

As this was already true for $\lambda = 1$, this signifies that the character -istic polynomial and the minimal polynomial of M are equal (up to the sign).

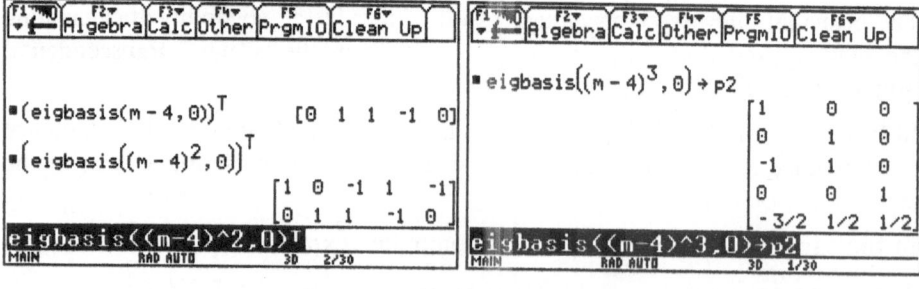

By juxtaposing the two matrices P_1 and P_2 obtained before, we construct a transition matrix P to a basis of $\mathbb{R}^4$ adopted from the direct sum of $\mathbb{R}^4$ into the two spectral subspaces. Of course, we observe that M is similar to a block diagonal matrix B with this change of basis. (We use the letter B to retain the notation at the beginning of this section.)

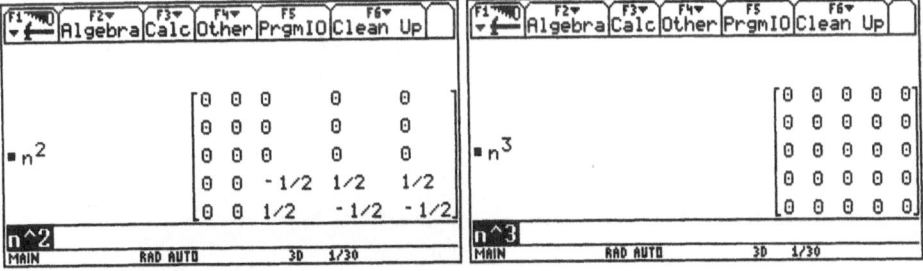

The matrix B decomposes into $B = D + N$ (D diagonal and N nilpotent):

The fact that N is nilpotent is confirmed by the following calculations.
The two blocks are nilpotent with respective indices 2 and 3 which agrees with
the observations made earlier about the spectral subspaces.

$$\begin{cases} B = P^{-1}MP \\ B = D + N \end{cases} \Rightarrow M = PBP^{-1} = P(D+N)P^{-1} = PDP^{-1} + PNP^{-1}.$$

• The matrix $\Delta = PDP^{-1}$ is diagonalizable (since it is similar to D, which is diagonal).

• The matrix $H = PNP^{-1}$ is nilpotent: $H^3 = PN^3P^{-1} = P0P^{-1} = 0$.

We have thus performed the Dunford decomposition of $M = \Delta + H$ of M.

Here is how to calculate the matrices Δ and H.

| F1▼ F2▼ F3▼ F4▼ F5 F6▼ |
| Algebra Calc Other PrgmIO Clean Up |

$\bullet\, p\cdot d\cdot p^{-1}\to\delta$
$$\begin{bmatrix} 1 & -3 & 6 & 3 & -6 \\ 3 & 1 & 3 & 0 & 0 \\ 3 & 3 & -2 & -3 & 6 \\ 3 & -3 & 3 & 4 & 0 \\ 3 & 3 & -6 & -3 & 10 \end{bmatrix}$$

`p*d*p^-1→δ`
MAIN   RAD AUTO   3D   1/30

| F1▼ F2▼ F3▼ F4▼ F5 F6▼ |
| Algebra Calc Other PrgmIO Clean Up |

$\bullet\, p\cdot n\cdot p^{-1}\to h$
$$\begin{bmatrix} -3 & 0 & 1 & 1 & -3 \\ -2 & 1 & -1 & 0 & 0 \\ 3 & 1 & -3 & -2 & 5 \\ -3 & -1 & 3 & 2 & -5 \\ 3 & 0 & -1 & -1 & 3 \end{bmatrix}$$

`p*n*p^-1→h`
MAIN   RAD AUTO   3D   1/30

For each integer r, $M^r = \sum_{k=0}^{2} \binom{r}{k} H^k \Delta^{r-k}$ (we limit the sum to $k = 2$ since after that $H^k = 0$), or again $M^r = PB^rP^{-1} = P\left(\sum_{k=0}^{2} \binom{r}{k} N^k D^{r-k}\right)P^{-1}$.

To calculate the symbolic powers of the diagonal matrix D (whose list of diagonal coefficients has been placed in the variable vp), it suffices to do the exponentiation of the list vp, then to use the diag instruction.

Here is how to define a function pd to calculate the powers of D, then how to calculate the r-th power of M. (The result is too long to be displayed on the calculator screen):

| F1▼ F2▼ F3▼ F4▼ F5 F6▼ |
| Algebra Calc Other PrgmIO Clean Up |

$\bullet\, \text{diag}(ev^r) \to pd(r)$ Done

$\bullet\, pd(r)$
$$\begin{bmatrix} 1 & 0 & 0 & 0 & 0 \\ 0 & 1 & 0 & 0 & 0 \\ 0 & 0 & 4^r & 0 & 0 \\ 0 & 0 & 0 & 4^r & 0 \\ 0 & 0 & 0 & 0 & 4^r \end{bmatrix}$$

`pd(r)`
MAIN   RAD AUTO   3D   2/30

| F1▼ F2▼ F3▼ F4▼ F5 F6▼ |
| Algebra Calc Other PrgmIO Clean Up |

$\bullet\, P\cdot\sum_{k=0}^{2} \text{ncr}(r,k)\cdot n^k\cdot pd(r-k)\cdot p^{-1}$

$$\begin{array}{cc} \dfrac{-r\cdot 4^r}{4} - 2\cdot r + 1 & 1 - 4^r \\ \left[\dfrac{r^2}{32} - \dfrac{r}{32} + 1\right]\cdot 4^r - 2\cdot r - 1 & \dfrac{r\cdot 4^r}{4} + 1 \end{array}$$

`...ncr(r,k)*n^k*pd(r-k),k,0,2)*...`
MAIN   RAD AUTO   3D   1/1

In the same manner,
$$\exp M = \exp H \exp \Delta$$
$$= \left(\sum_{k=0}^{2} \frac{1}{k!}H^k\right)P\exp D P^{-1}$$
and $\exp D$ may be calculated by taking the exponential of the diagonal coefficients.

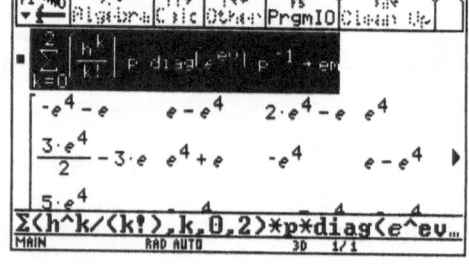

| F1▼ F2▼ F3▼ F4▼ F5 F6▼ |
| Algebra Calc Other PrgmIO Clean Up |

$\bullet\, \sum_{k=0}^{2} \dfrac{h^k}{k!}\cdot P\cdot \text{diag}(e^{ev})\cdot p^{-1} \to en$

$$\begin{array}{ccc} -e^4 - e & e - e^4 & 2\cdot e^4 - e \quad e^4 \\ \dfrac{3\cdot e^4}{2} - 3\cdot e \quad e^4 + e & -e^4 & e - e^4 \\ \dfrac{5\cdot e^4}{} & & \end{array}$$

`Σ(h^k/(k!),k,0,2)*p*diag(e^ev...`
MAIN   RAD AUTO   3D   1/1

We have obtained the *symbolic* form of exponential of M.

We compare with the result furnished by the built-in exponential function (which only works in the "approx " mode).

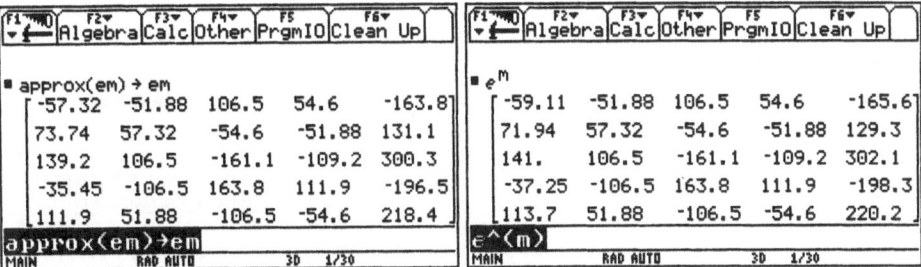

We observe an important difference in the precision in the columns 1 and 5. In fact, our result is correct! No doubt a rounding error in the eigenvalues of the matrix M caused the calculator to regard M as diagonalizable, which it is not.

The preceding calculations motivate us to program a function which calculates the symbolic form of the exponential of a matrix M. The method is copied from the example which we treated.

```
:myexp(m)
:Func
:Local n,s,ss,λ,i,j,t,p,h
:rowDim(m)→n:mzeros(det(m-θθ xx_),θθ xx_)→s
:If dim(s)<n:Return "error":augment(s,{∞})→ss
:For i,1,n: i→j:ss[i]→ λ
:   While when(ss[i+1]=λ,true,false,false):i+1→i:EndWhite
:   (m-λ)^(i-j+1)→t:basisvp(t,0)→t
:   If j=1 Then:t→p:Else:augment(p,t)→p:EndIf
:EndFor
:m-p*diag(s)*p^(-1)→h
:∑(h^j/(j!),j,0,n)*p*diag(e^s)*p^(-1)
:EndFunc
```

The matrix A is diagonalizable (the two eigenvalues are 1 and 4).
The exponential function gives $\exp A$ in approximate format.
Our function myexp calculates $\exp A$ in exact fashion. The conversion to approximate format gives exactly the first result.

The square matrix A of order 3 which follows has two eigenvalues, 1 (simple) and 2 (double).

The fact that $(A - I_3)(A - 2I_3)$ is not the zero matrix shows that A is not diagonalizable.

We calculate the exponential of A with the built-in function (an approximate result), then with our function myexp.

We could verify with a conversion "approx " format that the two results are almost identical.

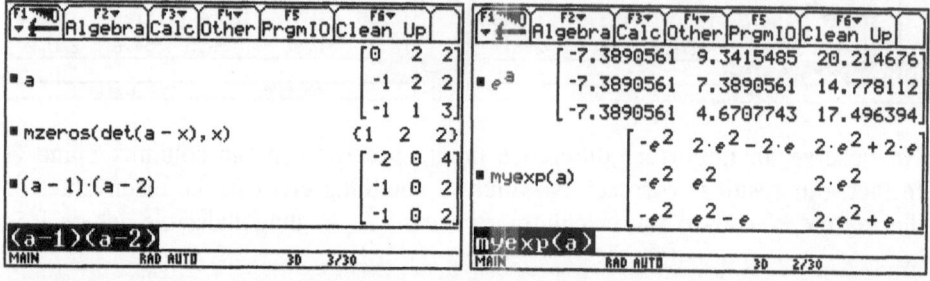

Now here is a non-diagonalizable square matrix A of order 4. (It is upper triangular with diagonal coefficients all equal to 1: if it were diagonalizable, it would be similar to I_4 and thus equal to I_4).

It is so evident that A is not diagonalizable that the built-in exponential function realizes it and refuses to calculate $\exp A$.

On the other hand, for our function myexp, this is not a problem!

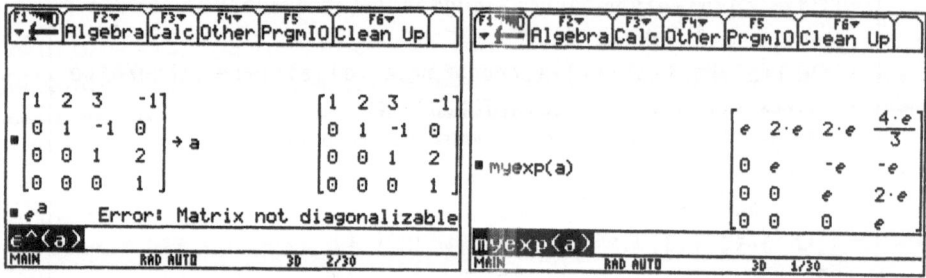

Here again is an example of the services which may be rendered by the function myexp.

The matrix A in fact contains the formal variable, which causes a "Data type" error if we try to calculate $\exp A$ directly.

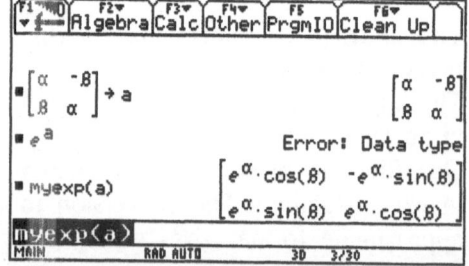

The calculation of the exponential of a matrix turns out to be useful, notably in solution of a system of linear differential equations with constant coefficients $X'(t) = AX(t)$. (Here X is a function with values in $\mathbb{K}^n$). In fact, the general solution of such a system is: $X'(t) = \exp(tA)X_0$, where X_0 is an arbitrary vector of $\mathbb{K}^n$ representing the value of the function X at $t = 0$.

With the same matrix A, here is the solution of the differential system

$$\begin{cases} x'(t) = \alpha x(t) - \beta y(t) \\ y'(t) = \beta x(t) + \alpha y(t) \end{cases}.$$

(x_0, y_0) is the initial position at $t = 0$ of the moving point $M(t) = \big(x(t), y(t)\big)$.

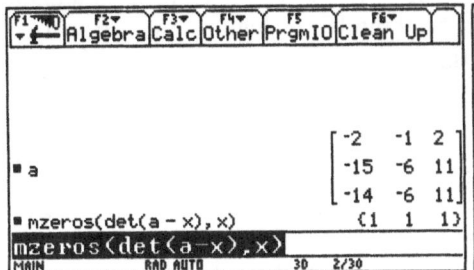

Here is a non-diagonalizable square matrix A of order 3. (It has only one eigenvalue, $\lambda = 1$, of multiplicity 3.) The second screen shows the solution of the differential system $X'(t) = AX(t)$ which passes through $(1, 1, 3)$ at the instant $t = 0$.

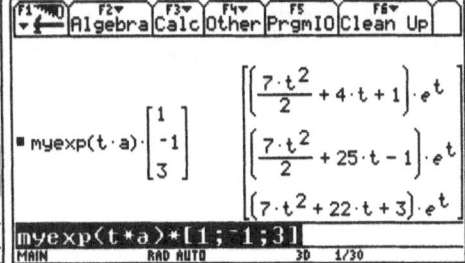

Last application: we may calculate the cosine of a matrix A by taking the real part of $\exp(iA)$ (with the condition that A has real coefficients).
Here is how to calculate, in exact fashion, the cosine fo the matrix $A = \begin{pmatrix} 0 & -1 \\ 1 & 0 \end{pmatrix}$.

The reader may wish to revisit the chapter on differential equations, especially the two cases covered in section **4.2.**

3. Localization of eigenvalues

Up to now we have been privileged to take a theoretical point of view in developing the utilities to calculate in exact fashion the eigenvalues or eigenvectors of reduced diagonal or triangular matrices — and especially of reduced Jordan matrices).

The Dunford decomposition likewise allowed us to calculate in exact fashion the exponential of a square matrix.

All the same we concede that the basis of these utilities is in finding the factorization of the characteristic polynomial by means of the instruction cSolve, and that we have limited our experiences to cases where this symbolic factorization is possible.

How will we do without these rather advanced utilities?

The eigenvalues and eigenvectors are not only of interest to the mathematician (who may be quite content with abstract situations which are a little idealistic) but also to the physicist (who must often face data which is tainted by imprecision) or even to the economist (whose finance methods may use matrices which are too big to use symbolic methods).

The search for eigenvalues and eigenvectors must thus usually be made approximately. The problem is sufficiently important that there is an abundant literature already available. The numeric methods are legion, and we will be interested only in the most classical of them which operate within the limits of our calculator, which will always be called upon to illustrate the methods.

To give honor where honor is due, we are going to begin by considering two built-in instruction of the TI-92+ (or the TI-89).

As for all built-in functions, they furnish rapid and immediately usable results, but without clarifying the methods and the underlying algorithms used.

We could nonetheless use them to test the quality of the methods which we study in what follows.

3. 1 The instructions eigVl and eigVc

The role of these two functions should be apparent from their names:

- The function eigVl calculates an approximation of the eigenvalues of a square matrix A. This must contain only real or complex numbers (in either exact or approximate form) and thus no formal variables. The result is a list in "approx " format.

- The function eigVc returns a matrix whose column vectors form a basis of the direct sum of the different eigenspaces. The conditions imposed on A are the same. The eigenvalues are evidently obtained in the "approx " format, and they are unit vectors.

Here are some examples of use of eigVl on matrices of order 2. Likewise, in the simplest cases (especially the third) the result is given in real format, as indicated by the presence of a decimal point.

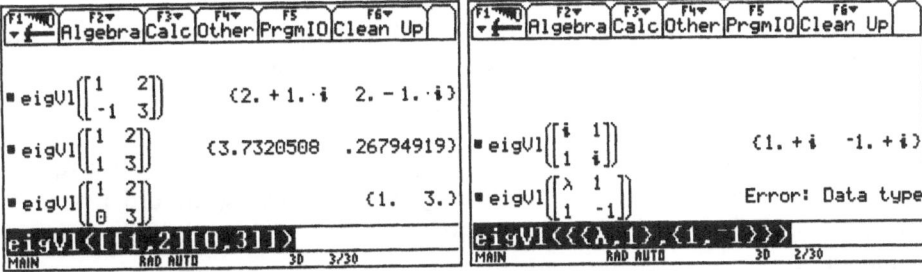

Here now is a square matrix of order 3 which is diagonalizable (three simple eigenvalues: $\alpha = 1$, $\beta = 2$ and $\gamma = 0$).

The function eigVl indeed gives us the eigenvalues in approximate fashion, then eigVc provides a transition matrix P to a basis of eigenvectors.

As has been said, the column vectors are unit vectors, but we see clearly that the eigenspaces are the vector lines generated respectively by the vectors $(1, -1, -1)$, $(0, 1, 1)$, and $(1, -1, 0)$.

The second screen lets us verify that the eigenvectors are given in the same order as that provided by eigVl, and the calculation of $P^{-1}AP$ illustrates the "approximate diagonalization " of A.

Here is an elementary Jordan matrix, constructed with the function ejordan from section **2.5**, which is not diagonalizable (only one eigenvalue of multiplicity 5, and the eigenspace is reduced to a vector line).

The function eigVc returns however a square matrix P of order 5, all of whose columns are equal (up to insignificant roundoff errors). The matrix P is considered by the calculator to be non-invertible.

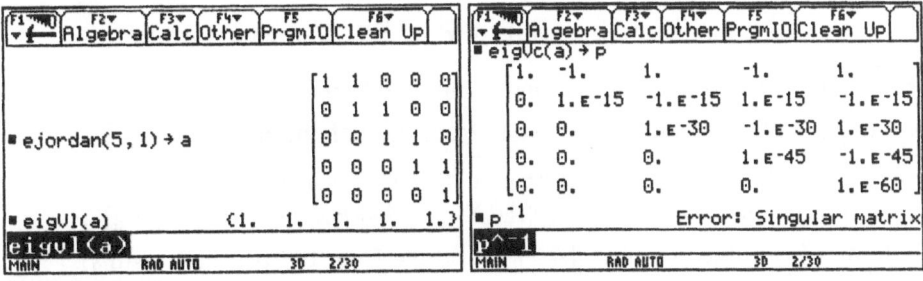

3. 2 Matrix norms and the spectral radius

Let A be a square matrix of order n with coefficients in $\mathbb{K}$. We call the maximum modulus of eigenvalues of A the *spectral radius* of A. We denote this by $\rho(A)$. Thus, $\rho(A) = \sup\limits_{\lambda \in Sp(A)} |\lambda|$.

Before trying to approximate the eigenvalues λ (a priori complex) of a square matrix A, we might more modestly be content with knowing the order of their magnitudes. For example, it is useful to find an upper bound of the spectral radius of A.

We will first recall some classical results. For convenience of notation, we will identify an element X of $\mathbb{K}^n$ with the $\mathcal{M}_{n,1}(\mathbb{K})$ row-matrix and the $\mathcal{M}_{1,n}(\mathbb{K})$ column matrix with which they correspond.

The vector space $\mathbb{K}^n$ may be equipped with a norm in several ways, but certain ones are more in use than others. Thus, for every vector $X = (x_1, x_2, \ldots, x_n)$, we put:

$$\|X\|_1 = \sum_{k=1}^{n} |x_k|, \quad \|X\|_2 = \left(\sum_{k=1}^{n} |x_k|^2 \right)^{1/2}, \quad \|X\|_\infty = \sup_{1 \le k \le n} |x_k|$$

For real $p \ge 1$, we may generalize these definitions by putting:

$$\|X\|_p = \left(\sum_{k=1}^{n} |x_k|^p \right)^{1/p} \text{ (we verify that } \lim_{p \to \infty} \|X\|_p = \|X\|_\infty)$$

For every vector norm $X \mapsto \|X\|$ on $\mathbb{K}^n$, we define a matrix norm $A \mapsto \|A\|$ sur $\mathcal{M}_n(\mathbb{K})$, said to be *subordinate* to the first, in one of the following entirely equivalent ways:

$$\|A\| = \sup_{\|X\|=1} \|AX\| = \sup_{\|X\| \le 1} \|AX\| = \sup_{X \ne 0} \frac{\|AX\|}{\|X\|}$$

In the same manner we define the subordinate norms on $\mathcal{L}(\mathbb{K}^n)$. For practical reasons we will take the matrix approach. The reader will without difficulty carry over the properties which follow to statements about operators.

The third definition shows that for every X in $\mathbb{K}^n$, $\|AX\| \le \|A\| \|X\|$.

One could characterize a matrix norm subordinate by:

$$\forall A \in \mathcal{M}_n(\mathbb{K}), \|A\| = \min\{\mu \in \mathbb{R}^+, \forall X \in \mathbb{K}^n, \|AX\| \le \mu \|X\|\}$$

Here are some properties of a matrix norm subordinate.

- $\|I\|_n = 1$ (This results immediately from the definition.)
- $\forall A, B \in \mathcal{M}_n(\mathbb{K}), \|AB\| \le \|A\| \|B\|$, and $\forall k \in \mathbb{N}, \|A^k\| \le \|A\|^k$.
  (a consequence of: $\forall X \in \mathbb{K}^n, \|ABX\| \le \|A\| \|BX\| \le \|A\| \|B\| \|X\|$).
- If A is invertible, $1 \le \|A\| \|A^{-1}\|$ (resulting from the two first results).

- There is a unit vector X such that $\|AX\| = \|A\|$: this is a consequence of the first definition of $\|A\|$, and of the continuity of the map $X \mapsto AX$ on the unit sphere S_1, which is compact since it is of finite dimension.

Remark:

There is another classical norm on $\mathcal{M}_n(\mathbb{K})$, called the *Schur norm* (or the *Frobenius norm*), which is defined by :

$$\|A\|_s = \left(\mathrm{Tr}(^\top \overline{A}A)\right)^{1/2} = \left(\sum_{i,j=1}^n |a_{i,j}|^2\right)^{1/2}$$

(denoting by $a_{i,j}$ the general coefficient of A)

This quantity likewise represents the Euclidean norm of A if we identify A with an element of $\mathbb{K}^{n^2}$. We observe that $\|I_n\|_s = \sqrt{n}$. This result proves that the Schur norm is not subordinate to any vector norm. The function norm of the calculator calculates the Schur norm of a matrix.

Classical examples

To each of the "usual " norms on $\mathbb{K}^n$ corresponds a matrix norm subordinate.

Let $A = (a_{i,j})$ be an arbitrary matrix of $\mathcal{M}_n(\mathbb{K})$ with $\|A\|_1$, $\|A\|_2$, $\|A\|_\infty$ the subordinate norms for the three classical norms on $\mathbb{K}^n$.

We denote by $L_i = \sum_{j=1}^n |a_{i,j}|$ and $C_j = \sum_{i=1}^n |a_{i,j}|$ the sum of the moduli of row i (respectively of column j) of A.

With this notation, one proves the following results:

- $\|A\|_1 = \max_{1\le j\le n} C_j$. We call this the *column norm* of A.

  It corresponds to the function colNorm of the calculator.

- $\|A\|_\infty = \max_{1\le i\le n} L_i$. We call this the *row norm* of A.

  It corresponds to the function rowNorm of the calculator.

- $\|A\|_2 = \sqrt{\rho(^\top \overline{A}A)}$ (square root of the spectral radius of $B = {}^\top \overline{A}A$).

  If the matrix A commutes with its adjoint $^\top \overline{A}$, notably if A is symmetric with real coefficients, then $\|A\|_2 = \rho(A)$.

The function norm calculates the Schur norm of a square matrix.

The *underscore* character allows us to specify that a formal variable represents a complex number, and we see the consequences on $\|A\|_s$.

Finally, rowNorm and colNorm don't accept formal matrices.

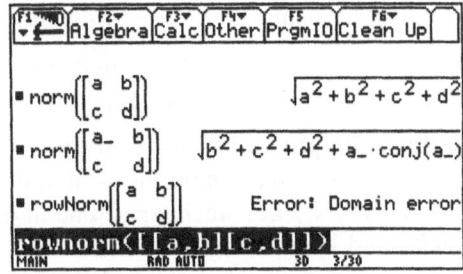

We see here the calculation of the three classical matrix norms for a matrix A of order 3.

- $\|A\|_\infty = 20$ (second row of A)
- $\|A\|_1 = 14$ (third column of A)
- $\|A\|_2 \simeq 13.77$
- $\|A\|_s = \sqrt{237}$

Here is how a transition matrix allows modification of a vector norm and what results for the associated matrix norm subordinate.

Proposition 1: Let $X \mapsto \|X\|$ be a norm on $\mathbb{K}^n$, and let $A \mapsto \|A\|$ be the matrix norm subordinate. Let Q be an invertible square matrix of order n.
We define a norm on $\mathbb{K}^n$ by putting $\|X\|_Q = \|QX\|$.
The matrix norm subordinate is then defined by $\|A\|_Q = \|QAQ^{-1}\|$.

Proof: The fact that $X \mapsto \|X\|_Q = \|QX\|$ is a norm on $\mathbb{K}^n$ is evident. For every matrix A of $\mathcal{M}_n(\mathbb{K})$, and by putting $X = Q^{-1}Y$ in $\mathbb{K}^n$:

$$\|A\|_Q = \sup_{X\neq 0} \frac{\|AX\|_Q}{\|X\|_Q} = \sup_{X\neq 0} \frac{\|QAX\|}{\|QX\|} = \sup_{Y\neq 0} \frac{\|(QAQ^{-1})Y\|}{\|Y\|} = \|QAQ^{-1}\|$$

Here is the first localization of the eigenvalues of a matrix A.

Proposition 2: For every matrix A of $\mathcal{M}_n(\mathbb{K})$, and for every matrix norm subordinate, we have the inequality $\rho(A) \leq \|A\|$.

Proof: Let λ be an eigenvalue of A, and let Y be an associated eigenvector. The equation $AY = \lambda Y$ gives $\|AY\| = |\lambda|\,\|Y\|$, then $\lambda = \frac{\|AY\|}{\|Y\|} \leq \sup_{X\neq 0} \frac{\|AX\|}{\|X\|}$.
We may conclude that $\forall \lambda \in \mathrm{Sp}(f)$, $|\lambda| \leq \|A\|$, and finally that $\rho(A) \leq \|A\|$.

Now we will see that for every matrix A, we may find a matrix norm subordinate which approximates very precisely the spectral radius of A.

Proposition 3: For every matrix A of $\mathcal{M}_n(\mathbb{K})$ (with character -istic polynomial splitting over $\mathbb{K} = \mathbb{R}$) and for every $\varepsilon > 0$, there exists a matrix norm subordinate on $\mathcal{M}_n(\mathbb{K})$ such that $\|A\| \leq \rho(A) + \varepsilon$.

Proof: The demonstation uses the Jordan decomposition of A.

Let $\varepsilon > 0$. We introduce the diagonal matrix $D = \mathrm{diag}(1, \varepsilon, \varepsilon^2, \ldots, \varepsilon^{n-1})$.

If $B = (b_{i,j}) \in \mathcal{M}_n(\mathbb{K})$, the general term of $C = D^{-1}BD$ is $c_{i,j} = \varepsilon^{j-i}b_{i,j}$.

In fact, the product by D^{-1} on the left divides row i by ε^{i-1}, and the product by D on the right multiplies column j by ε^{j-1}. In particular, the diagonal coefficients $b_{i,i}$ are unchanged and those immediately above the diagonal are multiplied by ε.

Let P be a transition matrix to the Jordan reduction J of A: $J = P^{-1}AP$ (see paragraph **2.6**), and let $K = D^{-1}JD = (PD)^{-1}A(PD)$. The matrices A and K are similar. They thus have the same eigenvalues and also the same spectral radius: $\rho(A) = \rho(K)$.

On the other hand, since J is a Jordan matrix, $K = D^{-1}JD$ is only distinguished from J by the non-zero coefficients just above the diagonal, and which have value ε instead of 1. In particular, the diagonal coefficients of K are those of J, that is, they are the eigenvalues of A.

Thus: $\|K\|_\infty = \max\{|\lambda| + \alpha, \lambda \in \mathrm{Sp}(A), \alpha \in \{0, \varepsilon\}\} \le \rho(A) + \varepsilon$.

We put $Q = PD$, and use the matrix norm $M \mapsto N(M) = \|QMQ^{-1}\|_\infty$, which we know is subordinate to the vector norm $X \mapsto \|QX\|_\infty$.

By definition of K, $N(A) = \|QAQ^{-1}\|_\infty = \|K\|_\infty \le \rho(A) + \varepsilon$.

Now we are going to see the importance of the spectral radius in the behavior of the sequence of successive powers of a matrix A.

We know that a geometric sequence $(q^n)_{n \ge 0}$ with ratio q converges to 0 if and only if $|q| < 1$. The following result shows that, in work with successive powers of A, the spectral radius plays the same role as q.

Proposition 4: Let $A \in \mathcal{M}_n(\mathbb{C})$. We have the equivalence $\lim\limits_{k \to \infty} A^k = 0 \Leftrightarrow \rho(A) < 1$.

Proof: First, we note that the norm used when we say that $\lim A^k = 0$ is of no particular importance: all the norms on $\mathcal{M}_n(\mathbb{C})$ are equivalent. It thus suffices that this be true for one of them, for example, for a matrix norm subordinate).

We suppose that $\lim\limits_{k \to \infty} A^k = 0$.

Let λ be an eigenvalue of A, and let X be an associated eigenvector. For every integer k, $A^k X = \lambda^k X$. The hypothesis implies $\lim\limits_{k \to \infty} A^k X = 0$ and thus $\lim\limits_{k \to \infty} \lambda^k = 0$, that is, $|\lambda| < 1$. We conclude that $\rho(A) < 1$.

Conversely, we suppose $\rho(A) < 1$.

By choosing ε "very small", we may find a matrix norm subordinate such that $\|A\| \le \rho(A) + \varepsilon < 1$.

We may conclude that $\lim\limits_{k \to \infty} \|A\|^k = 0$ and then $\lim\limits_{k \to \infty} A^k = 0$ (since $\|A^k\| \le \|A\|^k$).

Proposition 5: For every matrix A of $\mathcal{M}_n(\mathbb{C})$, for every matrix norm subordinate, and for every integer k, we have $\rho(A) \le \|A^k\|^{1/k}$ and $\rho(A) = \lim\limits_{k \to \infty} \|A^k\|^{1/k}$.

Proof: We know that the eigenvalues of A^k are the λ^k, where $\lambda \in \mathrm{Sp}(A)$.

Consequently: $\rho(A)^k = \rho(A^k) \le \|A^k\|$, then $\rho(A) \le \|A^k\|^{1/k}$.

Let $\varepsilon > 0$ and $B = \dfrac{1}{\rho(A) + \varepsilon} A$. We have $\rho(B) = \dfrac{1}{\rho(A) + \varepsilon} \rho(A) < 1$, thus $\lim\limits_{k \to \infty} B^k = 0$.

In particular: $\exists m \in \mathbb{N}, \forall k \ge m, \|B^k\| \le 1$, that is, $\|A^k\| \le (\rho(A) + \varepsilon)^k$.

We conclude: $\forall k \geq m$, $\rho(A) \leq \|A^k\|^{1/k} \leq \rho(A) + \varepsilon$ then $\rho(A) = \lim_{k \to \infty} \|A^k\|^{1/k}$.

Remark:

Contrary to the inequality $\rho(A) \leq \|A^k\|^{1/k}$, the equation $\rho(A) = \lim_{k \to \infty} \|A^k\|^{1/k}$ is true for every matrix norm, and not only those which are subordinate to vector norms. This is a consequence of the equivalence of norms in a vector space of finite dimension).

We thus recall the preceding results that, for every matrix norm subordinate, the quantity $\|A\|$ is an upper bound of the modulus of all the eigenvectors of A, and that an even better bound is given by $\|A^k\|^{1/k}$ (and the more so for larger k).

In the example below, the matrix A has four real eigenvalues and its spectral radius is roughly equal to 12.25. We have calculated the quantities $\|A^k\|^{1/k}$, for $1 \leq k \leq 5$ and for the column and row norms. The results confirm what is called convergence "from above " (but not very rapidly) to the spectral radius.

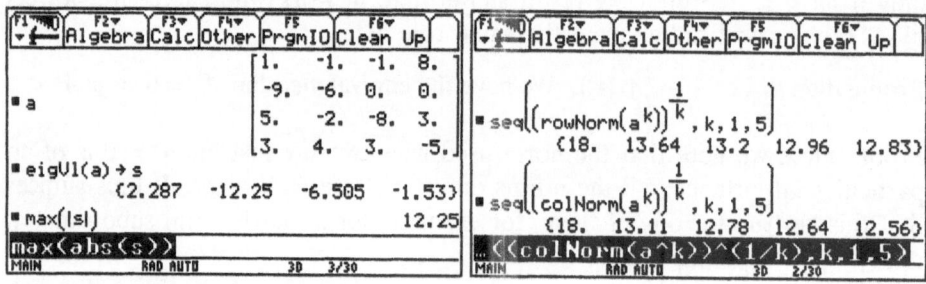

3. 3 Gershgorin Disks

Now we are going to refine the localization of the eigenvalues of a square matrix $A = (a_{i,j})$ from $\mathcal{M}_n(\mathbb{K})$.

We put: $\forall i \in \{1,\ldots,n\}$, $D_i = \{z \in \mathbb{C}, |z - a_{i,i}| \leq \sum_{j \neq i} |a_{i,j}|\}$.

Similarly, $\forall j \in \{1,\ldots,n\}$, $\Delta_j = \{z \in \mathbb{C}, |z - a_{j,j}| \leq \sum_{i \neq j} |a_{i,j}|\}$.

Definition 1: The disks D_i and Δ_j are called the Gershgorin disks of A.

Proposition 6: With the preceding notation, every eigenvalue is A is located in the union of the $(D_i)_{1 \leq i \leq n}$, and in that of the $(\Delta_j)_{1 \leq j \leq n}$.

Proof: Let λ be an eigenvalue of A, and let $X = (x_1, x_2, \ldots, x_n)$ be the associated eigenvector. Let i be an index such that $\|X\|_\infty = |x_i| > 0$.

We project the inequality $AX = \lambda X$ on the coordinate with index i:

$$\sum_{j=1}^{n} a_{i,j} x_j = \lambda x_i \Rightarrow (\lambda - a_{i,i}) x_i = \sum_{j \neq i} a_{i,j} x_j \Rightarrow |\lambda - a_{i,i}| \, |x_i| \leq \sum_{j \neq i} |a_{i,j} x_j|$$

$$\Rightarrow |\lambda - a_{i,i}| \, |x_i| \leq \left(\sum_{j \neq i} |a_{i,j}| \right) |x_i| \Rightarrow |\lambda - a_{i,i}| \leq \sum_{j \neq i} |a_{i,j}| \Rightarrow \lambda \in D_i$$

The eigenvalues of $^{\top}A$ are those of A. The preceding calculations thus show that λ is in one of the "D_j" of $^{\top}A$. Hence the disks D_j of $^{\top}A$ are the disks Δ_j of A. This finishes the proof.

Each D_i is contained in the disk with center O and with radius $\sum_{j=1}^{n} |a_{i,j}|$.

The union of the D_i is thus contained in the disk with center O and with radius $\|A\|_{\infty} = \max\limits_{1 \leq i \leq n} \sum_{j=1}^{n} |a_{i,j}|$. We thus see that the spectral radius $\rho(A)$ of A is less than or equal to $\|A\|_{\infty}$, but the precision obtained is much greater.

The program gersh graphs for a square matrix A the disk D_{∞} with center O and radius $\|A\|_{\infty}$ (we know that this contains all eigenvalues), then all the disks D_i of A.

Using "ZoomSqr" and centering the graph window on the origin, is chosen to best display the disk D_{∞}.

```
:gersh(a)
:Prgm:Local n,i,r,c
:rowDim(a)→n:rowNorm(a)→r
:ZoomStd:0→xscl:0→yscl
:-r→ymin:r→ymax:ZoomSqr:Circle 0,0,r
:For i,1,n: a[i,i]→c
:   Σ(abs(a[i,j]),j,1,n)-abs(c)→r
:   Circle real(c),imag(c),r
:EndFor
:EndPrgm
```

Here is an example using the program gersh with a square matrix A of order 3. We clearly see that the circle with center 0 and radius $\|A\|_{\infty}$ contains the three disks D_i of Gershgorin.

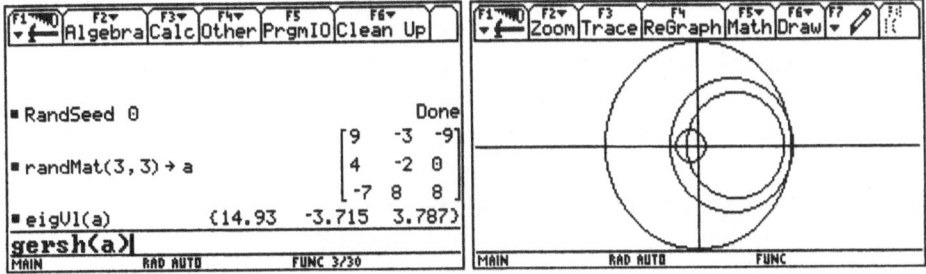

For a diagonal matrix A, the radius of the Gershgorin disks is zero!

If A is approximately a diagonal matrix, we may thus hope that use of the disks D_i will allow good separation of the eigenvalues.

Here is an example illustrating this situation. We see that the Gershgorin disks of A are pairwise disjoint.

Each of them thus contains an eigenvalue and only one. (In fact, one may show that when the union of k of the disks D_i is disjoint from that of the $n-k$ others, then it contains exactly k eigenvalues of A).

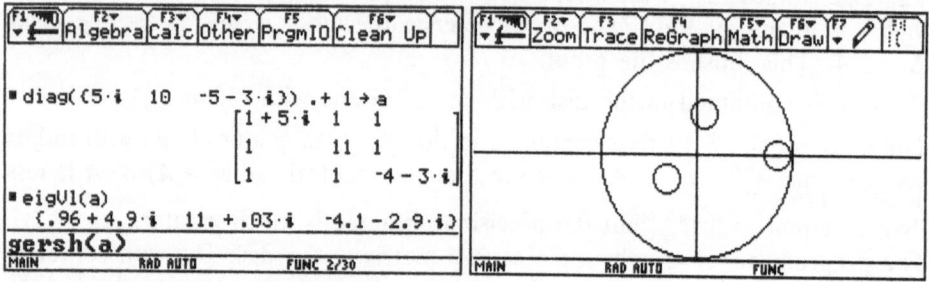

4. Power methods of finding eigenvalues

The preceding section allows us to locate rather grossly the eigenvalues of a square matrix M. But the bound of the spectral radius (by any subordinate matrix norm) is of mainly theoretical interest, and use of the Gershgorin disks doesn't furnish a very precise result. Even though certain improvements may be made, we won't develop them here.

We must thus tackle more effective methods, if we want to approximate more accurately at least one eigenvalue of M.

4. 1 Direct iterations (favorable case)

Let M be a matrix of $\mathcal{M}_n(\mathbb{K})$ (with $\mathbb{K} = \mathbb{R}$ or $\mathbb{C}$) representing an operator in the standard basis. We suppose (if $\mathbb{K} = \mathbb{R}$) that the character -istic polynomial of M splits. We recall that $\rho(M)$ (the spectral radius of M) is the maximum modulus of eigenvalues of M: $\rho(M) = \sup_{\mu \in \mathrm{Sp}(M)} |\mu|$.

In order that the method that we are going to present will be effective, we are going to add two additional hypotheses (which may appear strange since we know nothing a priori about the eigenvalues of M). These hypotheses constitute what we call the "favorable case ":

- The first is is that there exists a unique eigenvalue λ such that $\rho(M) = |\lambda|$. We say then that λ is a *dominant eigenvalue* of M.

- We will suppose in addition that the dimension $d(\lambda)$ of the eigenspace $E(\lambda)$ is equal to the multiplicity $m(\lambda)$ of this eigenvalue.

This second hypothesis is automatically realized if $m(\lambda) = 1$, that is, if λ is a simple eigenvalue – one with multiplicity 1. More generally, this signifies that the characteristic subspace $S(\lambda)$ associated with λ (see **2.7**) is equal to the eigenspace $E(\lambda)$ (then also that $E(\lambda) \subset S(\lambda)$).

The decomposition of $\mathbb{K}^n$ is a direct sum of different characteristic sub-spaces of f which is here written: $\mathbb{K}^n = E(\lambda) \underset{\mu \in \mathrm{Sp}(f), \mu \neq \lambda}{\bigoplus} S(\mu)$.

To simplify, we denote by G the direct sum of all the characteristic subspaces $S(\mu)$ of f, with $\mu \neq \lambda$. We of course assume that G does not reduce to $\{0\}$. (Otherwise, $\mathbb{K}^n = E(\lambda)$, $f = \lambda Id$, and $M = \lambda I_n$: to find the eigenvalues of M is then not very interesting).

The preceding hypotheses thus lead us to $\mathbb{K}^n = E(\lambda) \oplus G$, decomposition of $\mathbb{K}^n$ into two subspaces invariant under f such that:

• The restriction of f to $E(\lambda)$ is the dilation with respect to λ.

• The eigenvalues of the restriction g of f to G are those of f which are distinct from λ. In particular, $\rho(g) = \max\{|\mu|, \mu \in \mathrm{Sp}(f), \mu \neq \lambda\}$ is the second largest modulus among the different eigenvalues of f. It is strictly less than $\rho(f)$, that is, to $|\lambda|$.

Let u be an arbitrary vector of $\mathbb{K}^n$, and let $u = x + y$ be its decomposition under the direct sum $\mathbb{K}^n = E(\lambda) \oplus G$.

In general, the vector u is chosen at random. We thus suppose that it is neither an element of G (thus $x \neq 0$) nor an element of the eigenspace $E(\lambda)$ (thus $y \neq 0$). The component y is thus not an element of $E(\lambda)$ which leads to $g(y) \neq \lambda y$ (a useful fact a little later on).

We then define a sequence $(u_k)_{k \geq 0}$ of $\mathbb{K}^n$ by putting: $\forall k \in \mathbb{N}, u_k = f^k(u)$. The vectors u_k are the iterates of u ($u_0 = u$, $u_1 = f(u)$, $u_2 = f \circ f(u)$, etc.) In terms of matrices, if we denote by U_k and U the associated column vectors to u_k and u, we may thus write: $U_k = M^k U$.

The operator $h = \dfrac{1}{\lambda} g$ of G has for its spectral radius $\rho(h) = \dfrac{\rho(g)}{\lambda} = \dfrac{\rho(g)}{\rho(f)} < 1$.

We conclude that for every vector y of G, $\lim\limits_{k \to \infty} h^k(y) = \overrightarrow{0}$ (cf. **3.2**).

To be more precise, let $w \to \|w\|$ be a norm on $\mathbb{K}^n$ and let $\varphi \to \|\varphi\|$ be the subordinate norm on $\mathcal{L}(\mathbb{K}^n)$. We know that $\rho(h) = \lim\limits_{k \to \infty} \|h^k\|^{1/k}$.

In particular: $\forall r \in]\rho(h), 1[, \exists m \in \mathbb{N}$, such that $k \geq m \Rightarrow \|h^k\|^{1/k} \leq r \Rightarrow \|h^k\| \leq r^k$.

Under these conditions: $k \geq m \Rightarrow \|h^k(y)\| \leq \|h^k\| \|y\| \leq r^k \|y\|$.

The convergence of $h^k(y)$ to $\overrightarrow{0}$ is thus of exponential type.

Thus, with the preceding notation, and for every integer k:

$$u_k = f^k(x+y) = f^k(x) + g^k(y) = \lambda^k x + g^k(y) = \lambda^k\left(x + h^k(y)\right) \sim \lambda^k x$$

We see that the direction of the vector u_k gets closer to that of the vector x as k increases, that is, to that of an eigenvector of f for the dominant eigenvalue λ.

To arrive numerically at the vector x with good precision might take a large number of iterations. The game in numerical work thus risks doing damage by exceeding the calculator's capacity for the real numbers (overflow if $|\lambda| > 1$ and underflow if $|\lambda| < 1$).

To counter this type of risk, we normalize the vectors u_k at each step.

Thus, for every integer k we put: $v_k = \dfrac{1}{\|u_k\|} u_k$.

Since the vectors u_j are defined by the recurrence relation $u_{j+1} = f(u_j)$, the linearity of f means that, to calculate the vector v_k, everything happens as if we normalized u_k at the last step.

We may thus write, for every integer k:

$$v_k = \frac{1}{\|u_k\|} u_k = \left(\frac{\lambda}{|\lambda|}\right)^k \frac{1}{\|x + h^k(y)\|} (x + h^k(y)) \sim \left(\frac{\lambda}{|\lambda|}\right)^k \frac{1}{\|x\|} x$$

The preceding calculation shows that the sequence $(v_k)_{k \geq 0}$ is not convergent unless $\lambda > 0$, but this matters little since only the directions of the v_k matter. It is more interesting on the other hand to note that, since the v_k tend to become collinear with the eigenvector x (for each, one has $f(x) = \lambda x$), the vectors $f(v_k)$ and λv_k must also approach them.

More precisely, for every integer k:

$$f(v_k) - \lambda v_k = \frac{1}{\|u_k\|} \big(f(u_k) - \lambda u_k\big) = \frac{1}{\|u_k\|} (u_{k+1} - \lambda u_k)$$

$$= \frac{1}{\|u_k\|} \left[\lambda^{k+1}(x + h^{k+1}(y)) - \lambda^{k+1}(x + h^k(y))\right] = \frac{\lambda^{k+1}}{\|u_k\|} h^k\big(h(y) - y\big)$$

We conclude that, by putting $z = h(y) - y = \dfrac{1}{\lambda}(g(y) - \lambda y) \neq 0$:

$$\forall k \in \mathbb{N}, \ \|f(v_k) - \lambda v_k\| = \frac{|\lambda| \, \|h^k(z)\|}{\|x + h^k(y)\|} \sim \frac{|\lambda| \, \|h^k(z)\|}{\|x\|}$$

Since the spectral radius of h is less than 1, it follows that: $\displaystyle\lim_{k \to \infty} \|f(v_k) - \lambda v_k\| = 0$.

To normalize the vectors u_k, we may use any vector norm of $\mathbb{K}^n$, for example, $w \to \|w\|_1$ or the Euclidean norm $w \to \|w\|_2$.

We will choose the last one since it permits retrieving the eigenvalue λ by forming the scalar product of the vectors v_k and $f(v_k)$.

In fact, the vectors v_k are unit vectors:

$$|< v_k, f(v_k) > -\lambda| = |< v_k, f(v_k) - \lambda v_k >| \leq \|f(v_k) - \lambda v_k\|$$
$$\Rightarrow \lim_{k \to \infty} \ < v_k, f(v_k) > = \lambda$$

The speed of convergence of this sequence to λ is a direct function of that of the sequence $\big(f(v_k) - v_k\big)$ to 0, which depends geometrically on the spectral radius of h as we saw before.

The convergence is thus even more rapid when $\rho(h)$ is much less than 1, that is, when the dominant eigenvalue λ of f is much larger, in modulus, than the other eigenvalues of f.

The program iterdir will help us to illustrate the direct power method. It takes a matrix M as argument, chooses a vector u randomly, then displays the successive vectors v_k (using the preceding notation).

The scalar products $< v_k, f(v_k) >$ follow, which are supposed to converge to the dominant eigenvalue λ.

```
:iterdir(m)
:Prgm:Local v,u,λ,k:0→k:ClrIO
:approx(randMat(rowDim(m),1))→u
:Loop
: k+1→k:unitV(u)→v
: m*v→u:dotP(u,v)→ λ
: Disp "v["&string(k)&"]="&string(vᵀ)
: Pause "approx(λ)="&string(λ)
:EndLoop
:EndPrgm
```

Here is how to construct a good example

- We form a "simple " transition matrix P (with integer coefficients, with a determinant equal to ± 1 for which P^{-1} also has integer coefficients).

- We choose the eigenvalues (here $10, 2, 1$), then the reduced diagonal D.

- We next create the matrix $M = PDP^{-1}$.

In this example, M is thus diagonalizable, its dominant eigenvalue is $\lambda = 10$, and an associated eigenvector is $x = (1, 1, 3)$.

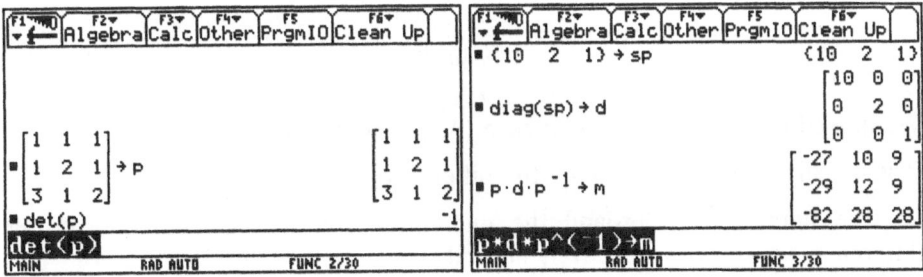

We next call the program iterdir to evaluate iterdir(m). We see that the eigenvalue $\lambda = 10$ is obtained after a dozen iterations.

At the same time, the sequence of vectors v_k converges to a unit vector which is clearly proportional to $x = (1, 1, 3)$.

But note: this convergence is here due to the fact that λ is a real positive number.

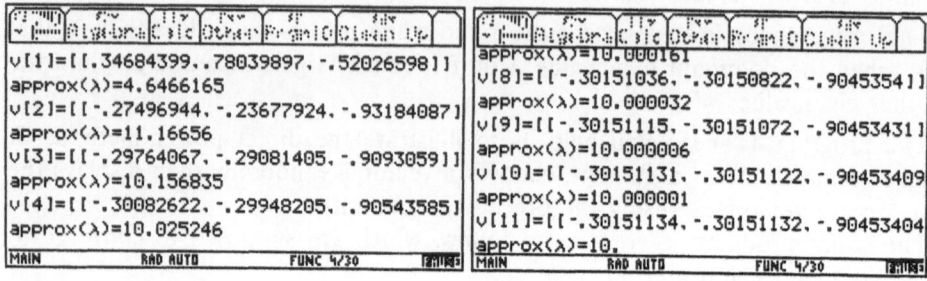

Now here is how to illustrate the method for symbolic calculation:

- We form the matrix $M^k = PD^kP^{-1}$, D^k, using the instruction diag applied to the list of k-th powers of eigenvalues.

- We create the vector $u_k = f^k(u)$ starting from a vector u chosen at random (here the vector $u = (3, -1, 7)$).

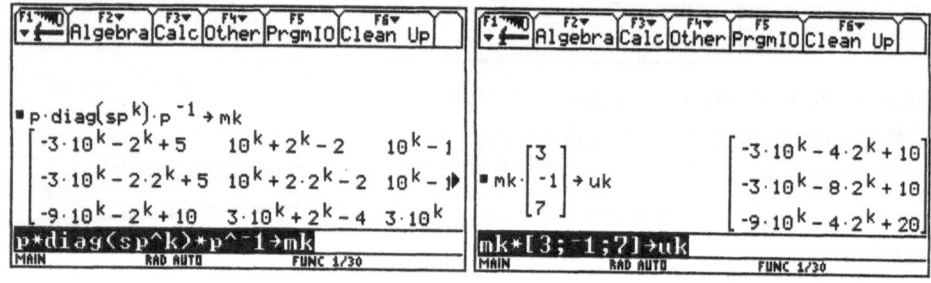

We next form the sequence of scalar products:

$$\lambda_k = <v_k, f(v_k)> = \frac{<u_k, f(u_k)>}{\|u_k\|^2} = \frac{<u_k, f(u_k)>}{<u_k, u_k>}$$

We indeed observe that the sequence $(\lambda_k)_{k\geq 0}$ converges to $\lambda = 10$.

It is equally interesting to study the difference $\lambda_k - 10$.

We see here how to expand the numerator and the denominator of this difference, which are respectively equivalent to $-576 \cdot 4^k \cdot 5^k$ and to $99 \cdot 4^k \cdot 52^k$, as k tends to $+\infty$.

The last result is confirmed by showing that $\lambda_k - 10 \sim -\frac{64}{11}5^{-k}$ when $k \to \infty$.

Taking the notation which served in the demonstration, the factor $1/5$ which appears in the equivalent of $\lambda - 10$ is none other than the spectral radius of the operator h. This is the ratio between 2, the second largest eigenvalue of f in modulus, and the dominant eigenvalue, 10.

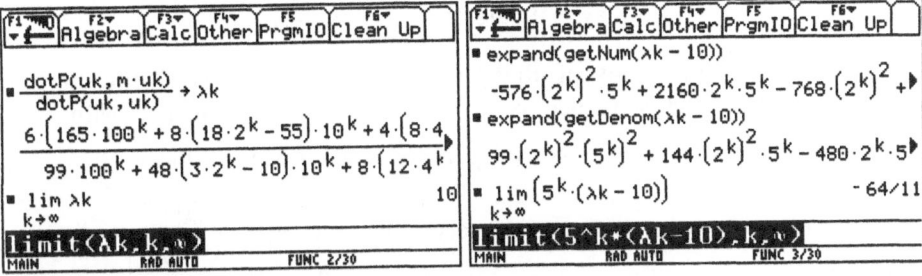

Now we resume use of the program iterdir with a matrix M with eigenvalues $10, 9, 1$ (the diagonal matrix P staying the same as in the first example, and with it the eigenvectors of M).

The novelty here is that the dominant eigenvalue 10 has approximately the same modulus as the second one. The convergence of the sequence of vectors v_k and of the sequence of scalar products λ_k is thus much slower.

We confirm this by observing that at the one hundredth iteration only 5 decimals have stabilized. The convergence is still geometric, but with the ratio $q = 9/10$, instead of $q = 2/10$ in the first example.

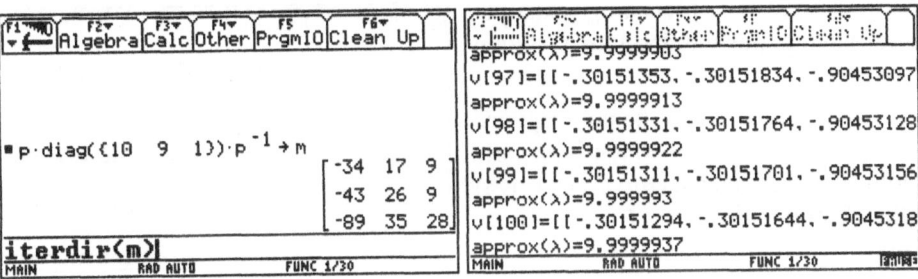

We modify the preceding example to a new one by giving to the matrix M the eigenvalues $10 + i$, 2, and 1.

The eigenvalue $\lambda = 10 + i$ is well separated from the two others, but it is not real and positive: the sequence of vectors v_k is thus no longer convergent. On the other hand, the convergence of the sequence $(\lambda_k)_{k \geq 0}$ is once more very rapid.

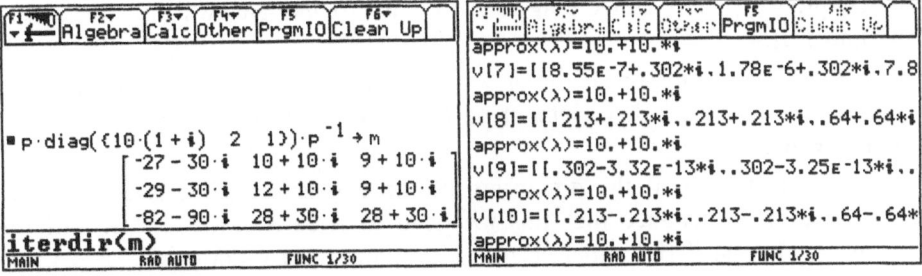

The theoretical study has shown that the vectors v_k satisfy: $v_k \sim \left(\dfrac{\lambda}{|\lambda|}\right)^k \dfrac{1}{\|x\|} x$.
When k tends to ∞, we may estimate that $v_{k+1} \approx \dfrac{\lambda}{|\lambda|} v_k$.

To take up the conditions of the preceding example, it is significant that the term (v_k) of the sequence is that of a geometric sequence with ratio $\dfrac{\lambda}{|\lambda|} = \dfrac{1+i}{|1+i|} = \exp\dfrac{i\pi}{4}$, and thus as a periodic sequence of period 8.

To reconsider when the sequence converges, we put: $w_k = \left(\dfrac{|\lambda|}{\lambda}\right)^k v_k$.

Indeed, since the exact value of λ being unknown at the moment when we define the v_k, we will replace it by the approximate value of $\lambda_k = \, < v_k, f(v_k) >$.
Then in the program iterdir, it suffices to replace the row

```
:Disp "v["&string(k)&"]="&string(vᵀ)
```

by the row

```
:v*(abs(λ)/λ)^k→v
```

We resume the preceding example after having thus modified the program iterdir.

Of course, we observe the convergence of the (v_k) to a vector which is proportional to the eigenvector $x = (1, 1, 3)$ (the display is in mode Float 2 to be more readable).

```
v[1]=[[ -.5-.48*i..5+.48*i..14+.14*i ]]
approx(λ)=18.+18.*i
v[2]=[[.24+.19*i..24+.21*i..71+.56*i ]]
approx(λ)=11.+10.*i
v[3]=[[.21+.21*i..21+.21*i..64+.64*i ]]
approx(λ)=10.+10.*i
v[4]=[[.21+.21*i..21+.21*i..64+.64*i ]]
approx(λ)=10.+10.*i
```
MAIN   RAD AUTO   FUNC 0/30

Up to now, we have treated the example where the dominant eigenvalue λ is simple, of multiplicity 1. The eigenspace $E(\lambda)$ is then a vector line. The study of the vectors v_k suffices to determine it entirely.

It is going to be different when the multiplicity of λ is greater than 1.

If we retain the notation used in the proof, the choice of an initial vector u (which decomposed into $u = x + y$ under the direct sum $\mathbb{K}^n = E(\lambda) \oplus G$) leads to a sequence $(v_k)_{k \geq 0}$ converging to x (this is just like $\mathbb{K}v_k$ which converges to the line $\mathbb{K}x$).

If we begin with another vector, $u' = x' + y'$, the sequence v'_k will lead to the component x', some other element of $E(\lambda)$, but a priori linearly independent of the vector x.

More generally, and if the dimension of the eigenspace $E(\lambda)$ is equal to m, there is a strong chance that this will lead to a basis $(x_1, x_2, \ldots, x_m)$ of the subspace, at least if the initial vectors u are chosen at random.

Here is a function iterdir2 (a simplified version of the program iterdir) which will allow us to illustrate this situation. The syntax is iterdir2(m,n), where M is a square matrix and n is the number of iterations to be done. The result is the list formed by, first, the approximation λ_k of the eigenvalue λ, then the components of the last v_k obtained.

```
:iterdir2(m,n):Func:Local u,v,λ,k:→k
:approx(randMat(rowDim(m),1))→u
:For k,1,n:unitV(u)→v:m*v→u:dotP(u,v)→ λ:EndFor
:mat▷list(unitV(u))→v:augment({λ},v)
:EndFunc
```

With a diagonalizable matrix M with spectrum $10, 10, 1$, three calls of the function iterdir2 give three results $a_k = \{\lambda_k = 10, v_k\}$.

Using ref shows in fact that v_1, v_2 and v_3 are in the same plane, the eigenspace for the dominant eigenvalue 10.

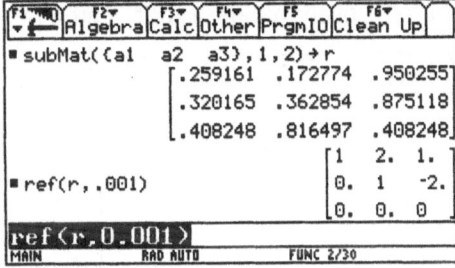

4. 2 Direct iterations (unfavorable case)

In the preceding section, we considered the favorable situation for the operator f of $\mathbb{K}^n$ (with matrix M in the standard basis):

- Existence of a single dominant eigenvalue λ.
- For λ, equality between the eigenspace and the characteristic sub-space.

These hypotheses allowed us to write $\mathbb{K}^n = E(\lambda) \oplus G$, by expressing G as the direct sum of all the characteristic sub-spaces of f, for the eigenvalues other than λ.

We have likewise supposed that the vector u (decomposed into $u = x + y$ in this direct sum) and which initiated the sequence of iterations was not in $E(\lambda)$ ($y \neq 0$, otherwise u is already a eigenvector of f), nor in G.

We then observed the "convergence" of the sequence of vector lines to the $\mathbb{K}v_k$, and that of the scalar products $< v_k, f(v_k) >$ to λ.

Now it is well known that mathematicians are just as interested in the unusual particular cases as in the general ones (if not more). Now we are going to examine some situations which are a little more problematic.

First possible problem:

The initial vector u is in the subspace G, invariant under f.

in this case, the iterates of u remain in G.

The sequences (v_k) and $(\lambda_k = < v_k, f(v_k) >)$ in general lead to a vector line for an eigenvalue and to the dominant eigenvalue of the restriction g of f to G.

To illustrate this situation by exact calculations, we are going to redo the first example used in the preceding paragraph.

The matrix M thus has three eigenvalues $10, 2, 1$ and it is diagonalizable by the transition matrix P leading to the eigenvectors $w = (1,1,3)$ for $\lambda = 10$, $w' = (1,2,1)$ for $\lambda = 2$, and $w'' = (1,1,2)$ for $\lambda = 1$.

We reuse the expression of M^k, which was stored in the variable mk.

On the other hand, we will choose as the initial vector u, a vector in the subspace G generated by w' and w''.

Specifically, we will choose $u = w' + w'' = (2,3,3)$.

We calculate $u_k = f^k(u)$, then the limit of the vectors v_k (normalized from u_k). We effectively obtain a unit vector proportional to $w' = (1,2,1)$.

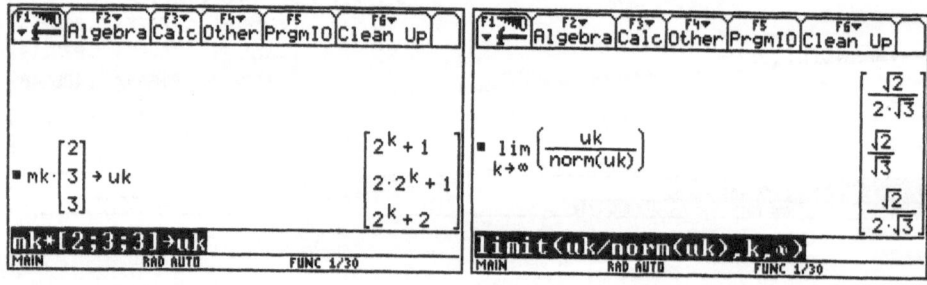

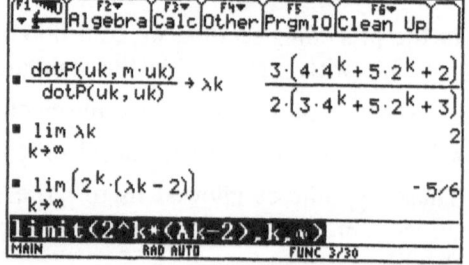

We see that the $\lambda_k = < v_k, f(v_k) >$ tend to the eigenvalue 2 (the second largest eigenvalue of f in modulus).

We observe however that $\lambda_k - 2 \sim -\frac{5}{6}2^{-k}$, the factor $1/2$ coming from the quotient of the eigenvalues 1 and 2.

Now we are going to replay exactly the same example, but in "approx" mode. To do this, we have modified the program iterdir so that it will request the initial u (rather than choosing it at random).

In the program iterdir, we have thus replaced the row:

$$:approx(randMat(rowDim(m),1))\rightarrow u$$

by the row:

$$:Input\ "Vector\ u",u:approx(u)\rightarrow u$$

So here is the reprise of the preceding example in "approx " mode. At first we observe that the (λ_k) seem to converge to 2, the v_k tending to a vector proportional to the eigenvector $w' = (1,2,1)$.

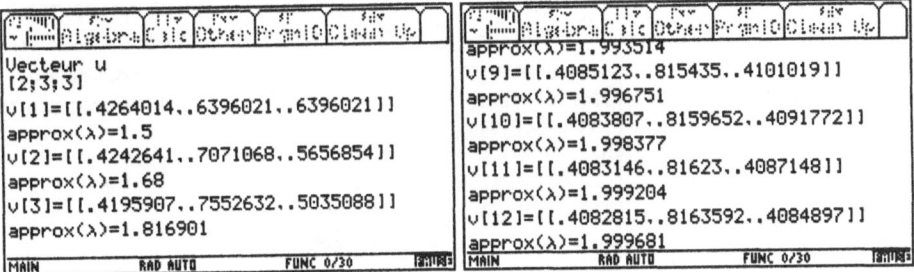

But pretty quickly (λ_k) exceeds the value 2 then "escapes " to a definite limit 10 (the dominant eigenvalue of f).

At the same time, the (v_k) tend to a vector proportional to the vector $w = (1,1,3)$, an eigenvector for $\lambda = 10$.

All this happens as if we had begun with a vector u not appearing in the plane generated by the vectors w' and w''.

The explanation is simple. The initial vector $u = (2,3,3)$ which we entered in the previous example is first normalized to give the vector v_1.

This normalization is done in approximate mode, and the inevitable rounding errors mean that v_1 is probably not entirely in the plane generated by w' and w''.

And even if it does lie there numerically, or within the precision allowed by the calculator, then it is almost certain that the vectors v_k at the end are going to escape the plane.

In order to have a clear conscience, we have calculated by hand the vectors v_1, v_2 and v_3.
The products by P^{-1} allow calculation of their coordinates in the basis w, w', w''.
We observe that v_3 is the first not to be numerically in the plane generated by w' and w''.

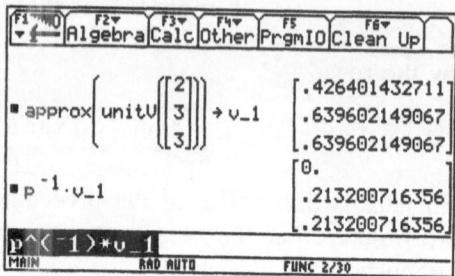

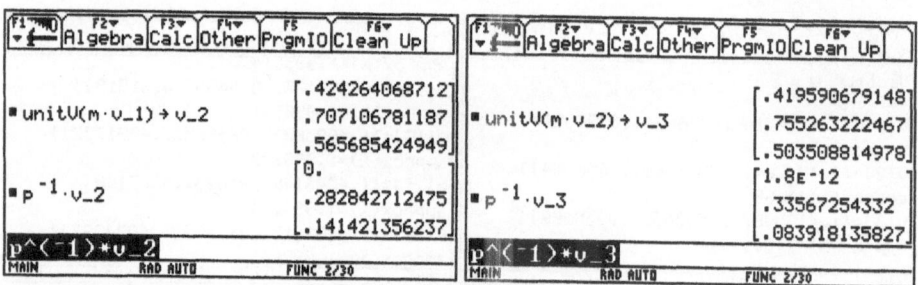

The following program, inspired by the function iterdir2, displays the list of approximations λ_k of the eigenvalue λ as the graph of a set of points in xyline mode. (The λ_k must be real).

```
:iterdir4(m,n,u)
:Prgm:Local u,v,k:approx(u)→u:{} → θλ
:For k,1,n:unitV(u)→v:m*v→u: augment(θλ,{dotP(u,v)})→ θλ:EndFor
:seq(k,k,1,n)→ θk:NewPlot 1,2,θk,θλ:ZoomData:DispG
:EndPrgm
```

We redo the previous example and see clearly that the sequence of (λ_k) does converge to 2, before converging to 10.

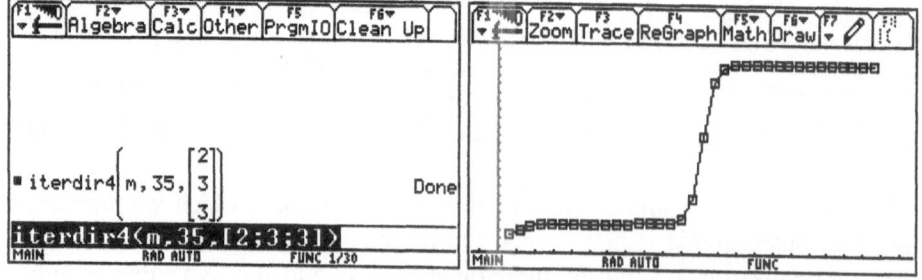

The second possible problem:

The matrix M has several distinct eigenvalues λ such that $|\lambda| = \rho(M)$ (the spectral radius of M).

This situation is not so uncommon: it happens if M has real coefficients and if the equation $\rho(M) = |\lambda|$ is true for a non-real eigenvalue λ. (The eigenvalues of M are then pairwise conjugate: $\overline{\lambda}$ is also an eigenvalue.)

To fix the ideas, suppose that there are two dominant eigenvalues λ and λ', and that for each of them, the eigenspace coincides with the characteristic sub-space.

Let u be any vector of E which is decomposed into $u = x + x' + y$ under the direct sum $\mathbb{K}^n = E(\lambda) \oplus E(\lambda') \oplus G$, where G is the sum of the characteristic sub-spaces of f for the eigenvalues other than λ and λ'. In order that we don't fall into a particular case which doesn't terminate, we will suppose that the vectors x, x' and y are non-zero. We put $\lambda' = \omega\lambda$, where $|\omega| = 1$.

As we have already done in the favorable case, we will call g the restriction of f to G, and we will put $g = \lambda h$. (h is thus an operator on G whose spectral radius is strictly less than 1.)

We have then, for every integer k:

$$u_k = f^k(u) = \lambda^k x + \lambda'^k x' + g^k(y) = \lambda^k(x + \omega^k x' + h^k(y)) \sim \lambda^k(x + \omega^k x')$$

We conclude by putting $z_k = x + \omega^k x'$:

$$\lambda_k = <v_k, f(v_k)> = \frac{1}{\|u_k\|^2} <u_k, u_{k+1}> \sim \frac{\lambda}{\|z_k\|^2} <z_k, z_{k+1}>$$

Imagine for example that ω is a p-th root of unity. Then the sequence (z_k) is p-periodic, and thus so is the sequence (λ_k). The values between those which evolve to λ_k are not generally eigenvalues of f.

So the sequence (λ_k) does not converge to any of the eigenvalues of f.

We treat here the example of a matrix M whose eigenvalues are 10, -10 and 1 (with the same transition matrix as in preceding examples).

Use of the program iterdir4 lets us follow the evolution of the sequence $(\lambda_k)_{k \geq 0}$, which "tends" to become 2-periodic (with the notation of the preceding example, $w = -1$). The two values of the cycle of length 2 depend on the initial vector u, chosen here at random.

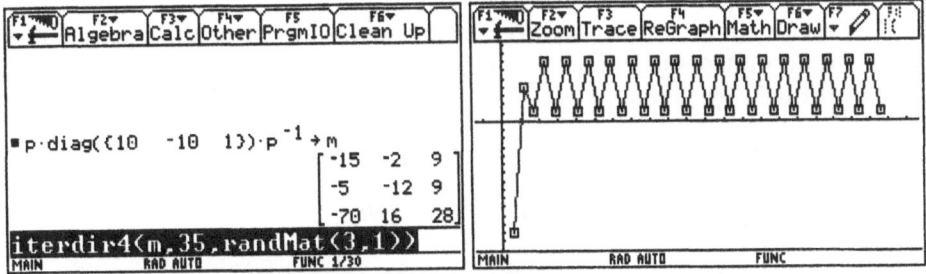

Third possible problem:

There is a unique dominant eigenvalue λ , but the eigenspace $E(\lambda)$ is strictly included in the characteristic sub-space $N(\lambda)$.

To simplify the calculations, we will suppose that $N(\lambda) = \mathrm{Ker}(f - \lambda Id)^2$. In other words, $\{\vec{0}\}$ is strictly included in $E(\lambda) = \mathrm{Ker}(f - \lambda Id)$ which is itself strictly included in $N(\lambda)$ and, for every $k \geq 2$: $\mathrm{Ker}(f - \lambda Id)^k = N(\lambda)$.

Let u be any vector of E and let $u = x + y$ be its decomposition under the direct sum $\mathbb{K}^n = N(\lambda) \oplus G$, where G is the direct sum of the characteristic sub-spaces of f for the eigenvalues other than λ.

As we have already done in the "favorable case", let g be the restriction of f to G and let h be the operator on G defined by $g = \lambda h$: the spectral radius of h is strictly less than 1.

We suppose, in order that we don't fall into a particular case (which would bring us back to the favorable case) that the component x of u in $N(\lambda)$ is not in the eigenspace $E(\lambda)$, that is, that $x' = f(x) - \lambda x \neq 0$.

Since the restriction of $(f - \lambda Id)^2$ to $N(\lambda)$ is zero, this implies that the vector $x' = f(x) - \lambda x$ is an eigenvector of f for λ.

We may express the iterates $f^k(x)$ as a function of x and of x'.

In fact, let $X^k = Q(X)(X - \lambda)^2 + \alpha X + \beta$, the division of X^k by $(X - \lambda)^2$.
We find $\lambda^k = \alpha\lambda + \beta$ then (after differentiation) $k\lambda^{k-1} = \alpha$. Thus, $\beta = (1 - k)\lambda^k$.

We conclude, for every integer $k \geq 1$:

$$f^k(x) = Q(f) \circ (f - \lambda Id)^2(x) + \alpha f(x) + \beta x = \alpha f(x) + \beta x$$
$$= k\lambda^{k-1} f(x) + (1 - k)\lambda^k x = k\lambda^{k-1}\big(f(x) - \lambda x\big) + \lambda^k x = \lambda^{k-1}(kx' + \lambda x)$$

The k-th iterate of u may thus be written:

$$u_k = f^k(u) = f^k(x) + g^k(y) = \lambda^{k-1}\big(kx' + \lambda x + \lambda h^k(y)\big) \sim k\lambda^{k-1}x'$$

We conclude that $u_{k+1} \sim (k + 1)\lambda^k x' \sim \lambda u_k$, then:

$$\lambda_k = <v_k, f(v_k)> = \frac{1}{\|u_k\|^2} <u_k, u_{k+1}> \sim \lambda$$

The sequence $(\lambda_k)_{k \geq 0}$ thus converges anew to to the dominant eigenvalue λ, as in the favorable case. The new evil is that this convergence is now much slower. In fact, by putting $z = h(y) - y$:

$$u_{k+1} - \lambda u_k = \lambda^k\big((k+1)x' + \lambda x + \lambda h^{k+1}(y)\big) - \lambda^k\big(kx' + \lambda x + \lambda h^k(y)\big)$$
$$= \lambda^k\big(x' + \lambda h^k(z)\big) \sim \lambda^k x' \sim \frac{\lambda}{k} u_k$$

Thus: $<v_k, f(v_k)> -\lambda = <v_k, f(v_k) - \lambda v_k> = \dfrac{<u_k, u_{k+1} - \lambda u_k>}{\|u_k\|^2} \sim \dfrac{\lambda}{k}$

The sequence of coefficients $\lambda_k = <v_k, f(v_k)>$ thus tends to λ, bu as $1/k$ tends to 0 (while in the "favorable" case this convergence is that of a geometric sequence with a ratio less than 1)!

To illustrate this situation, we are going to construct a M having the eigenvalues 10 (double) and 1 (simple), so M is not diagonalizable.

For this, we begin by forming the Jordan matrix J (see paragraph **2.5** for the functions mjordan and pjordan).

We conclude that the matrix M is similar to J (with the same transition matrix as in the preceding example).

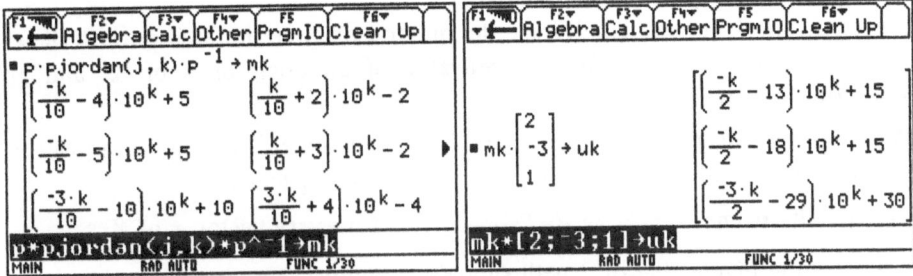

The function pjordan then lets us calculate the matrix J^k and thus the matrix $M^k = PJ^kP^{-1}$. The result is stored in the variable mk.
Starting with the vector u chosen at random (here $u = (2, -3, 1)$), we construct sequentially the k-th iterates $u_k = f^k(u)$.

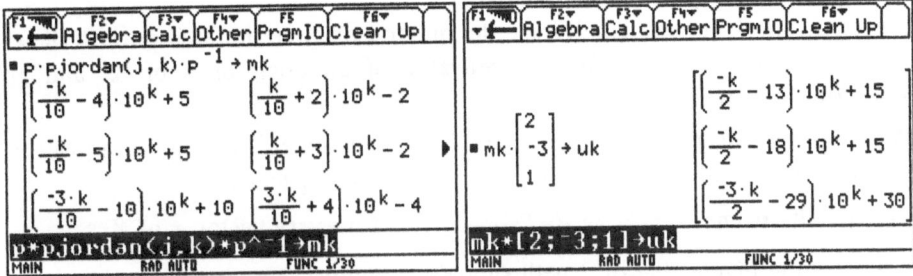

We now observe that the sequence of v_k (the normalized u_k vectors), indeed converge to a vector proportional to $x = (1, 1, 3)$ (which is the direction of an eigenvector line of M for the dominant eigenvalue 10).

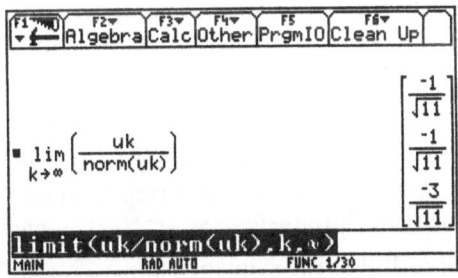

With the last screen, we calculate the scalar product $\lambda_k = <v_k, f(v_k)>$.
We see that the sequence of λ_k converges to the eigenvalue 10, but that the difference $\lambda_k - 10$ is equivalent to $10/k$ when $k \to \infty$ (which confirms the theoretical calculations).

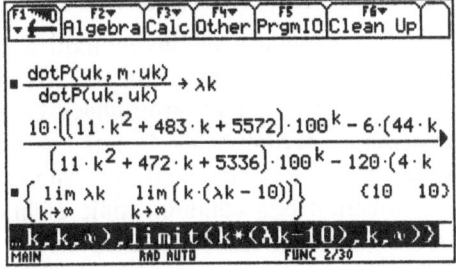

Here is a numeric confirmation of the problem posed by M.
We call the program iterdir, transmitting this matrix to it.
The first screen shows the beginning of the calculation, and the second illustrates
the slowness of the convergence: at iteration one hundred, we are still far from
the dominant eigenvalue 10 and from a vector proportional to $x = (1,1,3)$.

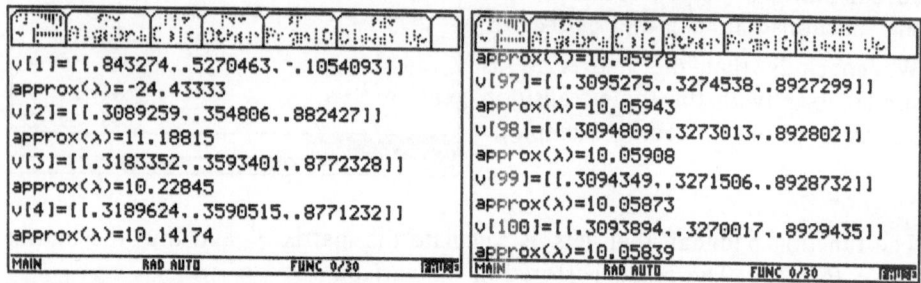

4. 3 Iterated inverse powers

Under certain favorable conditions, the method of direct iterated powers allows
approximation of the dominant eigenvalue of an operator f of $\mathbb{K}^n$ and an
associated eigenvector.

We suppose that f is invertible. The eigenvalues of f^{-1} are then the inverses
of those of f (with equality of the respective eigenspaces).

This results from: $f(x) = \lambda x \Leftrightarrow f^{-1}(x) = \dfrac{1}{\lambda}x$, for every vector x of $\mathbb{K}^n$.

Let λ be the eigenvalue of f with smallest modulus. The dominant eigenvalue
of f^{-1} is $1/\lambda$. The method of direct iterated powers applied to f^{-1} allows
approximation of $1/\lambda$ (and thus λ) and so also that of an associated eigenvector
x. We then speak of the method of iterated inverse powers.

In this method, as in the preceding one, there certainly are unfavorable
situations, which it is not necessary to review.

We thus start with a vector u of $\mathbb{K}^n$, and we construct the sequence of iterates
$u_{k+1} = f^{-1}(u_k)$. At each step we form the unit vector v_k from u_k. The sequence
of scalar products $< v_k, f^{-1}(v_k) >$ converges then to $1/\lambda$.

Recall that the sequence $(v_k)_{k \geq 0}$ is only convergent if λ is a positive real number
(but in any case, the vector line $\mathbb{K}v_k$ "converges " to an eigenvector line of f).

The inconvenience of this method is evidently the necessity of calculating the
inverse of the matrix M of f. We will entrust this work to the calculator,
but supposing that we are obliged to do it ourselves, the general strategy used
consists of decomposing M into the form $M = LU$ (L is lower triangular with
diagonal coefficients equal to 1, and U is upper triangular). Such so-called LU
factorization is a common topic of linear algebra. We study it in more detail
in the next section of this chapter.

In this manner, the equation $[u_{k+1}] = M^{-1}[u_k]$ is equivalent to $LU[u_{k+1}] = [u_k]$, and we find $[u_{k+1}]$ by solving two successive triangular systems.

The function iterinv very simply draws upon the function iterdir2. The syntax is the same.

The evaluation of iterinv(m,p) returns, after p iterations, a list $\{\lambda, x_1, \ldots, x_n\}$, where λ is the eigenvalue of M with smallest modulus, and where $x_1, \ldots, x_n$ are the components of an associated eigenvector x.

```
:iterinv(m,n)
:Func:Local r
: iterdir2(m^(-1),n)→r
: 1/r[1]→r[1]:r
:EndFunc
```

We resume with the matrix M used in the preceding example, which we know has the eigenvalues 10 (double) and 1 (simple).

The instruction iterinv(m,10) returns the eigenalue $\lambda = 1$ after 10 iterations and to the components of a vector proportional to $x = (1, 1, 2)$.

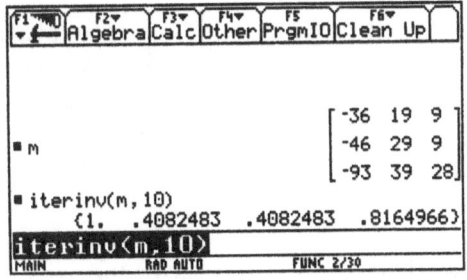

Remark:

If we denote by μ the eigenvalue of f with minimum modulus among all the other eigenvalues besides λ, the speed of convergence to λ by the method of inverse powers is that of the geometric sequence with ratio $q = \left|\frac{\lambda}{\mu}\right| < 1$.

This convergence is thus all the more rapid according as the modulus of λ is much less than all the other eigenvalues of f. (All this is under the assumption that the "favorable" conditions are satisfied: uniqueness of the eigenvalue with smallest modulus, and for which the eigenspace and the characteristic sub-space are equal).

4. 4 Iterations to an arbitrary eigenvalue

Let f be an operator on $\mathbb{K}^n$. We suppose that we know a first approximation λ_0 of an eigenvalue λ of f (that is, of the matrix M of f in the standard basis).

More precisely we assume that the eigenvalue of f closest to λ_0 is λ. As well, we suppose that λ_0 is not exactly an eigenvalue of f. Thus, the operator $f - \lambda_0 Id$ and its matrix $M - \lambda_0 I_n$ are not invertible.

The spectrum of $f - \lambda_0 Id$ is formed of $\mu - \lambda_0$, where μ runs through the spectrum of f.

The hypotheses made on λ_0 signify that $\lambda - \lambda_0$ is, among all the eigenvalues of $f - \lambda_0 Id$, which is the smallest in modulus.

The idea is thus to apply the method of iterated inverse powers of the map $g = f - \lambda_0 Id$, that is, with the matrix $M - \lambda_0 I_n$.

This method must lead to the eigenvalue $\lambda - \lambda_0$ of g, thus, to the eigenvalue λ of f) and to an eigenvector x of g for $\lambda - \lambda_0$: this vector x is then an eigenvector of f for λ.

The function itereig is then very simple to write. This uses the function iterinv.

The instruction itereig(m,λ_0,n) performs n iterations, starting from an estimation λ_0 of an eigenvalue of M. The result may be written $\{\lambda, x_1, \ldots, x_n\}$, where λ is the eigenvalue sought (at least, we hope to get a better approximation than that for λ_0!), and where $x_1, \ldots, x_n$ are the components of an associated eigenvector x.

```
:itereig(m,λ,n)
: Func:Local r
: iterinv(m-λ,n)→r
: r[1]+λ →r[1]:r
:EndFunc
```

We here form (always with the same transition matrix P!) a matrix M of eigenvalues 9, 7 and 1.

- Starting from $\lambda_0 = 7.1$, we find that $\lambda = 7$ and $x = (1, 2, 1)$ after 5 iterations.
- Starting from $\lambda_0 = 8.7$, we find $\lambda = 9$ and $x' = (1, 1, 3)$ after 10 iterations.

```
F1    F2    F3    F4    F5    F6
  Algebra Calc Other PrgmIO Clean Up

■ p·diag({9  7  1})·p⁻¹→m    ⎡ -29  14   8 ⎤
                             ⎢ -36  21   8 ⎥
                             ⎣ -78  30  25 ⎦
■ itereig(m,7.1,5)
            {7.  .408248  .816497  .408248}
■ itereig(m,8.7,10)
     {9.  -.301511  -.301511  -.904534}
itereig(m,8.7,10)
MAIN        RAD AUTO        3D   3/30
```

If we begin with an approximation λ_0 which is not much closer to any particular eigenvalue of M than to the others, the convergence is really less rapid, as we see with this example.

With the same matrix M, we start with $\lambda_0 = 8.1$ (almost midway between the eigenvalues 7 and 9).

The fact that λ_0 is slightly closer to 9 implies convergence, although somewhat slowly, to this eigenvalue.

```
F1    F2    F3    F4    F5    F6
  Algebra Calc Other PrgmIO Clean Up

■ itereig(m,8.1,10)
     {8.92099  .311679  .334519  .889356}
■ itereig(m,8.1,20)
     {8.99024  -.302741  -.305452  -.9028}
■ itereig(m,8.1,30)
     {8.99916  .301617  .30185  .904386}
■ itereig(m,8.1,40)
     {9.  -.301511  -.301511  -.904534}
itereig(m,8.1,40)
MAIN        RAD AUTO        3D   4/30
```

The method described here seems rather seductive as long as we know a correct estimation of the eigenvalue λ sought. In fact, not only does one very rapidly obtain a good approximation of λ, but the algorithm leads to another eigenvector for λ.

The defect of this method is that we must start with a vague idea of an eigenvalue of f. This is why it is often associated with another technique which does this first phase of approximation.

4. 5 Improvement of the method

We pick up the preceding method and the search for an eigenvalue λ of f (or of the matrix M) starting with the estimation of λ_0.

The algorithm consists of constructing the sequence of vectors U_k, starting from an arbitrary U, by: $U_{k+1} = (M - \lambda_0 I_n)^{-1} U_k$.

Then, at each step, we form the normalized vectors $V_k = \dfrac{1}{\|U_k\|} U_k$.

We know that the sequence of vector lines of the $\mathbb{K}V_k$ converge to an eigenvector line for the eigenvalue λ of M. The sequence of scalar products $\lambda_k = \ <V_k, MV_k>$ thus converges to λ.

We then replace the equation $U_{k+1} = (M - \lambda_0 I_n)^{-1} U_k$ by $U_{k+1} = (M - \lambda_k I_n)^{-1} U_k$.

In other words, at each step, we use λ_k as a new initial approximation of λ. If, as we hope, λ_k is closer to λ than was λ_0, the convergence must accelerate.

The defect of this method (largely compensated by the acceleration) is that at each step we must invert the matrix $(M - \lambda_k I_n)$, which becomes less and less invertible. If it is not invertible at all, then λ_k is the eigenvalue we sought!

The program itereig2 puts to work this improvement. The syntax is itereig2(m, λ_0) (λ_0 is an estimate of an eigenvalue λ of M).

As with the program itereig, the successive vectors v_k are displayed, as well as the approximations λ_k of λ.

```
:itereig2(m,λ)
:Prgm:Local v,u,k:0→k:ClrIO
:approx(randMat(rowDim(m),1))→u
:Loop
:  k+1→k:unitV(u)→v
:  (m-λ)^(-1)*v→u:dotP(m*v,v)→ λ
:  v*(abs(λ)/λ)^k→v
:  Disp "v["&string(k)&"]="&string(vᵀ)
:  Pause "approx(λ)="&string(λ)
:EndLoop
:EndPrgm
```

Look at the previous example with the estimate 8.1 for an eigenvalue of M, a very bad choice.

We observe that the eigenvalue $\lambda = 9$ is exactly detected at the ninth iteration, where forty steps were not sufficient formerly.

The acceleration in this case is thus spectacular.

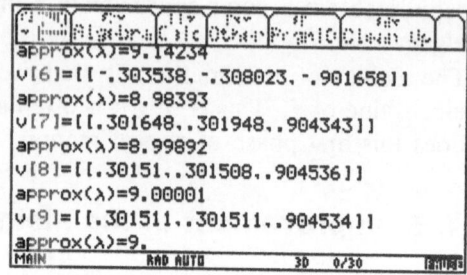

5. Other iterative methods

We will touch on three methods whose common point is to construct a sequence of matrices A_k similar to the initial matrix A and which converge to a diagonal or triangular matrix.

The diagonal of A_k thus converges to the eigenvalues of A.

5. 1 Use of the LU factorization

The LU instruction of the calculator does a decomposition or factorization of a square matrix A for which "$PA = LU$":

- L is a lower triangular matrix with ones on the diagonal, and U is an upper triangular matrix.
- P is a permutation matrix: each row and each column contains only one non-zero coefficient which has value 1.
- We have the equality $PA = LU$.

The syntax of the instruction LU is: LU A,L,U,P, where A is the matrix to be factored and where L, U, P are the names of the global variables which are to receive the three matrices. We may choose any names we wish, but the order is important.

Here is the factorization of a particular matrix A.

At first we displayed P, L, U, before verifying the equation $PA = LU$.

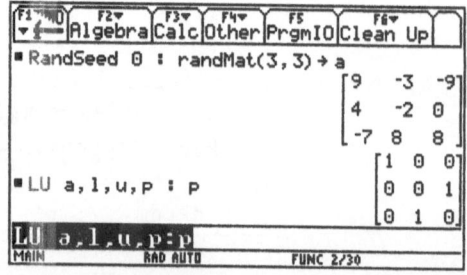

$$
\begin{array}{ll}
\bullet\ 1 & \begin{bmatrix} 1 & 0 & 0 \\ -7/9 & 1 & 0 \\ 4/9 & -2/17 & 1 \end{bmatrix} \\[6mm]
\bullet\ u & \begin{bmatrix} 9 & -3 & -9 \\ 0 & 17/3 & 1 \\ 0 & 0 & 70/17 \end{bmatrix}
\end{array}
\qquad
\begin{array}{ll}
\bullet\ p \cdot a & \begin{bmatrix} 9 & -3 & -9 \\ -7 & 8 & 8 \\ 4 & -2 & 0 \end{bmatrix} \\[6mm]
\bullet\ 1 \cdot u & \begin{bmatrix} 9 & -3 & -9 \\ -7 & 8 & 8 \\ 4 & -2 & 0 \end{bmatrix}
\end{array}
$$

Let $PA = LU$ be the decomposition of a matrix A. We put $B = UP^{-1}L$.

$PA = LU$ may also be written $AU^{-1} = P^{-1}L$.

Thus, $B = UAU^{-1}$ is a matrix similar to A.

Both thus have the same eigenvalues. We verify this as an extension of the previous example.

$$
\begin{array}{l}
\bullet\ u \cdot p^{-1} \cdot 1 \to b \quad \begin{bmatrix} 44/3 & -\dfrac{147}{17} & -3 \\[3mm] 47/27 & 1/3 & 17/3 \\[3mm] -\dfrac{490}{153} & 70/17 & 0 \end{bmatrix} \\[14mm]
\bullet\ \text{eigVl}(a) \quad (14.9285 \quad -3.71504 \quad 3.78651) \\
\bullet\ \text{eigVl}(b) \quad (14.9285 \quad 3.78651 \quad -3.71504)
\end{array}
$$

We generalize this idea by constructing a sequence of matrices $(A_k)_{k \geq 0}$ defined by $A_0 = A$ and, for every integer k: $A_k = U_k P_k^{-1} L_k$, denoting by $P_k A_{k-1} = L_k U_k$ the LU factorization of A_{k-1}. All the matrices A_k are similar to the initial matrix A (and thus have the same eigenvalues).

We will suppose that A has real coefficients, with a character -istic polynomial which splits over $\mathbb{R}$. One then proves that the sequence $(A_k)_{k \geq 0}$ converges to an upper triangular matrix T. In particular, the list of diagonal coefficients of the matrix A_k converges to the list of eigenvalues of A.

The proof of this result is difficult and strays considerably from the setting of this work. We will be content to experiment with the method considered here as a curiosity. The program iterlu (with the syntax iterlu(m,n)) does n iterations and displays the last matrix (A_n in the preceding notation), stored in the variable θm. The program likewise uses the global variable θl, θu and θp.

```
:iterlu(m,n)
:Prgm:Local k:approx(m)→ θm
:For k,1,n:LU θm,θl,θu,θp:θu*θp^(-1)*θl→ θm:EndFor
:Disp "θm=",θm
:EndPrgm
```

Here is an example of use of the program iterlu, with a square matrix A of order 4 whose eigenvalues are -2, 1, 5 and 9.

We have displayed the result of the first iteration, and of the twentieth.

We observe apparent convergence to an upper triangular matrix. The eigenvalues of A begin to appear with good precision on the diagonal of the matrices obtained.

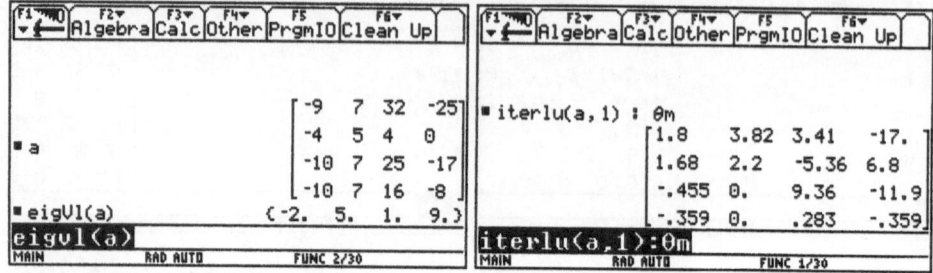

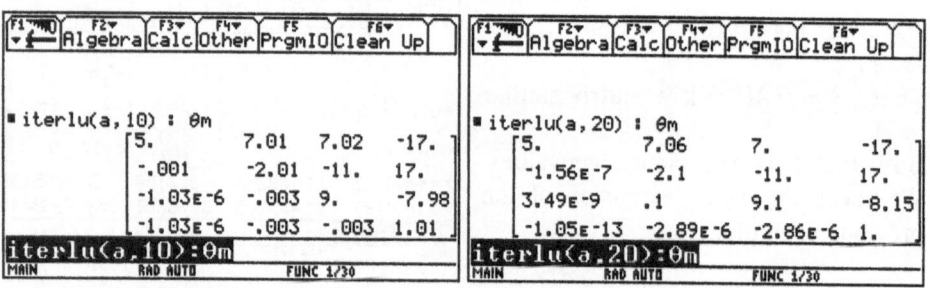

5. 2 Use of the QR factorization

In section **6** of the *Orthogonality* chapter, we covered extensively the QR factorization of a matrix, its significance, and the means of obtaining it.

Let $A \in \mathcal{M}_n(\mathbb{R})$. Recall that A may be written uniquely as $A = QR$, where Q is orthogonal ($^{\top}Q = Q^{-1}$) and R is upper triangular with positive elements on the diagonal. This last condition may be weakened, but at the cost of the uniqueness of the decomposition.

The built-in instruction QR does this factorization (with the obvious syntax QR A,Q,R, where Q and R are the names of global variables).

We construct a sequence $(A_k)_{k \geq 0}$ by putting $A_0 = A$ and, for every integer k, $A_k = R_k Q_k$, where $A_{k-1} = Q_k R_k$ is the QR factorization of A_{k-1}.

We observe that $A_k = R_k Q_k = Q_k^{-1} A_{k-1} Q_k$. The matrices A_k are thus pairwise similar and thus have the same eigenvalues as A.

Under certain hypotheses (when the character -istic polynomial splits in $\mathbb{R}$, and the eigenvalues are all of different modulus), one may proves that the matrices A_k tend to upper triangular matrices. Their diagonal then converges to the set of eigenvalues of A. One likewise may show that the convergence is all the more rapid as the eigenvalues of A are more separated in modulus.

This method obviously is analogous with the one in which we used the LU decomposition.

The program iterqr (syntax iterqr(m,n)) does n iterations and displays the last matrix, A_n, stored in the variable θm. The program also uses the global variable θq and θr.

```
:iterqr(m,n)
:Prgm:Local k:approx(m)→ θm
:For k,1,n:QR θm,θq,θr:θr*θq→ θm:EndFor
:Disp "θm=",θm
:EndPrgm
```

We reconsider here the matrix A of the preceding example.

We observe that the matrices A_k seem to tend to upper triangular matrices with the diagonal coefficients becoming very close to the eigenvalues of A, which are 5, 9, -2 and 1.

F1	F2 Algebra	F3 Calc	F4 Other	F5 PrgmIO	F6 Clean Up

```
■ iterqr(a,10) : θm
  ⎡5.        12.1    -6.03     -48.4⎤
  ⎢-6.36ᴇ-4  -2.01    6.75      22.3 ⎥
  ⎢3.63ᴇ-7    .002    9.        13.8 ⎥
  ⎣-2.09ᴇ-7  -.001    8.51ᴇ-4   1.01⎦
iterqr(a,10):θm
MAIN         RAD AUTO          FUNC 1/30
```

F1	F2 Algebra	F3 Calc	F4 Other	F5 PrgmIO	F6 Clean Up

```
■ iterqr(a,30) : θm
  ⎡5.        -1.21     13.5     -48.4⎤
  ⎢1.07ᴇ-5    9.        6.74     -23.4⎥
  ⎢-1.62ᴇ-11 -1.06ᴇ-5  -2.       11.8 ⎥
  ⎣0.         0.       -1.54ᴇ-9   1.  ⎦
iterqr(a,30):θm
MAIN         RAD AUTO          FUNC 1/30
```

5.3 The Jacobi method for real symmetric matrices

The method which we are going to present now only applies to symmetric matrices with real coefficients.

We might say that it only performs in a particular case, at least for a mathematician. But this situation presents itself very often in applications.

The Jacobi method makes use of rotation matrices such as those used in paragraph **6.3** of the Orthogonality chapter.

Recall that the method of Givens gives the QR factorization of a matrix A. It uses successive rotations $G_1, G_2, \ldots, G_p$, to annihilate the coefficients under the diagonal of A, in a manner which yields the upper triangular matrix R.
Thus, the orthogonal matrix $G = G_p G_{p-1} \ldots G_1$ satisfies $GA = R$.
It remains then to write $A = QR$, with $Q = G^{-1} = {}^\top G$.

We point out that this method is not useful in finding the spectrum of A. If the matrix R is really upper triangular, it is in fact not similar to A and thus doesn't have the same eigenvalues.

On the other hand, if G is a rotation matrix (or, more generally, an orthogonal matrix), then the matrices A and $B = {}^\top GAG$ are similar.

Better, if A is symmetric, so is B. If G is chosen in a way to annihilate a "sub-diagonal" element of A, then two coefficients are annihilated at one blow!

We may hope that a succession of matrices G_k allow the annihilation of all the non-diagonal coefficients of A, leading to a diagonal matrix D which is similar to A (and which thus gives the eigenvalues of this matrix...).

But things don't always work out that way. The problem comes from the fact that it is not possible to keep the preceding coefficients zero, contrary to the QR factorization by Givens' method.

But we will see all this in a while.

A) INTRODUCTION TO THE METHOD

The matrix A of coefficients $a_{i,j}$ is thus real and symmetric. This implies that A is diagonalizable in $\mathbb{R}$ and all of the eigenvalues are real.

Let p and q be two indices such that $1 \leq p < q \leq n$.

Our approach is to "attack" the coefficient $a_{p,q} = a_{q,p}$ of A (which coefficient is assumed to be non-zero).

Let $G = G(n, p, q, \theta) \in \mathcal{M}_n(\mathbb{R})$ whose general term $g_{k,l}$ is given by:

- If $k \notin \{p, q\}$, $g_{k,k} = 1$; $g_{p,p} = g_{q,q} = \cos \theta$.
- $g_{p,q} = \sin \theta$, and $g_{q,p} = -\sin \theta$.
- In all the other cases: $g_{k,l} = 0$.

If $\mathbb{R}^n$ is represented in its standard basis $(e_1, e_2, \ldots, e_n)$, the matrix G represents the rotation by an angle θ in the plane defined by e_q and e_p.

The following function constructs a rotation matrix.

```
:rot(n,p,q,θ)
:Func:Local m:identity(n)→m
:cos(θ)→m[p,p]:cos(θ)→m[q,q]
:sin(θ)→m[p,q]:-sin(θ)→m[q,p]:m
:EndFunc
```

Here is an example of such a matrix.

It represents, in $\mathbb{R}^6$, the rotation by an angle θ in the vector plane oriented by the vectors e_4 and e_2.

We put $B = {}^{\mathsf{T}}GAG$, with general term $b_{i,j}$. We want to choose G in a way that the coefficient $b_{p,q}$ is zero. A priori, this coefficient satisfies:

$$b_{p.q} = [B]_{p.q} = \sum_{k,l=1}^{n} [{}^{\top}G]_{p.k}[A]_{k.l}[G]_{l.q} = \sum_{k,l=1}^{n} g_{k.p}\, a_{k.l}\, g_{l.q} = \sum_{k,l \in \{p.q\}} g_{k.p}\, a_{k.l}\, g_{l.q}$$

The sum reduces to $k,l \in \{p,q\}$ since the other terms are zero.
We conclude that:

$$
\begin{aligned}
b_{p.q} &= \cos\theta(a_{p.p}\sin\theta + a_{p.q}\cos\theta) - \sin\theta(a_{q.p}\sin\theta + a_{q.q}\cos\theta)\\
&= (a_{p.p} - a_{q.q})\sin\theta\cos\theta + a_{p.q}(\cos^2\theta - \sin^2\theta) \quad \text{(by symmetry, } a_{p.q} = a_{q.p})\\
&= \frac{1}{2}(a_{p.p} - a_{q.q})\sin 2\theta + a_{p.q}\cos 2\theta
\end{aligned}
$$

Thus, the coefficient $b_{p.q}$ is zero if we put $\operatorname{cotan} 2\theta = \dfrac{a_{q.q} - a_{p.p}}{2a_{p.q}}$.

This equation defines a unique value of θ in $\left]-\frac{\pi}{4},0\right[\cup\left]0,\frac{\pi}{4}\right]$.

In fact, the map $\varphi \mapsto \operatorname{cotan}\varphi$ is a bijection of $\left]-\frac{\pi}{2},0\right[\cup\left]0,\frac{\pi}{2}\right]$ onto $\mathbb{R}$.

B) STUDY OF AN EXAMPLE

The following example has been chosen so that the value of θ is simple.
We propose to annihilate the coefficients with indices $(1,3)$ and $(3,1)$ in the symmetric matrix A.
We have calculated the quotient q, whose value must be given to θ, defined by the equation $\operatorname{cotan} 2\theta = q$.

We see that here $\theta = \dfrac{1}{2}\operatorname{Arctan}\dfrac{2a_{1.3}}{a_{3.3} - a_{1.1}} = \dfrac{1}{2}\operatorname{Arctan}\dfrac{1}{q} = \dfrac{1}{2}\operatorname{Arctan}\dfrac{-24}{7}$.

To calculate the exact value of $\cos\theta$ and of $\sin\theta$, we must first evaluate $\tan\theta$ and deduce a new expression $\theta = -\operatorname{Arctan}\dfrac{3}{4}$.

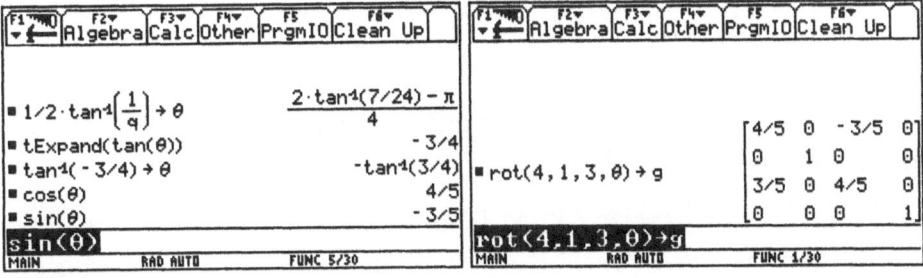

After having formed the rotation matrix G, we observe that the transition from A to $B = {}^TGAG$ accomplishes the annihilation of the coefficients with indices $(1,3)$ and $(3,1)$.

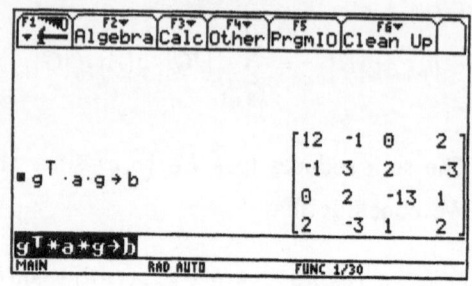

C) RETURN TO THE GENERAL CASE

With the value obtained for θ, we are going to calculate the other coefficients of B. To simplify the notation, we put: $c = \cos\theta$, $s = \sin\theta$, and $t = \tan\theta$.

With this notation: $(a_{q.q} - a_{p.p})sc = a_{p.q}(c^2 - s^2) \Rightarrow (a_{q.q} - a_{p.p})t = a_{p.q}(1 - t^2)$.

For every pair (i,j): $b_{i.j} = \displaystyle\sum_{k.l=1}^{n} g_{k.i}\, a_{k.l}\, g_{l.j}$.

We note first that if $i \notin \{p,q\}$ and $j \notin \{p,q\}$, then $b_{i.j} = a_{i.j}$.

In fact, in this case and using the notation of the Kronecker delta, $g_{k.i} = \delta_{k.i}$ and $g_{l.j} = \delta_{l.j}$.

Thus, we suppose that $i = p$: For every j of $\{1,\dots,n\}$,

$$b_{p.j} = \sum_{l=1}^{n}(g_{p.p}\, a_{p.l}\, g_{l.j} + g_{q.p}\, a_{q.l}\, g_{l.j}) = \sum_{l=1}^{n}(a_{p.l}\, c - a_{q.l}\, s)g_{l.j}$$

In particular, if $j \notin \{p,q\}$: $b_{p.j} = a_{p.j}\, c - a_{q.j}\, s$.

If $j = p$ (always with $i = p$), we may conclude:

$$b_{p.p} = (a_{p.p}\, c - a_{q.p}\, s)c - (a_{p.q}\, c - a_{q.q}\, s)s = a_{p.p}\, c^2 + a_{q.q}\, s^2 - 2\, s\, c\, a_{p.q}$$

$$= a_{p.p} + (a_{q.q} - a_{p.p})s^2 - 2\, s\, c\, a_{p.q} = a_{p.p} + sc\big((a_{q.q} - a_{p.p})t - 2a_{p.q}\big)$$

$$= a_{p.p} - s\, c\, a_{p.q}(1 + t^2) = a_{p.p} - t a_{p.q}$$

If $j = q$ (always with $i = p$), we know that $b_{p.q} = 0$.

Now we suppose that $i = q$: For every j of $\{1,\dots,n\}$,

$$b_{q.j} = \sum_{l=1}^{n}(g_{p.q}\, a_{p.l}\, g_{l.j} + g_{q.q}\, a_{q.l}\, g_{l.j}) = \sum_{l=1}^{n}(a_{p.l}\, s + a_{q.l}\, c)g_{l.j}$$

In particular, if $j \notin \{p,q\}$: $b_{q.j} = a_{p.j}\, s + a_{q.j}\, c$.

If $j = p$ (always with $i = q$), we know that $b_{q.p} = b_{p.q} = 0$.

If $j = q$ (always with $i = q$), we find:

$$b_{q.q} = (a_{p.p}\, s + a_{q.p}\, c)s + (a_{p.q}\, s + a_{q.q}\, c)c = a_{p.p}\, s^2 + a_{q.q}\, c^2 + 2\, s\, c\, a_{p.q}$$

$$= a_{q.q} - (a_{q.q} - a_{p.p})s^2 + 2\, s\, c\, a_{p.q} = a_{q.q} + t a_{p.q}$$

Summarizing:

- If $i \notin \{p,q\}$ and $j \notin \{p,q\}$, $b_{i.j} = a_{i.j}$.
- If $j \notin \{p,q\}$, $b_{p.j} = b_{j.p} = a_{p.j}\, c - a_{q.j}\, s$, and $b_{q.j} = b_{j.q} = a_{p.j}\, s + a_{q.j}\, c$.
- $b_{p.q} = b_{q.p} = 0$.
- $b_{p.p} = a_{p.p} - t a_{p.q}$ and $b_{q.q} = a_{q.q} + t a_{p.q}$.

D) USE OF THE SCHUR NORM

The preceding equations show that all the coefficients of rows and columns of A with index of p and q are affected in the passage from A to B.

This implies that it is futile to want to annihilate the non-diagonal coefficients one by one: when one works row by row or column by column, the coefficients made zero by the preceding operation can't be held to that value.

As a consolation, the possibility remains to annihilate those terms which have the maximum modulus at each step.

But to be sure that this is really a good idea, again we must verify that the modulus of the non-diagonal terms tends to decrease at each step.

To see this, we are going to consider the Schur norm of A and of B (see **3.2** for a reminder about matrix norms.)

We recall that for every matrix M of $\mathcal{M}_n(\mathbb{R})$, with general term $m_{i.j}$:

$$\|M\|_s^2 = \mathrm{Tr}(^\top M\, M) = \sum_{i.j=1}^{n} m_{i.j}^2$$

If Ω is an orthogonal matrix, then the matrices ΩM and $M\Omega$ have the same Schur norm. For example,

$$\|\Omega M\|_s^2 = \mathrm{Tr}\left(^\top(\Omega M)\, \Omega M\right) = \mathrm{Tr}(^\top M\, ^\top\Omega \Omega M) = \mathrm{Tr}(^\top M I_n M) = \|M\|_s^2$$

A consequence of this property is that the matrices A and $B = {}^\top G A G$ have the same Schur norm.

We denote by D (resp. Δ) the diagonal matrix whose diagonal coefficients are those of A (resp. B), and we let $A' = A - D$ (resp. $B' = B - \Delta$).

By construction: $\|A\|_s^2 = \|A'\|_s^2 + \|D\|_s^2$ and $\|B\|_s^2 = \|B'\|_s^2 + \|\Delta\|_s^2$.

First, we are going to compare the sum of the squares of the diagonal terms of A and of B, that is, the $\|D\|_s^2$ and the $\|\Delta\|_s^2$.

$$\sum_{j=1}^{n} b_{j.j}^2 = b_{p.p}^2 + b_{q.q}^2 + \sum_{j \notin \{p.q\}} b_{j.j}^2 = (a_{p.p} - t a_{p.q})^2 + (a_{q.q} + t a_{p.q})^2 + \sum_{j \notin \{p.q\}} a_{j.j}^2$$

$$= 2a_{p.q}\left(t(a_{q.q} - a_{p.p}) + a_{p.q}t^2\right) + \sum_{j=1}^{n} a_{j.j}^2$$

$$= 2a_{p.q}\left(a_{p.q}(1 - t^2) + a_{p.q}t^2\right) + \sum_{j=1}^{n} a_{j.j}^2 = 2a_{p.q}^2 + \sum_{j=1}^{n} a_{j.j}^2$$

We have thus obtained: $\|\Delta\|_s^2 = 2a_{p.q}^2 + \|D\|_s^2$.

We conclude that: $\|B'\|_s^2 = \|B\|_s^2 - \|\Delta\|_s^2 = \|A\|_s^2 - \|D\|_s^2 - 2a_{p.q}^2 = \|A'\|_s^2 - 2a_{p.q}^2$.

E) ANNIHILATING THE MAXIMUM NON-DIAGONAL COEFFICIENT

We suppose now that $a_{p.q}$ (the coefficient to be annihilated) has, among all the non-diagonal coefficients of A, the largest modulus.

We have then: $\|A'\|_s^2 = \sum_{i \neq j} a_{i.j}^2 \leq n(n-1)a_{p.q}^2$.

And we deduce the inequality: $\|B'\|_s^2 = \|A'\|_s^2 - 2a_{p.q}^2 \leq \left(1 - \dfrac{2}{n(n-1)}\right)\|A'\|_s^2$.

Remark: Since $|\theta| \leq \frac{\pi}{4}$, $t = \tan\theta$ satisfies that $|t| \leq 1$.

We conclude that $|b_{p.p} - a_{p.p}| \leq |a_{p.q}|$ and $|b_{q.q} - a_{q.q}| \leq |a_{p.q}|$.

More generally (and because all the diagonal coefficients other than $a_{p.p}$ and $a_{q.q}$ remain unchanged):$\forall j \in \{1, \ldots, n\}$, $|b_{j.j} - a_{j.j}| \leq |a_{p.q}|$.

Hence $a_{p.q}$ has the largest modulus of the coefficients of A'. But then again it is one of these coefficients so this verifies that $|a_{p.q}| \leq \|A'\|_s$.

We conclude the result which will be used in the sequel:

$$\forall j \in \{1, \ldots, n\}, \ |b_{j.j} - a_{j.j}| \leq \|A'\|_s$$

F) SETTING UP THE ALGORITHM

Let A be a symmetric matrix of order n with real coefficients.

We define a sequence $(A_k)_{k \geq 0}$ in the following fashion, denoting $a_{i.j}^{(k)}$ as the general term of A_k. We first put $A_0 = A$.

Let $k \in \mathbb{N}$. Suppose the matrix A_k is known, and let $a_{p.q}^{(k)}$ be its non-diagonal coefficient of maximum modulus.

If $a_{p.q} = 0$ (that is, if A_k is diagonal), we put $A_{k+1} = A_k$ (in this case the sequence of matrices $(A_k)_{k \geq 0}$ becomes stationary).

Otherwise, we put $A_{k+1} = {}^{\mathsf{T}}G_k A_k G_k$, where G_k is the rotation matrix allowing annihilation of (p, q). In other words, we pass from A_k to A_{k+1} exactly as we passed from A to B in E).

All the matrices A_k are real symmetric and similar to A. They thus have the same eigenvalues as A.

For every integer k, we put $A_k = A_k' + D_k$, where D_k is the diagonal matrix whose diagonal coefficients are those of A_k.

We also note that $q = \sqrt{1 - \frac{2}{n(n-1)}}$. The real number q satisfies $0 \leq q < 1$.

G) CONVERGENCE OF THE ALGORITHM

The preceding calculations show that for every integer k: $\|A_{k+1}'\|_s \leq q\|A_k'\|_s$. (This inequality is also valuable if the matrix A_k is diagonal since then the matrices A_k' and A_{k+1}' are zero.)

We may conclude: $\forall k \in \mathbb{N}$, $\|A'_k\|_s \leq q^k \|A'_0\|_s$. Thus, $\lim\limits_{k \to \infty} \|A'_k\|_s = 0$.

This result is already very important: it signifies that the non-diagonal coefficients of the matrices A_k tend to 0 when k tends to infinity.

Stated another way, matrices A_k "tend " to become diagonal.

It remains to understand how the diagonal of the matrices A_k evolve. After the remark of section **E**): $\forall j \in \{1, \ldots, n\}$, $\left| a_{j,j}^{(k+1)} - a_{j,j}^{(k)} \right| \leq \|A'_k\|_s \leq q^k \|A'_0\|_s$.

We conclude that, for every integer m and r:

$$\left| a_{j,j}^{(m+r)} - a_{j,j}^{(m)} \right| \leq \sum_{k=m}^{m+r-1} \left| a_{j,j}^{(k+1)} - a_{j,j}^{(k)} \right| \leq \|A'_0\|_s \sum_{k=m}^{m+r-1} q^k \leq \|A'_0\|_s \frac{q^m}{1-q}$$

Since $0 \leq q < 1$, this result shows that the sequence of general terms $a_{j,j}^{(k)}$ (the j-th diagonal coefficient A_k) is a Cauchy sequence. It is thus convergent.

Consequence: the sequence of diagonal matrices $(D_k)_{k \geq 0}$ converges to a diagonal matrix D. As we set up, for every integer k, $A_k = A'_k + D_k$, and as we know already that the sequence of matrices A'_k converges to zero, we deduce the convergence of the sequence $(A_k)_{k \geq 0}$ to the diagonal matrix D.

On the other hand, we recall that the matrices A_k are similar to the initial matrix A. The eigenvalues of A are thus those of A_k, and thus they are those of the limit matrix D. These are the diagonal coefficients of D.

This ends the proof.

We now know how, starting with a real symmetric matrix A, to create a sequence of real symmetric matrices A_k which converges to a diagonal matrix D carrying the eigenvalues of A. We know from the start that A has real eigenvalues.

G) PROGRAMMING THE ALGORITHM

The function jacobi1 applies the Jacobi transformation to a square symmetric matrix A and returns the matrix obtained by annihilating the non-diagonal terms of maximum modulus. With the preceding notation, it thus allows passing from the matrix A_k to the matrix A_{k+1}.

It will be possible to calculate (with the formula that we now know) the coefficients of A_{k+1} as a function of those of A_k. It seems more simple to us to create the rotation matrix G_k using the function rot and to calculate the $A_{k+1} = {}^{\top}G_k A_k G_k$ directly.

```
:jacobi1(a)
:Func:Local m,n,i,j,p,q,t,g
:0→m:rowDim(a)→n
:For i,2,n:For j,1,i-1:abs(a[i,j])→t
:  If t>m Then:t→m:i→p:j→q:EndIf
:EndFor:EndFor
:If m=0:Return a
:approx((a[q,q]-a[p,p])/(2*a[p,q]))→t
:when(t=0,π/4,tan⁻¹(1/t)/2)→t
:rot(n,p,q,t)→t:tᵀ*a*t
:EndFunc
```

Here are a real symmetric matrix A of order 3 and its eigenvalues. After a first iteration, we see how the coefficient -16, with index $(1,3)$, has been annihilated (up to rounding errors) by this operation.

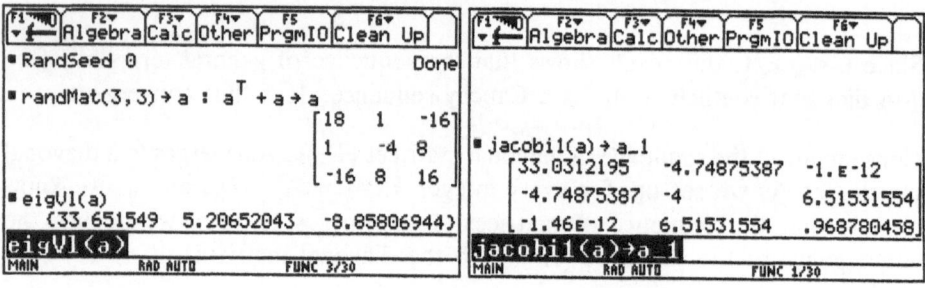

In the following iteration, we see that the preceding "annihilated " coefficient is not held to 0. At the tenth iteration, A_{10} is "almost diagonal " and gives the eigenvalues of A with very good precision.

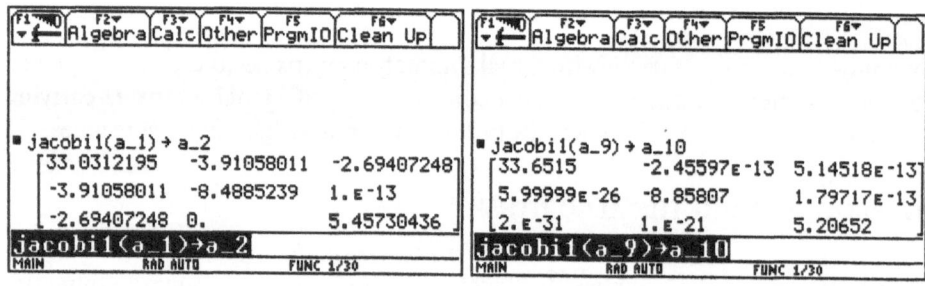

6. Symmetric operators

We considered symmetric matrices in the preceding section in our consideration of the iterative Jacobi method. The notion of a symmetric matrix (or operator) is sufficiently important to spend a little more time with it. The theoretical results are numerous and interesting, as are the numerical methods which relate to them.

6. 1 Review of theory

Let E be a vector space on $\mathbb{R}$ equipped with a scalar product.

Definition 1: We say that an operator f on E is symmetric if for all vectors u and v of E it satisfies the equation: $< f(u), v > = < u, f(v) >$. We denote by $S(E)$ the set of symmetric operators of E.

Here are some properties whose proofs are easy:

- The zero map and the identity map Id_E are obviously symmetric operators of E.
- $S(E)$ is a sub-space of $\mathcal{L}(E)$. This results from:
  $$< (\alpha f + \beta g)(u), v > = \alpha < f(u), v > + \beta < g(u), v >$$
  $$= \alpha < u, f(v) > + \beta < u, g(v) > = < u, (\alpha f + \beta g)(v) >$$
- If $f \in S(E)$ and if f is an isomorphism, then $f^{-1} \in S(E)$. In fact:
  $$< f^{-1}(u), v > = < f^{-1}(u), f \circ f^{-1}(v) > = < f \circ f^{-1}(u), f^{-1}(v) > = < u, f^{-1}(v) >.$$
- Let F be a finite dimensional sub-space of E. The orthogonal projection onto F and the projection onto the orthogonal complement of F are symmetric operators. In fact, if u and v decompose into $u = u' + u''$ and $v = v' + v''$ under the direct sum $E = F \oplus F^{\top}$:
  $$< p(u), v > = < u', v' + v'' > = < u', v' > = < u' + u'', v' > = < u, p(v) >$$
  Thus, $p \in S(E)$. Similarly, $s = 2p - Id_E \in S(E)$.
- If $f \in S(E)$, then $\operatorname{Ker} f = (\operatorname{Im} f)^{\top}$. In fact:
  $$v \in (\operatorname{Im} f)^{\top} \Leftrightarrow \forall u \in E \ < v, f(u) > = 0 \Leftrightarrow \forall u \in E \ < f(v), u > = 0 \Leftrightarrow f(v) = 0.$$
  In particular, a symmetric vector projection is necessarily an orthogonal vector projection (the converse of the preceding property).
- Let $f \in S(E)$, and let F be a sub-space of E, invariant under f. Then $F^{\top}$ is invariant under f. In fact, $\forall u \in F$, $\forall v \in F^{\top}$,
  $$< u, f(v) > = < f(u), v > = 0 \ (\operatorname{car} \ f(u) \in F).$$
  The restrictions of f to F and $F^{\top}$ are certainly symmetric operators.

The following definition generalizes the notion of symmetric operator to the case of a preHilbert $\mathbb{C}$–vector space. (A preHilbert space is a Hilbert space which is not necessarily complete and is also known as an inner product space. The definitions may vary slightly.)

Definition 2: Let E be a vector space over $\mathbb{C}$, equipped with a scalar product. We say that $f \in \mathcal{L}(E)$ is Hermitian if $\forall u, v \in E$: $< f(u), v > = < u, f(v) >$. We denote by $\mathcal{H}(E)$ the se of all Hermitian operators of E.

The preceding properties remain true for the Hermitian operators (except for the structure of a sub-space: we could however say that $f, g \in \mathcal{H}(E) \Rightarrow \alpha f + \beta g \in \mathcal{H}(E)$ with the condition that α and β are real numbers).

The following result is a first step toward the reduction of Hermitian operators (or symmetric operators).

Proposition 1: Let f be a Hermitian operator of E. Its eigenvalues are all real, and its eigenspaces are pairwise orthogonal.

Proof: Let λ be an eigenvalue of f, and let u be an associated eigenvector. Recall that the scalar product is semi-linear on the left and linear on the right.

$$< f(u), u > = < u, f(u) > \Rightarrow \overline{\lambda} \|u\|^2 = \lambda \|u\|^2 \Rightarrow \lambda = \overline{\lambda} \Rightarrow \lambda \in \mathbb{R}$$

Let μ be an eigenvalue of f, distinct from λ. $\forall u \in E_\lambda$, $\forall v \in E_\mu$:

$$< f(u), v > = < u, f(v) > \Rightarrow \lambda < u, v > = \mu < u, v > \Rightarrow < u, v > = 0$$

Remark: The preceding is again valid for the symmetric operators of a preHilbert $\mathbb{R}$-vector space E (at least in matters of the orthogonality of eigenspaces, since the question about the eigenvalues of f are real doesn't arise).

6. 2 Symmetric or Hermitian matrices

Now we are going to limit ourselves to finite dimensions. The two following definitions are probably familiar (at least the first!).

Definition 3: A matrix M of $\mathcal{M}_n(\mathbb{R})$ is said to be symmetric if $^{\top}M = M$.
A matrix M of $\mathcal{M}_n(\mathbb{C})$ is called Hermitian if $^{\top}\overline{M} = M$.

The matrix M with general term $m_{i,j}$ is thus symmetric (resp. Hermitian) if for all indices i, j, we have the equation $m_{j,i} = m_{i,j}$ (resp. $m_{j,i} = \overline{m_{i,j}}$).

The symmetric matrices are particular cases of Hermitian matrices, and the properties established for the latter also hold for both cases.

The following two simple functions allow creation of real or Hermitian symmetric matrices for small random matrices.

In both cases it suffices to indicate the order of the matrix which we want to create.

```
:randsym(n):Func:Local m:randMat(n,n)→m:m+mᵀ:EndFunc
```

```
:randherm(n):Func:Local m:
:randMat(n,n)+i*randMat(n,n)→m:m+mᵀ
:EndFunc
```

Here is the classic result which connects the vector point of view with that of the matrix approach.

Proposition 2: Let E be a $\mathbb{K}$-vector space of dimension $n \geq 1$, equipped with a scalar product, and let (e) be a orthonormal basis of E. Let f be an operator on E with matrix M in the basis (e). Then f is Hermitian (resp. symmetric) if and only if M is Hermitian (resp. symmetric)

Proof: (in the complex case)

Denote by X and Y the column matrices of the coordinates of $x, y \in E$, in (e). Since this basis is orthonormal, the scalar product may be written $< x, y > = {}^T\overline{X}Y$.

Under these conditions:

$$f \in \mathcal{H}(E) \Leftrightarrow \forall x, y \in E, < f(x), y > = < x, f(y) > \Leftrightarrow \forall X, Y, \; {}^T\overline{MX}\,Y = {}^T\overline{X}MY$$

$$\Leftrightarrow \forall X, Y, \; {}^T\overline{X}\,{}^T\overline{M}Y = {}^T\overline{X}MY \Leftrightarrow {}^T\overline{M} = M$$

The demonstration is identical (up to conjugation) to the real case.

We may draw from the preceding result that the eigenvalues of a Hermitian matrix are all real.

Here is an illustration with a random matrix of order 3. Up to roundoff errors, we indeed see that the imaginary parts of the eigenvalues are zero.

Every real symmetric matrix A of order n is a particular case of Hermitian matrix and thus represents in the standard basis of $\mathbb{C}^n$ (orthonormal for the usual scalar product) a Hermitian operator f.

The eigenvalues of A in $\mathbb{C}$ (of which there are n, each counted as many times as its multiplicity) are those of f and are thus all real.

In other words, the character -istic polynomial of A splits in $\mathbb{R}$.

Conversely, if E is a finite dimensional Euclidean (vector space over $\mathbb{R}$ equipped with a scalar product), and if f is a symmetric operator of E, its character -istic polynomial is that of every matrix representing f, in particular in an orthonormal basis.

This character -istic polynomial thus splits in $\mathbb{R}$.

We have shown the following result.

Proposition 3: Let A be a real symmetric matrix of order n (resp. f a symmetric operator of a Euclidean space of dimension n) with $n \geq 1$.

The characteristic polynomial of A (resp. that of f) splits in $\mathbb{R}$.

A and f thus have n real eigenvalues (each counted as many times as its multiplicity).

6. 3 Diagonalization in an orthonormal basis

We have arrived at the principal result of this section.

Proposition 4: Let f be a Hermitian operator of a $\mathbb{C}$-vector space E with dimension $n \geq 1$. Then there exists in E an orthonormal basis of eigenvectors of f. In particular f is diagonalizable.

Here is the analog for the real case.

Proposition 5: Let f be a symmetric operator of a $\mathbb{R}$-vector space E of dimension $n \geq 1$. Then there exists in E an orthonormal basis of eigenvectors of f. In particular f is diagonalizable.

Proof: (by induction on n). We denote $\mathbb{K} = \mathbb{R}$ or $\mathbb{C}$. For $n = 1$, this is evident. Suppose the property holds for rank $(n-1)$.
Let λ_1 be an eigenvalue of f, and let u_1 be an associated eigenvector (the existence of λ_1 is assured since the character -istic polynomial of f splits in $\mathbb{K}$). The vector line $\mathbb{K}u_1$ is invariant under f, so this is also the case for $G = (\mathbb{K}u_1)^\top$ (cf the beginning of this paragraph), and the restriction g of f to G is a Hermitian operator (if $\mathbb{K} = \mathbb{C}$) of symmetric (if $\mathbb{K} = \mathbb{R}$) of G.
Since $\dim G = n - 1$, there exists in G an orthonormal basis $u_2, u_3, \ldots, u_n$ formed of eigenvectors of g (thus of f).
The family $u_1, u_2, \ldots, u_n$ constitutes then an orthonormal basis of E, and it is composed of eigenvectors of f. This accomplishes the induction.

Recall that a matrix M of $\mathcal{M}_n(\mathbb{C})$ is called unitary if it is invertible and if $M^{-1} = {}^\top\overline{M}$. We know (cf section **5.5** of the Orthogonality chapter) that the unitary matrices are the matrices to change between orthonormal bases de $\mathbb{C}^n$. The Hermitian matrices of order n form a group under the product operation, called the unitary group of order n and denoted $\mathcal{U}(n)$.

Likewise, a matrix M of $\mathcal{M}_n(\mathbb{R})$ is called orthogonal if it is invertible and if $M^{-1} = {}^\top M$. The orthogonal matrices are the transition matrices between orthonormal bases of $\mathbb{R}^n$. They form a group under the matrix product called the orthogonal group of index n, and denoted $\mathcal{O}(n)$.

The Hermitian matrices (resp. symmetric) are the matrices of the Hermitian operators (resp. symmetric) of $\mathbb{C}^n$ (resp. of $\mathbb{R}^n$) in the standard basis (which is orthonormal for the usual scalar product).

Here are the matrix equivalents of propositions **4** and **5**.

Proposition 6: Let $M \in \mathcal{M}_n(\mathbb{C})$ be a Hermitian matrix.
then M is diagonalizable. There exists a unitary matrix P such that $P^{-1}MP = {}^\top\overline{P}MP$ is diagonal. We express this property by saying that M is diagonalizable in in the unitary group.

Proposition 7: Let $M \in \mathcal{M}_n(\mathbb{R})$ be a symmetric matrix. Then M is diagonalizable over $\mathbb{R}$. There exists an orthogonal matrix P such that $P^{-1}MP = {}^\top PMP$ is diagonal. We express this property by saying that M is diagonalizable in the orthogonal group.

With the function randherm, we create a Hermitian matrix A of order 4.

Then we calculate the eigenvalues of A. We use real to avoid the negligible imaginary parts which only come from roundoff errors.

```
F1▼ F2▼ F3▼ F4▼ F5 F6▼
  Algebra Calc Other PrgmIO Clean Up
■ RandSeed 0 : randherm(4)→a
  ⎡16        -8·i      16·i-5    -5-11·i⎤
  ⎢8·i       12        8·i-15    2-10·i ⎥
  ⎢-5-16·i   -15-8·i   8         18·i+6 ⎥
  ⎣11·i-5    10·i+2    6-18·i    16     ⎦
■ real(eigVl(a))→vp
           {51.652  -18.03  4.3857  13.993}
real(eigVl(a))→vp
MAIN          RAD AUTO        FUNC 2/30
```

The function cround will be useful to pursue this example. It fills in a defect of the built-in function round which doesn't apply to complex numbers.

```
:cround(z,n):round(real(z),n)+i*round(imag(z),n)
```

The built-in function eigVc returns the transition matrix P for a basis of unitary eigenvectors of A.

The eigenspaces (we know that they are pairwise orthogonal) here are (eigenvalues distinct) vector lines. The eigenvectors obtained thus form an orthonormal basis of $\mathbb{C}^n$.

In other words, the matrix P is unitary.

We verify this by verifying that the product $^\top\overline{P}P$ is indeed equal to the identity matrix I_4. (Recall: the built-in "transposition" function in fact does a conjugate transpose or transconjugation. Hence, where we read $P^\top$, we must understand $^\top\overline{P}$.). The function cround lets us obtain an exact result by rounding to the precison 10^{-10}.

```
F1▼ F2▼ F3▼ F4▼ F5 F6▼
  Algebra Calc Other PrgmIO Clean Up
■ eigVc(a)→P
  ⎡-.22002+.43139·i   .04487-.36466·i  ⎤
  ⎢-.4294+.05233·i    .19342-.33778·i  ⎥▶
  ⎢.54962             .71857           ⎥
  ⎣-.04851-.52339·i   -.3486+.27502·i  ⎦
eigVc(a)→p
MAIN          RAD AUTO        FUNC 1/30
```

```
F1▼ F2▼ F3▼ F4▼ F5 F6▼
  Algebra Calc Other PrgmIO Clean Up
                              ⎡1.  0.  0.  0.⎤
                              ⎢0.  1.  0.  0.⎥
■ cround(p^T·p,10)            ⎢0.  0.  1.  0.⎥
                              ⎣0.  0.  0.  1.⎦
cround(p^Tp,10)
MAIN          RAD AUTO        FUNC 1/30
```

By evaluating $^\top\overline{P}AP = P^{-1}AP$, we verify that the unitary matrix P diagonalizes A.

In fact we obtain a diagonal matrix D, carrying the eigenvalues (all real) of A.

```
F1▼ F2▼ F3▼ F4▼ F5 F6▼
  Algebra Calc Other PrgmIO Clean Up
■ cround(p^T·a·p,10)→d
  ⎡51.652  0.      0.      0.    ⎤
  ⎢0.      -18.03  0.      0.    ⎥
  ⎢0.      0.      4.3857  0.    ⎥
  ⎣0.      0.      0.      13.993⎦
cround(p^T*a*p,10)→d
MAIN          RAD AUTO        FUNC 1/30
```

Now we treat the example of a symmetric real matrix A of order 5 (and formed by our function randsym).

The function eigVc gives us a transition matrix P to a basis of unitary eigenvectors of A.

F1 ▾ ﹍	F2 ▾ Algebra	F3 ▾ Calc	F4 ▾ Other	F5 PrgmIO	F6 ▾ Clean Up	

■ RandSeed 0 : randsym(5)→a

$$\begin{bmatrix} -18 & 0 & -13 & -10 & 7 \\ 0 & 2 & -11 & 2 & 8 \\ -13 & -11 & -10 & 6 & -1 \\ -10 & 2 & 6 & -18 & 12 \\ 7 & 8 & -1 & 12 & 16 \end{bmatrix}$$

RandSeed 0:randsym(5)→a
MAIN   RAD AUTO   FUNC 1/30

F1 ▾ ﹍	F2 ▾ Algebra	F3 ▾ Calc	F4 ▾ Other	F5 PrgmIO	F6 ▾ Clean Up	

■ eigVc(a)→p

$$\begin{bmatrix} -.1497 & .7334 & -.5322 & -.1786 & .3529 \\ -.4164 & .1262 & .7056 & -.2874 & .4799 \\ .1806 & .3215 & .2022 & -.6613 & -.6211 \\ -.2018 & .5308 & .3139 & .6579 & -.3823 \\ -.8549 & -.2473 & -.2818 & -.1238 & -.3365 \end{bmatrix}$$

eigVc(a)→p
MAIN   RAD AUTO   FUNC 1/30

We observe that P is an orthogonal matrix ($^TPP = I_5$, the roundoff errors being corrected by a call to the function round). This signifies that the basis of eigenvectors obtained by eigVc is orthonormal.

A last verification: the matrix P diagonalizes A ($^TPAP = P^{-1}AP = I_5$). We observe that the eigenvalues of A are distinct.

The fact that the eigenvectors obtained are orthonormal is thus foreseeable when we know that they are unit vectors. This is the reason that the function function eigVc always returns a matrix whose columns are unit eigenvectors.

F1 ▾ ﹍	F2 ▾ Algebra	F3 ▾ Calc	F4 ▾ Other	F5 PrgmIO	F6 ▾ Clean Up	

■ round$\left(p^T \cdot p, 10\right)$

$$\begin{bmatrix} 1. & 0. & 0. & 0. & 0. \\ 0. & 1. & 0. & 0. & 0. \\ 0. & 0. & 1. & 0. & 0. \\ 0. & 0. & 0. & 1. & 0. \\ 0. & 0. & 0. & 0. & 1. \end{bmatrix}$$

round(pᵀ*p,10)
MAIN   RAD AUTO   FUNC 1/30

F1 ▾ ﹍	F2 ▾ Algebra	F3 ▾ Calc	F4 ▾ Other	F5 PrgmIO	F6 ▾ Clean Up	

■ round$\left(p^T \cdot a \cdot p, 10\right)$

$$\begin{bmatrix} 24.17 & 0. & 0. & 0. & 0. \\ 0. & -33.3 & 0. & 0. & 0. \\ 0. & 0. & -3.458 & 0. & 0. \\ 0. & 0. & 0. & -24.45 & 0. \\ 0. & 0. & 0. & 0. & 9.036 \end{bmatrix}$$

round(pᵀ*a*p,10)
MAIN   RAD AUTO   FUNC 1/30

One may logically ask whether, for a given symmetric matrix (or Hermitian) A, the function eigVc always returns an orthogonal matrix (or unitary) matrix.

The two preceding examples shouldn't give this illusion, since a problem arises when at least one eigenvalue is multiple.

If different eigenspaces are obtained and are mutually orthogonal, the following example shows that, within a certain eigenspace (of dimension greater than 1), the function eigVc returns a basis of certain unit eigenvectors which are not orthogonal!

Here is a real symmetric matrix A of order 5. Its eigenvalues are 6 (simple) and 1 (quadruple). The function $\texttt{eigVc}$ returns the transition matrix P to a basis of unit eigenvectors.

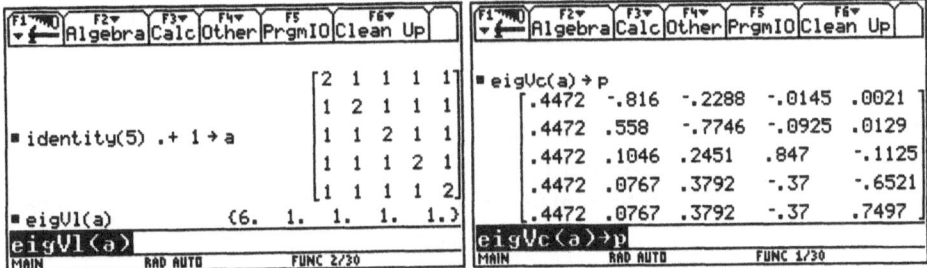

The first column vector of P is proportional to $(1, 1, 1, 1, 1)$: It forms a basis of the vector line E_6 of A for $\lambda = 6$.

The other column vectors of P form a basis of eigenspace E_1 of A for the eigenvalue $\lambda = 1$.

The calculation of $P^{\top}P$ shows that P is not orthogonal.
The diagonal coefficients of $P^{\top}P$ are equal to 1, which confirms that the eigenvectors are unit vectors.
The first row and the first column are those of I_5, which leads to the orthogonality of E_6 and E_1.

We know that the instruction QR decomposes a matrix A into a product of an orthogonal matrix Q and a triangular matrix R, proceeding from the orthogonalization of the columns of A. If the column vectors of A are obtained by juxtaposition of bases of subspaces which are pairwise orthogonal, the orthonormalization effected "respects" these subspaces.

To return to our example, this signifies that the instruction QR transforms P (which is the transition matrix to a basis of eigenvectors, the eigenspaces being pairwise orthogonal) into Q, a transition matrix to an orthonormal basis of eigenvectors.

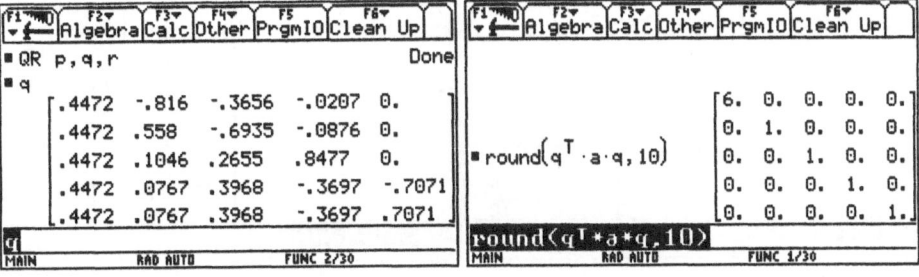

The following experience shows that the eigenvectors could be given in any order. After the change of columns 1 and 3 of P, the direction vector u of E_6 is found in the vectors of a basis of E_1.

The instruction QR furnishes an orthogonal matrix Q, which is still a transition matrix to an orthonormal basis of eigenvectors.

Only the order of these vectors has changed: the vector u is found in the third column, which is represented on the diagonal of $Q^{\mathsf{T}}AQ$.

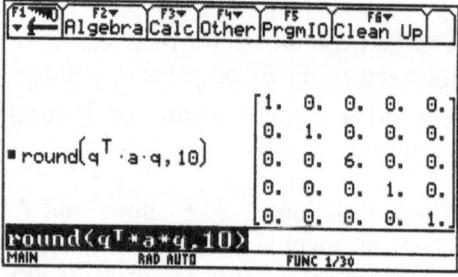

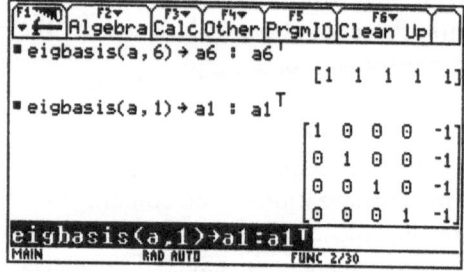

We have diagonalized A in the orthogonal group, but the recourse to the eigVc function constrains us to the "approx " mode.

Here is how we could do the same diagonalization, but in symbolic mode.

The function eigbasis (see section 2.2) first furnishes a basis of the two eigenspaces of A.

An orthonormal basis of the eigenspace of A for $\lambda = 6$ is easily chosen since it suffices to normalize the vector $(1, 1, 1, 1, 1)$. For the eigenvalue 1 of multiplicity 4, we again use the built-in function QR. In fact, if it operates on exact matrices, it returns an exact result.

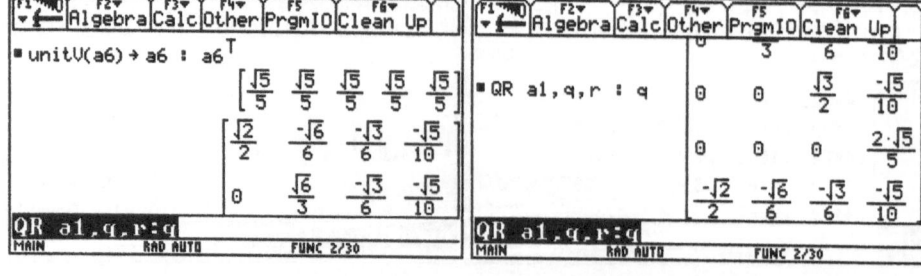

We adjoin the column matrix a6 and the matrix q to a new matrix. The result, always saved in the variable q, is so large that two screens are necessary.

We see that it is only necessary to call the round instruction in order to avoid rounding errors: the matrix Q is exactly orthogonal, and exactly diagonalizes the matrix A.

The function orthdiag automates the preceding calculations. It takes as an argument a matrix M and returns a transition matrix P to a basis of eigenvectors of M (orthonormal within each eigenspace).

If the matrix M is symmetric (resp. Hermitian), the matrix P is thus orthogonal (resp. unitary).

```
:orthdiag(m)
:Func:Local n,s,i,j,b,t,c,d,q
:cZeros(det(m-θθ_),θθ_)→s
:newMat(rowdim(m),1)→q
:For i,1,dim(s)
:   eigbasis(m,s[i])→b
:   1→t:(unitV(bᵀ[1]))ᵀ →c:c→d
:   For j,2,colDim(b):t-c*cᵀ →t
:      unitV(t*(bᵀ[j])ᵀ)→c:augment(d,c)→d
:   EndFor
:   augment(q,d)→q
:EndFor:subMat(q,1,2)
:EndFunc
```

In the listing of the function orthdiag, we recognize a call to our eigbasis function (which finds an eigenvector for a given eigenvalue). The inner loop "For j " orthonormalizes this basis by the Gram-Schmidt process (see the function qrsch in section **6.1** of the *Orthogonality* chapter).

Here is a square Hermitian matrix A of order 3. We see that the eigenvalues are 1, $1 - \sqrt{3}$ and $1 + \sqrt{3}$.

The function orthdiag returns a matrix Q which is verified to be unitary $(^T\overline{Q}Q = I_3)$, and which diagonalizes the matrix A.

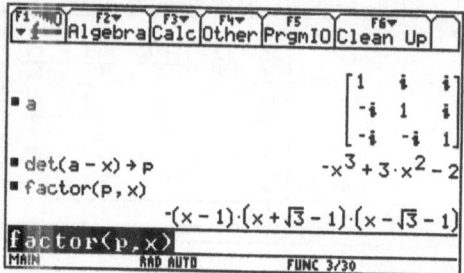

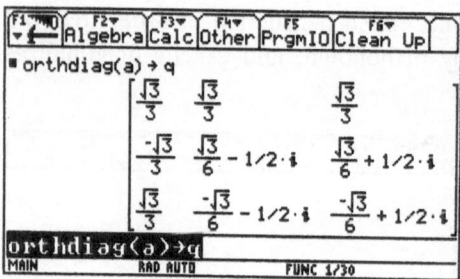

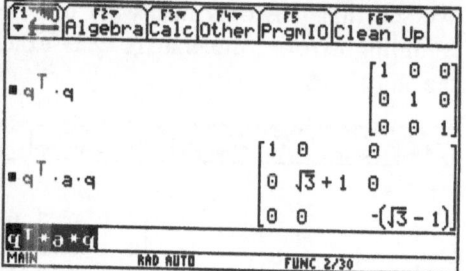

In the second example, A is a real symmetric matrix of order 4, whose eigenvalues are -4 (double), 4 (simple) and 8 (simple).

The function orthdiag converts A into the matrix Q.

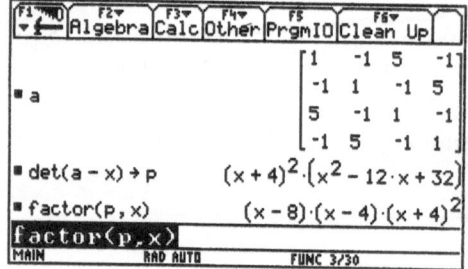

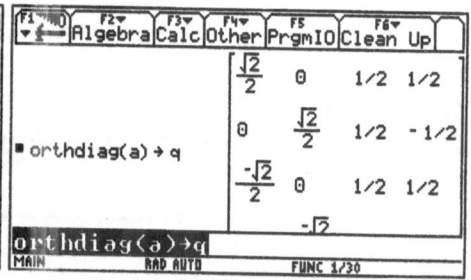

We verify finally that the matrix Q is an orthogonal matrix which diagonalizes the matrix A.

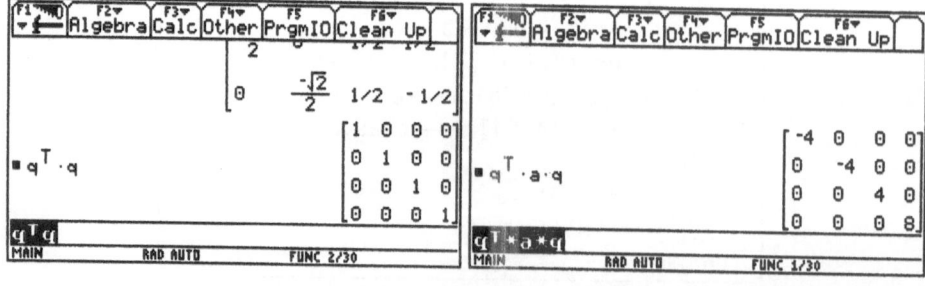

6. 4 Orthogonal polynomials: a return visit

Let I be an interval of $\mathbb{R}$ (not reducing to a point) and let $x \mapsto w(x)$ be a continuous map which has strictly positive values on I (possibly vanishing at isolated points). Let E be the vector space of continuous maps $f : I \to \mathbb{R}$ such that $x \mapsto f^2(x)w(x)$ is integrable on I. (We suppose that this vector space contains the polynomials with real coefficients.)

The map $(f, g) \mapsto \int_I f(x)g(x)w(x)\,dx$ defines a scalar product on E. The function $w(x)$ is called a weight function.

We know from the Orthogonality chapter, section 3, that there is a sequence of pairwise orthogonal polynomials $(P_n)_{n \geq 0}$, arranged by degree ($\forall n \in \mathbb{N}$, $\deg P_n = n$), and unique up to a non-zero multiplicative constant.

The uniqueness of P_n is often assured by additional constraints.

The most standard particular cases lead to the classical sequences $(P_n)_{n \geq 0}$ of (Legendre, Chebyshev, Laguerre, etc.). Finally, the existence of such a scalar products brings us to raise the question of symmetric operators in this setting.

A) A FIRST EXAMPLE

We define a scalar product on $\mathbb{R}[X]$ by: $< P, Q > = \int_{-1}^{1} P(x)Q(x)\,dx$.

This leads to the family $(L_n)_{n \geq 0}$ of Legendre polynomials. We are going to retrieve these as eigenvectors of a symmetric operator of $\mathbb{R}[X]$.

We consider the map f defined on E par: $f(P) = \dfrac{d}{dx}\big((x^2 - 1)P'(x)\big)$.

The linearity of f is evident. On the other hand, if P is a non-constant polynomial, $f(P) = (x^2 - 1)P''(x) + 2xP'(x)$ is a polynomial of the same degree as P. (The constant polynomials have as their image the zero polynomial).

We conclude that f is an operator on $\mathbb{R}[X]$, and also that, for every n, of the sub-space $\mathbb{R}_n[X]$ of polynomials of degree less than or equal to n.

For all polynomials P and Q, and by use of integration by parts:

$$< f(P), Q > = \int_{-1}^{1} f(P)(x)Q(x)\,dx = \int_{-1}^{1} \frac{d}{dx}\big((x^2 - 1)P'(x)\big)Q(x)\,dx$$

$$= \Big[(x^2 - 1)P'(x)Q(x)\Big]_{-1}^{1} - \int_{-1}^{1} (x^2 - 1)P'(x)Q'(x)\,dx$$

$$= \int_{-1}^{1} (1 - x^2)P'(x)Q'(x)\,dx$$

P and Q play the same role: $< f(P), Q > = < f(Q), P > = < P, f(Q) >$.

In other words, f is a symmetric operator of $E = \mathbb{R}[X]$. It is the same, for every n, for the restriction f_n of f to $\mathbb{R}_n[X]$.

If a polynomial P is of degree k, and if its leading term is $a_k x^k$, then the leading term of $f(P) = (x^2 - 1)P''(x) + 2xP'(x)$ is $k(k + 1)a_k x^k$.

The matrix of f_n in the standard basis $1, X, X^2, \ldots, X^n$ is thus of the form shown here. The eigenvalues of f_n are the $\lambda_k = k^2 + k$, with $0 \leq k \leq n$. They are all of multiplicity 1, which confirms that f_n is diagonalizable.

$$\begin{pmatrix} 0 & \star & \cdots & \cdots & \cdots & \star \\ 0 & 2 & \star & & & \vdots \\ \vdots & & \ddots & \ddots & \ddots & \vdots \\ \vdots & & & \ddots & k^2+k & \star & \vdots \\ \vdots & & & & \ddots & \ddots & \star \\ 0 & \cdots & \cdots & \cdots & 0 & n^2+n \end{pmatrix}$$

More generally, the eigenvalues of f are the $\lambda_k = k^2 + k$, with $k \in \mathbb{N}$, and all the eigenspaces $E(\lambda_k)$ pairwise orthogonal vector lines for the scalar product $(P, Q) \mapsto \int_{-1}^{1} P(x)Q(x)\, dx$.

Let P be an eigenvector of f. We will show that $P(1) \neq 0$. For this, we prove that the multiplicity m of 1 as a root of P is equal to 0.

There is a polynomial Q such that $P = (x-1)^m Q$, with $Q(1) \neq 0$. Thus:

$$\begin{aligned} f(P) = \lambda P &\Rightarrow \lambda(x-1)^m Q = (x^2-1)\big((x-1)^m Q\big)'' + 2x\big((x-1)^m Q\big)' \\ &\Rightarrow \lambda(x-1)Q = (x+1)\big((m^2-m)Q + 2m(x-1)Q' + (x-1)^2 Q''\big) \\ &\quad + 2x\big(mQ + (x-1)Q'\big) \text{ (simplifying by } (x-1)^{m-1}) \end{aligned}$$

With $x = 1$, we find: $0 = 2(m^2 - m)Q(1) + 2mQ(1) \Rightarrow m^2 Q(1) = 0 \Rightarrow m = 0$.

In the vector line $E(\lambda_n)$, there is thus a unique polynomial L_n such that $L_n(1) = 1$. This is the Legendre polynomial of index n.

This new definition contains the essential properties of the Legendre polynomials ($\deg L_n = n$ and pairwise orthogonality of L_n) but which are mainly expressed in the differential equation:

$$f(L_n) = \lambda_n P_n \Rightarrow (x^2 - 1)L_n''(x) + 2xL_n'(x) = n(n+1)L_n(x)$$

B) GENERALIZATION

We may succinctly extend the preceding point of view to the general case of a scalar product defined, at least for $\mathbb{R}[X]$, by $< P, Q > = \int_I P(x)Q(x)\omega(x)\, dx$.

We suppose that ω remains strictly positive on the interior of I, and that $\varphi : I \to \mathbb{R}$ is a function such that $\varphi\omega$ vanishes at the extremities of I.

Then we define a map f of $\mathbb{R}[X]$ by: $f(P) = \dfrac{1}{\omega}(\varphi\omega P')'$.

Finally, we suppose that if $\deg P = n \geq 1$, then $f(P)$ is a polynomial of degree n. These hypotheses may seem artificial, but in every classical case there exists such a function φ. (In the "Legendre case", $\omega = 1$ and $\varphi = x^2 - 1$).

Thus: $< f(P), Q > = \int_I (\varphi\omega P')'Q = \big[\varphi\omega P'Q\big]_I - \int_I \varphi\omega P'Q' = -\int_I \varphi\omega P'Q'$,

which proves as before that f is symmetric.

The diagonalization of the restriction f_n of f to $\mathbb{R}_n[X]$ leads to a family of orthogonal polynomials P_n, the equation $f(P_n) = \lambda_n P_n$ gives the differential equation satisfied by the P_n.

6. 5 Tridiagonalization of symmetric matrices

We saw in paragraph section **5.3** that the Jacobi method allows one to start with a real symmetric matrix A and construct a sequence of symmetric matrices $(A_k)_{k \geq 0}$ (with $A_0 = A$), all similar to A and which converge to a diagonal matrix D (the diagonal elements of which are the eigenvalues of A).

We know that each A_k may be constructed from A_{k-1} by the equation $A_k = {}^T G_k A_{k-1} G_k$, where G_k is a rotation matrix, which is therefore orthogonal matrix.

Thus, $A_k = {}^T \Omega_k A \Omega_k$, where Ω_k is the orthogonal matrix $\Omega_k = G_1 G_2 \ldots G_k$.

We may thus write: $D = \lim\limits_{k \to \infty} A_k = \lim\limits_{k \to \infty} {}^T \Omega_k A \Omega_k$

If we have proved the convergence of the sequence of matrices A_k, we haven't done so for the sequence of matrices Ω_k. We could nonetheless consider the iterative method of Jacobi as a technique of diagonalization "in the limit " in the orthogonal group.

It is not unreasonable to hope that such a method allows diagonalization of an arbitrary real symmetric matrix A by means of an orthogonal transition matrix and a finite number of steps.

But such a method doesn't exist.
The most we may hope for in a finite number of orthogonal transformations is to pass from A to a symmetric tridiagonal matrix of the form shown opposite:

$$T = \begin{pmatrix} a_1 & b_1 & 0 & \cdots & & 0 \\ b_1 & a_2 & b_2 & \ddots & & \vdots \\ 0 & b_2 & a_3 & \ddots & & 0 \\ \vdots & \ddots & \ddots & \ddots & & b_{n-1} \\ 0 & \cdots & 0 & b_{n-1} & & a_n \end{pmatrix}$$

There are two "orthogonal tridiagonalizations " of a symmetric matrix. One uses orthogonal symmetries with respect to hyperplanes (Householder's method) and the other uses rotation matrices (Givens' method).

A) THE METHOD OF GIVENS

We won't give the details of this technique, which repose in large part in the calculations in section **5.3**.

The only change with respect to the Jacobi method, where each rotation matrix G_k was chosen in a way to annihilate the non-diagonal coefficient with maximum modulus in the matrix A_{k-1}, is that we are going to successively annihilate all the coefficients not on the three diagonals.

We leave to the reader to see how this is possible by proceeding on each column from top to bottom, that is, on column j, from row $i = j + 2$ to row $i = n$, and from the first to the next to last column. The crucial point is to verify that the treatment of a non-tridiagonal coefficient does not "unannihilate " the coefficients already set to zero.

The function `tridiagg` tridiagonalizes a real symmetric matrix A.

This function displays the successive steps A_k, then saves them in the variable θt and θq respectively the symmetric tridiagonal matrix T and the orthogonal matrix Q such that $^TQAQ = T$.

```
:tridiagg(a)
:Prgm:Local n,i,j,k,r,c,s,g
:ClrIO:rowDim(a)→n:0→k
:identity(n)→ θq
:For j,1,n-2:For i,j+2,n
:   √(a[j,j+1]^2+a[j,i]^2)→r
:   If r≠0 Then
:      a[j,j+1]/r→c:a[j,i]/r→s
:      identity(n)→g
:      c→g[i,i]:c→g[j+1,j+1]
:      s→g[i,j+1]:-s→g[j+1,i]
:      θq*g→ θq:g^T*a*g→a:clrIO
:      k+1→k:Disp "A["&string(k)&"]="
:      Pause a
:   EndIf
:EndFor:EndFor
:a→ θt
:EndPrgm
```

The following screens show the tridiagonalization of a symmetric matrix A of order 4. The program `tridiagg` displays the steps A_1, A_2 and A_3 of the decomposition. The non-tridiagonal coefficients are progressively annihilated.

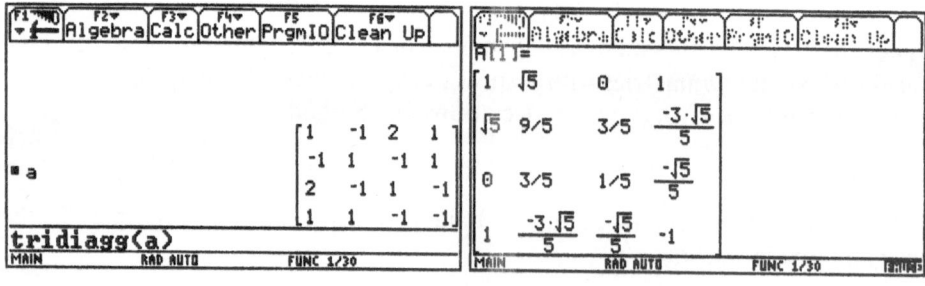

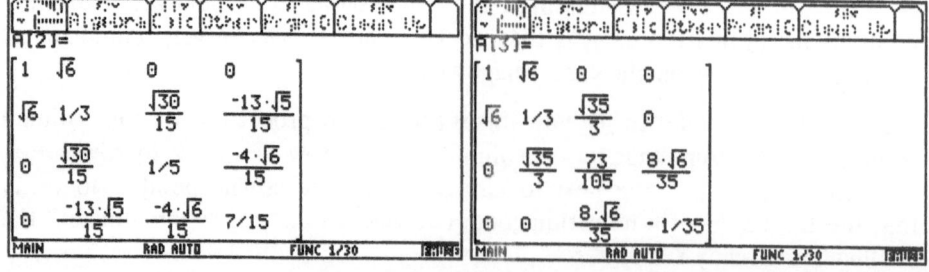

The symmetric tridiagonal matrix final T is stored in the variable θt, and the orthogonal matrix Q transforming A into T is stored in θq.

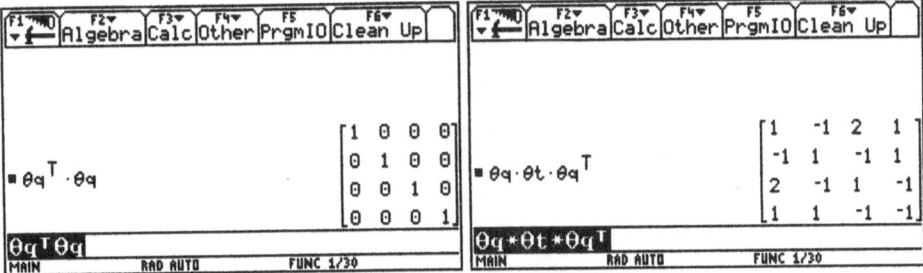

We verify that Q is indeed an orthogonal matrix, and equation $QT^\top Q = A$ holds.

B) THE METHOD OF HOUSEHOLDER

To tridiagonalize a real symmetric matrix A by means of orthogonal transformations, we could just as well use Househoder matrices. We refer to sections **6.4** and **6.5** of the chapter on *Orthogonality*.

The Householder matrix associated with v is $H = I_n - \dfrac{2}{\|v\|^2}[v]^\top[v].$

It represents, in $\mathbb{R}^n$, the orthogonal symmetry h with respect to the hyperplane orthogonal to the vector v.

Let $u = (a_1, a_2, \ldots, a_n)$ be a vector of $\mathbb{R}^n$, and j be an index between 1 and n. We denote by $u' = (0, \ldots, 0, a_j, \ldots, a_n)$ and $v = (0, \ldots, 0, a_j + \varepsilon \|u'\|, a_{j+1}, \ldots, a_n)$, where $\varepsilon = \pm 1$ is the sign of the coefficient a_j.

The orthogonal symmetry h with respect to the plane orthogonal to v sends the vector u onto $h(u) = (a_1, \ldots, a_{j-1}, -\varepsilon \|u'\|, 0, \ldots, 0)$. In particular, it annihilates all the coordinates of u to starting with the $(j+1)$-th. It is important to observe that h leaves invariant the first $j - 1$ vectors of the standard basis.

These remarks are at the basis of the QR decomposition by the method of Householder: The successive products of the square matrix A of order n by the matrices $H_1, H_2, \ldots, H_{n-1}$ allow the progressive annihilation of all the subdiagonal coefficients of A (from the first to the next to last column, the matrix H_k treating the k-th column).

We thus arrive at $\Omega A = R$, where $\Omega = H_{n-1} H_{n-2} \cdots H_1$ (R being upper triangular), then $A = QR$ with $Q = H_1 H_2 \cdots H_{n-1}$.

This method is of no interest if we want to find the eigenvalues of A, since the two matrices A and R are not similar.

The solution consists of using the transformation $A \mapsto B = HAH$ with Householder matrices H (recall that $H = {}^{T}H = H^{-1}$), the idea always being to annihilate the subdiagonal coefficients of A.

But if A is real symmetric, B is also. Every annihilation of a subdiagonal coefficient also does that to a "superdiagonal " coefficient.

However, if we know to annihilate the subdiagonal coefficients of a column of A by passing from A to HA, what happens when we pass from A to HAH?

The function hh will allow us to illustrate this possibility. It constructs a Householder matrix H starting from a vector u and an index j, in a manner to annihilate all the coefficients u beginning with the $(j+1)$-th. This is the matrix of the hyperplane symmetry h mentioned at the beginning of this section.

```
:hh(u,j)
:Func:Local i,ε
:mat▷list(u)→u
:For i,1,j-1:0→u[i]:EndFor
:when(u[j]>0,1,-1,-1)→ ε
:u[j]+ε√(dotP(u,u))→u[j]
:list▷mat(u)→u
:1-2*uᵀ*u/(norm(u))^2
:EndFunc
```

We are going to use A, a real, symmetric matrix of order 5 as a guinea pig. The first time we calculated the eigenvalues of A. We then form the Housholder matrix H intending to annihilate all the subdiagonal coefficients of A.

Conforming to what was previewed, the subdiagonal coefficients of the first column of of HA are zero.

Sadly, the eigenvalues of the matrix HA have nothing to do with those of A.

If we want to conserve the eigenvalues
of A, we must form the matrix HAH.

The result is very deceiving since
all the coefficients annihilated by the
operation $A \mapsto HA$ are "unannihilated
" by the transformation $A \mapsto HAH$.

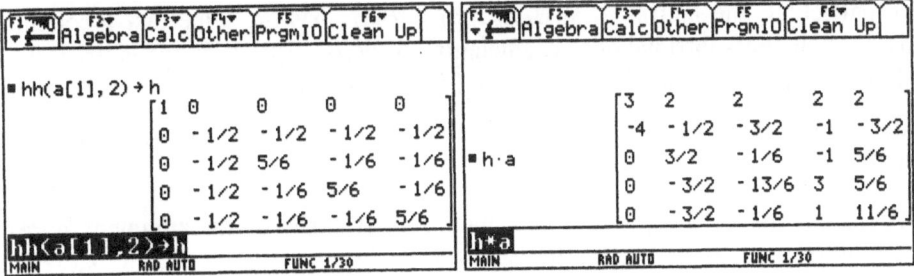

We now form the Householder matrix H which allows the annihation of all
the coefficients of the first column of A, not starting from the second (as
before), but starting with the third. We indeed observe that the "subdiagonal
" coefficients of the first column of HA are zero.

It is remarkable to observe that the transformation $A \mapsto B = HAH$ holds to 0
the "subtridiagonal " coefficients of the first column!

Since B is symmetric, there are other coefficients above the diagonal this time,
which are likewise now zero. Finally, the matrix B is similar to A. We check
that the eigenvalues are the same.

We are going to generalize the preceding observation and show that one may
tridiagonalize A in a finite number of steps.

We thus construct, starting with the real symmetric matrix A of order n, a
finite sequence $A_0 = A$, $A_1 = H_1 A_0 H_1, \ldots$, $A_{n-2} = H_{n-2} A_{n-3} H_{n-2}$: each matrix
H_k is charged with "treating " the k-th column of A_{k-1} by annihilating the
"subtridiagonal " coefficients.

We have to show that A_{n-2} is symmetric tridiagonal.

Suppose we know that the matrix A_{k-1} is symmetric and tridiagonal for the columns and rows with indices 1 to $k-1$. (These conditions hold for $k = 1$, with the symmetric matrix $A_0 = A$).

Then we form the Householder matrix K_k, in order to annihilate subtridiagonal coefficients of the k-th column of A_{k-1} (and thus from row $k+2$ to row n).

By definition of H_k, the subtridiagonal coefficients of the first k columns of $B_k = H_k A_{k-1}$ are thus zero.

On the other hand, H_k is the matrix (in the standard basis $e_1, e_2, \ldots, e_n$ of $\mathbb{R}^n$) of an orthogonal hyperplane symmetry leaving invariant the vectors $e_1, \ldots, e_k$.

Thus, $H_k = \begin{pmatrix} I_r & 0 \\ 0 & H' \end{pmatrix}$, where H' is a Householder matrix of order $n - r$.

In particular, the matrix $A_k = H_k A_{k-1} H_k = B_k H_k$ has the same first k columns as B_k: all the subtridiagonal coefficients are zero. But A_k is symmetric: it is thus tridiagonal for its first k rows and its first k columns.

By a finite induction, we have deduced that A_{n-2} is a symmetric tridiagonal matrix (and indeed it is similar to A).

The program tridiagh does this method. It takes as an argument a real symmetric matrix A, displays the steps A_k of the decomposition, and stores in the variable θt and θq respectively the reduced tridiagonal symmetric matrix T and the orthogonal matrix Q such that $^\top QAQ = T$.

The instruction approx(a)→a forces the "approx " mode. We could just as well eliminate this instruction, but one must know that with this method the calculations in exact mode quickly become very complicated.

Display of the steps A_k is done with rounding to the precision 10^{-10}. In this way, we better see the progressive annihilation of the appropriate coefficients.

```
:tridiagh(a)
:Prgm:Local n,h,k
:approx(a)→a
:rowDim(a)→n:identity(n)→ θq
:For k,1,n-2
:   hh(a⊤[k],k+1)→h:θq*h→ θq
:   h*a*h→a
:   ClrIO:Disp "A["&string(k)&"]="
:   Pause round(a,10)
:EndFor
:a→ θt
:EndPrgm
```

We reconsider the matrix A used above.

Here are three steps of its tridiagonalization.

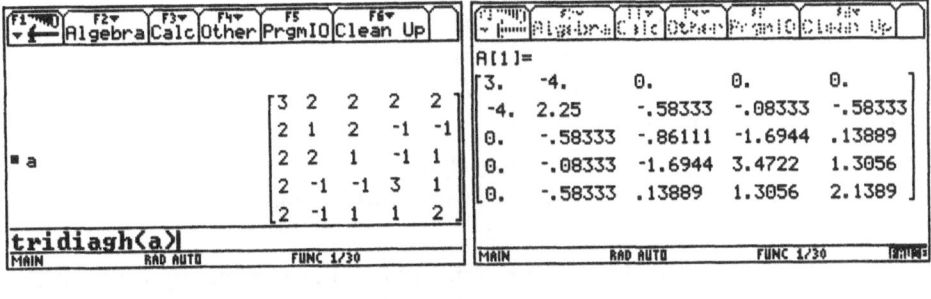

The matrix $T = A_3$, the last step of the decomposition, is symmetric and tridiagonal. It is stored in the variable θt. Here is the orthogonal matrix Q which tridiagonalizes A and also a verification of the equation $A = QT^{\top}Q$.

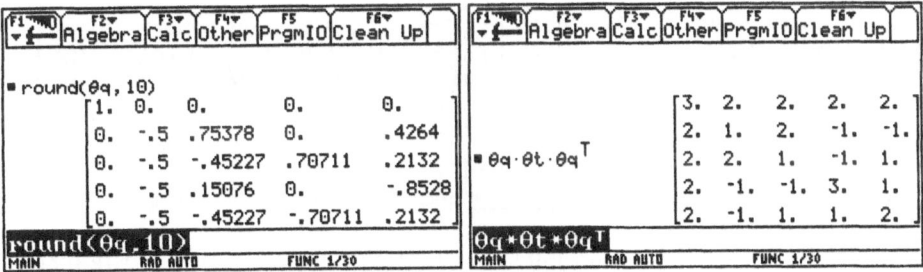

B) REDUCTION TO THE HESSENBERG FORM

In the listing of the function tridiagh, we read the instruction hh(a$^{\top}$[k],k+1)→h.

This use of the transpose of A is a little surprising since we know that A (and all the matrices A_k which is deduced from it, and which are in turn stored in the local variable a) are symmetric.

In fact, we remember that the Householder matrices constructed by the program tridiagh are known to annihilate the subtridiagonal coefficients of A, column by column.

To address the k-th column of A, it is simplest to transpose A and to extract the k-th row of $^{\top}A$. (The syntax is: a$^{\top}$[k]).

The actual listing of the function tridiagh thus permits us to treat real non-symmetric matrices A.

If we look over the preceding proof, especially the passage of A_{k-1} to A_k, we observe that the subtridiagonal coefficients of the first k columns of $B_k = H_k A_{k-1}$ are zero, and it is the same for $A_k = B_k H_k = H_k A_{k-1} H_k$, which has the same first k columns as B_k.

Thus, the application of the preceding algorithm leads from a square matrix A with real coefficients to a matrix T whose subtridiagonal coefficients are zero. We say that T is a Hessenberg matrix.

Here are the successive steps of the reduction to the Hessenberg form, in the orthogonal group, of an arbitrary square matrix A of order 5.

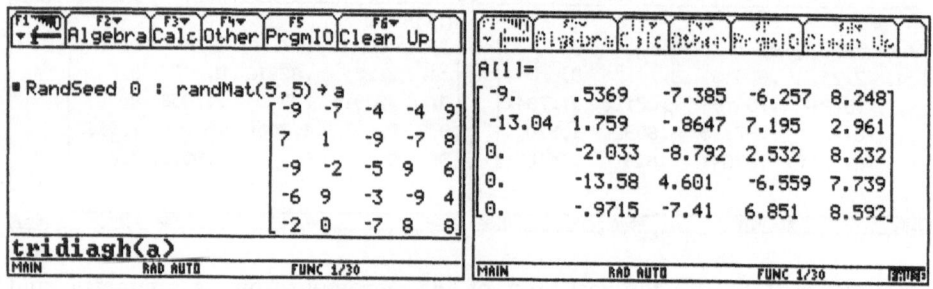

The matrix $T = A_3$ in the last step of the decomposition is a Hessenberg matrix. It is stored in the variable θt.

Here is the orthogonal matrix Q which allows this decomposition, and the verification of the equation $A = Q T^{\top} Q$.

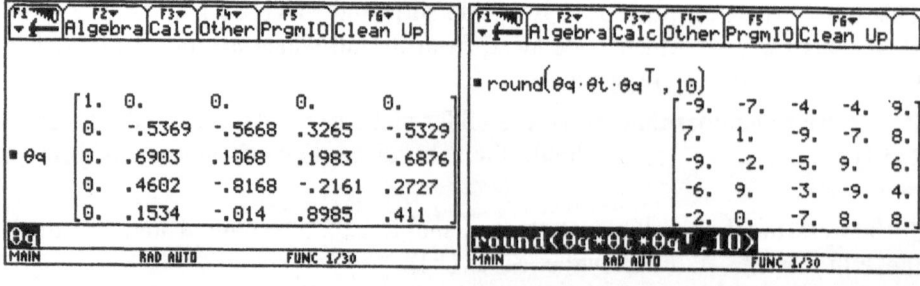

7. Positive symmetric matrices

The preceding section was committed to symmetric operators of a real preHilbert space. In finite dimensions, their study must be confined to real symmetric matrices.

Now we are going to consider particular real symmetric matrices: those whose eigenvalues are all positives.

7. 1 Preliminary theory

Let E be a $\mathbb{R}$-vector space of dimension $n \geq 1$, equipped with a scalar product. Let f be a symmetric operator on E ($f \in \mathcal{S}(E)$).
We define a map $\varphi : E \times E \to \mathbb{R}$, by: $(u,v) \mapsto \varphi(u,v) = \,< f(u),v>$.

The map φ is clearly bilinear (using the linearity of f and bilinearity of the scalar product) and symmetric (the equation $\varphi(u,v) = \varphi(v,u)$ carrys the symmetry of f).

In order that φ defines a scalar product on E, it must be the case that:

- φ is positive: $\forall x \in E$, $\varphi(x,x) = \,< f(x),x> \geq 0$.
- φ is definite: $\forall x \in E$, $< f(x),x> = 0 \Rightarrow x = \overrightarrow{0}$.

We know that there exists in E an orthonormal basis $(\varepsilon) = (\varepsilon_1, \varepsilon_2, \ldots, \varepsilon_n)$ of eigenvectors of f. Let λ_k be the eigenvalue associated with the vector ε_k.

If $x = \displaystyle\sum_{k=1}^{n} x_k \varepsilon_k$, then $f(x) = \displaystyle\sum_{k=1}^{n} \lambda_k x_k \varepsilon_k$ and: $\varphi(x,x) = \,< f(x),x> = \displaystyle\sum_{k=1}^{n} \lambda_k x_k^2$.

In particular, for every k of $\{1, \ldots, n\}$, $\varphi(\varepsilon_k, \varepsilon_k) = \lambda_k$.

Under these conditions:

- φ is positive if and only if the eigenvalues λ_k of f are positive or zero. We express this situation by saying that f is a positive symmetric operator of E.
- φ is positive definite if and only if the eigenvalues λ_k of f are strictly positive. We express this situation by saying that f is a positive definite symmetric operator of E.

Let S be a real symmetric matrix of order n. It represents an symmetric operator f of $\mathbb{R}^n$ (equipped with the usual scalar product) in the standard basis of $\mathbb{R}^n$. The eigenvalues λ_k of S, all real, are those of f.

We say that S is a symmetric matrix positive if f is a positive symmetric operator, that is, if the λ_k are elements of $\mathbb{R}^+$.

We say that S is a symmetric matrix positive definite if f is positive definite, that is, if the λ_k are strictly positive.

Let S be a real symmetric matrix of order n.

- S is positive $\Leftrightarrow \forall X \in \mathbb{R}^n$, $< SX, X> = \,^{\top}XSX \geq 0$.
- S is positive definite if moreover: $\forall X \in \mathbb{R}^n$, $X \neq \overrightarrow{0} \Rightarrow \,^{\top}XSX > 0$.

Recognizing "by sight" that a matrix S is symmetric is easy! It is completely otherwise to tell whether S is positive or positive definite.

This symmetric matrix S has three strictly positive eigenvalues.

It is thus positive definite.

However, the coefficients of the matrix S are not all positive!

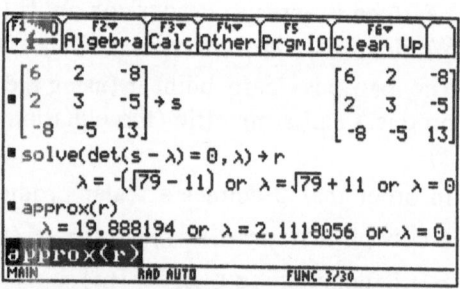

Here is another symmetric matrix S of order 3 whose coefficients are again of arbitrary sign.

Two eigenvalues are strictly positive, and the third is zero.

S is thus positive, but it is not positive definite.

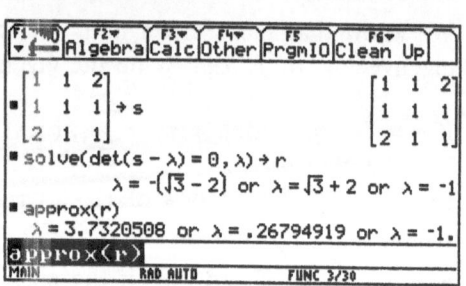

Finally here is a symmetric matrix of order 3, all of whose coefficients are strictly positive.

However, we see that one of the eigenvalues of S is strictly negative.

The matrix S is thus not positive (nor a fortiori positive definite).

These three examples show that while the symmetric nature of a matrix S may be easy to see, it is not evident whether S is positive or positive definite.

All the same, we may make the two following remarks:

- If S is symmetric and positive, and if X represents the k-th vector of the standard basis of $\mathbb{R}^n$, then $<SX, X>> 0$. But this scalar product is also the k-th diagonal coefficient of S. The diagonal coefficients of a positive definite symmetric matrix S are thus strictly positive. The converse is false, as the third example above shows.

- If the real symmetric matrix $S = (a_{i,j})$ has positive diagonal coefficients and, at the same time, has a strictly positive dominant diagonal element (that is, for every i, $a_{i,i} > \sum_{j \neq i} |a_{i,j}|$), then it is positive definite.
In fact, use of the Gershgorin disks of S (cf section **3.3**) shows that each eigenvalue λ is in a disk with center $a_{i,i}$ and of radius $r_i = \sum_{j \neq i} |a_{i,j}|$. Since λ is real and $0 \leq r_i < a_{i,i}$, λ is strictly positive.

7. 2 Square root of a symmetric matrix positive

Let A be a square matrix of order n with real coefficients. To determine the square root of A, that is, the matrices B such that $B^2 = A$, is not an easy thing: this problem may not admit a solution, or indeed may possess a finite number, or even an infinite number.

In fact everything depends on whether the matrix M is diagonalizable or not, whether its eigenvalues are distinct or not, etc., and certainly, one must decide whether to start the search for solutions B in $\mathcal{M}_n(\mathbb{R})$ or in $\mathcal{M}_n(\mathbb{C})$).

Rather than develop this problem in general fashion, which would be a little ambitious, we are going to restrict ourselves to symmetric matrices positive for which there exists an interesting result.

We begin with an example: we will extract the square roots (it will have four) of a symmetric matrix positive definite of order 2. The matrix A is meticulously chosen in order that the calculations yield only whole numbers. For example, its eigenvalues are 1 and 9).

We create the matrix A. We then form an unknown matrix B of order 2, then the equation $B^2 = A$, which we convert into a list of equations.
We then solve this system with cSolve (using Solve will lead to the same result: all the solutions are real).

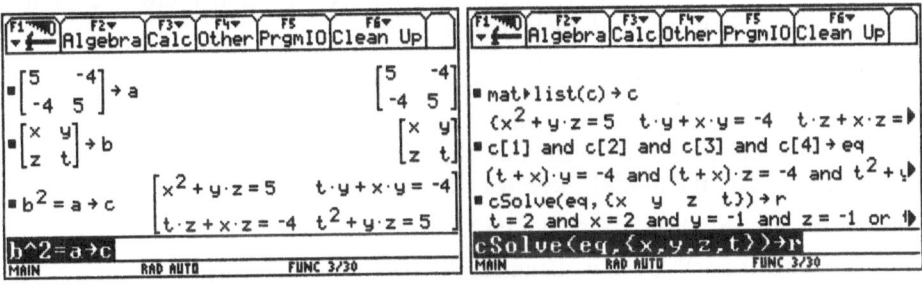

We may verify that the result r is of the form r_1 or $\left(r_2 \text{ or } (r_3 \text{ or } r_4)\right)$, calling the four solutions r_1, r_2, r_3, r_4. This shows the use of the function part to isolate each of them.
We may then exhibit the different matrices B such that $B^2 = A$.

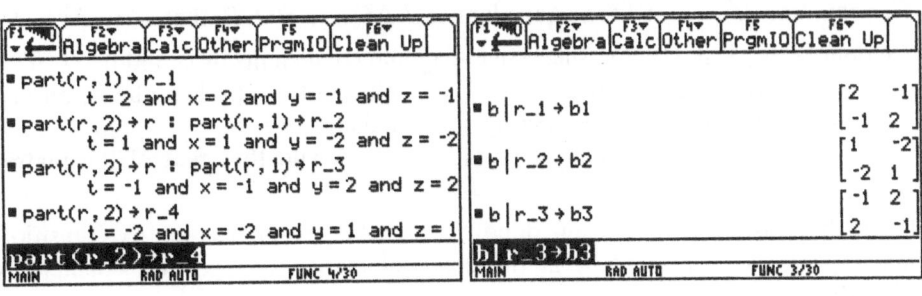

Here is the last solution B_4, then the list of eigenvalues of B_1, B_2, B_3, B_4.

We observe that the four matrices are symmetric, but that only B_1 is positive definite.

```
F1 ▾      F2 ▾   F3 ▾  F4 ▾  F5      F6 ▾
  ◆ ⁻ ⁻ Algebra Calc Other PrgmIO Clean Up

■ b|r_4 → b4                        [ -2   1 ]
                                    [  1  -2 ]
■ eigVl(b1)                            (1.   3.)
■ eigVl(b2)                            (-1.  3.)
■ eigVl(b3)                            (1.  -3.)
■ eigVl(b4)                            (-1.  -3.)
eigVl(b4)
MAIN          RAD AUTO        FUNC 5/30
```

On looking at the previous example, we must guard against hasty generalizations of believing that all the square roots of a symmetric matrix positive (or positive definite) A are symetric. However, this is true if the eigenvalues of A are distinct, as was the case in the preceding example.

For example, $B = \begin{pmatrix} 1 & 0 \\ 1 & -1 \end{pmatrix}$ is not symmetric, but its square is the identity matrix (which is certainly symmetric and positive definite!).

Here is a result which has no exception and for which the previous example is an illustration.

Proposition 1: Let A be a symmetric matrix positive. Then there is a unique symmetric matrix positive B such that $B^2 = A$.

Proof: We show first the uniqueness if the matrix B exists.

Let X be an eigenvector of B for the eigenvalue μ. The equation $BX = \mu X$ and the hypothesis $B^2 = A$ implies that $B^2 X = \mu BX$ which is to say that $AX = \mu^2 X$: X is thus an eigenvector of A for $\lambda = \mu^2$.

Thus, with the usual notation, $E_\mu(B) \subset E_\lambda(A)$.

But (by hypothesis) B is diagonalizable in an orthonormal basis (ε). The preceding inclusions show that (ε) is likewise an orthonormal basis of eigenvectors of A.

Let P be the transition orthogonal matrix from the standard basis to the base (ε).

The matrices $D = {}^T\!PAP$ and $\Delta = {}^T\!PBP$ are both diagonal matrices.
The equation $B^2 = A$ is then equivalent to $P\Delta^2 {}^T\!P = PD{}^T\!P$, that is, to $\Delta^2 = D$.

We denote by μ_k and λ_k the successive diagonal coefficients of Δ and by D. The μ_k are the eigenvalues of B and the λ_k are those of A. By hypothesis, all are real positive numbers. It necessarily follows that: $\forall k \in \{1, \ldots, n\}$, $\mu_k = \sqrt{\lambda_k}$, which establishes the uniqueness of the matrix Δ and thus that of the matrix B.

Now, the existence: We again denote by P the orthogonal matrix transition matrix from A to D, and by Δ the diagonal matrix with positive coefficients such that $\Delta^2 = D$.

The matrix $B = P\Delta^{\top}P$ is positive symmetric and satisfies:

$$B^2 = \left(P\Delta^{\top}P\right)^2 = P\Delta^{\top}PP\Delta^{\top}P = P\Delta^{2\top}P = PD^{\top}P = A$$

The function sqrtsym calculates the square root of the symmetric matrix positive A. (No test is done to verify that A satisfies these hypotheses.) Most of the work is done by the function orthdiag (see section **6.3**).

```
:sqrtsym(a)
:Func:Local q,s
:orthdiag(a)→q
:mzeros(det(a-θθ_),θθ_)→s
:q*diag(√(s))*qᵀ
:EndFunc
```

Here is a real symmetric matrix A of order 5. Its eigenvalues are 0, 16 and 8 (double). A is thus positive.
We calculate its symmetric positive square root B: its eigenvalues are 0, 4 and $2\sqrt{2}$ (double).
Finally, we verify that $B^2 = A$.

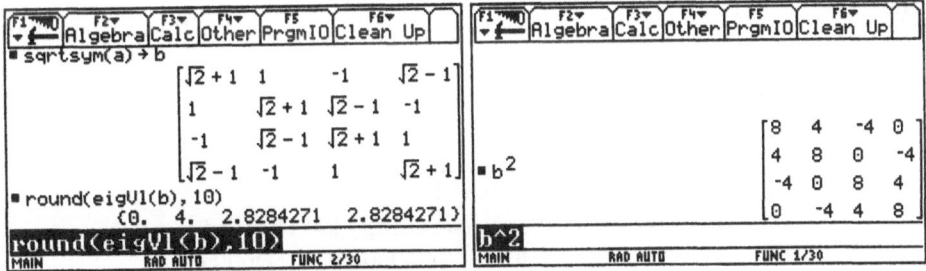

We could generalize what preceded with Hermitian matrices positive matrices A of $\mathcal{M}_n(\mathbb{C})$ satisfying $^{\top}\overline{A} = A$ and all of whose eigenvalues are positive or zero. The formulation of Proposition **1** becomes:

Every Hermitian matrix positive A is the unique square of a Hermitian matrix positive B.

Our function sqrtsym again lets us pass from A to its square root B. Here is an illustration. It remains to verify that the eigenvalues of B are positive.

(They are the square roots of those of A).

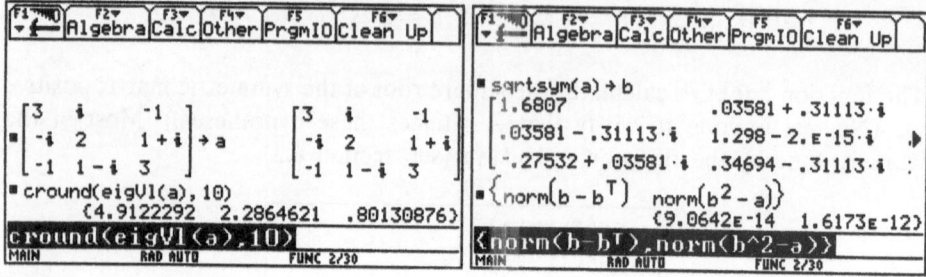

7. 3 Polar decomposition

We give an application of the existence of the symmetric square root of a symmetric matrix positive definite.

Proposition 2: Let A be a square invertible matrix of order n with real coefficients. Then there exists a unique pair (Q, S), where Q is orthogonal and S is symmetric and positive definite such that $A = QS$.

Proof: Suppose that this factorization exists.
Then $^{T}A = {}^{T}S\,{}^{T}Q = SQ^{-1}$ so $^{T}AA = SQ^{-1}QS = S^2$.
The matrix ^{T}AA is symmetric and positive definite. S is thus its symmetric positive square root, which assures the uniqueness of S, and thus of $Q = AS^{-1}$. We define S as the unique symmetric positive definite matrix such that $S^2 = A$ and we put $Q = AS^{-1}$.
We observe that $^{T}QQ = {}^{T}(S^{-1})\,{}^{T}AAS^{-1} = S^{-1}S^2S^{-1} = I_n$ (Q is thus orthogonal) and $QS = AS^{-1}S = A$.

The program poldec takes an invertible matrix A as an argument, then calculates and displays S and Q (stored in the variables θs and θq).

```
:poldec(a):Prgm:Local b
:aᵀa→b:sqrtsym(b)→ θs:Disp "S=":Pause θs
:a∗θs^(-1)→ θq:Disp "Q=":Pause θq :EndPrgm
```

We decompose a random matrix A of order 3. We then verify that Q is orthogonal and that the product QS indeed produces the matrix A.
The matrix S is visibly symmetric, so it remains to verify that its eigenvalues are strictly positive, which is easy with eigVl.

S=
$$\begin{bmatrix} 9.3898812 & -4.0412657 & -6.4419176 \\ -4.0412657 & 6.6416126 & 4.0690482 \\ -6.4419176 & 4.0690482 & 9.3244058 \end{bmatrix}$$
Q=
$$\begin{bmatrix} .65177196 & .35537373 & -.67000211 \\ .75498997 & -.22016099 & .61767248 \\ -.07199624 & .90842647 & .41179836 \end{bmatrix}$$

• round$(\theta q \cdot \theta q^T, 10)$
$$\begin{bmatrix} 1. & 0. & 0. \\ 0. & 1. & 0. \\ 0. & 0. & 1. \end{bmatrix}$$

• round$(\theta q \cdot \theta s, 10)$
$$\begin{bmatrix} 9. & -3. & -9. \\ 4. & -2. & 0. \\ -7. & 8. & 8. \end{bmatrix}$$

round(θq*θs,10)

Remark: There is an analogous result in $\mathcal{M}_n(\mathbb{C})$: An invertible matrix A from $\mathcal{M}_n(\mathbb{C})$ may be uniquely decomposed into $A = QS$, where Q is unitary $(^T\overline{Q} = Q^{-1})$ and S is Hermitian $(^T\overline{S} = S)$ and positive definite (eigenvalues > 0). The program poldec may again be used to effect this decomposition.

7. 4 The Choleski decomposition

We know that it is not simple to recognize at a glance whether a symmetric matrix S is positive or positive definite. However, it is very easy to construct such matrices, as the following result shows.

Proposition 3: Let M be a matrix of $\mathcal{M}_n(\mathbb{R})$. Then $S = M^TM$ is positive. If M is invertible, then S is positive definite.

Proof: We remark first that $\det S = (\det M)^2$. The two matrices M and S are thus invertible at the same time.

To begin, S is symmetric: $^TS = {}^T(M^TM) = {}^T(^TM)^TM = M^TM = S$.

S is positive: $\forall X \in \mathbb{R}^n$, $^TXSX = {}^TXM^TMX = {}^T(^TMX)^TMX = \|^TMX\|^2 \geq 0$.

Finally, if M (and thus ^TM) is invertible: $^TXSX = 0 \Rightarrow {}^TMX = 0 \Rightarrow X = \overrightarrow{0}$.

It is remarkable to observe that the converse of the preceding result is true. We thus obtain a characterization of the positive or positive definite symmetric matrices.

Proposition 4: Let S be a symmetric matrix positive of order n. There exists a matrix M such that $S = M^TM$. If S is positive definite, M is invertible.

Proof: There is nothing to prove since we know already that this problem has at least one solution symmetric positive M ($S = M^TM$ becomes $S = M^2$).

Nevertheless, we may easily observe that the equation $S = M^TM$ in general has an infinity of solutions.
In fact, if M is one of them and if Q is an orthogonal matrix, then $N = MQ$ again is another since $N^TN = MQ^TQ^TM = M^TM = S$.

The following result (which will be used to prove the existence of the Choleski decomposition) shows that one symmetric matrix positive definite may always be concealed in another.

Proposition 5: Let S be a symmetric matrix positive definite of order n, and let k be an integer between 1 and n. Let S_k be the submatrix of S formed at the intersection of the first k rows and columns. (We call this S_k principal submatrix of order k of S). Then S_k is positive definite.

Proof: We write $S = \begin{pmatrix} S_k & {}^{\top}R_k \\ R_k & P_k \end{pmatrix}$.

Let $y = (y_1, y_2, \ldots, y_k)$ be a vector of $\mathbb{R}^k$ and let $x = (y_1, y_2, \ldots, y_k, 0, \ldots, 0)$ be the vector of $\mathbb{R}^n$ obtained by extending y by zero coefficients.

Let X, Y be the column vectors representing x and y. We denote in the same manner the scalar product scalaire on $\mathbb{R}^n$ and on $\mathbb{R}^k$.

With this notation:

$$< SX, X > = {}^{\top}XSX = ({}^{\top}Y \; 0) \begin{pmatrix} S_k & {}^{\top}R_k \\ R_k & P_k \end{pmatrix} \begin{pmatrix} Y \\ 0 \end{pmatrix}$$

$$= ({}^{\top}Y \; 0) \begin{pmatrix} S_k Y \\ R_k Y \end{pmatrix} = {}^{\top}Y S_k Y = < S_k Y, Y >$$

We may conclude that $< S_k Y, Y >$ is always strictly positive (as is $< SX, X >$) unless X is zero (that is, only if Y is zero). The matrix S_k (which is evidently symmetric) is thus positive definite.

As an example, here is a symmetric positive definite matrix S, created from a random matrix M of order 5.

We verify that the eigenvalues of S are all strictly positive, a simple consequence of the fact that M is invertible. S is thus symmetric and positive definite. We then verify that it is the same for the principal submatrices S_k. (We dispensed with S_1!).

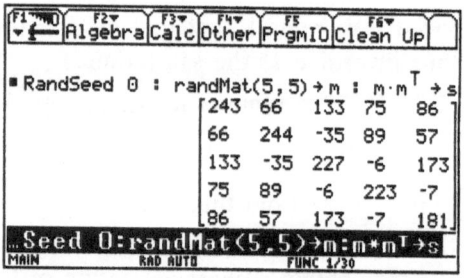

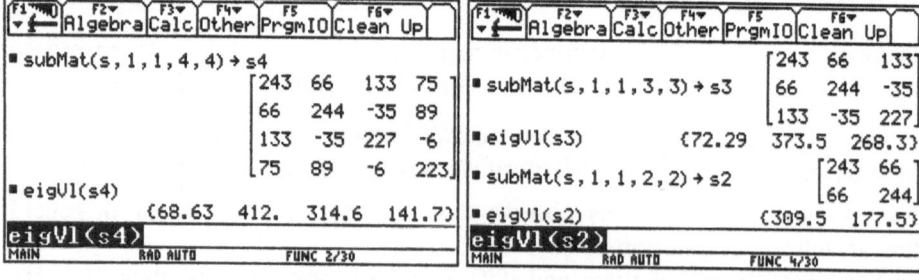

The decomposition $S = M^{\top}M$ of a positive (definite) symmetric matrix is not unique. To arrive at uniqueness, there must be a little more effort exerted on matrix M. On arrive then at the following result, known as the Choleski decomposition.

Proposition 6: Let S be a symmetric matrix positive definite of order n. Then there exists a unique matrix L, lower triangular with diagonal coefficients strictly positive, such that $S = L^{\mathsf{T}}L$.

Proof: By induction on the order n of S.

If $n = 1$, this is evident: $S = (\lambda)$, with $\lambda > 0$, and the only solution is $L = (\sqrt{\lambda})$.

We suppose that the property is true for rank $n \geq 1$.

Let S be a symmetric matrix positive definite of order $n + 1$.

We must show the existence and uniqueness of L, lower triangular with diagonal coefficients strictly positive, of order $n + 1$ such that: $S = L^{\mathsf{T}}L$.

We could write $S = \begin{pmatrix} S_n & {}^{\mathsf{T}}R_n \\ R_n & a \end{pmatrix}$ and find L in the form $L = \begin{pmatrix} L_n & 0 \\ T_n & \lambda \end{pmatrix}$.

In this notation:

- S_n is the principal submatrix of order n of S.
- L_n is lower triangular with diagonal coefficients strictly positive, of order n.
- R_n and T_n are two row vectors of size n.
- a and λ are two strictly positive real numbers.

We must prove the existence and the uniqueness of L_n, T_n and λ.

With the preceding notation:

$$L^{\mathsf{T}}L = S \Leftrightarrow \begin{pmatrix} L_n & 0 \\ T_n & \lambda \end{pmatrix}\begin{pmatrix} {}^{\mathsf{T}}L_n & {}^{\mathsf{T}}T_n \\ 0 & \lambda \end{pmatrix} = \begin{pmatrix} S_n & {}^{\mathsf{T}}R_n \\ R_n & a \end{pmatrix} \Leftrightarrow \begin{cases} L_n{}^{\mathsf{T}}L_n = S_n & (1) \\ T_n{}^{\mathsf{T}}L_n = R_n & (2) \\ T_n{}^{\mathsf{T}}T_n + \lambda^2 = a & (3) \end{cases}$$

We know that S_n is symmetric and positive definite (Proposition 4). By the induction hypothesis, equation (1) thus has a unique solution L_n (lower triangular with diagonal coefficients strictly positive).

Such a matrix L_n is evidently invertible.

Equation (2) defines then row T_n in unique fashion: $T_n = R_n({}^{\mathsf{T}}L_n)^{-1}$.

Equation (3) remains, which may be written: $\lambda^2 = a - T_n{}^{\mathsf{T}}T_n$ (T_n is now known). For the moment we could again consider that λ is complex, and the choice of one of the two square roots of the scalar $a - T_n{}^{\mathsf{T}}T_n$.

With such a tentative λ, we effectively have the equation $S = L^{\mathsf{T}}L$.

The determinant of S is a strictly positive real number, the product of the eigenvalues of S, which are all strictly positive by hypothesis.

The matrix L is block triangular, so $\det L = (\det L_n)\lambda$.

Thus, $\det S = \det(L^{\mathsf{T}}L) = (\det L)^2 = (\det L_n)^2\lambda^2$. We may conclude that $\lambda^2 > 0$, which proves that λ may be chosen to be strictly positive and in a unique way.

This proves the property for rank $n + 1$ and completes the induction argument.

We now come to the programming of the Choleski decomposition.

We denote by $s_{i,j}$ and $\ell_{i,j}$ the coefficients of the matrices S and L.

We are going to see that the equality $L^\top L = S$ allow, by identification, the calculation of the coefficients $\ell_{i,j}$ in explicit fashion.

$$S = L^\top L \Leftrightarrow \forall i,j \in \{1,\ldots,n\}, \; s_{i,j} = \sum_{k=1}^{n} [L]_{i,k} [{}^\top L]_{k,j} = \sum_{k=1}^{n} \ell_{i,k}\ell_{j,k} = \sum_{k=1}^{\min(i,j)} \ell_{i,k}\ell_{j,k}$$

The sum is limited above by $\min(i,j)$ since L is lower triangular.

On the other hand, we know that the matrix S is symmetric. The identification in $S = L^\top L$ may thus be limited (without loss of any generality) to the case $i \geq j$ (diagonal and subdiagonal coefficients).

For every fixed integer j between 1 and n, we find:

- (1): For $i = j$: $\displaystyle\sum_{k=1}^{j} \ell_{j,k}^2 = s_{j,j} \Rightarrow \ell_{j,j}^2 = s_{j,j} - \sum_{k=1}^{j-1} \ell_{j,k}^2$

- (2): For $j+1 \leq i \leq n$: $\displaystyle\sum_{k=1}^{j} \ell_{i,k}\ell_{j,k} = s_{i,j} \Rightarrow \ell_{i,j} = \frac{1}{\ell_{j,j}}\left(s_{i,j} - \sum_{k=1}^{j-1} \ell_{i,k}\ell_{j,k}\right).$

The coefficients of L may thus be calculated column by column.

More precisely, here is the calculation of the j-th column of L:

- Phase (1) gives the j-th diagonal coefficient $\ell_{j,j} > 0$ (If the second member is not strictly positive, then S is not positive definite.) We observe that $\ell_{j,j}$ is obtained as a function of $s_{j,j}$ and of coefficients already calculated in L in the same row, in preceding columns.

- Phase (2) gives the subdiagonal in terms of the j-th column of L. Each $\ell_{i,j}$ is obtained as a function of $s_{i,j}$ (at the same position in S), of the diagonal coefficient $\ell_{j,j}$ (which is to be calculated), and of the terms already known in L (in rows i,j, but in columns with index less than j).

Thus here is the function choleski, which takes the matrix S as an argument, and which returns the matrix L. It is to be verified that S is real and symmetric. (The calculations use the diagonal or subdiagonal coefficients of S).

When phase (1) poses a problem (if the second member is negative or zero), the function choleski returns the message "not positive definite".

```
:choleski(s)
:Func:Local n,L,j,t,k,i
:rowDim(s)→n:newMat(n,n)→L
:For j,1,n:s[j,j]-∑(L[j,k]^2,k,1,j-1)→t
:   If when(t≤0,true,false,false)
:   Return "not positive definite "
:   √(t)→t:t→L[j,j]
:   For i,j+1,n
:      (s[i,j]-∑(L[i,k]*L[j,k],k,1,j-1))/t→L[i,j]
:EndFor:EndFor:L
:EndFunc
```

Here is an example of decomposition of a symmetric positive definite matrix S.

F1 ▾▾	F2 ▾	F3 ▾	F4 ▾	F5	F6 ▾
▾ ✐	Algebra	Calc	Other	PrgmIO	Clean Up

■ s
$$\begin{bmatrix} 1 & -2 & 4 \\ -2 & 13 & -11 \\ 4 & -11 & 21 \end{bmatrix}$$

■ eigVl(s)
{.22773095 5.3779542 29.394315}

`eigVl(s)`

MAIN   RAD AUTO   FUNC 2/30

F1 ▾▾	F2 ▾	F3 ▾	F4 ▾	F5	F6 ▾
▾ ✐	Algebra	Calc	Other	PrgmIO	Clean Up

■ choleski(s) → l
$$\begin{bmatrix} 1 & 0 & 0 \\ -2 & 3 & 0 \\ 4 & -1 & 2 \end{bmatrix}$$

■ $l \cdot l^T$
$$\begin{bmatrix} 1 & -2 & 4 \\ -2 & 13 & -11 \\ 4 & -11 & 21 \end{bmatrix}$$

`L*L^T`

MAIN   RAD AUTO   FUNC 2/30

Calculator Guide

These pages will allow the reader to review the principal instructions and functions of the TI-92, TI-92 Plus, and TI-89, at least those which have allowed us to illustrate the mathematics in this book.

It is impossible in a few pages to "review" everything. We have thus made some subjective choices. The syntax in the examples and in the screen copies will make clear how to use a function, and the results shown will help one to understand their application. We have grouped instructions by themes rather than in alphabetical order. We assume that the reader has a copy of the respective TI GUidebook at hand to refer to.

1. Expansion, factorization

1.1 Expansion

The main utility for expansion is the function expand ([F2]2).

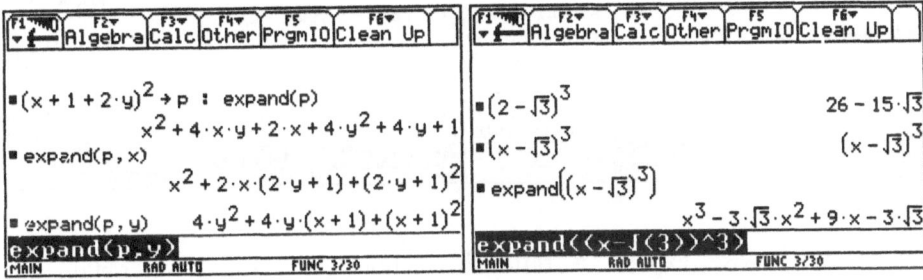

For example, we may expand a polynomial expression: if we specify the name of a variable as a second argument the expansion will be arranged according to decreasing powers of that variable.

Some expansions are automatic, notably when the expression contains only constants, while others require an explicit expand.

In order to expand logarithms one may sometimes require a hypothesis about the sign.

The expansion of $\ln(x\_^2y)$ is not completed since the content of $x\_$ is a complex number by default while the content of x is a real number.

■ expand(ln(x·y))	ln(x·y)
■ expand(ln(x·y), x) | x > 0	ln(x) + ln(y)
■ expand(ln(x²·y))	ln(x²) + ln(y)
■ expand(ln(x_²·y))	ln(x_²·y)
■ expand(ln(xᵐ)) | x > 0	m·ln(x)

`expand(ln(x^m))|x>0`

Here are some other examples of expansions with power functions.

Again, some hypotheses about the sign of the argument may be necessary to obtain expected results.

We recall that on the TI-92 the character | may be obtained by [2nd]K. On the TI-89, it is on the keyboard.

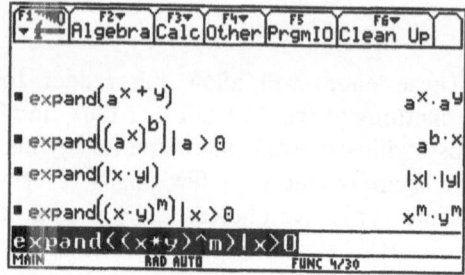

In order to obtain trigonometric expansions, we may use tExpand ([F2]91).

The functions expand and tExpand allow expansion of expressions containing hyperbolic trigonometric functions, but the results are sometimes expressed in an unexpected form.

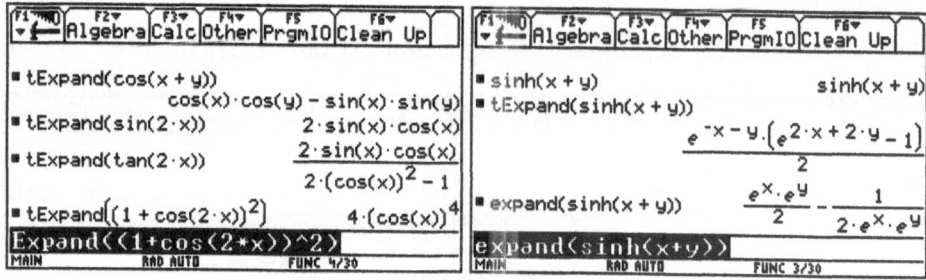

We may likewise expand a rational fraction R. Thus, we obtain the decomposition into simple factors of R: partial fraction decomposition.

We note that if R has complex coefficients, the answer is arranged in terms of its real and imaginary parts. An artifice to avoid this is to replace the complex i by the letter i to retain the complex factors.

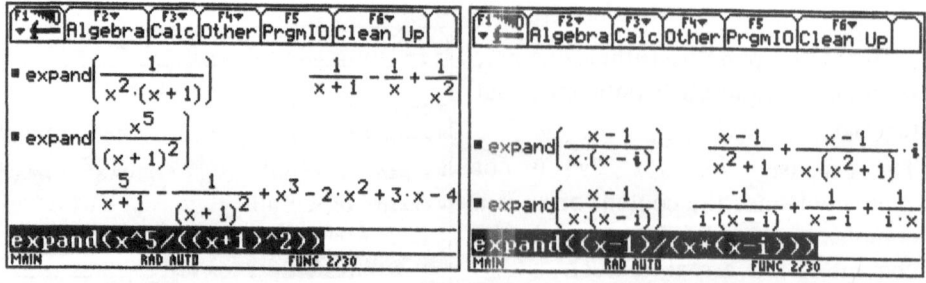

`propFrac` develops a rational fraction R as $R = P + S$, where P is a polynomial and S is a rational fraction of negative total degree (sum of the degrees in the numerator and the denominator).

The second, optional, parameter indicates the variable with respect to which the degree is calculated.

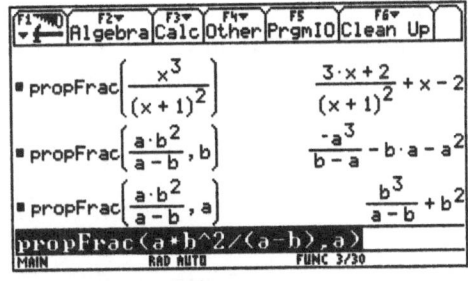

1. 2 Factorization

The instructions covered here are `factor` ([F2]2) and `cFactor` ([F2]A2).

If we don't specify the name of the variable, `factor` works in terms of rational numbers.

If we do indicate the name of the variable, it will find factors with irrational roots.

Finally, using `cFactor` allows us to get complex roots.

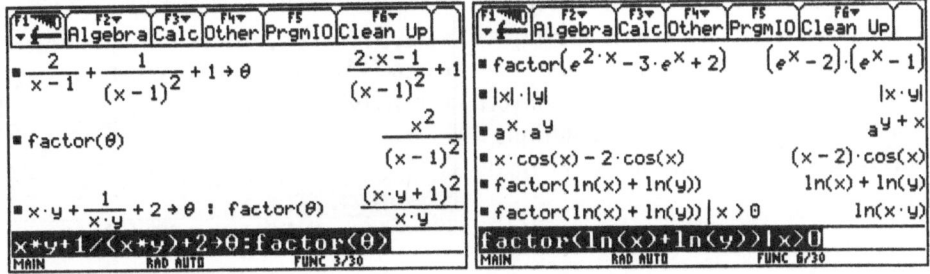

The examples which follow show that the instruction `factor` allows factorizations of non-polynomial expressions, those with rational fractions, for example.

In certain cases, the automatic procedures of the calculator suffice to obtain a factored expression.

The instruction tCollect ([F2]92) plays an inverse role to that of tExpand: it linearizes the powers of trigonometric functions.
One probably shouldn't call this factorization.
tCollect does not accept optional arguments.

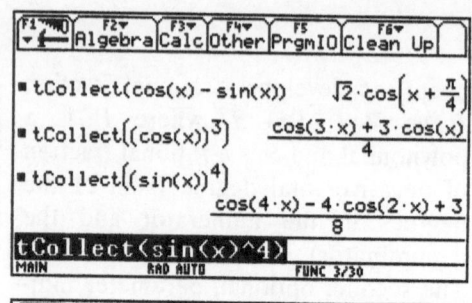

The instruction comDenom ([F2]6) reduces an expression to a single or common denominator but without finding its factors, contrary to factor, which may be the desired effect.

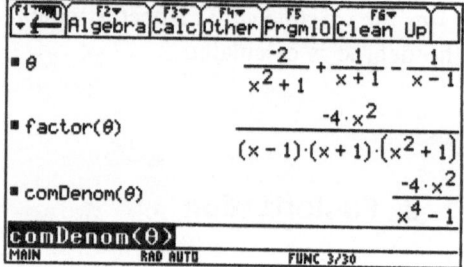

2. Equations and systems

The connections between factoring an expression $f(x)$ and finding the exact solutions of $f(x) = 0$ are as expected for equations of "reasonable" degree.

But if the degree is too high, or for non-polynomial expressions, finding roots no longer may proceed by factorization. Then one must seek a numerical solution. In this case the calculator may only give us an approximate solution which depends on an initial estimate of the solution.

2. 1 The equation f(x)=0

The instructions solve ([F2]1) and cSolve ([F2]A1) try to solve an equation or a system of equations in symbolic fashion over $\mathbb{R}$ or $\mathbb{C}$.

If there are no solutions, they respond false. If there are many, the equations are separated by or.
Exp▷list then allows one to convert such an answer to a list of solutions.
Solve will also handle linear inequalities

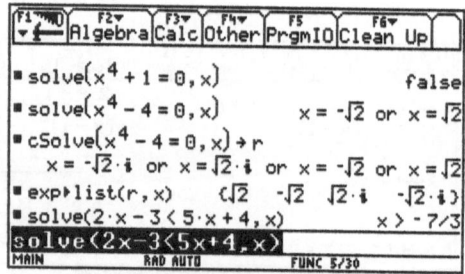

If symbolic solution is impossible, and if the calculator is in AUTO mode (see the MODE menu) then solve and cSolve automatically pass into the "approx " mode.

One may always impose this mode by pressing ◇[Enter] instead of [Enter].

When there is an infinity of solutions, the calculator uses integer symbols @n1, @n2, etc. to designate them.

In some cases, the calculator announces that there may be other solutions in addition to those it finds:

Warning: More solutions may exist.

The instructions zeros ([F2]4) and cZeros ([F2]A3) are cousins related to solve and cSolve. The second argument is always the name x of the variable for which we wish to find a zero in the expression or inequality placed in the first argument. The result is then a list of solutions rather than a logical expression containing the equations of the solutions separated by or.

Here are some examples, essentially the same ones used with solve and cSolve above.

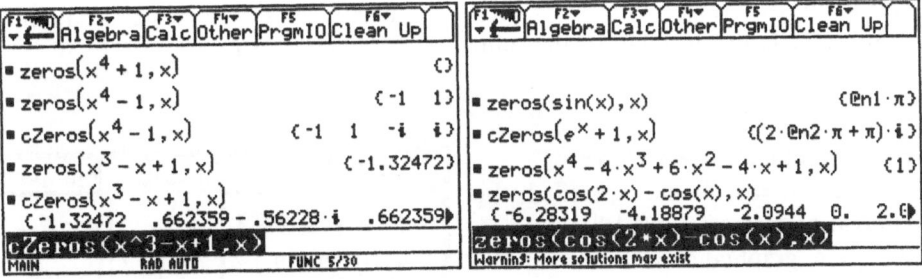

When we look for particular solutions to the equation $f(x) = 0$ or for zeros of the expression $f(x)$, we may "inform " the functions solve, cSolve, zeros and cZeros with the "given that " operator, the | obtained by [2nd]K on the TI-92 and which appears on the keyboard on the TI-89. We thus express the conditions which selected solutions must satisfy.

We may likewise replace the name of the variable by an equation expressing a first estimate of the solution expected, but in this case the search will be made in "approx " mode. We remark that this method is not always infallible with trigonometric equations.

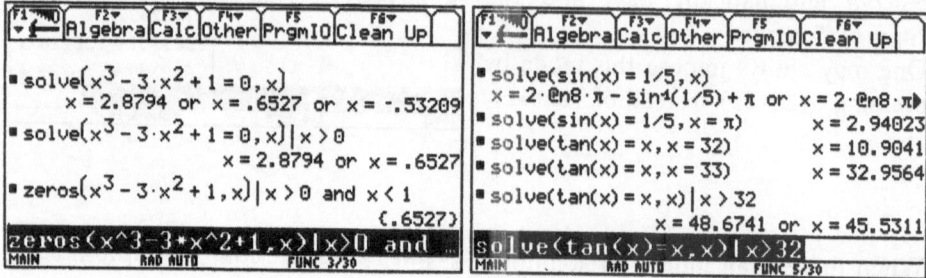

The instruction nSolve ([F2]8) finds a real numeric value of the expression in the first argument.

We may specify an initial estimate or conditions to be fulfilled by the solution sought, but trigonometric equations again may yield some surprises.

2. 2 The numeric solver

Accessible by ([Apps]A), this is very easy to use. One specifies the equation to be solved, and it analyzes the names which appear in it.

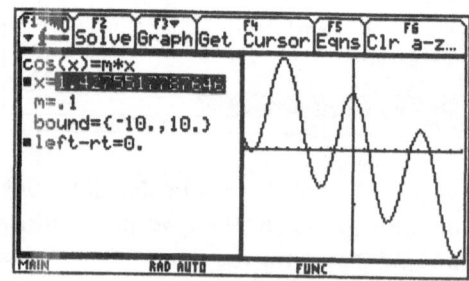

Here we entered $\cos x = mx$.

We put $m = 0.1$ and specified the interval of interest $[-10, 10]$.

We then graphed $\cos x - mx$ in the ZoomFit mode.

Then we placed the cursor on the line $x=$ and launched the search by pressing F2.

To find another solution, we have se-
lected the graph window ([2nd]Apps),
positioned the cursor at the intersection of the curve with the axis Ox, and
retrieved the value at the cursor by
[F4]; then we relaunched the search
for a solution with [F2].

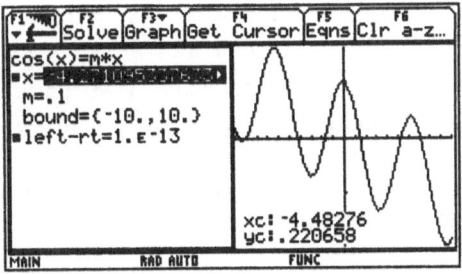

If the equation uses many variable
names, we may search with respect
to any one of them by specifying the
values of the others. Here we have
put $x = 1$, positioned the cursor on
the line $m =$, then launched the search
for m by [F2].

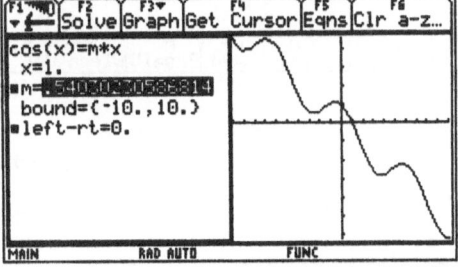

The interactive solver remembers the
last equations which have been inves-
tigated.

It is thus possible to resume with
the preceding equations studied by
pressing [F5].

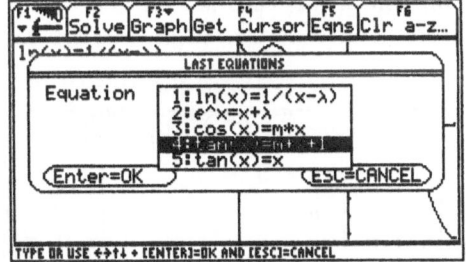

2. 3 Systems of equations

The functions solve et cSolve allow us to solve systems with several equations.
It suffices to separate the equations in the first argument by the logical connective
and. The second argument then contains the list of variables with respect to
which it must solve the system.

Here we solve $\begin{cases} x^2 + 3x - y^2 = 0 \\ x^2 + y^2 = 5 \end{cases}$,

beginning in $\mathbb{R}$ with solve (two pairs of solutions) then in $\mathbb{C}$ with cSolve (four pairs of solutions).
Exp▷list allows separation of the different results.

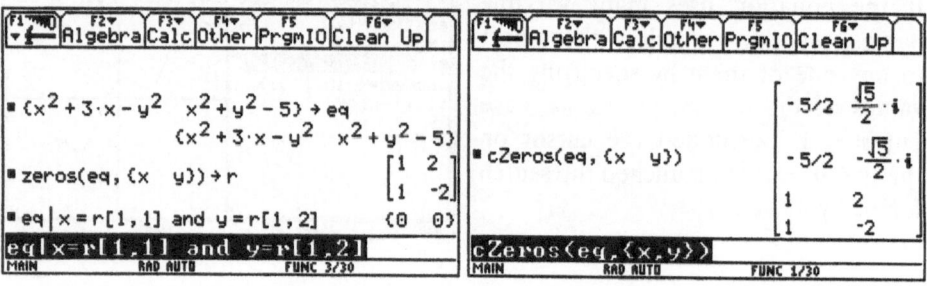

The functions zeros and cZeros also allow solving systems, but the syntax is different. The first argument is the list of equations, and the result is returned in the form of a matrix, each line of which is a solution of the system.

We resume the preceding example with zeros then cZeros.

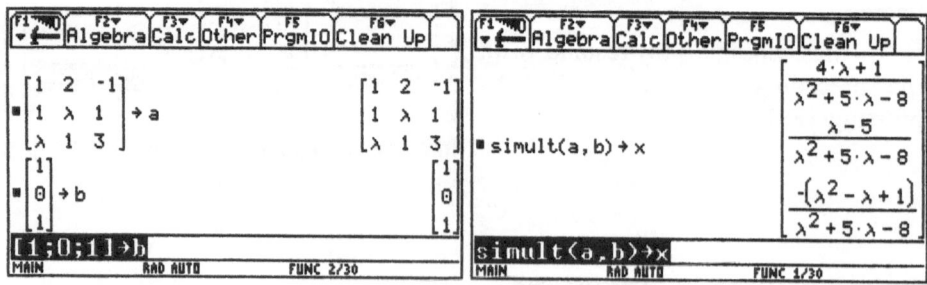

We may likewise solve the linear system $AX = B$ with simult, with the condition that the system has an invertible (thus, square) coefficient matrix A. The unique solution in this case is returned by simult(A,B).
Here is an example. Mind the two values of λ for which $\lambda^2 + 5\lambda - 8$ vanishes.

2. 4 Differential equations

The instruction deSolve ([F3] C) allows solution of certain differential equations in exact form. These are mostly linear differential equations of order 1 or 2.

The syntax is deSolve(*eqn*,*x*,*y*) if we want to solve the differential equation *eqn*, where y is to be found as a function of x. The character indicating a derivative may be obtained by [2nd]B on the TI-92 and [2nd]= on the TI-89.

Here are three examples of differential equations of order 1. In the third (non-linear) case the solution is given implicitly.

In each case the solution depends on a parameter @ which is indexed by successive integers. The indexing resets to 1 when we erase the screen.

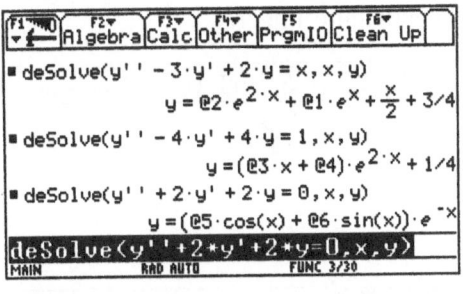

Here are three differential equations of order 2, each linear with constant coefficients. Other than in this simple but very common situation, there are few chances that the calculator will find a solution.

We may include initial conditions or boundary conditions.

Here we have obtained the solutions of $y'' - 3y' + 2y = 2$, then those which have the value 1 when $x = 0$, and then the only solution which satisfies $y(0) = 1$ and $y'(0) = -1$.

Here, in the DIFF EQUATIONS graph mode, is how the solution of $y'' + y' + 2y = 0$ which satisfies $y(0) = 1$ and $y'(0) = 5$ is represented.

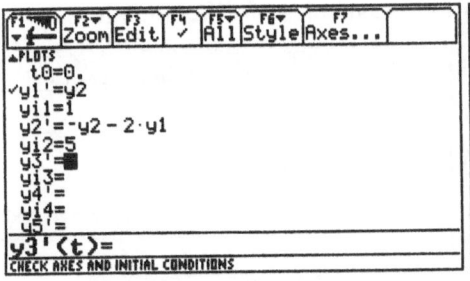
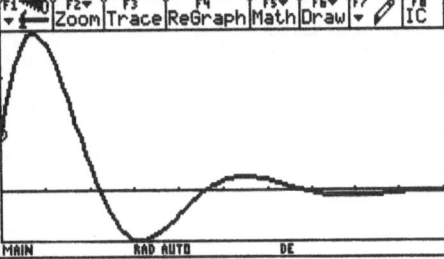

3. Differentiation and integration

Differentiation and integration are two activities which we are glad to assign to our calculator. In these two essential domains of analysis, it serves very well.

3. 1 Symbolic differentiation

Symbolic differentiation is the purview of the keyboard instruction d ([2nd] 8), which must not be confused with the letter "d ".

Here are some examples of differentiation of the same expression f (first derivative, then second derivative, and finally the fifth derivative). To evaluate a derivative at a point, we may use the | "given that " operator.

If the order of differentiation is negative, we obtain a primitive or anti-derivative or indefinite integral.

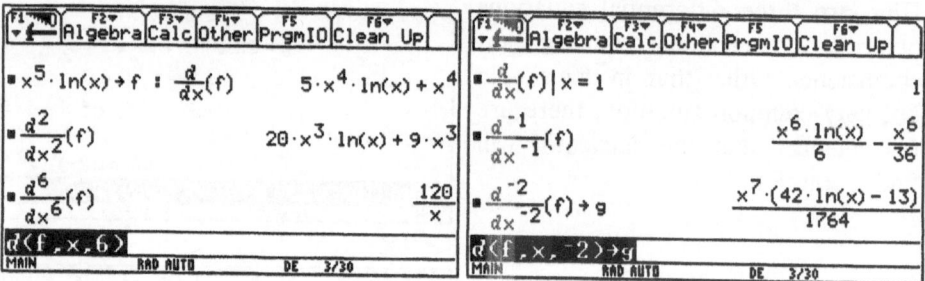

In the example below, we note that: $y\dfrac{\partial f}{\partial x} - x\dfrac{\partial f}{\partial y} = \dfrac{y}{x}$ et $\dfrac{\partial^2 f}{\partial y \partial x} = \dfrac{\partial^2 f}{\partial x \partial y}$.

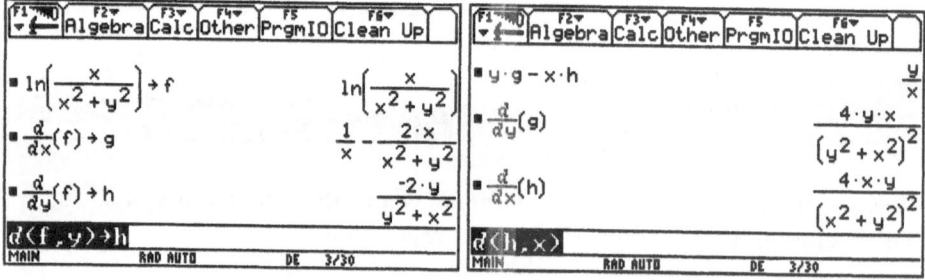

Here we see that the calculator knows the usual rules of differentiation:

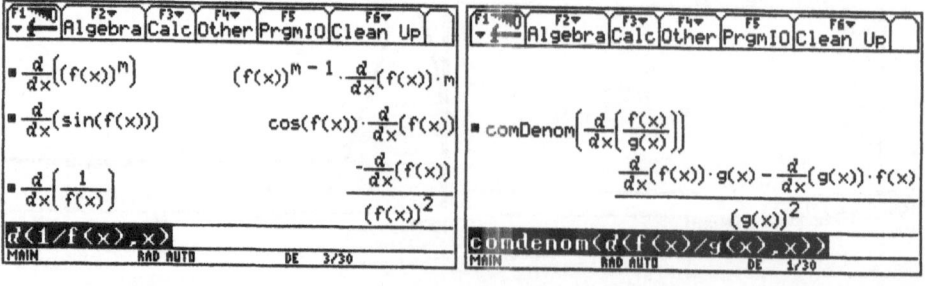

3. 2 Numerical differentiation

The instruction nDeriv ([F3]A) calculates an approximation of the derivative of an expression at a point. The syntax is nDeriv(f,x).
The calculation is much faster than the symbolic differentiation.

By default this approximation is $f'(x) \simeq \dfrac{f(x+.001) - f(x-0.001)}{0.002}$.

A third, optional, argument h allows use of $f'(x) \simeq \dfrac{f(x+h) - f(x-h)}{2h}$.

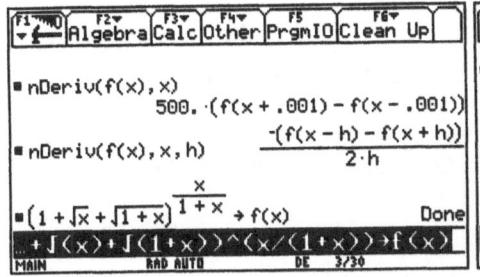

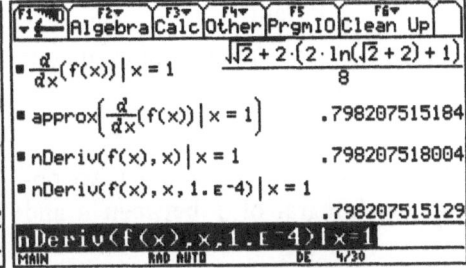

As we see here, decreasing the value of h a little may lead to better precision. But an excessive decrease may introduce significant roundoff errors and a very approximate result.

3. 3 Taylor's formula

The function Taylor allows calculation of the expansion of the same name of an expression f depending on a variable x, of order n, and at a point a:

$$f(x) = f(a) + f'(a)(x-a) + \frac{f''(a)}{2!}(x-a)^2 + \cdots + \frac{f^{(n)}(a)}{2!}(x-a)^n + o\big((x-a)^n\big)$$

The function Taylor gives the polynomial part of this expansion.
The syntax is Taylor(f,x,n,a). If a is omitted, the expansion is at 0 (a Maclaurin expansion).

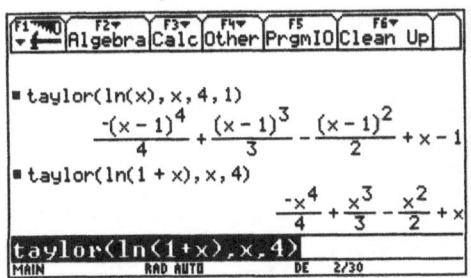

3. 4 Integration

Here we use the instruction $\int$ (accessible on the calculator by [F3] 2).

$\int(f,x)$ gives a primitive or anti-derivative of f, with respect to the variable x.

If we use a third argument, for example λ, this is added to the result as the "arbitrary constant ".

Thus, we obtain the indefinite integral of f.

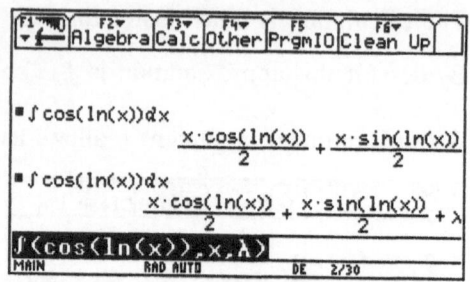

With the syntax $\int(f,x,a,b)$ we obtain the integral of f between a and b.

We may likewise calculate integrals on unbounded intervals or for unbounded functions.

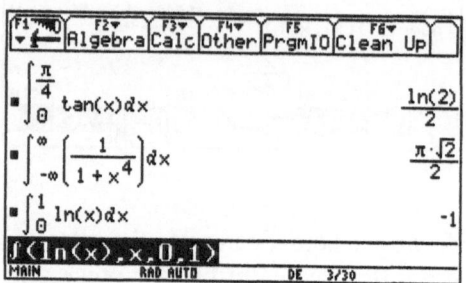

If the exact calculation fails, and if the calculator is in Auto mode, then an appoximate result is returned.

We may force the approx mode by pressing ◇Enter rather than Enter.

We obtain the same result, but more rapidly, with the numerical integration function nInt ([F3]B).

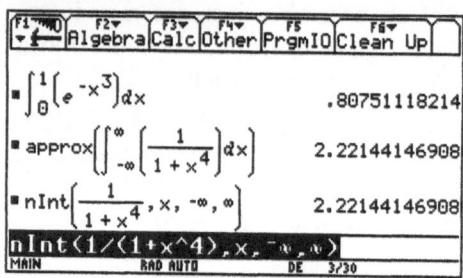

The calculator allows us to study integrals depending on their limits, as testified by the following definition of the function $f : x \mapsto \int_{x}^{x^2} \dfrac{dt}{\ln t}$.

If we want to represent the curve $y = f(x)$, it is better to use the second method, which plots the graph much faster.

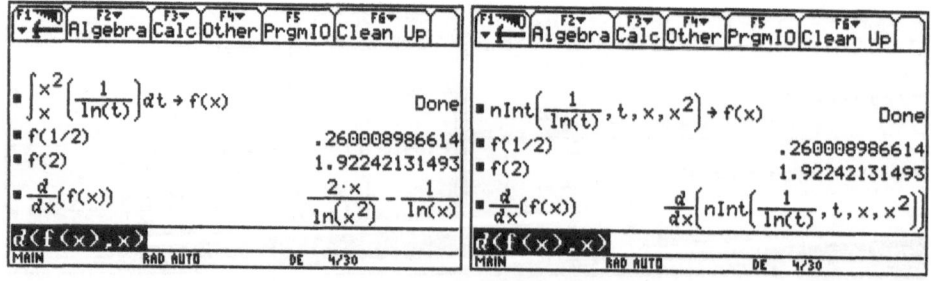

4. Matrices and lists

Matrices and lists represent two ways to group objects. The elements of a matrix or a list may be numbers, names, expressions, or even strings of characters.

4. 1 Matrices

For the calculator, a matrix M is always an array with two dimensions. Each element is referred to with a row index and a column index.

We begin by evaluating the instruction
$[[4, 1, 8, 2][9, 1, 5, 0][3, 7, 4, 2]] \rightarrow m$,
or $[4, 1, 8, 2; 9, 1, 5, 0; 3, 7, 4, 2] \rightarrow m$,
creating a matrix of type 3×4.

$m[2, 1]$ isolates the term in position $(2, 1)$.

$m[2] \rightarrow v$ puts the second row of m into v: this is again a matrix, but of type 1×4.

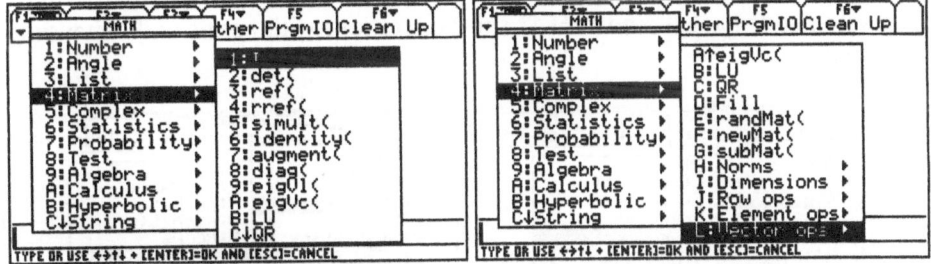

We may read the element of a matrix M with indices (i, j) (respectively its row i) by evaluating $M[i, j]$ (resp. $M[i]$).
We may modify the element with indices (i, j) by $something \rightarrow M[i, j]$.
Most operations on matrices are grouped under the menu [Math]4.

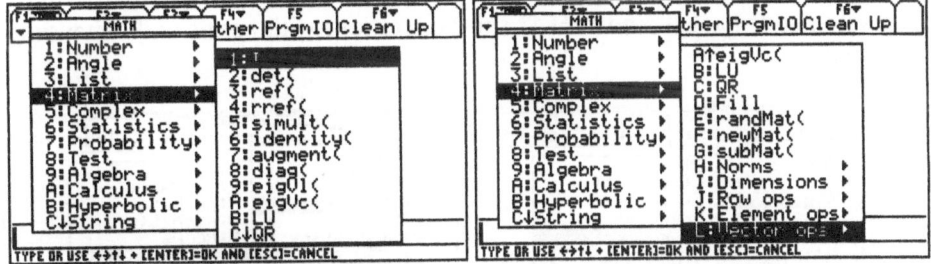

- $M^{\top}$, the "conjugate transpose " of M: the term with indices (i, j) of the new matrix is the conjugate of $M_{j,i}$. If M has real coefficients, this is the transpose.

- det(M) calculates the determinant of M. The syntax det(M, tol) considers scalars whose absolute value is less than tol to be zero. This is useful in order that a singular matrix won't be considered invertible because of rounding errors.

- ref(M) and rref(M) give the upper or row echelon form (reduced in the case of rref) of M. We may also use the optional argument tol here.

- simult(M,B) returns $M^{-1}B$ (that is, the solution of the system $MX = B$).
- identity(n) returns the identity matrix of order n.
- augment(M,N) places the matrices M and N side by side. They must have the same number of rows.
- diag($[\lambda_1, \lambda_2, \ldots, \lambda_n]$) creates a diagonal matrix whose diagonal coefficients are the λ_k.
- eigVl(M) and eigVc(M) return respectively the list of eigenvalues of M and a transition matrix to a basis of eigenvectors. The matrix M must contain only constants. The result is in "approx " form and the eigenvectors are column unit vectors.
- LU M,L,U,P creates a lower triangular matrix L with 1's on the diagonal, an upper triangular matrix U, and a matrix P such that $PA = LU$. A *tol* argument may be added.
- QR M,Q,R creates an orthogonal matrix Q and an upper triangular matrix R such that $M = QR$.
  One may also add a tolerance *tol*.
- Fill λ, M fills the matrix M with the coefficient λ.
- randMat(n,p) generates a random $n \times p$ matrix with integer coefficients between -9 et 9.
- newMat(n,p) returns a zero $n \times p$ matrix.
- subMat(M,n_1,p_1,n_2,p_2) extracts a sub-matrix from M formed by rows n_1 to n_2 and columns p_1 to p_2. By default p_1 and p_2 are equal to the number of columns of M and n_2 to its number of rows.

The sub-menus Math4H and Math4I show operations with norms and dimensions of a matrix.

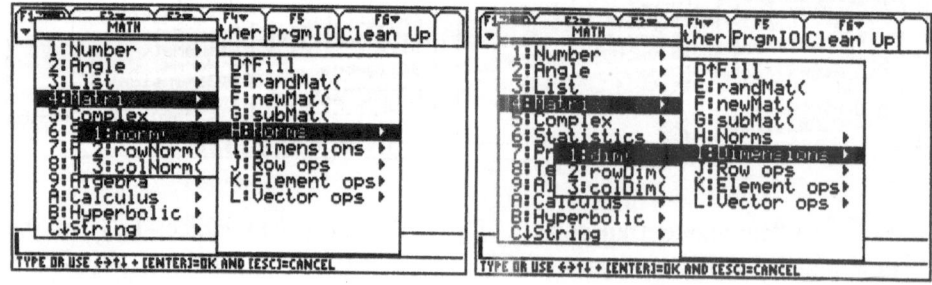

- norm(M), rowNorm(M) and colNorm(M) calculate respectively the Schur norm of M (the square root of the sum of the squares of the coefficients), its row norm (the largest sum of moduli of a row), or its column norm (the maximum sum of the moduli of a column).
- If M is an $n \times p$ matrix, then dim(M), rowDim(M) and colDim(M) return respectively $\{n,p\}$, n and p.

The sub-menus Math4J and Math4K open to operations which work on the rows of a matrix M, or which effect the same operation on all elements of M.

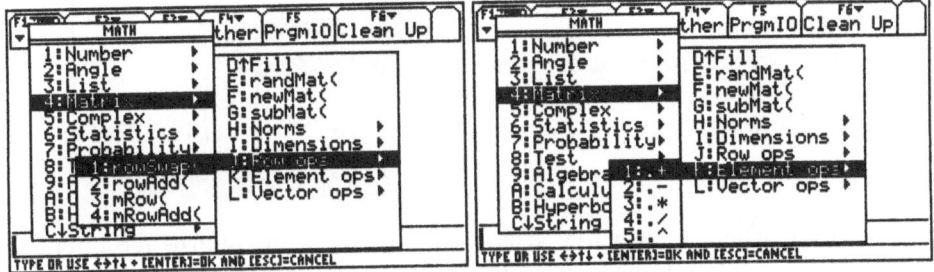

- rowSwap(M,j,k) returns the matrix obtained by exchanging the rows with index j and k: it is the operation coded $L_j \leftrightarrow L_k$

- rowAdd(M,j,k) returns the matrix obtained by adding in M the row j to the row k:it is coded $L_k \leftarrow L_k + L_j$.

- mRow(λ,M,j) returns the matrix obtained by multiplying row j in M by the coefficient λ: it is coded $L_j \leftarrow \lambda L_j$.

- mRowAdd(λ,M,j,k)returns the matrix obtained by adding to row k in M the product by λ of row j : it is coded $L_k \leftarrow L_k + \lambda L_j$.

We may always work on columns of a matrix by a double transposition.

We see here how to swap two columns of a matrix M.

- The expressions $M.+\lambda$, $M.-\lambda$, $M.*\lambda$, $M./\lambda$ and $M.\hat{\ }\lambda$ do the same operation involving the coefficient λ (for example $m_{i,j} \leftarrow m_{i,j} + \lambda$ for the first) on all elements of $m_{i,j}$ de M.

In these instructions, λ may likewise be a matrix N, or M may be a coefficient: the operations are all done term by term.

Here are three examples. In the third don't forget the space after 2.

The sub-menu Math4L opens on vector operations.

Some of these are reserved for row or column vectors $[a, b]$ or $[a, b, c]$ (dimension 2 or 3).

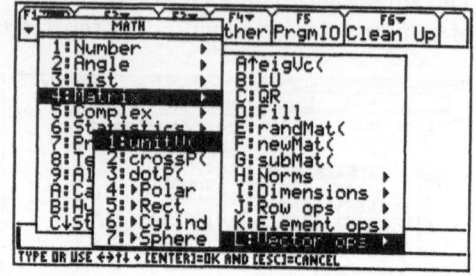

- unitV(V), where V is a row vector or column vector, returns the quotient of V divided by its Euclidean norm. We obtain a unit vector, the "direction " of V.

- crossP(U, V) forms the cross product of U and V (in dimension 2 or 3). The result is a vector of dimension 3.

- dotP(U, V) calculates the scalar product of vectors U and V, either rows or columns. The formula used is $\sum \overline{u_k} v_k$.

- ▷Polar, ▷Rect, ▷Cylind, and ▷Sphere convert the expression of a vector to be displayed to polar coordinates (dimension 2), rectangular, cylindrical or spherical coordinates.

Here are three conversion examples.

We may choose the default display format in the Mode menu: here we are in Rectangular mode.

The result is purely "visual ": in fact, the vector is always remembered in rectangular format.

We remark that we may do a large number of operations on matrices (sometimes they must be square):

- If M, N are matrices, and n is an integer, the operations $M + N$, $M - N$, $M * N$, M^n are possible (with the condition that the dimensions of M and of N allow). We may also calculate the polynomial of a square matrix M by polyEval($\{a_n, \ldots, a_1, a_0\}$, M).

- If λ is a scalar and M is a square matrix of order n, we may evaluate the expressions $M + \lambda$ or $M - \lambda$, the coeffcient λ first being converted to the matrix λI_n so that dimensions match.

- If M is a square numerical matrix (without formal variables) we may calculate $\cos M$, $\sin M$, $\exp M$, etc. The results are in "approx " format.

- If M is a matrix, we may evaluate expand(M), factor(M), etc. The result is the matrix derived from M by application of the functions expand, factor, etc. to each of the elements of M.

4. 2 Lists

For our calculator, a list is a finite ordered sequence of objects, delimited by the characters { and }.

The objects may be expressions, numbers, or strings of characters.

A list L may contain lists, but with the condition that the n elements of L are all lists with the same length p: L is then converted to an $n \times p$ matrix.

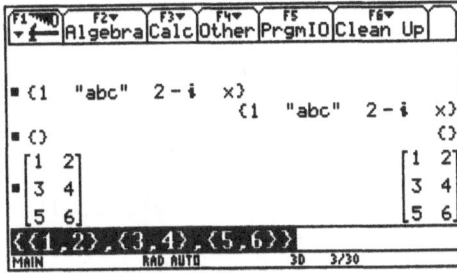

We may read the element with index i in a list L evaluating $L[i]$, and it may be modified by $something \rightarrow L[i]$. If L has length n, the instruction $something \rightarrow L[n+1]$ adjoins an element to L.

Many instructions on lists are grouped in the menu [Math]3.

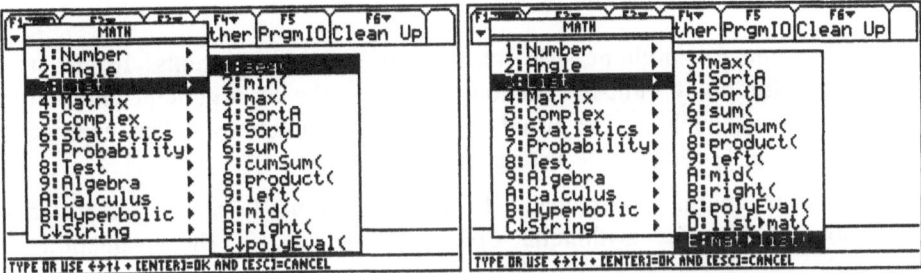

The instruction seq is the usual way to create a sequence whose general term is given by a formula. The syntax is seq($expr,var,beg,end,step$), where $expr$ is an expression depending on the index var varying from beg to end with the specified $step$ (an optional argument whose default value is 1).

Two seq instructions may be "nested " to construct a matrix whose general term follows a given formula. To form the $n \times p$ matrix M whose general term is $m_{i.j}$, we thus evaluate an expression such as seq(seq($M_{i.j},j,1,p),i,1,n$).

Here are three examples using seq. The order of elements in a list is important, and one list may contain the same element several times.

- $\min(\{a_1, a_2, \ldots, a_n\})$ and $\max(\{a_1, a_2, \ldots, a_n\})$ respectively return the minimum or the maximum element of the list $\{a_1, a_2, \ldots, a_n\}$.

- SortA L and SortD L arrange the contents of the list saved in the global variable L in ascending (respectively, descending) order.
  After L, one may adjoin names designating other lists: these are then subject to parallel modifications with L.

- sum(L) and product(L) respectively return the sum and product of the elements contained in the list L.

- cumSum($\{a_1, a_2, \ldots, a_n\}$) returns the list $\{a_1, a_1 + a_2, a_1 + a_2 + a_3, \ldots\}$.

- left(L, n) (resp. right(L, n)) returns the list of the n leftmost (resp. rightmost) elements of the list L and mid(L, n, k) gives the sub-list of L formed of k elements beginning with that in position n.

- polyEval($\{a_n, \ldots, a_1, a_0\}, x$) returns the expression $a_n x^n + \cdots + a_1 x + a_0$.
  Here x may be an expression, a constant, a square matrix, etc.

- list▷mat(L, n) converts the list L into a matrix with n columns. If n is omitted, the result is a matrix with one row.

- mat▷list(M) converts the matrix M into a list.

Lists allow one to simultaneously calculate on several arguments. In fact, this is the case for many built-in functions f which take an argument and return an image $f(x)$, such as the usual mathematical functions. The expression $f(\{a_1, a_2, \ldots, a_n\})$ returns the list of images $\{f(a_1), f(a_2), \ldots, f(a_n)\}$.

In the same manner, for many binary operations $(x, y) \mapsto g(x, y)$, and and notably for the arithmetic operations $+, -, *, /, \hat{\ }$, the evaluation of $g(\{a_1, a_2, \ldots, a_n\}, \{b_1, b_2, \ldots, b_n\})$ leads to $\{g(a_1, b_1), \ldots, g(a_n, b_n)\}$.

We can't review all the possibilities nor all the exceptions for the use of the calculator. We will be content with some examples.

F1▾	F2▾ Algebra	F3▾ Calc	F4▾ Other	F5 PrgmIO	F6▾ Clean Up
▪ {2 × -1 π} → a			{2 × -1 π}		
▪ √a			{√2 √x i √π}		
▪ cos(a)		{cos(2) cos(x) cos(1) -1}			
▪ a²			{4 x² 1 π²}		
▪ d/dx(a)			{0 1 0 0}		
d(a,x)					
MAIN RAD AUTO 3D 5/30					

F1▾	F2▾ Algebra	F3▾ Calc	F4▾ Other	F5 PrgmIO	F6▾ Clean Up
▪ {2 × -1 π} → a			{2 × -1 π}		
▪ {1 2 3 4} → b			{1 2 3 4}		
▪ a + b			{3 x+2 2 π+4}		
▪ a · b			{2 2·x -3 4·π}		
▪ aᵇ			{2 x² -1 π⁴}		
▪ a/b			{2 x/2 -1/3 π/4}		
MAIN RAD AUTO 3D 6/30					

Finally, we note that in a number of cases, those where a function operates on a row or vector column, it is possible to give a list as an argument.
To take only one example, diag($[a_1, a_2, \ldots, a_n]$) and diag($\{a_1, a_2, \ldots, a_n\}$) give the same result (a diagonal matrix).

The instructions sum, product and cumSum may also be applied to matrices, but they work column by column.

5. Sums, products, limits

We know of sum and product, which add or multiply the elements of a list, but we may also calculate sums or products (even infinite!) by means of the instructions $\sum$ and $\prod$ (menu [F3]):

$\sum(expr,k,beg,end)$ and
$\prod(expr,k,beg,end)$
respectively calculate the sum and product of values of $expr$, this expression depending on the index k running from beg to end.

The examples which follow show that in some cases it is possible to calculate a sum or a product on an arbitrary interval.

Finally, the calculator is capable of recognizing certain classical series, notably all the series obtained by integrating or differentiating the geometric series, the series of terms $1/k!$, etc.)

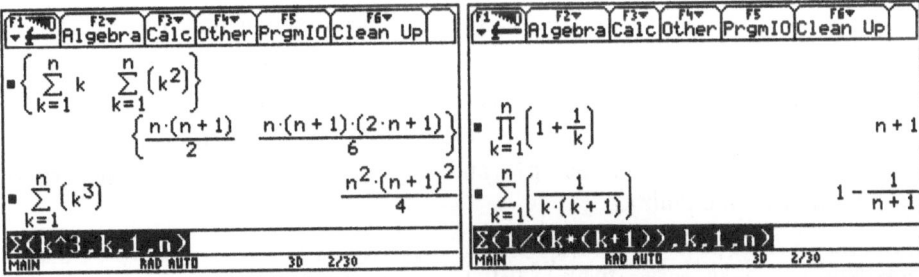

To calculate a limit at a point x_0, the syntax is $\text{limit}(expr, var, x_0, dir)$, where *expr* designates an expression considered in terms of the variable *var*. The argument *dir* is optional: a positive value indicates a limit from the right, and a limit from the left if it is negative.

Here are examples:

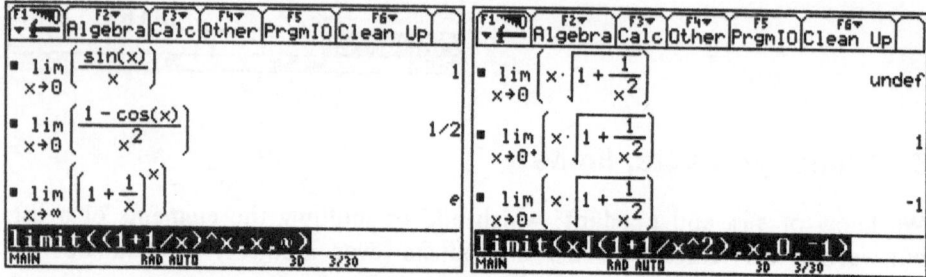

6. Expressions and sub-expressions

An algebraic expression (other than possibly the names of variables and numbers which are only "atoms ") is a combination of operations and arguments which together could be expressions. Such a tree structure has a root (the operator which gives the final result) with some number of branches (the arguments of this operator).

The instruction part allows the dissection of an expression:

• part(*expr*) gives the number of arguments of the root operator of *expr*.

• part(*expr*,0) gives the root operator of *expr*, as a string.

• part(*expr*,n), where $n \geq 1$, gives the n-th argument of this root operator.

Below we see how the instruction part allows analysis of the expression f.

Likewise we dispose of the instructions getNum and getDenom which extract the numerator or the denominator of an expression, after reduction to a common denominator. The instructions left and right return the left or right parts of an equation or inequality.

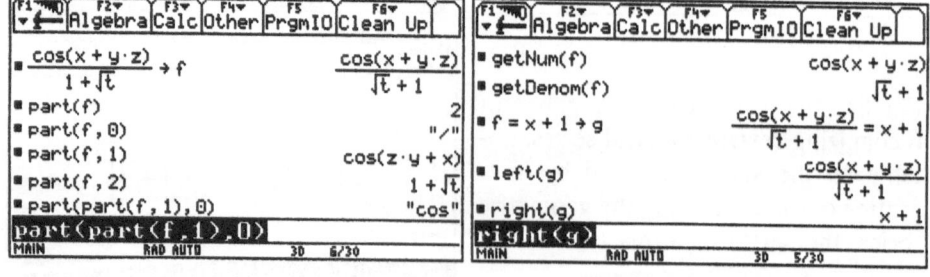

Bibliography

Much of the mathematics in this book is "classical", so references abound.
Here is a brief list of readily available and useful background material.

E. Aboufadel , S. Schlicke
Discovering Wavelets
Wiley-Interscience New York, 1999; ISBN 0471331937

D.M. Bressoud
A Radical Approach to Real Analysis
Mathematical Association of America, Washington, D.C.,1994; ISBN 0-88385-701-4

R. L. Devaney
An Introduction to Chaotic Dynamical Systems 2/e
Addison-Wesley (Perseus Press), Reading, MA, 1989; ISBN 0201130467

C. E. Edwards , D. E. Penney
Differential Equations and Boundary Value Problems: Computing and Modeling
Prentice-Hall, 1999; ISBN 0-13-079770-7

G. Hämmerlin, K.H. Hoffmann
Numerical Mathematics
Springer Verlag New York, 1991; ISBN 0-387-97494-6

M. Misiurewic
Remarks on Sharkovskii's Theorem
Amer. Math. Monthly, **104, (9)**,846-847 (1997)

H.O. Peitgen, H. Jürgens, D. Saupe
Chaos and Fractals
Springer Verlag, New York, 1993; ISBN 3-540-97903-4

J. Stoer, R. Bulirsch
Introduction to Numerical Analysis 2nd ed.
Springer Verlag, New York, 1996; ISBN 3-540-97878-X

G. Strang
Introduction To Linear Algebra (2/E).
Wellesley-Cambridge Press, Wellesley MA, 1998

G.P. Tolstov (Translation by E. Silverman)
Fourier Series
Dover Publications, New York, ISBN 0-486-63317-9

TI-89 Guidebook, TI-92 Guidebook, TI-92+ Guidebook
Texas Instruments, Dallas, TX, 1998

J. A. Yorke, K. Alligood, T. Sauer
Chaos : An Introduction to Dynamical Systems
Springer Verlag, New York, 1996; ISBN 0-387-94677-2

List of the programs used in the book

Commands and functions of the TI-89/92/92Plus calculator are used throughout this text to illustrate many computational and theoretical details. Most are also covered elsewhere in this book. Here are names and brief descriptions of additional programs and functions used. We have retained the names from the French version of the book.

Name	Brief description	Location
cycles	determine limit cycles of dynamical system	1.1.3
bifurc	plot Feigenbaum set	1.1.3
iter	used in feigb	1.1.5
feigb	determine Feigenbaum constant	1.1.5
racines	Newton's method as dynamical system	1.2
julia	plot Julia set for $z \mapsto z^2$	1.4.1
mandelb	plot Mandelbrot set	1.4.2
racinec	Newton's method for complex plane as dynamical system	1.4.3
euler	Euler's method for ODEs	2.7.1
eultest	a test of Euler's method for ODEs	2.7.1
fourierf	generate Fourier coefficients of f	3.1
fouriern	compute partial sums for Fourier series of f	3.1.1
cesaro	compute Cesaro partial sums for Fourier series of f	3.1.3
fft	Fast Fourier Transform of f	3.5.2
tcheb	interpolatingpolynomial by FFT and Chebishev polynomials	3.5.2
ond	compute Haar wavelet coefficients	3.6.3
wal	waveletcompression example	3.6.3
lagr	compute Lagrange interpolating polynomial	4.1.1
interpol	compute Vandermonde interpolating polynomial	4.1.2
dd	compute divided differences by recursion	4.1.3
ddi	compute divided differences byiteration	4.1.3
newt	compute Newton interpolating polynomial	4.1.3
neville	compute Neville interpolating polynomial	4.1.4
graphnev	plot Neville interpolating polynomial	4.1.4

Symbols used in the book

Throughout the text, we shall use a number of symbols, briefly listed here:

$\binom{n}{k}$	the binomial coefficient; $\frac{n!}{k!(n-k)!}$
χ_f	the characteristic polynomial
$\chi_{[-a,a]}$	the characteristic function of the interval [-a,a]
$\times$	Cartesian product of sets
$X^{\perp}$	the orthogonal complement of a set X
$\delta_{n,m}$	Kronecker delta; 1 if $m = n$ and 0 otherwise
$y^{(n)}$	the n-th derivative of $y(x)$
$\exists$	there exists; there is
$\forall$	for each; for every
$\Longleftrightarrow$	if and only if
$\in$	is an element of; belongs to
$\subseteq$	contained in; included in
$]a, b[$	an open interval on the real line
$[a, b]$	a closed interval on the real line
$x \mapsto f(x)$	x is assigned to $f(x)$; used to show a function as a map
$\mathrm{Ann}(f)$	the subset of all polynomial annihilators of f
$\mathbb{C}$	the set of complex numbers
$C^n, C^n(I, \mathbb{R})$	the set of real valued functions on an interval I, which have n continuous derivatives
$E_{\lambda}(f)$	the eigenspace of the eigenvalue λ
$\mathcal{GL}(E)$	the group of the automorphisms of E, the linear group of E
$\mathrm{Inv}(f)$	the set of vectors invariant under f
$\mathbb{K}[f]$	the set of all the polynomials of an operator f
$\mathrm{Ker}(f)$	the kernel of a map f; the preimage of 0
$\mathcal{M}_n(\mathbb{R})$	the set of $n \times n$ matrices with real entries
N or $\mathbb{N}^{>0}$	the set of natural numbers or positive integers
$O(E)$	the set of orthogonal operators on E
$\mathrm{Opp}(f)$	the set vectors which image under f is equal to its opposite.
$\mathcal{L}(E)$	the set of all linear operators on the vector space E
$\mathcal{P}_n$	the vector space of trigonometric polynomials
$\mathbb{R}^n = \mathbb{R} \times \ldots \times \mathbb{R}$	the set of real n-tuples
$\mathbb{R}^{>0}$	the set of positive real numbers
$\mathbb{R}^{\geq 0}$	the set of non-negative real numbers
$\mathbb{R}^{<0}$	the set of negative real numbers
$\mathbb{R}[X]$	the space of polynomials with coefficients in $\mathbb{R}$
$\mathbb{R}_n[X]$	the space of polynomials of degree $\leq n$
$\mathrm{Span}(u_1, u_2, \ldots, u_n)$	the vector space generated by these vectors or the set of all linear combinations of these vectors
$\mathrm{Sp}(f)$	the spectrum of f
$\mathrm{tr}(A)$	the trace of the matrix A; the sum of its diagonal elements
$\mathbb{Z}$	the set of real integers

Index